VOLUME SIX HUNDRED AND FIFTY SEVEN

METHODS IN ENZYMOLOGY

Photoacoustic Probes for In Vivo Imaging

METHODS IN ENZYMOLOGY

VOLUME SIX HUNDRED AND FIFTY SEVEN

METHODS IN ENZYMOLOGY

Photoacoustic Probes for In Vivo Imaging

Edited by

JEFFERSON CHAN

Department of Chemistry and Beckman Institute for Advanced Science and Technology, University of Illinois at Urbana–Champaign, Urbana, IL, United States

Academic Press is an imprint of Elsevier
50 Hampshire Street, 5th Floor, Cambridge, MA 02139, United States
525 B Street, Suite 1650, San Diego, CA 92101, United States
The Boulevard, Langford Lane, Kidlington, Oxford OX5 1GB, United Kingdom
125 London Wall, London, EC2Y 5AS, United Kingdom

First edition 2021

Notices
Knowledge and best practice in this field are constantly changing. As new research and experience broaden our understanding, changes in research methods, professional practices, or medical treatment may become necessary.

Practitioners and researchers must always rely on their own experience and knowledge in evaluating and using any information, methods, compounds, or experiments described herein. In using such information or methods they should be mindful of their own safety and the safety of others, including parties for whom they have a professional responsibility.

To the fullest extent of the law, neither the Publisher nor the authors, contributors, or editors, assume any liability for any injury and/or damage to persons or property as a matter of products liability, negligence or otherwise, or from any use or operation of any methods, products, instructions, or ideas contained in the material herein.

ISBN: 978-0-323-85530-3
ISSN: 0076-6879

For information on all Academic Press publications
visit our website at https://www.elsevier.com/books-and-journals

Publisher: Zoe Kruze
Developmental Editor: Federico Paulo S. Mendoza
Production Project Manager: James Selvam
Cover Designer: Alan Studholme

Typeset by SPi Global, India

Contents

Contributors

Jefferson Chan
Department of Chemistry and Beckman Institute for Advanced Science and Technology, University of Illinois at Urbana–Champaign, Urbana, IL, United States

Junjie Chen
State Key Laboratory of Luminescent Materials and Devices, Guangdong Provincial Key Laboratory of Luminescence from Molecular Aggregates, College of Materials Science and Engineering, South China University of Technology, Guangzhou, China

Weijie Chen
Key Laboratory of Pesticide and Chemical Biology, Ministry of Education, International Joint Research Center for Intelligent Biosensing Technology and Health, College of Chemistry, Central China Normal University, Wuhan, PR China

Liang Cheng
Institute of Functional Nano & Soft Materials (FUNSOM), Jiangsu Key Laboratory for Carbon-Based Functional Materials and Devices, Soochow University, Suzhou, JS, China

Penghui Cheng
School of Chemical and Biomedical Engineering, Nanyang Technological University, Singapore, Singapore

Chengchao Chu
State Key Laboratory of Molecular Vaccinology and Molecular Diagnostics, Center for Molecular Imaging and Translational Medicine, School of Public Health; State Key Laboratory of Cellular Stress Biology, Innovation Center for Cell Biology, School of Life Sciences, Xiamen University; Amoy Hopeful Biotechnology Co., Ltd., Xiamen, China

Wei Du
Hefei National Laboratory of Physical Sciences at Microscale, Department of Chemistry, University of Science and Technology of China, Hefei, AH, China

Amanda K. East
Department of Chemistry and Beckman Institute for Advanced Science and Technology, University of Illinois at Urbana–Champaign, Urbana, IL, United States

Yue Fei
CAS Center for Excellence in Nanoscience, CAS Key Laboratory for Biomedical Effects of Nanomaterials and Nanosafety, National Center for Nanoscience and Technology (NCNST), University of Chinese Academy of Sciences, Beijing, China

Mingyuan Gao
State Key Laboratory of Radiation Medicine and Protection, School for Radiological and Interdisciplinary Sciences (RAD-X) and Collaborative Innovation Center of Radiation Medicine of Jiangsu Higher Education Institutions, Soochow University, Suzhou, PR China

Vipul Gujrati
Center for Translational Cancer Research (TranslaTUM), School of Medicine, Technical University of Munich, Munich; Institute of Biological and Medical Imaging (IBMI), Helmholtz Zentrum München (GmbH), Neuherberg, Germany

Zijuan Hai
Key Laboratory of Structure and Functional Regulation of Hybrid Materials, Ministry of Education, Institutes of Physical Science and Information Technology, Anhui University, Hefei, AH, China

Kenjiro Hanaoka
Graduate School of Pharmaceutical Sciences, The University of Tokyo, Tokyo, Japan

Shuangyan Huan
State Key Laboratory of Chemo/Biosensing and Chemometrics, College of Chemistry and Chemical Engineering, Hunan University, Changsha, PR China

Wenting Huo
Department of Energy and Hydrocarbon Chemistry, Graduate School of Engineering, Kyoto University, Kyoto, Japan

Takayuki Ikeno
Graduate School of Pharmaceutical Sciences, The University of Tokyo, Tokyo, Japan

Naoto Imaizumi
Department of Energy and Hydrocarbon Chemistry, Graduate School of Engineering, Kyoto University, Kyoto, Japan

Xin Jin
CAS Center for Excellence in Nanoscience, CAS Key Laboratory for Biomedical Effects of Nanomaterials and Nanosafety, National Center for Nanoscience and Technology (NCNST), University of Chinese Academy of Sciences, Beijing, China

Kai Li
Shenzhen Key Laboratory of Smart Healthcare Engineering, Department of Biomedical Engineering, Southern University of Science and Technology (SUSTech), Shenzhen, China

Li-Li Li
CAS Center for Excellence in Nanoscience, CAS Key Laboratory for Biomedical Effects of Nanomaterials and Nanosafety, National Center for Nanoscience and Technology (NCNST), University of Chinese Academy of Sciences, Beijing, China

Yaxi Li
Shenzhen Key Laboratory of Smart Healthcare Engineering, Department of Biomedical Engineering, Southern University of Science and Technology (SUSTech), Shenzhen, China

Gaolin Liang
Hefei National Laboratory of Physical Sciences at Microscale, Department of Chemistry, University of Science and Technology of China, Hefei, AH; State Key Laboratory of Bioelectronics, School of Biological Sciences and Medical Engineering, Southeast University, Nanjing, JS, China

Gang Liu
State Key Laboratory of Molecular Vaccinology and Molecular Diagnostics, Center for Molecular Imaging and Translational Medicine, School of Public Health; State Key Laboratory of Cellular Stress Biology, Innovation Center for Cell Biology, School of Life Sciences, Xiamen University, Xiamen, China

Melissa Y. Lucero
Department of Chemistry and Beckman Institute for Advanced Science and Technology, University of Illinois at Urbana–Champaign, Urbana, IL, United States

Jingmei Ma
CAS Center for Excellence in Nanoscience, CAS Key Laboratory for Biomedical Effects of Nanomaterials and Nanosafety, National Center for Nanoscience and Technology (NCNST), University of Chinese Academy of Sciences, Beijing, China

Koji Miki
Department of Energy and Hydrocarbon Chemistry, Graduate School of Engineering, Kyoto University, Kyoto, Japan

Kanuj Mishra
Institute of Biological and Medical Imaging (IBMI), Helmholtz Zentrum München (GmbH), Neuherberg, Germany

Huiying Mu
Department of Energy and Hydrocarbon Chemistry, Graduate School of Engineering, Kyoto University, Kyoto, Japan

Van Phuc Nguyen
Department of Ophthalmology and Visual Sciences, University of Michigan, Ann Arbor, MI, United States; NTT-Hitech Institutes, Nguyen Tat Thanh University, Ho Chi Minh City, Vietnam

Jen-Shyang Ni
Department of Chemical and Materials Engineering, Photo-sensitive Material Advanced Research and Technology Center (Photo-SMART), National Kaohsiung University of Science and Technology, Kaohsiung, China

Kohei Nogita
Department of Energy and Hydrocarbon Chemistry, Graduate School of Engineering, Kyoto University, Kyoto, Japan

Vasilis Ntziachristos
Center for Translational Cancer Research (TranslaTUM), School of Medicine, Technical University of Munich, Munich; Institute of Biological and Medical Imaging (IBMI), Helmholtz Zentrum München (GmbH), Neuherberg, Germany

Masahiro Oe
Department of Energy and Hydrocarbon Chemistry, Graduate School of Engineering, Kyoto University, Kyoto, Japan

Kouichi Ohe
Department of Energy and Hydrocarbon Chemistry, Graduate School of Engineering, Kyoto University, Kyoto, Japan

Yannis M. Paulus
Department of Ophthalmology and Visual Sciences; Department of Biomedical Engineering, University of Michigan, Ann Arbor, MI, United States

Kanyi Pu
School of Chemical and Biomedical Engineering, Nanyang Technological University, Singapore, Singapore

Wei Qian
IMRA America Inc, Ann Arbor, MI, United States

Haibin Shi
State Key Laboratory of Radiation Medicine and Protection, School for Radiological and Interdisciplinary Sciences (RAD-X) and Collaborative Innovation Center of Radiation Medicine of Jiangsu Higher Education Institutions, Soochow University, Suzhou, PR China

Xiaoxiao Shi
State Key Laboratory of Molecular Vaccinology and Molecular Diagnostics, Center for Molecular Imaging and Translational Medicine, School of Public Health, Xiamen University, Xiamen, China

Guosheng Song
State Key Laboratory of Chemo/Biosensing and Chemometrics, College of Chemistry and Chemical Engineering, Hunan University, Changsha, PR China

Mariia Stankevych
Institute of Biological and Medical Imaging (IBMI), Helmholtz Zentrum München (GmbH), Neuherberg, Germany

Andre C. Stiel
Institute of Biological and Medical Imaging (IBMI), Helmholtz Zentrum München (GmbH), Neuherberg, Germany

Lihe Sun
State Key Laboratory of Luminescent Materials and Devices, Guangdong Provincial Key Laboratory of Luminescence from Molecular Aggregates, College of Materials Science and Engineering, South China University of Technology, Guangzhou, China

Rodrigo Tapia Hernandez
Department of Chemistry and Beckman Institute for Advanced Science and Technology, University of Illinois at Urbana–Champaign, Urbana, IL, United States

Yasuteru Urano
Graduate School of Pharmaceutical Sciences; Graduate School of Medicine, The University of Tokyo, Tokyo, Japan

Anna Wang
State Key Laboratory of Radiation Medicine and Protection, School for Radiological and Interdisciplinary Sciences (RAD-X) and Collaborative Innovation Center of Radiation Medicine of Jiangsu Higher Education Institutions, Soochow University, Suzhou, PR China

Hao Wang
CAS Center for Excellence in Nanoscience, CAS Key Laboratory for Biomedical Effects of Nanomaterials and Nanosafety, National Center for Nanoscience and Technology (NCNST), University of Chinese Academy of Sciences, Beijing, China

Xueding Wang
Department of Biomedical Engineering, University of Michigan, Ann Arbor, MI, United States

Yanpei Wang
State Key Laboratory of Chemo/Biosensing and Chemometrics, College of Chemistry and Chemical Engineering, Hunan University, Changsha, PR China

Yuqi Wang
State Key Laboratory of Analytical Chemistry for Life Science, Chemistry and Biomedicine Innovation Center (ChemBIC), School of Chemistry and Chemical Engineering, Nanjing University, Nanjing, China

Chengfan Wu
Hefei National Laboratory of Physical Sciences at Microscale, Department of Chemistry, University of Science and Technology of China, Hefei, AH, China

Shuizhu Wu
State Key Laboratory of Luminescent Materials and Devices, Guangdong Provincial Key Laboratory of Luminescence from Molecular Aggregates, College of Materials Science and Engineering, South China University of Technology, Guangzhou, China

Yinglong Wu
State Key Laboratory of Luminescent Materials and Devices, Guangdong Provincial Key Laboratory of Luminescence from Molecular Aggregates, College of Materials Science and Engineering, South China University of Technology, Guangzhou, China

Shuyu Xu
State Key Laboratory of Molecular Vaccinology and Molecular Diagnostics, Center for Molecular Imaging and Translational Medicine, School of Public Health, Xiamen University, Xiamen, China

Anuj K. Yadav
Department of Chemistry and Beckman Institute for Advanced Science and Technology, University of Illinois at Urbana–Champaign, Urbana, IL, United States

Guang-Fu Yang
Key Laboratory of Pesticide and Chemical Biology, Ministry of Education, International Joint Research Center for Intelligent Biosensing Technology and Health, College of Chemistry, Central China Normal University, Wuhan, PR China

Deju Ye
State Key Laboratory of Analytical Chemistry for Life Science, Chemistry and Biomedicine Innovation Center (ChemBIC), School of Chemistry and Chemical Engineering, Nanjing University, Nanjing, China

Fengying Ye
Key Laboratory of Pesticide and Chemical Biology, Ministry of Education, International Joint Research Center for Intelligent Biosensing Technology and Health, College of Chemistry, Central China Normal University, Wuhan, PR China

Zhifei Ye
State Key Laboratory of Chemo/Biosensing and Chemometrics, College of Chemistry and Chemical Engineering, Hunan University, Changsha, PR China

Baoli Yin
State Key Laboratory of Chemo/Biosensing and Chemometrics, College of Chemistry and Chemical Engineering, Hunan University, Changsha, PR China

Jun Yin
Key Laboratory of Pesticide and Chemical Biology, Ministry of Education, International Joint Research Center for Intelligent Biosensing Technology and Health, College of Chemistry, Central China Normal University, Wuhan, PR China

Ling Yin
Department of Chemistry and Chemical Engineering, Jining University, Qufu, PR China

Fang Zeng
State Key Laboratory of Luminescent Materials and Devices, Guangdong Provincial Key Laboratory of Luminescence from Molecular Aggregates, College of Materials Science and Engineering, South China University of Technology, Guangzhou, China

Menglei Zha
Shenzhen Key Laboratory of Smart Healthcare Engineering, Department of Biomedical Engineering, Southern University of Science and Technology (SUSTech), Shenzhen, China

Lele Zhang
Key Laboratory of Structure and Functional Regulation of Hybrid Materials, Ministry of Education, Institutes of Physical Science and Information Technology, Anhui University, Hefei, AH, China

Rui Zhang
Institute of Functional Nano & Soft Materials (FUNSOM), Jiangsu Key Laboratory for Carbon-Based Functional Materials and Devices, Soochow University, Suzhou, JS, China

Preface

Imaging modalities with deep tissue capabilities can provide researchers with a powerful lens to see what is invisible to the naked eye within its native in vivo environment. In this regard, photoacoustic imaging (also known as optoacoustic imaging) has emerged as a promising approach for basic scientific research, as well as for clinical applications in humans. By combining safe, nonionizing light irradiation with the generation of ultrasound waves that can travel through the body with minimal perturbation, photoacoustic imaging strikes the perfect balance between sensitivity, contrast, resolution, imaging depth, affordability, and ease of use. Beyond the initial application of endogenous pigments such as hemoglobin, oxy-hemoglobin, and melanin for label-free studies, the field has made considerable strides toward molecular imaging through the advent of targeted contrast agents and analyte-responsive imaging probes. One of the major highlights that characterize this impressive palette of chemical tools are new and innovative design approaches that have enabled the selective detection and tracking of a given biomarker using photoacoustic imaging. The goal of this *Methods in Enzymology* volume is to continue the longstanding tradition of this series by showcasing new advances by leaders in the field with the intent to inspire future developments.

In this volume, Hanaoka and Urano employ a spirocyclization design to disrupt the conjugated pi-system to access an activatable probe for hypochlorous acid (Chapter 1). Ye demonstrates the utility of a multicomponent design that involves a key reduction, proteolysis, and cyclization sequence that enables self-assembly of their probe to yield an activated nanoparticle for caspase-3 sensing in tumors (Chapter 2). Shi and Gao discuss the development of a multimodal probe with a ratiometric readout for MMP-2 detection (Chapter 3). Likewise, Miki and Ohe target the MMP-2 enzyme using a complementary approach that involves the aggregation of their MMP-2-activated probe, giving rise to a change in the photoacoustic signal at two wavelengths (Chapter 4). Liang utilizes an alkaline phosphatase-triggered self-assembly design to enable enhanced photoacoustic detection of this important cancer biomarker (Chapter 5). Similarly, Chu and Liu employ a self-assembly probe design that is augmented with multimodal imaging capabilities for sensing of glutathione and hydrogen peroxide, both important disease biomarkers (Chapter 6). To access deeper regions of the body, Chan describes the development of an NIR-II probe for nitric oxide sensing in multiple tumor models (Chapter 7). Likewise, Li discusses the fabrication of a semiconducting polymer that features an absorbance band that extends into the NIR-II region and its application for tracking hepatic tumor growth in a longitudinal study (Chapter 8). Yin and Yang introduce conjugated bridges and electron donors to improve the photoacoustic properties of their imaging agents in the NIR-II imaging window (Chapter 9). Hai demonstrates the utility of a mitochondria-targeted photoacoustic probe for detecting hydrogen peroxide in a murine model of inflammation (Chapter 10). Pu introduces an activatable photoacoustic agent with the potential to diagnose acute kidney injury using photoacoustic imaging (Chapter 11). Zheng and Wu employ multispectral optoacoustic tomography to detect a new alkaline phosphatase activatable probe they developed for in vivo imaging of drug-induced liver injury (Chapter 12). Wang discusses the development of a caspase-1-responsive probe for detecting bacterial infections in host cells that feature a self-assembly mechanism for photoacoustic signal enhancement upon activation (Chapter 13). Ntziachristos describes the development of genetically engineered bacterial outer membrane vesicles for optoacoustics-guided phototherapy applications (Chapter 14). Stiel discusses the advantages of photoswitchable imaging agents that enable their signal to be distinguished from nonphotoswitching background pigments (Chapter 15). Huan and Song describe the development of an oxygen-embedded quinoidalacene-based semiconducting nanoprobe and its application to guide tumor treatment (Chapter 16). Chan employs a general conformational restriction design to access a new dye scaffold with optimized photoacoustic properties for probe development (Chapter 17). Finally, Wang and Paulus detail the fabrication of colloidal gold nanoparticles and their application to image newly developed blood vessels in the retina (Chapter 18).

Together, the work presented in each of the 18 chapters of this volume showcase work from a vibrant international community of scientists who are leaders in their respective disciplines. Indeed, the authors collectively share the common goal of employing photoacoustic imaging to illuminate unseen features of the body as demonstrated by the impressive diversity of the probe designs and breadth of biomedical applications. While the current repertoire of imaging agents is notable, it will be exciting to see future developments as the field continues to advance. In the capacity of the volume editor, I thank all the authors for their contributions; this would not have been possible without their willingness to share their work or generosity with their time. I also acknowledge the *Methods in Enzymology* editors and Elsevier editorial and production teams. In addition, I thank Prof. David Christianson, Zoe Kruze, and Paulo Mendoza for their guidance and patience. Last but not least, I thank my family Melissa, Ashley, Ryan, and Leroy for supporting this endeavor.

CHAPTER ONE

Development of a small-molecule-based activatable photoacoustic probe

Takayuki Ikeno[a], Kenjiro Hanaoka[a,*,†], and Yasuteru Urano[a,b,*]

[a]Graduate School of Pharmaceutical Sciences, The University of Tokyo, Tokyo, Japan
[b]Graduate School of Medicine, The University of Tokyo, Tokyo, Japan
*Corresponding authors: e-mail address: khanaoka@keio.jp; uranokun@m.u-tokyo.ac.jp

Contents

Abstract

Photoacoustic (PA) imaging is an emerging imaging modality that combines the advantages of optical imaging and ultrasound imaging. In particular, activatable PA probes, which visualize the presence or the activity of target molecules in terms of a change of the PA signal, are useful tools for functional imaging. In this chapter, we describe the development of small-molecule-based activatable PA probes, focusing on the design and synthesis of PA-MMSiNQ, our recently developed activatable PA probe for HOCl. We also describe the protocols used for evaluation of PA-MMSiNQ with a UV–vis spectrometer and a PA imaging microscope.

† Current address: Graduate School of Pharmaceutical Sciences, Keio University, Tokyo, Japan.

Methods in Enzymology, Volume 657
ISSN 0076-6879
https://doi.org/10.1016/bs.mie.2021.06.041

1. Introduction

In vivo imaging techniques, including magnetic resonance imaging (MRI), computerized tomography (CT), positron-emission tomography (PET), ultrasound and fluorescence imaging, are now indispensable in clinical practice and biological research, because they can provide detailed anatomical and physiological information non-invasively (Weissleder & Pittet, 2008). Among these imaging techniques, photoacoustic (PA) imaging has recently been attracting much attention (Wang & Hu, 2012; Wang, Lin, Wang, Chen, & Huang, 2016). The mechanism of PA imaging is as follows. Firstly, the optical absorber absorbs a short laser pulse and the energy is converted to heat, which causes thermoelastic expansion and subsequent ultrasound generation. The PA image is constructed by detecting these ultrasound signals. Since PA imaging utilizes light for excitation of the optical absorber and ultrasound as the detection signal, it offers the advantages of both optical imaging and ultrasound imaging, *i.e.*, high contrast with selective photoexcitation in optical imaging and deep tissue penetration of ultrasound in ultrasound imaging. These advantages make it possible to obtain more detailed information in deeper tissue than is possible with fluorescence imaging.

PA imaging of endogenous chromophores such as hemoglobin (Bohndiek et al., 2015) is useful because it can provide physiological information without the need to administer an exogenous contrast agent. For example, PA imaging of hemoglobin provides not only structural information about blood vessels, but also information about oxygen saturation (Bohndiek et al., 2015). Although PA imaging of endogenous absorbers is useful, the biological phenomena that can be imaged in this way are limited. Therefore, exogenous contrast agents that can enable PA imaging of a wider range of biological phenomena are required (Luke, Yeager, & Emelianov, 2012; Weber, Beard, & Bohndiek, 2016). In particular, activatable PA probes (Li et al., 2021; Miao & Pu, 2016; Wang & Zhang, 2021), which can reveal the presence of the target biological phenomenon in terms of a change of PA properties, have been attracting great attention due to their high signal-to-noise ratio and wide range of targetable biological phenomena.

In this chapter, we focus on exogenous contrast agents for PA imaging, especially small organic molecule-based activatable PA probes. Firstly, we briefly summarize key points to consider in developing activatable PA probes. Then, we describe an activatable PA probe for HOCl, PA-MMSiNQ

(Ikeno et al., 2019), which we have recently developed. We present in detail the protocols used for its evaluation by UV–vis spectrometry and PA imaging microscopy.

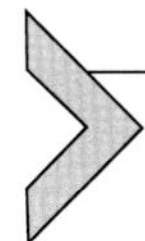

2. Development of small-molecule-based activatable PA probes

With the development of PA imaging technology, there is an increasing demand for exogenous contrast agents, such as carbon nanotubes (De La Zerda et al., 2008), gold nanoparticles (Cheng et al., 2014), polymer nanoparticles (Zha et al., 2013) and organic small molecules (Kim, Song, Gao, & Wang, 2010; Sreejith et al., 2015), that can broaden the range of detectable biological phenomena and provide a high signal-to-noise (S/N) ratio. Although these contrast agents have been applied for lymph vessel imaging (Nagaoka, Tabata, Yoshizawa, Umemura, & Saijo, 2018), tumor imaging (De La Zerda et al., 2012) and so on, most of them are based on passive or active accumulation at the target region. However, these accumulation-based PA probes ("always-on" type PA probes) need a relatively long time to generate a sufficiently high signal, and it is difficult to monitor molecular activities such as enzymatic activities. On other hand, activatable PA probes can exhibit a PA signal change upon binding to or reaction with target molecules, and also enable PA imaging with a high S/N ratio. So, the development of such probes has been actively studied in recent years (Li et al., 2021; Miao & Pu, 2016; Wang & Zhang, 2021).

Several properties are desirable for the development of activatable PA probes. Firstly, near-infrared region (NIR) absorption is important for PA probes because NIR light can penetrate deeper into tissues than visible light due to lower light scattering and the weak absorption of endogenous absorbers, such as hemoglobin, in this region. This high tissue penetration of NIR light is extremely desirable for *in vivo* imaging. Secondly, a high molar extinction coefficient is required for PA probes to efficiently absorb light energy in order to generate a strong ultrasound signal. Thirdly, a low fluorescence quantum yield is preferred for PA probes. Since the PA signal is derived from the nonradiative decay process of dyes after photoexcitation, the radiative decay process contributes to a reduction of the PA signal of dyes. Moreover, photostability is an important property for long-term PA imaging.

Activatable PA probes can be classified in two types: off/on type and ratiometric type. The off/on type activatable PA probes show a PA signal intensity change in the presence of target molecules, and require excitation at a suitable wavelength and an ultrasound transducer for signal detection. This simple equipment requirement is an advantage of off/on type activatable PA probes, while their signal can be dependent on the probe concentration and also can be affected by photodegradation. On the other hand, PA signals are obtained at two or more excitation wavelengths with ratiometric type activatable PA probes, and the change of their ratio is measured. This method can effectively cancel out the effects of differences in concentration, localization and photodegradation of the probe.

Various activatable small-molecule-based PA probes have been developed (Miao & Pu, 2016; Wang & Zhang, 2021), including, for example, activatable PA probes for reactive oxygen species (ROS) (Reinhardt, Zhou, Jorgensen, Partipilo, & Chan, 2018; Toriumi et al., 2019; Yin, Zhen, Fan, Huang, & Pu, 2017; Zhang et al., 2017), metal ion (Li, Zhang, Smaga, Hoffman, & Chan, 2015; Roberts et al., 2018), enzymatic activity (Dragulescu-Andrasi, Kothapalli, Tikhomirov, Rao, & Gambhir, 2013; Levi et al., 2010; Toriumi et al., 2019; Wang et al., 2019) and pH (Jo, Lee, Kopelman, & Wang, 2017). Reported activatable PA probes have been well reviewed elsewhere (Li et al., 2021; Miao & Pu, 2016; Wang & Zhang, 2021). Many activatable PA probes utilize the absorption spectral change caused by a change of the intramolecular charge transfer (ICT) process when they react with or bind to target molecules to generate the PA signal change (Wang & Zhang, 2021). For example, BODIPY-based activatable PA probes for NO (Reinhardt et al., 2018), Cu^{2+}(Li, Zhang, Smaga, Hoffman, & Chan, 2015), and hypoxia (Knox, Kim, Zhu, & Chan, 2018) have been developed, and these probes show ratiometric PA signal changes induced by the reaction with the target molecules. Activatable PA probes that utilize enzymatic cleavage as a PA signal switch have also been developed. Gambhir et al. reported an activatable PA probe for MMP-2 (Levi et al., 2010). This probe is composed of two dyes, BHQ3 and Alexa750, which are linked by an activatable cell-penetrating peptide. The probe can be cleaved by MMP-2 enzymatic activity, and the subsequent accumulation of the cell-penetrating part of the probe in cells leads to an increase of the subtraction PA signal (PA signal at 680 nm—PA signal at 750 nm). Besides MMP-2 (Levi et al., 2010; Yin et al., 2019), activatable PA probes for protease activity, such as furin (Dragulescu-Andrasi et al., 2013) and caspase-3 (Wang et al., 2019), have been developed by utilizing enzymatic cleavage of a peptide substrate as a PA signal switch.

In spite of recent progress in the molecular design strategies for activatable PA probes, the variety of such probes is still limited, probably because it is more difficult to synthesize NIR dyes than dyes working in the visible light region. In particular, the design strategies for absorption control of NIR dyes are less well developed, and therefore new design strategies are still required for NIR activatable PA probes, which would be applicable to various biological targets.

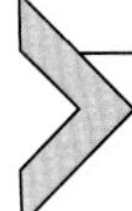

3. Development of an activatable PA probe for HOCl: PA-MMSiNQ

3.1 Design of 2-Me wsSiNQ660 as a platform dye scaffold for NIR rhodamine-based activatable PA probes

In the above context, we set out to establish a rational design strategy for activatable PA probes, and for this purpose, we focused on Si-rhodamine (Koide, Urano, Hanaoka, Terai, & Nagano, 2011a; Kushida, Nagano, & Hanaoka, 2015) (SiR) dyes. SiR is a group of rhodamine dyes in which the O atom at the 10 position of the xanthene structure is replaced by a Si atom, and these dyes show about 90 nm longer absorption wavelengths than conventional rhodamine dyes. Their absorption and fluorescence wavelengths can reach the near-infrared region (>650 nm). In addition, SiR is more resistant to photodegradation than cyanine dyes, which are commonly used as NIR dyes (Koide et al., 2011a). Thus, SiR should be suitable for PA imaging. So far, there have been few examples of the use of rhodamine dyes as scaffolds for activatable PA imaging probes. However, rhodamine dyes have been used as scaffolds for many activatable fluorescent probes, and some of the fluorescence control strategies can also be applied to activatable PA probes. For example, the absorption and fluorescence control strategy based on the intramolecular spirocyclization of rhodamine dyes is considered to be applicable to activatable PA probes. On the other hand, since SiRs are fluorescent dyes, their PA signal is relatively weak because of the efficient fluorescence emission process after photoexcitation. Taking these factors into account, we considered that nonfluorescent SiR would be a promising dye scaffold for activatable PA probes. Therefore, we designed and synthesized 2-Me wsSiNQ660 (Fig. 1A), a dye scaffold which is expected to emit a PA signal with high efficiency, taking advantage of our previous finding that SiNQs (Myochin et al., 2015) are essentially nonfluorescent (fluorescence quantum yield of <0.001) (Fig. 1B). Indeed, we found that 2-Me wsSiNQ660 showed a high light-to-PA signal conversion efficiency compared to fluorescent SiR dye or cyanine dye (Ikeno et al., 2019). Another key point in the molecular design of 2-Me wsSiNQ660 is that it contains two sulfone groups to improve its water solubility, so that

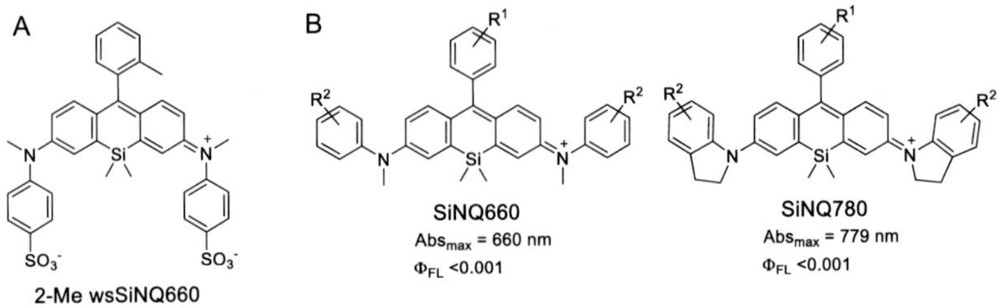

A

2-Me wsSiNQ660

B

SiNQ660
Abs_{max} = 660 nm
Φ_{FL} <0.001

SiNQ780
Abs_{max} = 779 nm
Φ_{FL} <0.001

Fig. 1 (A) Chemical structure of 2-Me wsSiNQ660. (B) Chemical structures and photophysical properties of SiNQs (Myochin et al., 2015). *Reprinted (adapted) with permission from* Anal. Chem. *(2019), 91(14), 9086–9092. Copyright (2019) American Chemical Society.*

a sufficient signal can be obtained. In general, PA imaging shows lower sensitivity than fluorescence imaging, so a high concentration of PA probe is needed to acquire a sufficiently high PA signal.

3.2 Design, synthesis and evaluation of PA-MMSiNQ

To examine whether 2-Me wsSiNQ660 is a promising scaffold for activatable PA probes, we set out to develop an activatable PA probe for hypochlorous acid (HOCl). HOCl is one of the ROS, which kills a variety of pathogens and plays an important role in the human immune system (Prokopowicz et al., 2010). Our molecular design of an activatable PA probe for HOCl, PA-MMSiNQ, is shown in Fig. 2. We employed the HOCl detection strategy of previously reported activatable fluorescent probes for HOCl, HySOx (Kenmoku, Urano, Kojima, & Nagano, 2007) and MMSiR (Koide, Urano, Hanaoka, Terai, & Nagano, 2011b). These probes adopt an intramolecular spirocyclic structure because of nucleophilic attack of the mercaptomethyl group at the 2′-position of the benzene moiety on the 9-position of the xanthene moiety. But, upon reaction with HOCl, the S atom of mercaptomethyl group is oxidized, and the probe molecules take an open form. This leads to the recovery of both absorption and fluorescence of the xanthene moiety. We considered that we could develop an activatable PA probe for HOCl by combining this HOCl detection strategy with the 2-Me wsSiNQ660 scaffold, and so we synthesized PA-MMSiNQ (Fig. 2). Then, we measured the UV–vis spectra of PA-MMSiNQ to investigate whether it would show a large and rapid increase of absorption at 660 nm upon reaction with HOCl. After confirming a suitable absorption spectral change, we investigated whether PA-MMSiNQ works as an activatable PA probe for HOCl in silicone tubes (*in vitro*), and further in mouse subcutis (*in vivo*). The synthetic scheme and the step-by-step protocols of these evaluations are described below.

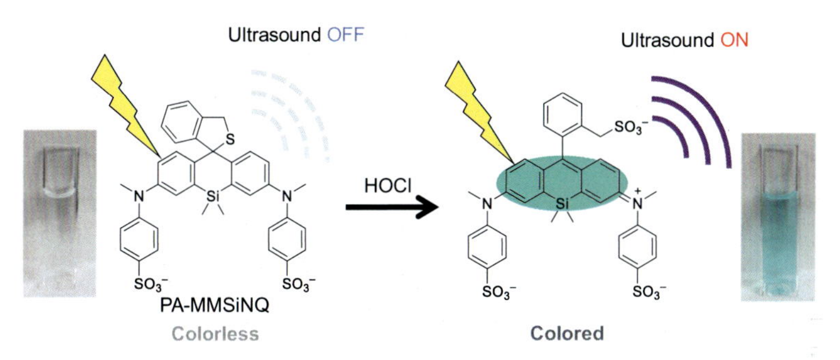

Fig. 2 Design of an activatable PA probe for HOCl. Photographs show cuvettes containing the reaction mixture before and after reaction with HOCl (Ikeno et al., 2019). *Reprinted (adapted) with permission from* Anal. Chem. *(2019), 91(14), 9086–9092. Copyright (2019) American Chemical Society.*

4. Protocols

4.1 Synthesis of PA-MMSiNQ

PA-MMSiNQ can be synthesized by the reaction of a lithiated benzene moiety bearing a mercaptomethyl group protected by a *tert*-butyl group with *N*-Ph Si-xanthone, followed by deprotection of the *tert*-butyl and isopropyl groups.

The synthetic scheme of PA-MMSiNQ is shown in Scheme 1.

4.1.1 N,N,N′,N′*-Tetraallyldiamino-Si-xanthone (5)*

Compound **3** is prepared as reported (Egawa et al., 2011) from 3-bromoaniline (**1**) in two steps.

1. Add compound **3** (5.33 g, 10.3 mmol) and tetrahydrofuran (THF) (30 mL) to a round-bottom flask equipped with a magnetic stirring bar.
2. Flush the flask with Ar, then cool to −78 °C in a mixture of acetone/dry ice.
3. Slowly add 1 M sec-BuLi (30.9 mmol) to the mixture, then stir it for 20 min.
4. At the same temperature, add $SiMe_2Cl_2$ (2.66 g, 20.6 mmol) slowly, then warm the mixture to r.t. Stir the mixture for a further 1 h.
5. Add 2 N HCl aq. to quench the reaction, then neutralize the mixture by adding sat. $NaHCO_3$ aq.
6. Extract the mixture with CH_2Cl_2, and wash the organic layer with brine. Dry the organic layer over Na_2SO_4, and evaporate it to dryness to obtain crude compound **4**.

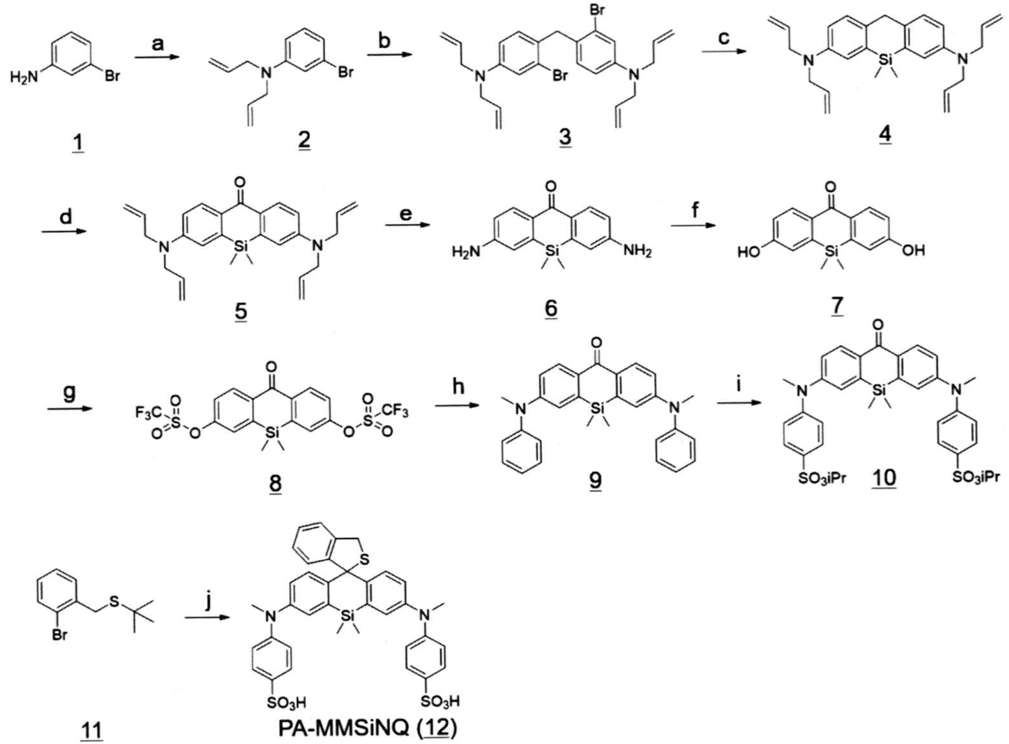

Scheme 1 Synthetic scheme for PA-MMSiNQ. (a) Allyl bromide, K_2CO_3/CH_3CN, 80 °C. (b) HCOH/AcOH, 60 °C. (c)—(i) *sec*-BuLi/THF, −78 °C, (ii) dichlorodimethylsilane/THF, −78 °C to r.t. (d) $KMnO_4/CH_3CN$, 0 °C. (e) $Pd(PPh_3)_4$, 1,3-dimethylbarbituric acid/CH_2Cl_2, 35 °C. (f) $NaNO_2/CH_3OH$, H_2SO_4aq., 0 °C to reflux. (g) *N*-Phenylbis(trifluoromethanesulufonimide), DIEA/DMF, r.t. (h) *N*-Methylaniline, Cs_2CO_3, $Pd_2(dba)_3$, xantphos/toluene, 100 °C. (i)—(i) Chlorosulfuric acid/CH_2Cl_2, 0 °C, (ii) Triisopropyl orthoformate/2-propanol, 55 °C. (j)—(i) *sec*-BuLi/THF, −78 °C, (ii) Compound (**10**)/THF, −78 °C to r.t., and (iii) 2 N HCl aq., CH3CN, reflux.

7. Dissolve crude <u>4</u> in CH_3CN (100 mL) in a round-bottom flask equipped with a magnetic stirring bar, and cool the mixture to 0 °C. To this solution, add $KMnO_4$ (4.93 g, 31.2 mmol) in small portions over a period of 2 h with stirring.
8. Stir the mixture for another 1 h at the same temperature, then dilute it in CH_2Cl_2. Filter the solution through a celite pad, and evaporate the filtrate to dryness.
9. Purify the residue by medium-pressure liquid chromatography (MPLC) (silica gel, CH_2Cl_2) to afford <u>5</u> (1.25 g, 2.92 mmol, 28% yield).

4.1.2 Diamino-Si-xanthone (<u>6</u>)

1. Dissolve compound <u>5</u> (549 mg, 1.28 mmol) and 1,3-dimethylbarbituric acid (1.01 g, 6.48 mmol) in CH_2Cl_2 (10 mL) in a Schlenk flask equipped with a magnetic stirring bar, and flush the flask with Ar.
2. Add $Pd(PPh_3)_4$ (231 mg, 0.200 mmol) to the mixture, and stir it at 35 °C under an Ar atmosphere for 26 h. Then cool it to r.t.
3. Add sat. $NaHCO_3$ aq. to the mixture, and extract the whole with CH_2Cl_2. Wash the organic layer with brine, then dry it over Na_2SO_4. Evaporate it to dryness.
4. Purify the residue by MPLC (silica gel, AcOEt/hexane = 50/50 to 71/29, linear gradient) to afford pure <u>6</u> (297 mg, 1.11 mmol, 87% yield).

4.1.3 Dihydroxy-Si-xanthone (<u>7</u>)

1. Dissolve compound <u>6</u> (297 mg, 1.11 mmol) in CH_3OH/6 N H_2SO_4 aq. (15 mL/30 mL) in a round-bottom flask equipped with a magnetic stirring bar, and cool the mixture to 0 °C.
2. Slowly add a solution of $NaNO_2$ (490 mg, 7.10 mmol) dissolved in H_2O (5 mL) to the mixture, and stir it at the same temperature for 30 min.
3. Slowly pour the mixture into boiling 1 N H_2SO_4 aq. (54 mL), then reflux it for another 15 min.
4. After cooling to 0 °C, extract the whole with CH_2Cl_2. Dry the organic layer over Na_2SO_4, and evaporate it to dryness.
5. Purify the residue by MPLC (silica gel, AcOEt/hexane = 30/70 to 55/45, linear gradient) to afford <u>7</u> (176 mg, 0.651 mmol, 59% yield).

4.1.4 3,6-Ditrifluoromethanesulfonate-Si-xanthone (<u>8</u>)

1. Dissolve compound <u>7</u> (176 mg, 0.651 mmol), *N*-phenylbis(trifluoromethanesulfonimide) (943 mg, 2.64 mmol) and *N,N*-diisopropylethylamine (DIEA) (675 mg, 5.23 mmol) in DMF (8 mL) in a round-bottom flask equipped with a magnetic stirring bar under an Ar atmosphere.

2. Stir the mixture at r.t. for 1.5 h. Then, add H_2O to the mixture to quench the reaction.
3. Extract the whole with CH_2Cl_2, then wash the organic layer with H_2O. Dry the organic layer over Na_2SO_4 and evaporate it to dryness.
4. Purify the residue by MPLC (silica gel, hexane/CH_2Cl_2=80/20 to 30/70, linear gradient) to afford **8** (278 mg, 0.520 mmol, 80% yield).

4.1.5 3,6-Bis(N-methylaniline)-Si-xanthone (9)

1. Dissolve compound **8** (235 mg, 0.440 mmol), *N*-methylaniline (282 mg, 2.64 mmol) and Cs_2CO_3 (867 mg, 2.66 mmol) in toluene (40 mL) in a Schlenk flask equipped with a magnetic stirring bar.
2. Flush the flask with Ar, then add $Pd_2(dba)_3$ (135 mg, 0.147 mmol) and xantphos (140 mg, 0.242 mmol) to the mixture under an Ar atmosphere.
3. Stir the mixture at 100 °C for 14 h.
4. Cool the mixture to r.t. Filter the mixture through paper, and extract the filtrate with CH_2Cl_2. Evaporate the organic layer to dryness.
5. Purify the residue by MPLC (silica gel, hexane/AcOEt=97/3 to 74/26, linear gradient) to afford **9** (109 mg, 0.242 mmol, 55% yield).

4.1.6 3,6-Bis(4-i-PrSO$_3$-N-methylaniline)-Si-xanthone (10)

1. Dissolve compound **9** (137 mg, 0.305 mmol) in CH_2Cl_2 (15 mL) in a round-bottom flask equipped with a magnetic stirring bar, and cool it to 0 °C.
2. Add chlorosulfuric acid (600 μL) dropwise to the mixture, and stir it for 4 h at the same temperature.
3. Add H_2O to quench the reaction, and evaporate the mixture to remove CH_2Cl_2. Roughly purify the residue on a Sep-Pak vac 35 cc (10 g) C18 cartridge (wash with H_2O, then elute with CH_3OH).
4. Evaporate the eluent to remove CH_3OH, then lyophilize it.
5. Dissolve the residue and triisopropyl orthoformate (6.5 mL, 29 mmol) in 2-propanol (13 mL) in a round-bottom flask equipped with a magnetic stirring bar, and stir the mixture for 5 h at 55 °C under Ar.
6. Cool the mixture to r.t., and evaporate it to dryness.
7. Add H_2O to the residue, and extract the mixture with CH_2Cl_2. Dry the organic layer over Na_2SO_4, and evaporate it to dryness. Purify the residue by MPLC (silica gel, hexane/AcOEt=73/27 to 52/48, linear gradient) to afford **10** (140 mg, 0.202 mmol, 65% yield).

4.1.7 PA-MMSiNQ (12)

(2-Bromobenzyl)(*tert*-butyl)sulfane (**11**) can be prepared as reported (Yan & RajanBabu, 2000).

1. Add compound **11** (570 mg, 2.20 mmol) and anhydrous THF (5 mL) to a flame-dried flask equipped with a magnetic stirring bar, and flush the flask with Ar.
2. Cool the mixture to −78 °C, and slowly add 1 M *sec*-BuLi (1.65 mmol) to it with stirring.
3. Stir the mixture for 15 min. Then, add compound **10** (38.1 mg, 0.0551 mmol) dissolved in anhydrous THF (3 mL) to the mixture.
4. Continue to stir for a further 1 h at r.t., then add 2 N HCl aq. to quench the reaction.
5. Add 2 N HCl aq. (15 mL) and CH_3CN (10 mL) to the mixture, reflux it for 19 h, then cool it to r.t.
6. Evaporate the mixture to dryness. Purify the residue by high-performance liquid chromatography (HPLC) (eluent, A/B = 70/30 → 0/100, 40 min; A: H_2O containing 0.1% TFA (v/v), B: CH_3CN/H_2O = 80/20 containing 0.1% TFA (v/v)) to afford PA-MMSiNQ (**12**) (17.0 mg, 0.0238 mmol, 43% yield).

4.2 Evaluation of PA-MMSINQ with a UV–vis spectrometer

4.2.1 Equipment

1. UV–vis spectrometer (Shimadzu UV-1850)
2. Disposable plastic cuvette (light path length is 1 cm)
3. Micropipette and tips

4.2.2 Reagents

1. PBS (pH 7.4) (Gibco)
2. 1 mM PA-MMSiNQ solution (dissolved in DMF)
3. 10 mM HOCl solution (dissolved in 0.1 N NaOHaq.)

4.2.3 Procedure

1. Pour 3 mL of PBS (pH 7.4) into a plastic cuvette.
2. Do baseline correction (500–800 nm) with the UV–vis spectrometer.
3. Add 3 μL of 1 mM PA-MMSiNQ to the cuvette (final: 1 μM). Then, measure the absorption spectrum before reaction with HOCl (500–800 nm).

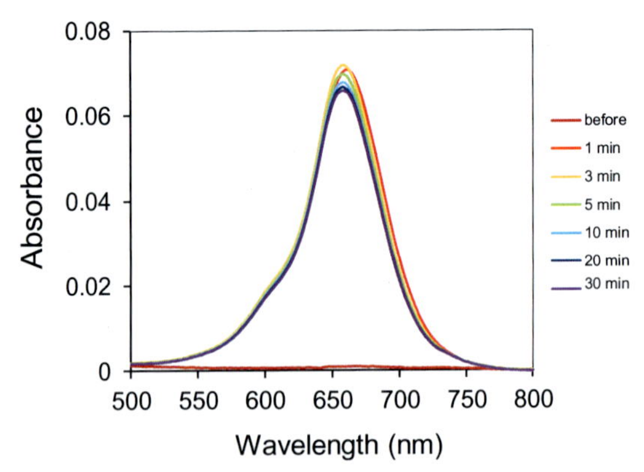

Fig. 3 Time-dependent absorption spectral change of 1 μM PA-MMSiNQ before and at 1, 3, 5, 10, 20, and 30 min after addition of 9 μM HOCl in PBS (pH 7.4) containing 0.1% DMF as a cosolvent. *Reprinted (adapted) with permission from* Anal. Chem. *(2019),* 91*(14), 9086–9092. Copyright (2019) American Chemical Society.*

4. Add 2.7 μL of 10 mM HOCl (final: 9 μM) to the solution, and start absorption spectral measurement to obtain the time-dependent absorption spectral change (500–800 nm, for 30 min every 1 min).

If PA-MMSiNQ works properly, a large absorption increase around 660 nm can be observed (Fig. 3).

4.3 *In vitro* evaluation of PA-MMSiNQ with a PA microscope

4.3.1 Equipment

1. PA microscope (Fig. 4C)
2. 1.5 mL Eppendorf tubes
4. Silicone tubes (inner diameter: 1 mm)
5. Glass plate
6. Clay
7. Micropipette and tips

NOTE: Clay is used to fix the silicone tubes filled with the probe solution to a glass plate.

4.3.2 Reagents

1. PBS (pH 7.4) (Gibco)
2. 10 mM PA-MMSiNQ solution (dissolved in DMSO)
3. 100 mM HOCl solution (dissolved in 0.1 N NaOH aq.)

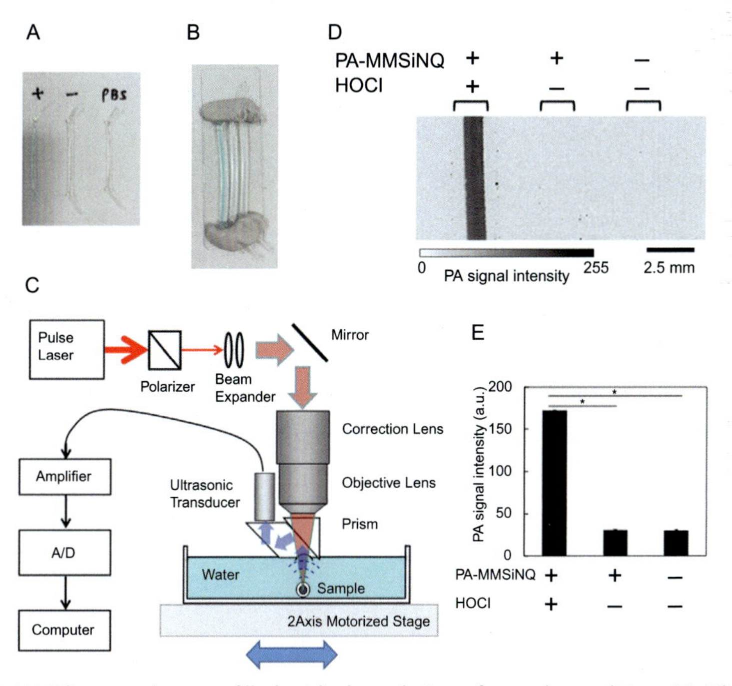

Fig. 4 (A) Silicone tubes are filled with the solutions for each condition. (B) Silicone tubes are fixed on a glass plate with clay. (C) The experimental setup for PA imaging within a silicone tube (inside diameter: 1 mm)(Hu, Maslov, & Wang, 2011). We employed a Q-switched pulse laser system (QL671, CrystaLaser, wavelength 671 nm, pulse width 15 ns, repetition rate 10 kHz). The laser pulse generated by the laser system is attenuated by a polarizer, and the beam diameter is adjusted by a beam expander. The laser pulse passes through a correction lens, an objective lens (LMPLFLN10X, Olympus), a prism, and water. The correction lens corrects spherical aberration produced in the prism and water. The prism acts as a PA wave collector and reflector. The space between the prism and a sample is filled with water. The sample is inserted into a silicone tube (internal diameter 1 mm). The laser pulse is focused into the sample through the objective lens and a lens part of the prism (total optical NA 0.21). A PA wave emanates from the sample, and is gathered by the lens part of the prism (acoustic NA 0.46) and transmitted within the prism to an ultrasonic transducer (C110, Olympus, center frequency 5 MHz). The ultrasonic transducer converts mechanical energy into electrical energy and the electrical energy is amplified by an amplifier (PR5900, Olympus, 54 dB) and measured by an analog-to-digital converter (U1071, Keysight Technologies). To generate 1 pixel of image data, we averaged 100 pulses, so the imaging speed is 100 pixels/s (10 KHz/100). Pixel size is 0.05 mm/pixel. Pulse energies vary from image to image. The sample is moved by a 2-axis motorized stage to acquire a PA image. (D) PA image of 100 μM PA-MMSiNQ after (left) and before (middle) reaction with 600 μM HOCl in PBS (pH 7.4), and PBS only (right). The excitation wavelength was 671 nm. (E) PA signal intensities of three ROIs (8×8 pixels) from each silicone tube in (D) were calculated with ImageJ and averaged. Data are shown as mean ± S.D. ($n = 3$ ROIs). * indicates $p < 0.05$ (one-sided Student's *t*-test). *Panel (C) Reprinted (adapted) with permission from* Anal. Chem. *(2019),* 91*(14), 9086–9092. Copyright (2019) American Chemical Society, Panel (D) Reprinted (adapted) with permission from* Anal. Chem. *(2019),* 91*(14), 9086–9092. Copyright (2019) American Chemical Society. Panel (E) Reprinted (adapted) with permission from* Anal. Chem. *(2019),* 91*(14), 9086–9092. Copyright (2019) American Chemical Society.*

4.3.3 Procedure

1. Prepare three 1.5 mL Eppendorf tubes for three conditions, (A): PA-MMSiNQ + HOCl, (B): PA-MMSiNQ only, (C): PBS (pH 7.4) only.
2. Add PA-MMSiNQ (100 μM) dissolved in PBS (pH 7.4) (500–1000 μL) to the Eppendorf tubes (A) and (B). Add only PBS (pH 7.4) to the Eppendorf tube (C).
3. Add HOCl (final concentration: 600 μM) to the Eppendorf tube (A). The reaction of PA-MMiNQ with HOCl starts. (The reaction usually ends quickly.)
4. Fill silicone tubes with each solution (A) ~ (C). In this process, firstly, prepare silicone tubes (the length is about 10 cm, and both ends are open). Then, slowly fill an adequate volume of the appropriate solution into each silicone tube. After filling the silicone tubes with the solution, tie both ends of the tube to prevent the solution from leaking out (Fig. 4A).
5. Fix the prepared three silicone tubes on a glass plate with clay (Fig. 4B), and then fill the plate with water.
6. Perform PA imaging with the PA microscope (Fig. 4C).
7. Analyze the obtained PA images with ImageJ, and compare the PA signal intensities of each silicone tube to examine whether or not PA-MMSiNQ works as an activatable PA probe for HOCl (Fig. 4D and E).

4.4 *In vivo* evaluation of PA-MMSiNQ with a PA microscope

4.4.1 Equipment

1. PA microscope (Fig. 5C)
2. A small animal anesthetizer, MK-A110 (Muromachi Kikai Co., Ltd)
3. Tape
4. Film
5. Absorbent cotton
6. Micropipette and tips
7. Disposable syringe with needle

4.4.2 Animals and reagents

1. Mouse (BALB/cAJc1-nu-nu, male, 6 weeks old, CLEA Japan)
2. Water
3. Saline (Otsuka Pharmaceutical Co. Ltd)
4. Isoflurane (Wako Pure Chemical)
5. 10 mM PA-MMSiNQ solution (dissolved in DMSO)
6. 100 mM HOCl solution (dissolved in 0.1 N NaOH aq.)

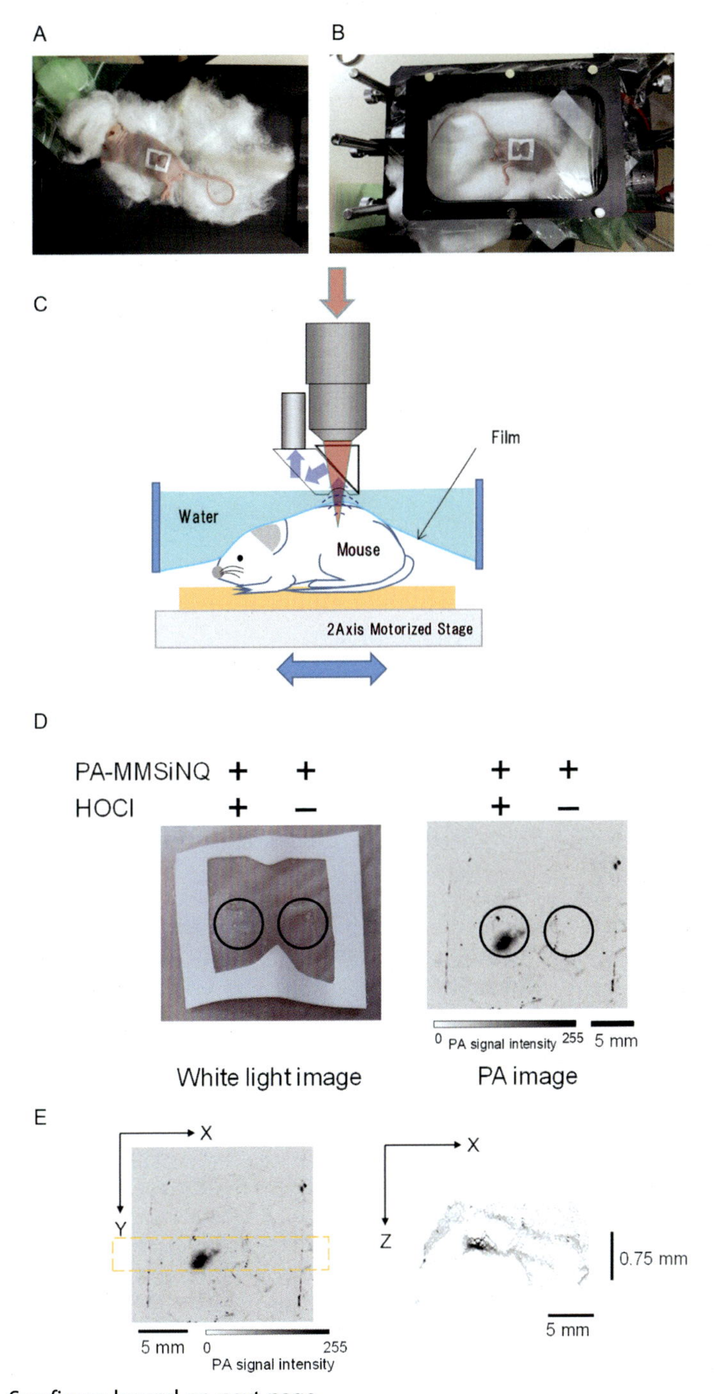

Fig. 5 See figure legend on next page.

4.4.3 Procedure

1. Anesthetize a mouse with the small animal anesthetizer using isoflurane. It is important to use appropriate anesthetic conditions so that the mouse remains alive during the experiment.
2. Lay the anesthetized mouse onto the stage on absorbent cotton. Mark the injection sites on the mouse's thigh with a tape (Fig. 5A).
3. Dissolve PA-MMSiNQ in saline so that the final concentration becomes 200 μM containing 2% DMSO as a cosolvent. Dissolve HOCl in saline so that the final concentration becomes 500 μM.
4. Inject PA-MMSiNQ dissolved in saline (20 μL) into two sites of the thigh of the mouse. Then, inject HOCl dissolved in saline (20 μL) into one of two sites where PA-MMSiNQ was injected.
5. Place film over the injected site, then add water on the upper side of the film (Fig. 5B and C).
6. Perform PA imaging with the PA microscope (Fig. 5C).

In addition to two-dimensional PA images, three-dimensional PA images can be constructed, because the probe position, *i.e.*, tissue depth, can be determined from the propagation time of ultrasound (Fig. 5E). We succeeded in obtaining three-dimensional PA images of PA-MMSiNQ in mouse subcutis.

Fig. 5 (A) Anesthetized mouse on absorbent cotton. (B) After the probe injection, a film is placed over the mouse. (C) The experimental setup for PA imaging of a mouse. Most parts are the same as in Fig. 4C. A mouse is covered with film, and the space between the film and the prism is filled with water. To obtain 1 pixel of an image, we averaged the PA signals obtained with 100 laser pulses. The imaging speed is 100 pixels/s. The pixel size is 0.1 mm/pixel (horizontal direction), 0.015 mm/pixel (vertical (depth) direction). (D) White-light image of mouse thigh and PA image of HOCl in mouse subcutis with PA-MMSiNQ. Left circle: 200 μM PA-MMSiNQ in 20 μL of saline containing 2% DMSO as a cosolvent was subcutaneously injected into the thigh of a mouse. Subsequently, 500 μM HOCl in 20 μL of saline was injected into the same location. Right circle: 200 μM PA-MMSiNQ in 20 μL of saline containing 2% DMSO as a cosolvent was subcutaneously injected into the thigh of a mouse. The white-light image was obtained just after the injection, and the PA image was obtained 40 min after the injection (this time difference was due to the time required for imaging). The excitation wavelength was 671 nm. (E) PA imaging in the XY direction (left) and the XZ direction (right). The XZ direction image, which corresponds to the orange square area in the XY-direction image, was also constructed. The XZ direction image was constructed from the PA signals whose waveform was correlated to the waveform from PA-MMSiNQ (correlation coefficient: $r \geq 0.4$). *Panel (C) Reprinted (adapted) with permission from* Anal. Chem. *(2019), 91(14), 9086–9092. Copyright (2019) American Chemical Society. Panel (D) Reprinted (adapted) with permission from* Anal. Chem. *(2019), 91(14), 9086–9092. Copyright (2019) American Chemical Society. Panel (E) Reprinted (adapted) with permission from* Anal. Chem. *(2019), 91(14), 9086–9092. Copyright (2019) American Chemical Society.*

5. Concluding remarks

PA imaging is a promising imaging technique which allows deep-tissue imaging with high spatial resolution in a non-invasive manner, and intensive researches have led to improvements in many aspects of PA imaging, such as PA microscopy, analysis, contrast agents, *etc.* To further broaden the range of detectable biological events and to improve the S/N ratios of PA images, the development of new contrast agents, in particular activatable PA probes, is needed. In this chapter, we have briefly summarized the molecular design strategy for small-molecule-based activatable PA probes, and described in more detail an activatable PA probe for HOCl, which we have recently reported. We anticipate that the 2-Me wsSiNQ660 dye scaffold used in this probe will be applicable to many other activatable PA probes in the future, and this will enable PA imaging to become an increasingly useful imaging modality in life science research and clinical medicine.

References

Bohndiek, S. E., et al. (2015). Photoacoustic tomography detects early vessel regression and normalization during ovarian tumor response to the antiangiogenic therapy trebananib. *Journal of Nuclear Medicine*, *56*, 1942–1947.

Cheng, K., et al. (2014). Construction and validation of nano gold tripods for molecular imaging of living subjects. *Journal of the American Chemical Society*, *136*, 3560–3571.

De La Zerda, A., et al. (2008). Carbon nanotubes as photoacoustic molecular imaging agents in living mice. *Nature Nanotechnology*, *3*, 557–562.

De La Zerda, A., et al. (2012). Family of enhanced photoacoustic imaging agents for high-sensitivity and multiplexing studies in living mice. *ACS Nano*, *6*, 4694–4701.

Dragulescu-Andrasi, A., Kothapalli, S. R., Tikhomirov, G. A., Rao, J., & Gambhir, S. S. (2013). Activatable oligomerizable imaging agents for photoacoustic imaging of furin-like activity in living subjects. *Journal of the American Chemical Society*, *135*, 11015–11022.

Egawa, T., et al. (2011). Development of a fluorescein analogue, TokyoMagenta, as a novel scaffold for fluorescence probes in red region. *Chemical Communications*, *47*, 4162–4164.

Hu, S., Maslov, K., & Wang, L. V. (2011). Second-generation optical-resolution photoacoustic microscopy with improved sensitivity and speed. *Optics Letters*, *36*, 1134–1136.

Ikeno, T., et al. (2019). Design and synthesis of an activatable photoacoustic probe for hypochlorous acid. *Analytical Chemistry*, *91*, 9086–9092.

Jo, J., Lee, C. H., Kopelman, R., & Wang, X. (2017). In vivo quantitative imaging of tumor pH by nanosonophore assisted multispectral photoacoustic imaging. *Nature Communications*, *8*, 471.

Kenmoku, S., Urano, Y., Kojima, H., & Nagano, T. (2007). Development of a highly specific rhodamine-based fluorescence probe for hypochlorous acid and its application to real-time imaging of phagocytosis. *Journal of the American Chemical Society*, *129*, 7313–7318.

Kim, C., Song, K. H., Gao, F., & Wang, L. V. (2010). Sentinel lymph nodes and lymphatic vessels: Noninvasive dual-modality in vivo mapping by using indocyanine green in rats—Volumetric spectroscopic photoacoustic imaging and planar fluorescence imaging. *Radiology*, *255*, 442–450.

Knox, H. J., Kim, T. W., Zhu, Z., & Chan, J. (2018). Photophysical tuning of N-oxide-based probes enables ratiometric photoacoustic imaging of tumor hypoxia. *ACS Chemical Biology*, *13*, 1838–1843.
Koide, Y., Urano, Y., Hanaoka, K., Terai, T., & Nagano, T. (2011a). Evolution of group 14 rhodamines as platforms for near-infrared fluorescence probes utilizing photoinduced electron transfer. *ACS Chemical Biology*, *6*, 600–608.
Koide, Y., Urano, Y., Hanaoka, K., Terai, T., & Nagano, T. (2011b). Development of an Si-rhodamine-based far-red to near-infrared fluorescence probe selective for hypochlorous acid and its applications for biological imaging. *Journal of the American Chemical Society*, *133*, 5680–5682.
Kushida, Y., Nagano, T., & Hanaoka, K. (2015). Silicon-substituted xanthene dyes and their applications in bioimaging. *Analyst*, *140*, 685–695.
Levi, J., et al. (2010). Design, synthesis, and imaging of an activatable photoacoustic probe. *Journal of the American Chemical Society*, *132*, 11264–11269.
Li, C., et al. (2021). Recent development of near-infrared photoacoustic probes based on small-molecule organic dye. *RSC Chemical Biology*, *2*, 743–758.
Li, H., Zhang, P., Smaga, L. P., Hoffman, R. A., & Chan, J. (2015). Photoacoustic probes for ratiometric imaging of copper(II). *Journal of the American Chemical Society*, *137*, 15628–15631.
Luke, G. P., Yeager, D., & Emelianov, S. Y. (2012). Biomedical applications of photoacoustic imaging with exogenous contrast agents. *Annals of Biomedical Engineering*, *40*, 422–437.
Miao, Q., & Pu, K. (2016). Emerging designs of activatable photoacoustic probes for molecular imaging. *Bioconjugate Chemistry*, *27*, 2808–2823.
Myochin, T., et al. (2015). Development of a series of near-infrared dark quenchers based on Si-rhodamines and their application to fluorescent probes. *Journal of the American Chemical Society*, *137*, 4759–4765.
Nagaoka, R., Tabata, T., Yoshizawa, S., Umemura, S.-I., & Saijo, Y. (2018). Visualization of murine lymph vessels using photoacoustic imaging with contrast agents. *Photoacoustics*, *9*, 39–48.
Prokopowicz, Z. M., et al. (2010). Hypochlorous acid: A natural adjuvant that facilitates antigen processing, cross-priming, and the induction of adaptive immunity. *The Journal of Immunology*, *184*, 824–835.
Reinhardt, C. J., Zhou, E. Y., Jorgensen, M. D., Partipilo, G., & Chan, J. (2018). A ratiometric acoustogenic probe for in vivo imaging of endogenous nitric oxide. *Journal of the American Chemical Society*, *140*, 1011–1018.
Roberts, S., et al. (2018). Calcium sensor for photoacoustic imaging. *Journal of the American Chemical Society*, *140*, 2718–2721.
Sreejith, S., et al. (2015). Near-infrared squaraine dye encapsulated micelles for in vivo fluorescence and photoacoustic bimodal imaging. *ACS Nano*, *9*, 5695–5704.
Toriumi, N., et al. (2019). Design of photostable, activatable near-infrared photoacoustic probes using tautomeric benziphthalocyanine as a platform. *Angewandte Chemie, International Edition*, *58*, 7788–7791.
Wang, L. V., & Hu, S. (2012). Photoacoustic tomography: In vivo imaging from organelles to organs. *Science*, *335*, 1458–1462.
Wang, S., Lin, J., Wang, T., Chen, X., & Huang, P. (2016). Recent advances in photoacoustic imaging for deep-tissue biomedical applications. *Theranostics*, *6*, 2394–2413.
Wang, S., & Zhang, X. (2021). Design strategies of photoacoustic molecular probes. *Chembiochem*, *22*, 308–316.
Wang, Y., et al. (2019). Photoacoustic imaging of tumor apoptosis via caspase-instructed macrocyclization and self-assembly. *Angewandte Chemie (Weinheim an der Bergstrasse, Germany)*, *131*, 4940–4944.

Weber, J., Beard, P. C., & Bohndiek, S. E. (2016). Contrast agents for molecular photoacoustic imaging. *Nature Methods*, *13*, 639–650.

Weissleder, R., & Pittet, M. J. (2008). Imaging in the era of molecular oncology. *Nature*, *452*, 580–589.

Yan, Y. Y., & RajanBabu, T. V. (2000). Highly flexible synthetic routes to functionalized phospholanes from carbohydrates. *The Journal of Organic Chemistry*, *65*, 900–906.

Yin, C., Zhen, X., Fan, Q., Huang, W., & Pu, K. (2017). Degradable semiconducting oligomer amphiphile for ratiometric photoacoustic imaging of hypochlorite. *ACS Nano*, *11*, 4174–4182.

Yin, L., et al. (2019). Quantitatively visualizing tumor-related protease activity in vivo using a ratiometric photoacoustic probe. *Journal of the American Chemical Society*, *141*, 3265–3273.

Zha, Z., et al. (2013). Biocompatible polypyrrole nanoparticles as a novel organic photoacoustic contrast agent for deep tissue imaging. *Nanoscale*, *5*, 4462–4467.

Zhang, J., et al. (2017). Activatable photoacoustic nanoprobes for in vivo ratiometric imaging of peroxynitrite. *Advanced Materials*, *29*, 1604764.

CHAPTER TWO

A caspase-3 activatable photoacoustic probe for in vivo imaging of tumor apoptosis

Yuqi Wang and Deju Ye*
State Key Laboratory of Analytical Chemistry for Life Science, Chemistry and Biomedicine Innovation Center (ChemBIC), School of Chemistry and Chemical Engineering, Nanjing University, Nanjing, China
*Corresponding author: e-mail address: dejuye@nju.edu.cn

Contents

Methods in Enzymology, Volume 657
ISSN 0076-6879
https://doi.org/10.1016/bs.mie.2021.06.021

Abstract

Photoacoustic (PA) imaging is an emerging imaging technique, which combines high spatial resolution and deep tissue penetration of ultrasound imaging with high sensitivity of fluorescence imaging. In the past few years, PA has shown promise for noninvasive imaging of biomolecules in vivo. In this chapter, we present the synthesis and application of a tumor targeting and caspase-3 activatable PA probe (**1-RGD**) for real-time and noninvasive imaging of tumor apoptosis. **1-RGD** can be efficiently delivered into tumor tissues and recognized by caspase-3, which triggered efficient proteolysis of DEVD substrate and subsequent intramolecular macrocyclization, followed by in situ self-assembly into nanoparticles, leading to prolonged retention in apoptotic tumors and enhanced PA signals. With **1-RGD**, high-resolution 3D PA images of tumor tissues can be obtained, allowing to report on the activity and distribution of caspase-3 within DOX-treated tumors, which was helpful for early monitoring of tumor response to therapy. We provide detailed protocols for the synthesis, in vitro characterization and in vivo applications of **1-RGD**.

1. Introduction

Caspase-3, which belongs to cysteine protease family, has been considered as a key effector enzyme in inducing cell apoptosis. Caspase-3 is present as an inactive pro-caspase in viable cells, which is activated during apoptosis, ultimately executing cell death (Brindle, 2008; Shalini, Dorstyn, Dawar, & Kumar, 2014). As tumor cells response to chemotherapy is mainly via cell apoptosis pathway and activation of caspase-3, caspase-3 has been recognized as an important early biomarker for evaluating chemotherapy-induced cell death (Pop & Salvesen, 2009; Ray, De, Patel, & Gambhir, 2008). However, due to the high heterogeneity of tumor microenvironment, chemotherapy-induced tumor apoptosis is likely heterogeneous as well; the caspase-3 is activated within specific regions of tumor regions response to therapy. Therefore, precise detection of caspase-3 activity and visualization of its distribution within a whole tumor tissue are of great importance for early evaluation of therapeutic efficacy. To date, many fluorescent probes have been developed to detect caspase-3 activity (Edgington et al., 2009; Huang et al., 2013; Maxwell, Chang, Zhang, Barnett, & Piwnica-Worms, 2009; Shi et al., 2012; Wang et al., 2011). However, fluorescence imaging suffers from low penetration depth and poor spatial resolution as a result of strong absorption and scattering of light by biological tissues, limiting the ability for noninvasive detection of caspase-3 in deep-seated tumors (Ntziachristos, 2010; Wang, 2009). Several positron emission tomography (PET) probes have also been reported to visualize caspase-3 activity in tumor tissues with high sensitivity (Hight et al., 2014; Nguyen et al., 2009; Su et al., 2013). However, PET probes also faces the limitations of low spatial resolution and risky radiation. Magnetic resonance imaging (MRI) contrast agents have the advantages of high spatial resolution and unlimited penetration depth for the detection of tumor apoptosis. However, it is required to accumulate high concentration of tracers in apoptotic regions due to the low sensitivity of MRI, which prohibits the sensitive detection of caspase-3 activity in early stages of therapy (Mizukami et al., 2008; Mizukami, Takikawa, Sugihara, Shirakawa, & Kikuchi, 2009; Yuan et al., 2015). Therefore, new imaging technologies and imaging probes capable of detecting caspase-3 activity in vivo with high sensitivity and high spatial resolution are urgently demanded.

Photoacoustic (PA) imaging is a new type of molecular imaging technology developed rapidly in recent years (Kim, Favazza, & Wang, 2010; Zhang, Smaga, Satyavolu, Chan, & Lu, 2017). Based on the detection of ultrasound waves generated from the thermoelastic expansion of an absorber upon light irradiation, PA imaging can produce high resolution 3D images with much deeper tissue penetration in relative to fluorescence imaging, showing promise for in vivo imaging of essential physiological and pathological processes (Miao & Pu, 2016; Pu, Chattopadhyay, & Rao, 2016; Razgulin, Ma, & Rao, 2011; Reinhardt & Chan, 2018; Wang & Hu, 2012). To achieve higher PA contrast, a number of PA probes capable of accumulating in disease sites through active or passive delivery have been developed (De La Zerda et al., 2008, 2012; Sun et al., 2016; Yang, Cai, Forrest, Cui, & Yang, 2011; Zhang et al., 2014). In addition, many activatable PA probes with their PA signals selectively switched from "off" to "on" in response to a biomolecule of interest, such as ions (Li, Zhang, Smaga, Hoffman, & Chan, 2015), reactive oxygen species (Chen, Zhen, & Pu, 2017; Pu et al., 2014; Yin, Zhen, Fan, Huang, & Pu, 2017), and enzymes (Levi et al., 2010; Miao & Pu, 2018; Wu, Zhang, Du, Cheng, & Liang, 2018) have also been devoted, enabling real-time detection of biomolecules in deep tissues with low background signal and high sensitivity. Though encouraging progresses have been achieved, a caspase-3 activatable PA probe capable of detecting caspase-3 activity and mapping apoptotic regions in tumor tissues are remaining elusive.

In this chapter, we present a caspase-3 activatable PA probe (**1-RGD**) based on our previously reported biocompatible reaction-mediated macrocyclization and self-assembly strategy (Ye, Liang, Ma, & Rao, 2011). This strategy has been proved to be effective to design fluorescence

(Ye et al., 2014), PET (Shen et al., 2013), and MRI (Liang et al., 2011; Nejadnik et al., 2015; Ye et al., 2014) probes for in vivo imaging. We outline detailed protocols for the synthesis and in vitro characterization of **1-RGD**. We demonstrated that **1-RGD** can be efficiently activated in the presence of caspase-3 and then self-assemble into nanoparticles, providing a significantly enhanced PA signal. We also provide a concise but detailed outline for the application of **1-RGD** for in vivo PA imaging of apoptotic tumor regions in living mice with DOX-treated tumors.

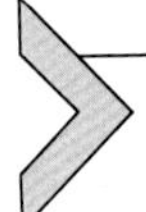

2. Design of tumor targeting and caspase-3 activatable probe, 1-RGD

1-RGD consists of five major components: (1) D-cysteine (D-Cys) and 2-cyano-6-hydroxyquinoline (CHQ) residues for biocompatible intramolecular cyclization, (2) a glutathione (GSH) reducible disulfide bond; (3) a caspase-3 cleavable peptide substrate (DEVD); (4) a clinically applicable near-infrared (NIR) dye (indocyanine green, ICG); and (5) a cyclic peptide (c-RGD) ligand to target the $\alpha_v\beta_3$ integrin receptor overexpressed on tumor cells (Fig. 1A). Upon intravenous (i.v.) injection, **1-RGD** can be delivered into tumor tissues through the targeting effect of c-RGD. After internalized into apoptotic tumor cells, thiol and amino groups of the D-Cys residue in **1-RGD** are uncaged by intracellular GSH and active caspase-3. Then, the fast intramolecular condensation reaction between free D-Cys and CHQ proceeds, generating a more rigid and hydrophobic cyclized product **1-cycl**, which can further self-assemble into ICG-containing nanoparticles assisted by the enhanced intermolecular interactions (e.g., hydrophobic interaction and π-π stacking). The emission of the nanoparticles containing ICG molecules greatly decreases owing to the aggregation-caused quenching (ACQ) effect, while the PA signal increases resulting from the augment of the nonradiative relaxation process (see Fig. 1B). Compared with small molecules, the molecular sizes of nanoparticles increase dramatically, facilitating to prolong the retention in apoptotic tumor tissues. With the unactivated small-molecule **1-RGD** washed out of viable tumor tissues, enhanced PA contrast to signal caspase-3 activity in apoptotic tumor regions can be achieved. Benefiting from the high-resolution of PA imaging, **1-RGD** can detect the apoptotic regions distributed in a whole tumor, allowing to evaluate the therapeutic efficacy against tumors at the early stage prior to the tumor size changes.

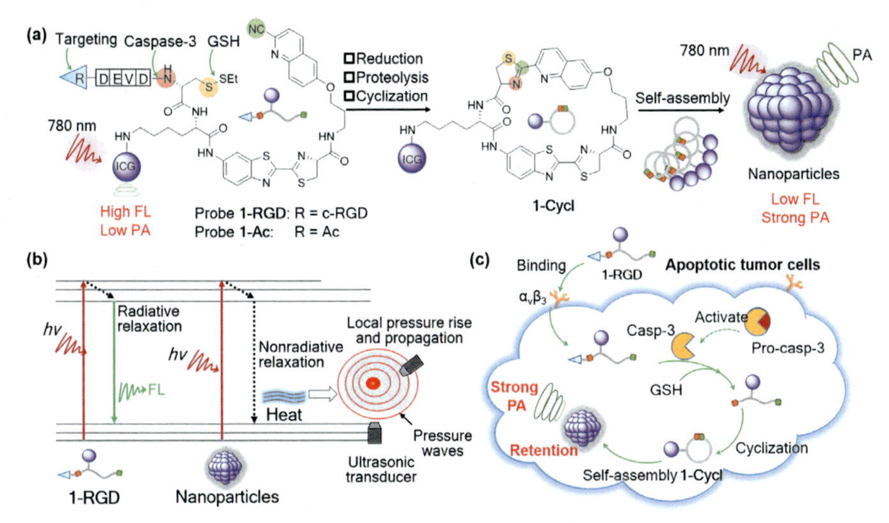

Fig. 1 General design of the caspase-3 activatable PA probes. (A) Proposed chemical conversion of **1-RGD** and **1-Ac** in the presence of caspase-3 and GSH. (B) Jablonski diagram presents the proposed mechanism to amplify PA signal by augmenting nonradiative relaxation of the excited ICG fluorophores within nanoparticles. (C) Schematic illustration of **1-RGD** for in vivo PA imaging of caspase-3 activity in apoptotic tumor cells. *Reproduced with permission from John Wiley and Sons (Wang Y., Hu X., Weng J., Li J., Fan Q., Zhang Y., et al. (2019). A Photoacoustic Probe for the Imaging of Tumor Apoptosis by Caspase-Mediated Macrocyclization and Self-Assembly.* Angewante Chemie International Edition, 58, *4886–4890. https://doi.org/10.1002/anie.201813748).*

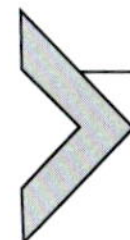

3. Synthesis of 1-RGD and control probe 1-Ac

3.1 Compound 2 and compound 3

Compound **2** and compound **3** were synthesized according to previously reported method (Ye et al., 2011).

3.2 Compound 4

1. Add **2** (0.2 mmol) to a 50 mL round bottom flask equipped with a magnetic stir bar. Seal the flask with a rubber plug
2. Fill the flask with dry nitrogen
3. Add N,N-diisopropylethylamine (DIPEA, 0.2 mmol), dichloromethane (DCM, 5 mL), and methanol (MeOH, 5 mL) to the flask via syringe
4. Add 5 mL of MeOH to a 10 mL centrifuge tube. Add **3** (0.22 mmol), tris(2-carboxyethyl) phosphine hydrochloride (TCEP, 0.2 mmol) and

DIPEA (1.0 mmol) to the tube successively. The mixture was vortexed until the solids were dissolved completely

5. Add the mixture of **3** to the 50 mL flask dropwise via syringe
6. Continue stirring the mixture at room temperature (25 °C) for 30 min or until the reaction is complete by TLC (DCM/MeOH = 20/1)
7. Remove the solvent using rotary evaporator and dissolve the residue with 20 mL of ethyl acetate (EA). Transfer the solution to a separating funnel. Wash the organic layer with water (2 × 20 mL) and brine (1 × 20 mL)
8. Dry the organic layer using anhydrous sodium sulfate and evaporate the solvent to obtain crude product
9. Add the crude product, triisopropylsilane (TIPSH, 0.4 mmol), trifluoroacetic acid (TFA, 5 mL), and DCM (5 mL) to a 50 mL round bottom flask equipped with a magnetic stir bar
10. Stir the reaction mixture at room temperature for 3 h or until the reaction is complete by TLC (DCM/MeOH = 20/1). Evaporate the solvent under reduced pressure
11. Add cold diethyl ether (Et_2O, 20 mL) to the flask, yielding a light yellow precipitate. Transfer the suspension to a 50 mL centrifuge tube. After centrifugation, remove the supernatant and add 20 mL of Et_2O to the tube to redisperse. Remove the supernatant after centrifugation. Dry the precipitate under vacuum and obtain the crude intermediate as a light yellow solid
12. Without further purification, add the intermediate and 10 mL of MeOH to a 50 mL round bottom flask equipped with a magnetic stir bar
13. Measure 2-(ethyldisulfanyl)pyridine (PySSEt, 0.3 mmol) and 5 mL of MeOH to a 10 mL centrifuge tube. Add the solution to the 50 mL round bottom flask within 10 min using a syringe
14. Continue stirring the mixture at room temperature (25 °C) for 30 min or until the reaction is complete by TLC (DCM/MeOH = 20/1)
15. Remove the solvent using rotary evaporator and use silica gel column chromatography to isolate the product (DCM/MeOH = 50/1 to DCM/MeOH = 20/1) as a light yellow solid in 33% yield (from compound **2**)

3.3 Compound 5

1. Prepare peptide Alkyne-Asp(O^{t}Bu)-Glu(O^{t}Bu)-Val-Asp(O^{t}Bu)-COOH following the standard procedure of solid phase peptide synthesis

2. Add compound **4** (0.2 mmol), Alkyne-Asp(O^tBu)-Glu(O^tBu)-Val-Asp(O^tBu)-COOH (0.22 mmol), and HBTU (0.22 mmol) to a 50 mL round bottom flask equipped with a magnetic stir bar. Seal the flask with a rubber plug
3. Fill the flask with dry nitrogen
4. Add 20 mL of anhydrous THF and DIPEA (0.4 mmol) to a 50 mL tube. After fully mixed, add the solution to the 50 mL flask via syringe
5. Continue stirring the mixture at room temperature for 2 h or until the reaction is complete by TLC (DCM/MeOH = 15/1). Evaporate the solvent under reduced pressure
6. Use DCM/MeOH (15/1) as the eluent to purify the residue with a short silica gel column
7. Add the intermediate to a 10 mL round bottom flask equipped with a magnetic stir bar. Add 5% piperidine (5 mL) to the flask
8. Stir the reaction solution at room temperature for 10 min. Place the flask into an ice-bath. Then, add 1 M HCl dropwise to adjust the pH to 5–7
9. Using semi preparative HPLC to isolate compound **5**. After lyophilization, compound **5** can be obtained as a light yellow powder (41% yield, two steps from compound **4**).

3.4 Compound 6

1. Add compound **5** (0.2 mmol) to a 50 mL round bottom flask equipped with a glass stopper and a magnetic stir bar
2. Measure triisopropylsilane (TIPSH, 0.6 mmol), trifluoroacetic acid (TFA, 19 mL), DCM (0.5 mL), and acetonitrile (ACN, 0.5 mL) to a tube. Add the mixture to the 50 mL flask after fully mixed
3. Stir the reaction mixture at room temperature for 3 h until all of the $O^{t-}Bu$ group were removed
4. Remove the solvent under reduced pressure and add cold Et_2O (~30 mL) to the flask, yielding a light yellow precipitate. Transfer the suspension to a 50 mL centrifuge tube. After centrifugation, remove the supernatant and add 30 mL of Et_2O to the tube to redisperse. Remove the supernatant after centrifugation. Dry the precipitate under vacuum to obtain compound **6** as a light yellow solid in 84% yield

3.5 Compound 7

1. Add **6** (0.1 mmol) and tert-Butanol (2.5 mL) to a 10 mL round bottom flask equipped with a stir bar

2. Add Tris(3-hydroxypropyltriazolylmethyl)amine (THPTA, 0.05 mmol), $CuSO_4$ (0.05 mmol), sodium L-Ascorbate (0.1 mmol), and deionized water (2.5 mL) to a 5 mL tube
3. Add the mixture prepared in step 2 and c-RGDfK(N_3) (0.11 mmol) to the flask successively. Stir the reaction mixture at room temperature for 30 min or until the reaction is complete by HPLC
4. Isolate product **7** using semi preparative HPLC. Compound **7** can be obtained as a white to light yellow powder in 81% yield after lyophilization

3.6 1-RGD

1. Add **7** (2.6 mg) to a 1.5 mL tube
2. Add anhydrous N,N-Dimethylformamide (DMF, 0.2 mL) and DIPEA (1.0 μL) to the tube to dissolve **7**. NOTE: Confirm that the solution is weakly alkaline. If the pH of the mixture is still below 7, more DIPEA should be added. This is very important because it is hard to confirm the pH once ICG-OSu (dark green) is added
3. Add ICG-OSu (1.0 mg) and DMF (0.1 mL) to the tube and then place the tube on a vortex mixer to vortex for 2 h or until the reaction is complete by HPLC
4. Use semi-preparative HPLC to isolate **1-RGD**. The final product should be a dark green fluffy powder (47% yield) (Scheme 1).

3.7 Compound 8

1. Prepare peptide Ac-Asp(O^{t}Bu)-Glu(O^{t}Bu)-Val-Asp(O^{t}Bu)-COOH following the standard procedure of solid phase peptide synthesis
2. Add compound **4** (0.2 mmol), Ac-Asp(O^{t}Bu)-Glu(O^{t}Bu)-Val-Asp(O^{t}Bu)-COOH (0.22 mmol), and HBTU (0.22 mmol) to a 50 mL round bottom flask equipped with a magnetic stir bar and then seal the flask with a rubber plug
3. Fill the flask with dry nitrogen
4. Add 20 mL of anhydrous THF and DIPEA (0.4 mmol) to a 50 mL tube. After fully mixed, add the solution into the 50 mL flask via syringe
5. Stir the reaction mixture at room temperature for 2 h or until the reaction is complete by TLC (DCM/MeOH = 15/1). Evaporate the solvent under reduced pressure

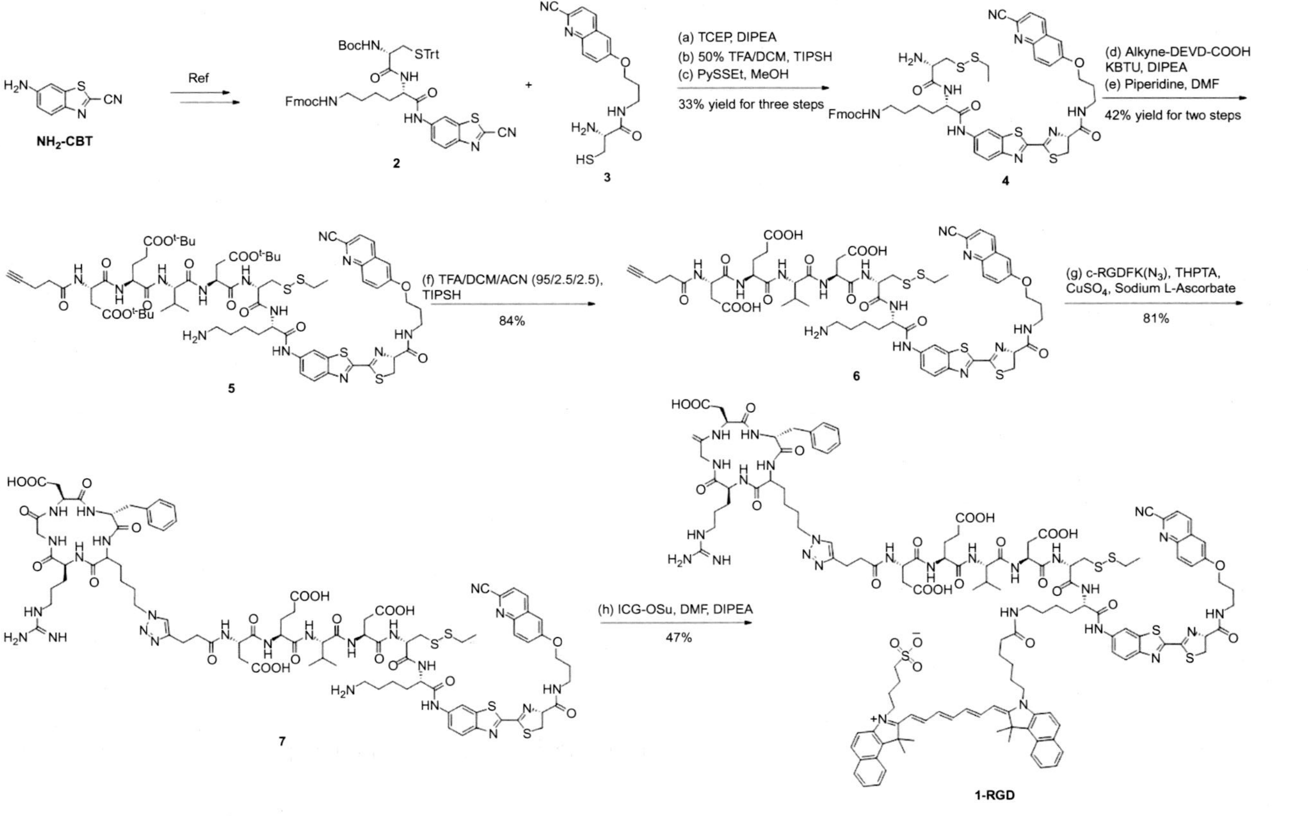

Scheme 1 Synthetic route of **1-RGD**.

6. Purify the residue with a short silica gel column using DCM/MeOH (15/1) as the eluent
7. Add the intermediate to a 10 mL round bottom flask equipped with a magnetic stir bar. Add 5% piperidine (5 mL) to the flask
8. Stir the reaction mixture at room temperature for 10 min. Put the flask into an ice-bath and add 1 M HCl dropwise to the flask to adjust the pH to 5–7
9. Using semi-preparative HPLC to isolate compound **8**. After lyophilization, compound **8** can be obtained as a light yellow powder in 39% yield (two steps from compound **4**)

3.8 Compound 9

1. Measure 0.2 mmol of compound **9** to a 50 mL round bottom flask equipped with a magnetic stir bar
2. Add triisopropylsilane (TIPSH, 0.6 mmol), trifluoroacetic acid (TFA, 19 mL), DCM (0.5 mL), and ACN (0.5 mL) to the 50 mL flask
3. Stir the reaction mixture at room temperature for 3 h until all of the O^{t-}Bu group were removed
4. Remove the solvent under reduced pressure and add cold Et_2O (~30 mL) to the flask, yielding a light yellow precipitate. Transfer the suspension to a 50 mL centrifuge tube. After centrifugation, remove the supernate and add 30 mL of Et_2O to the tube to redisperse. Remove the supernate and dry the precipitate under vacuum to obtain compound **9** as a light yellow solid in 82% yield

3.9 1-Ac

1. Add **9** (2.6 mg) to a 1.5 mL tube
2. Add dry DMF (0.2 mL) and DIPEA (1.0 μL) to the tube to dissolve **9**
3. Add ICG-OSu (1.0 mg) and DMF (0.1 mL) to the tube and then placed the tube on a vortex mixer to vortex for 2 h or until the reaction is complete by HPLC
4. Isolate **1-Ac** using semi-preparative HPLC. The final product should be a dark green fluffy powder in 43% yield (Scheme 2)

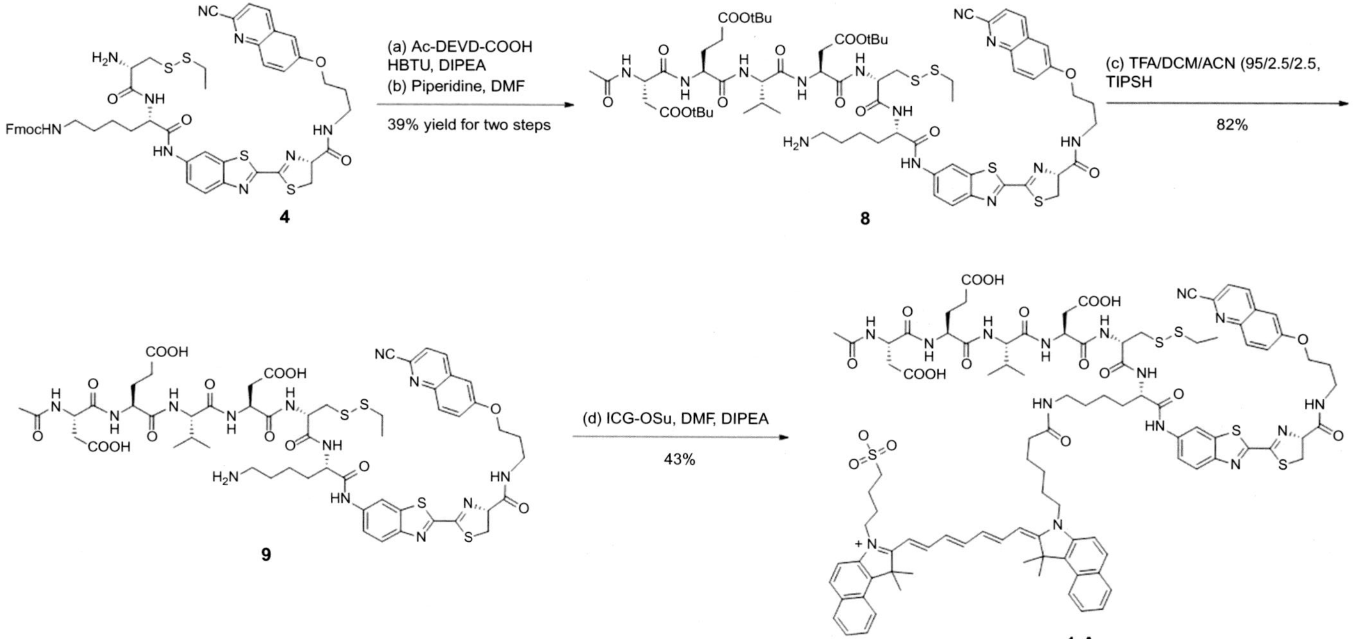

(a) Ac-DEVD-COOH
HBTU, DIPEA
(b) Piperidine, DMF

39% yield for two steps

(c) TFA/DCM/ACN (95/2.5/2.5,
TIPSH

82%

(d) ICG-OSu, DMF, DIPEA

43%

4

8

9

1-Ac

Scheme 2 Synthetic route of the control probe **1-Ac**.

4. In vitro characterization of 1-RGD and 1-Ac

After completing the synthesis of **1-RGD** and **1-Ac**, the following experiments are needed to evaluate the ability of **1-RGD** to detect caspase-3: (1) confirm the cyclized product and the self-assembled nanostructures of **1-RGD** upon incubation with caspase-3, (2) verify the fluorescence and photoacoustic signal of **1-RGD** before and after incubation with caspase-3, (3) confirm the stability, selectivity, sensitivity, and biocompatibility of **1-RGD**, (4) validate that **1-RGD** can detect caspase-3 activity in apoptotic U87MG cells.

In order to prevent the systematic error generated by weighing a small amount of **1-RGD**, we suggest that **1-RGD** can be prepared into high-concentration stock solutions and stored in −20 °C refrigerator. The concentration of the stock solution can be calibrated against a standard ICG solution by using UV–Vis spectrophotometry. The protocols for preparation of stock solution, preparation of standard ICG solution, and concentration calibration of stock solution are described step by step as below.

Equipment

(1) Analytical balance (0.01 mg readability)
(2) Centrifuge tube (1.5 and 50 mL)
(3) Micropipettes and tips (5 mL, 1000 μL, 200 μL, 10 μL, and 2.5 μL)
(4) Vortex mixer
(5) UV–Vis spectrophotometer (Ocean Optics)

Reagents

(1) **1-RGD**, **1-Ac**, and **ICG** (obtained from commercial source)
(2) Dimethyl sulfoxide (DMSO)
(3) MeOH or other organic solvent in which **1-RGD**, **1-Ac**, and ICG can be dissolved completely
(4) 1 M HEPES buffer
(5) 3-[(3-Cholamidopropyl)dimethylammonio] propanesulfonate (CHAPS)
(6) Glycerol
(7) Sodium chloride (NaCl)
(8) Ethylenediaminetetraacetic acid disodium salt (EDTA-2Na)
(9) Dithiothreitol (DTT)
(10) Deionized water
(11) Recombinant Human Caspase-3 (R&*D* systems)
(12) Caspase-3 storage buffer. NOTE: The composition of this storage buffer should follow the instruction offered by enzyme supplier

Procedure

Protocol for the preparation of standard ICG solution:

1. Place a 1.5 mL tube into the analytical balance to acquire the precise weight
2. Add an amount of ICG (>1 mg) solid to the tube and acquire the precise weight
3. Calculate the precise weight of ICG and add an appropriate volume of DMSO to generate a 10 mM solution
4. Place the tube on a vortex mixer and vortex for 1 min to dissolve the solid completely
5. Seal the tube with parafilm and store the ICG standard solution in −20 °C refrigerator

Protocol for the preparation and concentration calibration of **1-RGD** or **1-Ac** stock solutions:

1. Add a certain amount of **1-RGD** or **1-Ac** solid (weighted by analytical balance) to a 1.5 mL microcentrifuge tube
2. Calculate the volume of DMSO required to give a 10 mM solution. Add the appropriate volume of DMSO to the tube by using a micropipette
3. Place the tube on a vortex mixer and vortex for 1 min to dissolve the solid completely
4. Add 999 μL MeOH to a 1.5 mL microcentrifuge tube
5. Add 1 μL **1-RGD** or **1-Ac** stock solution to the tube and mix the solution thoroughly by using the vortex mixer
6. Add 1 mL MeOH to a quartz cuvette and baseline the spectrophotometer
7. Add 1 mL of **1-RGD** or **1-Ac** solution to the same quartz cuvette and acquire the absorbance value at 780 nm ($A_{\text{(diluted } \mathbf{1\text{-}RGD} \text{ or } \mathbf{1\text{-}Ac} \text{ solution)}}$).
8. Repeat step 5–6 to obtain the absorbance value at 780 nm of ICG solution ($A_{(10 \mu M \text{ ICG solution})}$).
9. Calculate the actual concentration of **1-RGD** or **1-Ac** stock solution by using the following equations:

$$C_{(\text{diluted } \mathbf{1-RGD} \text{ or } \mathbf{1-Ac} \text{ solution})} = 1.00\,\mu M_{(\text{ICG})} \times A_{(\text{diluted } \mathbf{1-RGD} \text{ or } \mathbf{1-Ac} \text{ solution})} / A_{(1\,\mu M \text{ ICG solution})}$$

$$C_{(\mathbf{1-RGD} \text{ or } \mathbf{1-Ac} \text{ stock solution})} = C_{(\text{diluted } \mathbf{1-RGD} \text{ or } \mathbf{1-Ac} \text{ solution})} \times 10{,}000$$

NOTE: We assume that the extinction coefficient (ε) of free ICG, **1-RGD** and **1-Ac** in methanol at low concentration

($<10\,\mu M$) was the same and the absorbance of the three diluted compounds in methanol follow the Beer's law.

10. Mark the actual concentration of the stock solution on the tube and seal the tube with parafilm. Store the solutions in a −20 °C refrigerator. NOTE: To avoid repeated freezing and thawing, we suggest that the stock solutions of **1-RGD** or **1-Ac** can be divided into small portions and stored in −20 °C refrigerator. When preparing for any experiment, warm the stock solution to room temperature and dilute to the required concentration with caspase assay buffer

All of the in vitro experiments that are related to caspase-3 should be proceeded in caspase assay buffer (50 mM HEPES, 100 mM NaCl, 1 mM EDTA, 10 mM DTT, 10% glycerol and 0.1% CHAPS, pH 7.4) otherwise specified. The protocol for preparation of caspase assay buffer is described as bellow:

1. Add 1 M HEPES buffer (2 mL) and deionized water (38 mL) to a 50 mL centrifuge tube (tube A) using a pipette. Mix the solution thoroughly by using vortex mixer to generate 50 mM HEPES buffer
2. Add 4 g of glycerol, 40 mg of CHAP, 13.5 mg of EDTA-2Na, and 233.8 mg of NaCl to another 50 mL centrifuge tube (tube B)
3. Add 36 mL of 50 mM HEPES buffer to tube B. Mix the solution thoroughly by using vortex mixer
4. Adjust the pH of caspase assay buffer to 7.4 with 1 M HCl and 1 M NaOH to generate caspase assay buffer. NOTE: We suggest that caspase assay buffer should be stored in −20 °C refrigerator in small portions, which should be taken as needed during each test and used after thawing. As the reducing agent (DTT) is easy to be oxidized, it should be added to caspase assay buffer before each experiment. After adding DTT, it is usually necessary to readjust the pH to 7.4

Protocol for the preparation of caspase-3 stock solution:

1. Dilute the original solution to $100\,\mu g\,mL^{-1}$ with storage buffer
2. Separate the caspase-3 stock solution into serval 1.5 mL tubes and stored in −80 °C refrigerator. NOTE: Repeated freezing and thawing may cause disfunction of the enzyme, so that the stock solution of caspase-3 should be divided into small portions and stored in −80 °C refrigerator. Because only very few stock solution will be used each time, so 1, 2, and 5 μL of caspase-3 stock solution in each

1.5 mL centrifuge tube should be reasonable. The whole procedures for the preparation of caspase-3 stock solution should proceeded in an ice-bath to avoid affecting the activity of the enzyme

4.1 Validation of caspase-3-mediated macrocyclization and self-assembly of 1-RGD

After synthesizing the probe, it is important to verify whether **1-RGD** can undergo the intramolecular cyclization reaction and self-assemble to form nanoparticles after incubation with caspase-3. Here, we use HPLC analysis and MALDI-TOF MS to monitor and confirm the formation of the cyclized product (**1-cycl**). By using dynamic light scattering (DLS) and transmission electron tomography (TEM), we demonstrated that incubation of **1-RGD** with caspase-3 can produce mono-disperse nanoparticles (as shown in Fig. 2).

Equipment

(1) Centrifuge tube (1.5 and 50 mL)
(2) Micropipettes and tips (1 mL, 200 μL, 10 μL, and 2.5 μL)
(3) Disposable single sealed cuvettes (50–2000 μL, Eppendorf)
(4) TEM grid (copper, coated with carbon film)
(5) High Performance Liquid Chromatography (Thermo Scientific Dionex Ultimate 3000 with 1‰ CF_3COOH CH_3CN/H_2O as eluents)

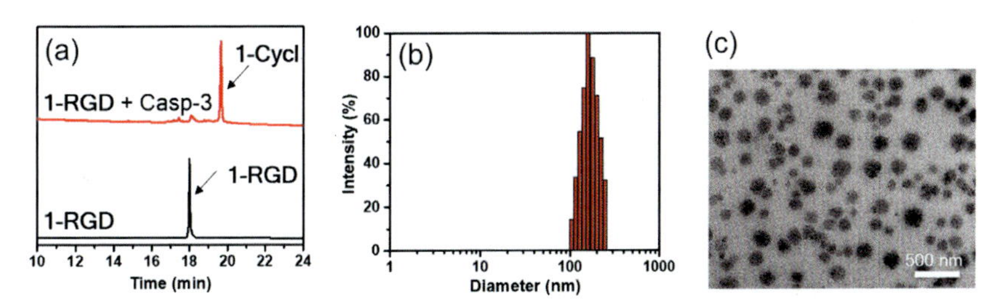

Fig. 2 Determination of the caspase-3-initiated macrocyclization and self-assembly of **1-RGD**. (A) HPLC analysis of **1-RGD** before (black) and after (red) incubation with caspase-3 for 5 h. DLS analysis (B) or TEM analysis (C) of **1-RGD** after incubation with caspase-3 overnight. *Reproduced with permission from John Wiley and Sons (Wang Y., Hu X., Weng J., Li J., Fan Q., Zhang Y., et al. (2019). A Photoacoustic Probe for the Imaging of Tumor Apoptosis by Caspase-Mediated Macrocyclization and Self-Assembly.* Angewante Chemie International Edition, 58, *4886–4890. https://doi.org/10.1002/anie.201813748).*

(6) MALDI-TOF mass spectrograph (ABSCIEX MALDI TOF-TOF 4800 plus)
(7) Transmission Electron Microscope (JEM 1011 Electron Microscope)
(8) Dynamic light scattering system (Brookhaven 90Plus)

Reagents

(1) Caspase assay buffer.
(2) Deionized water.
(3) Dithiothreitol (DTT)
(4) **1-RGD** stock solution (10 mM in DMSO)
(5) Caspase-3 stock solution (100 μg mL^{-1} in storage buffer)
(6) MeOH (chromatographic grade)

Procedure

4.1.1. Protocol for HPLC analysis of **1-RGD** before and after incubation with caspase-3

1. Warm a 10 mM stock solution of **1-RGD** and an appropriate volume of caspase assay buffer to room temperature
2. Add appropriate amount of DTT to prepare caspase assay buffer (containing 10 mM DTT)
3. Prepare a 10 μM solution of **1-RGD** in caspase assay buffer (containing 10 mM DTT), and incubate the solution at 37 °C for 10 min in water bath
4. Warm the tube that containing 1 μL of caspase-3 solution in an ice-bath. Add **1-RGD** solution which prepared in step 3 (100 μL) to the tube. Gently pipetting up and down to mix the solution evenly. NOTE: The final concentration of caspase-3 is 1 μg mL^{-1}
5. Add **1-RGD** solution prepared in step 3 (100 μL) to another 1.5 mL tube
6. Incubate each solution at 37 °C for 5 h in water bath
7. Add 450 μL of MeOH and 450 μL of deionized water to each tube. After thoroughly mixing, analyze the two solutions by HPLC respectively. Collect main peaks with 780 nm absorption for MALDI-TOF MS analysis

4.1.2. Protocol for TEM and DLS analysis of **1-RGD** before and after incubation with caspase-3

1. Warm a 10 mM stock solution of **1-RGD** and an appropriate volume of caspase assay buffer to room temperature
2. Add appropriate amount of DTT to the buffer to generate caspase assay buffer (containing 10 mM DTT). Use a 0.22 μm filter to remove dust

3. Prepare a 50 μM solution of **1-RGD** in caspase assay buffer (containing 10 mM DTT), and incubate the solution at 37 °C for 10 min in water bath
4. Warm the tube that containing 1 μL of caspase-3 solution in an ice-bath. Add **1-RGD** solution which prepared in step 3 (100 μL) to the tube. Gently pipetting up and down to mix the solution evenly. Incubate the solution at 37 °C overnight in water bath (Marked as solution A). In the meantime, add **1-RGD** solution which prepared in step 3 (100 μL) to another tube. Incubate the solution at 37 °C overnight in water bath (Marked as solution B)
5. After incubation, sonicate each solution for 10 min
6. Load 10 μL of a solution A onto the surface of a TEM grid (copper, coated with carbon film) using a 10 μL micropipette. After 10 min, remove the droplet using a pipette
7. Load 10 μL of deionized water (filtered by a 0.22 μm filter) carefully onto the TEM grid, and remove the droplets immediately afterwards
8. Repeat step 6–7 and dry the TEM grid under vacuum to prepare TEM sample A (Experimental group: **1-RGD** incubated with caspase-3)
9. Repeat steps 6–8 to prepare TEM sample B (Control group: **1-RGD** incubated in caspase assay buffer)
10. Acquire TEM images of each sample
11. Load the rest of solution A or solution B into a disposable single sealed cuvettes and acquire the DLS data of each solution

4.2 Photophysical characterization and photoacoustic signal validation of 1-RGD upon incubation with caspase-3

The above experiments have demonstrated that **1-RGD** can self-assemble into nanoparticles after incubation with caspase-3, it is necessary to evaluate the optical property and PA signal changes in subsequent experiments. Upon incubation with caspase-3 for 5 h, we found that the fluorescence intensity at 810 nm declined for ~22-fold whereas the PA signal increased ~3-fold. These signal changes could hindered by *Z*-VAD-fmk, a caspase-3 inhibitor (as shown in Fig. 3). The results demonstrated that caspase-3-instructed self-assembly of **1-RGD** was crucial to enhance the PA signal. The detailed protocols for absorption spectra and fluorescence spectra acquisition and PA signal measurement are described as below.

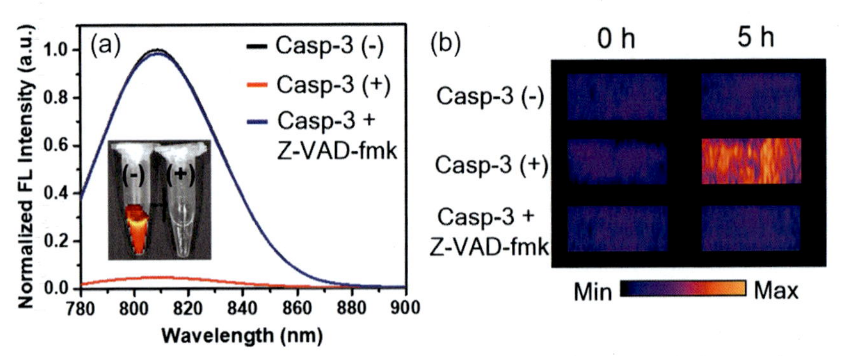

Fig. 3 Fluorescence spectra (A) or PA images (B) of **1-RGD** before and after incubation with caspase-3 or caspase-3 together with *Z*-VAD-fmk. *Reproduced with permission from John Wiley and Sons (Wang Y., Hu X., Weng J., Li J., Fan Q., Zhang Y., et al. (2019). A Photoacoustic Probe for the Imaging of Tumor Apoptosis by Caspase-Mediated Macrocyclization and Self-Assembly.* Angewante Chemie International Edition, 58, *4886–4890. https://doi.org/10.1002/anie.201813748).*

Equipment

(1) Centrifuge tube (1.5 and 50 mL)

(2) Micropipettes and tips (1 mL, 200 μL, 10 μL, and 2.5 μL)

(3) Quartz cuvette (0.35 mL, 10 mm path length, 4 transparent sides)

(4) Fine bore polythene tube (0.86 mm ID, 1.27 mm OD)

(5) UV–Visible spectrometer (Ocean Optics)

(6) Fluorescence spectrometer (HORIBA Jobin Yvon Fluoromax-4)

(7) Photoacoustic imaging system (Nexus128, Endra Life Sciences)

Reagents

(1) Caspase assay buffer.

(2) Deionized water.

(3) Dithiothreitol (DTT)

(4) **1-RGD** stock solution (10 mM in DMSO)

(5) Caspase-3 stock solution (100 μg mL^{-1} in storage buffer)

(6) *Z*-VAD-fmk stock solution (1 mg mL^{-1} in DMSO)

Procedure

1. Warm up the UV–Visible spectrometer and fluorescence spectrometer at least 15 min
2. Warm up the photoacoustic imaging system at least 30 min
3. Prepare a 10 μM solution and a 20 μM solution of **1-RGD** in caspase assay buffer (containing 10 mM DTT), and pre-warm the solutions at 37 °C for 10 min in water bath

4. Prepare three 1.5 mL tubes and marked as tube A, tube B, and tube C respectively
5. Add 300 μL of **1-RGD** solution (10 μM) to tube A. Add 300 μL of **1-RGD** solution (10 μM) and 3 μL of caspase-3 stock solution (100 $\mu g\,mL^{-1}$) to tube B. Add 150 μL of caspase assay buffer (containing 10 mM DTT), 3 μL of caspase-3 stock solution (100 $\mu g\,mL^{-1}$), and 0.6 μL of Z-VAD-fmk stock solution (1 $mg\,mL^{-1}$ in DMSO) to tube C. After incubation at 37 °C for 10 min in water bath, add 150 μL of **1-RGD** solution (20 μM) to tube C and pipette up and down to mix
6. Incubate all of the three solutions at 37 °C for 5 h in water bath
7. After incubation, sonicate each solution for 10 min
8. Load each solutions to quartz cuvette and acquire absorbance spectra (600–900 nm) and fluorescence spectra (λ_{ex} = 760 nm).
9. Load each solution into a length of fine bore polythene tube:
   a. Using a 1 mL syringe, load one of the three solutions into a fine bore polythene tube (about 30–40 mm in length). Check whether there are bubbles in the polythene tube, and if so, reload the solution
   b. Heat both ends of the tube and quickly pinch the heated end with a tweezer to seal the tube. Critical: If heated over an open flame, keeps flammable reagents away
   c. Check whether there are bubbles in the polythene tube, and if so, repeat steps 9a and 9b to remake the sample
10. Gently stick the sample tube on the surface of a piece of medical transparent tape. Stick the tape on the bottom of the bowl of the photoacoustic imaging system. Add deionized water to the bowl to completely cover the sample tube
11. Acquire PA data of each sample (λ_{ex} = 780 nm). Quantify the average PA signals of each sample after reconstruction of PA images using the data analyzing software (OrsiriX Lite)

4.3 Evaluation of enzyme kinetics, sensitivity, selectivity, and stability of 1-RGD

For a probe used to detect enzyme activities in vivo, fast enzymatic reaction rate, high selectivity and sensitivity to caspase-3, and good stability in physiological environment are extremely important. We evaluated the caspase-3 instructed activation kinetics of **1-RGD**. The catalytic constant k_{cat} and the Michaelis constant K_M were measured to be $0.44 \pm 0.01\,s^{-1}$ and $5.9 \pm 0.03\,\mu M$, respectively, the k_{cat}/K_M was then calculated to be $\approx 0.7 \times 10^5\,M^{-1}s^{-1}$. These parameters are comparable to that of other

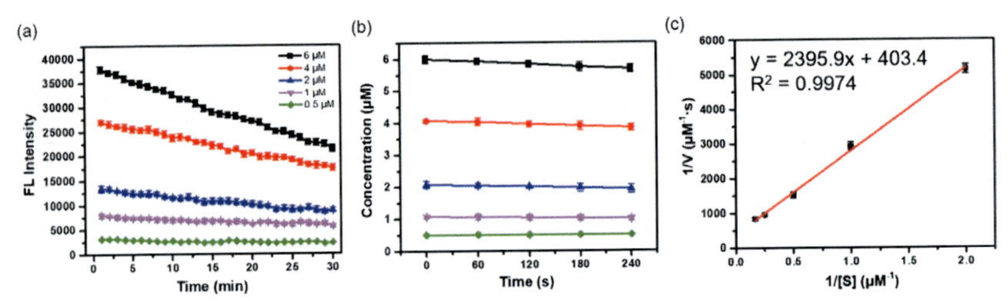

Fig. 4 Kinetics evaluation of **1-RGD**. (A) Plots of fluorescence intensity vs time under different probe concentrations (0.5, 1, 2, 4 and 6 μM). (B) Plots of probe concentrations vs time under different probe concentrations. (C) Lineweaver-Burk plots of 1/V vs 1/[S] for **1-RGD**. Values are mean $\pm$ SD (n = 3). *Reproduced with permission from John Wiley and Sons (Wang Y., Hu X., Weng J., Li J., Fan Q., Zhang Y., et al. (2019). A Photoacoustic Probe for the Imaging of Tumor Apoptosis by Caspase-Mediated Macrocyclization and Self-Assembly.* Angewante Chemie International Edition, 58, *4886–4890. https://doi.org/10.1002/anie.201813748).*

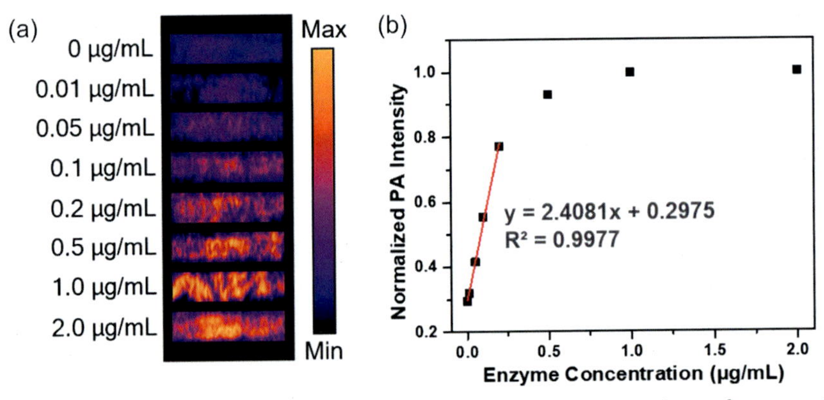

Fig. 5 (A) PA images of **1-RGD** incubated with different concentrations of caspase-3 (λ_{ex} = 780 nm). Liner relationship of PA intensity with enzyme concentration between 0.01 and 0.2 μg mL^{-1}. *Reproduced with permission from John Wiley and Sons (Wang Y., Hu X., Weng J., Li J., Fan Q., Zhang Y., et al. (2019). A Photoacoustic Probe for the Imaging of Tumor Apoptosis by Caspase-Mediated Macrocyclization and Self-Assembly.* Angewante Chemie International Edition, 58, *4886–4890. https://doi.org/10.1002/anie.201813748).*

reported casapase-3 imaging probes (Fig. 4). The detection limit was measured to be ~0.02 μg mL^{-1} (≈0.7 nM, signal to noise, S/N = 3) (Fig. 5). We also demonstrated that only caspase-3 could enhance the PA intensity of **1-RGD**. Negligible PA intensity enhancement was observed when incubated with other representative proteases in tumor cells including cathepsin B (CTB), γ-glutamyl transferase (GGT) and matrix metalloproteinase-2

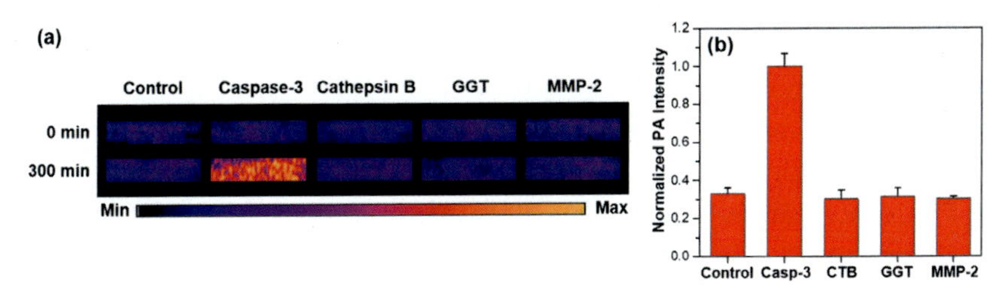

Fig. 6 (A) Photoacoustic images of **1-RGD** or **1-RGD** before and after incubation with Casp-3, CTB, GGT or MMP-2. $\lambda_{ex} = 780\,nm$. (B) Normalized PA intensity of **1-RGD** following treatment with conditions in (A). *Reproduced with permission from John Wiley and Sons (Wang Y., Hu X., Weng J., Li J., Fan Q., Zhang Y., et al. (2019). A Photoacoustic Probe for the Imaging of Tumor Apoptosis by Caspase-Mediated Macrocyclization and Self-Assembly.* Angewante Chemie International Edition, 58, *4886–4890. https://doi.org/10.1002/anie.201813748).*

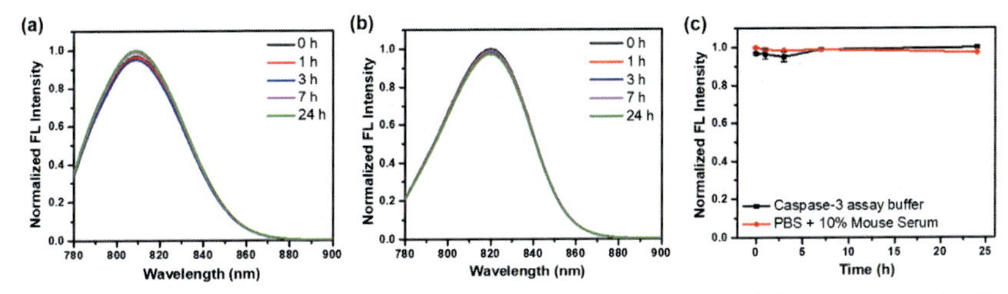

Fig. 7 Fluorescence spectra of **1-RGD** following incubation in (A) Caspase assay buffer or (B) 10% mouse serum containing PBS buffer (pH 7.4) for 0–24 h. $\lambda_{ex} = 760\,nm$. (C) Plot of the fluorescence intensity of **1-RGD** at 810 nm from (A) and (B). *Reproduced with permission from John Wiley and Sons (Wang Y., Hu X., Weng J., Li J., Fan Q., Zhang Y., et al. (2019). A Photoacoustic Probe for the Imaging of Tumor Apoptosis by Caspase-Mediated Macrocyclization and Self-Assembly.* Angewante Chemie International Edition, 58, *4886–4890. https://doi.org/10.1002/anie.201813748).*

(MMP-2) (Fig. 6). In addition, we confirmed that **1-RGD** exhibited good stability in complex physiological environments (e.g., 10% mouse serum) (Fig. 7). These results demonstrated that **1-RGD** had a good ability for PA imaging of caspase-3 activity. The relative protocols for each experiments are described as below.

Equipment

(1) Centrifuge tube (1.5 mL)

(2) Micropipettes and tips (1 mL, 200 μL, 10 μL, and 2.5 μL)

(3) 96-well black plate (Transparent bottom)

(4) Quartz cuvette (0.35 mL, 10 mm path length, 4 transparent sides)
(5) Fine bore polythene tube (0.86 mm ID, 1.27 mm OD)
(6) Microplate reader (SPARK 20 M, TECAN)
(7) Fluorescence spectrometer (HORIBA Jobin Yvon Fluoromax-4)
(8) Photoacoustic imaging system (Nexus128, Endra Life Sciences)

Reagents

(1) Caspase assay buffer
(2) Deionized water
(3) Dithiothreitol (DTT)
(4) **1-RGD** stock solution (10 mM in DMSO)
(5) Caspase-3 stock solution (100 μg mL^{-1} in storage buffer)
(6) Cathepsin B stock solution (100 μg mL^{-1} in storage buffer)
(7) γ-glutamyl transferase stock solution (100 μg mL^{-1} in storage buffer)
(8) Matrix metalloproteinase-2 stock solution (100 μg mL^{-1} in storage buffer)
(9) 1 × PBS (containing 10% mouse serum)

Procedure

4.3.1. Measurement of enzyme kinetic parameters of **1-RGD** toward caspase-3

1. Warm up the microplate reader for at least 15 min
2. Prepare different concentrations of **1-RGD** (6, 4, 2, 1, 0.5, and 0.2 μM) in 110 μL of caspase assay buffer (containing 10 mM DTT)
3. Load 100 μL of each solution into a 96-well black plate (Transparent bottom). Pre-warm the solutions at 37 °C for 10 min. Acquire the fluorescence intensity of each well (λ_{ex} = 770 nm, λ_{em} = 810 nm)
4. Plot the fluorescence intensity against the concentration of **1-RGD** to generate a calibration curve
5. Prepare solutions of **1-RGD** (12, 8, 4, 2, and 1 μM) in 60 μL of caspase assay buffer (containing 10 mM DTT).
6. Prepare the solution of caspase-3 (0.4 μg mL^{-1}) in 300 μL of caspase assay buffer (containing 10 mM DTT).
7. Load 50 μL of each **1-RGD** solution into five wells of a 96-well black plate (Transparent bottom). Load 60 μL of caspase-3 solution to another five wells. Place the plate into the microplate reader to pre-warm at 37 °C for 10 min

8. Using a multi-channel pipette to add 50 μL of pre-warmed caspase-3 to each **1-RGD** well and pipette up and down two times to mix
9. Right after step 8, begin to acquire fluorescence intensities at 810 nm every 1 min at 37 °C ($\lambda_{ex}=770$ nm)
10. Calculation of catalytic constant k_{cat}, the Michaelis constant K_M, and k_{cat}/K_M (as shown in Fig. 4):
    a. Take the data of the first 5 time points (0–300 s). Convert the fluorescence intensity into the concentration of **1-RGD** by using the calibration curve
    b. Plot scatter plot of the concentration of **1-RGD** against time under different **1-RGD** concentrations. Fit each scatter plot linearly to calculate the slope of each plot. The absolute value of each slope is the initial reaction velocity (V) related to corresponding **1-RGD** concentration ([S])
    c. Plot the Lineweaver-Burk plot of 1/V vs. 1/[S] for **1-RGD**. The x-intercept of the plot represents $-1/K_m$. The y-intercept of the plot is equivalent to the inverse of V_{max}. The value of k_{cat} can be calculated:

$$k_{cat} = V_{max}/[E] \text{ ([E] is the concentration of enzyme)}$$

4.3.2. Determination of the sensitivity of **1-RGD** toward caspase-3

1. Warm up the photoacoustic imaging system at least 30 min
2. Prepare a solution of **1-RGD** (20 μM) in 400 μL of caspase assay buffer (containing 10 mM DTT)
3. Prepare solutions of caspase-3 (4, 2, 1, 0.4, 0.2, 0.1, and 0.02 μg mL^{-1}) in 60 μL of caspase assay buffer (containing 10 mM DTT)
4. Add 50 μL of **1-RGD** (20 μM) to seven 1.5 mL centrifuge tubes respectively
5. Add 50 μL of different concentration of caspase-3 to each tube respectively and pipette up and down to mix
6. Incubate each solution at 37 °C for 5 h
7. Prepare samples for PA imaging according to the previously described procedures

8. Acquire PA data of each sample ($\lambda_{ex}=780$ nm). Quantify the average PA signals of each sample after reconstruction of PA images using the data analyzing software (OrsiriX Lite)
9. Plot scatter plot of PA signal against caspase-3 concentration. Fit scatter plot linearly to calculate the slope (as shown in Fig. 5)
10. Calculate the limit of detection (LOD) with the following equation:

$$\mathrm{LOD} = 3\,\sigma/k$$

where σ is the standard deviation of 11 blank measurements; k is the slope between the PA signal versus caspase-3 concentration (0.01–0.2 $\mu g\,mL^{-1}$). NOTE: Prepare at least 11 fine bore polythene tube filled with caspase assay buffer and acquire PA data of these samples. Calculate the standard deviation of average PA signals of these measurements to obtain σ.

4.3.3. Validation of the selectivity of **1-RGD** toward caspase-3

1. Warm up the photoacoustic imaging system at least 30 min
2. Prepare caspase assay buffer (containing 10 mM DTT), cathepsin B assay buffer (50 mM MES, 1 mM EDTA, 1 mM TCEP, 0.1% Triton X-100, pH=5.5), GGT assay buffer (1 × PBS, pH=7.4), and MMP-2 assay buffer (TCNB buffer, 150 mM NaCl, 10 mM $CaCl_2$, 100 mM Tris·HCl, 0.05% Brij 35, pH=7.4).
3. Dilute **1-RGD** stock solution to 10 μM by using different assay buffer respectively
4. Add 1 μL of caspase-3, CTB, GGT or MMP-2 to 100 μL of corresponding **1-RGD** solution respectively and pipette up and down to mix
5. Incubate all of the mixtures at 37 °C for 5 h
6. Prepare samples for PA imaging according to the previously described procedures
7. Acquire PA data of each sample ($\lambda_{ex}=780$ nm). Quantify the average PA signals of each sample after reconstruction of PA images (see Fig. 6) using the data analyzing software (OrsiriX Lite)

4.3.4. Accessing the stability of **1-RGD** in physiological conditions

1. Warm up the fluorescence spectrometer at least 15 min
2. Pre-warm **1-RGD** stock solution (10 mM) to room temperature

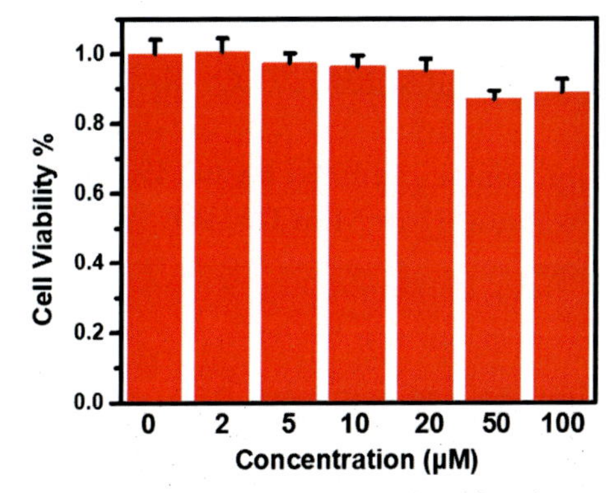

Fig. 8 Cell viability of U87MG cells. Cells were incubated with **1-RGD** at 0, 2, 5, 10, 20, 50 and 100 μM for 24 h. The cell viability was determined by the MTT assay. Error bars are standard deviation (n = 3). *Reproduced with permission from John Wiley and Sons (Wang Y., Hu X., Weng J., Li J., Fan Q., Zhang Y., et al. (2019). A Photoacoustic Probe for the Imaging of Tumor Apoptosis by Caspase-Mediated Macrocyclization and Self-Assembly.* Angewante Chemie International Edition, 58, *4886–4890. https://doi.org/10.1002/anie.201813748).*

3. Prepare two 1.5 mL microcentrifuge tubes. Add 1990 μL of caspase assay buffer (containing 10 mM DTT) or PBS buffer (containing 10% mouse serum, pH 7.4) to each tube respectively
4. Add 10 μL of **1-RGD** stock solution to each tube and pipette up and down to mix thoroughly
5. Load 200 μL of each solution to a quartz cuvettes and acquire initial fluorescence spectra (λ_{ex} = 760 nm)
6. Incubate the remaining **1-RGD** solutions at 37 °C. Take out 200 μL of the solutions with pipette and acquire fluorescence spectra at 1, 3, 7 and 24 h respectively (as shown in Fig. 7)

4.4 Validation that 1-RGD has good biocompatibility

For a probe used to detect enzyme activities in vivo, it is crucial to confirm that **1-RGD** is non-cytotoxic to living cells. Here we use the MTT assay to determine the survival rate of U87MG cells incubated with different concentrations of **1-RGD** for 24 h (see Fig. 8).

Equipment

(1) Centrifuge tube (1.5 mL)
(2) Micropipettes and tips (1 mL, 200 μL, 10 μL, and 2.5 μL)

(3) 96-well transparent plate
(4) Microplate reader (SPARK 20 M, TECAN)

Reagents

(1) DMEM medium (containing 10% FBS and 1%penicillin/streptomysin)
(2) **1-RGD** stock solution (10 mM in DMSO)
(3) 1 × MTT solution
(4) DMSO

Procedure

1. Seed approximately 5×10^4 U87MG cells (100 μL DMEM medium) in 42 wells of a 96-well plate
2. Incubate the plate for 12 h at 37 °C in an incubator
3. Dilute **1-RGD** stock solution in DMEM medium to generate solutions of 2, 5, 10, 20, 50, and 100 μM
4. Remove the medium in 96-well plate carefully and add 100 μL of each solution prepared in step 3 (six wells per concentration). Add 100 μL of DMEM medium to the remaining six wells (control group)
5. Incubate the plate at 37 °C for 24 h in the incubator
6. Add 50 μL of 1 × MTT to the 42 wells. Incubate the plate at 37 °C for 4 h in the incubator
7. Remove the medium carefully and add 150 μL of DMSO. Shake the plate by a plate shaker for 30 min
8. Acquire the absorbance at 490 nm of each well using a microplate reader
9. Calculate the percent viability

4.5 Fluorescence imaging and photoacoustic imaging of viable or apoptotic cells incubated with 1-RGD

Before in vivo applications, we first applied **1-RGD** for imaging of caspase-3 activity in staurosporine (STS)-induced apoptotic U87MG cells. As shown in Fig. 9, **1-RGD** could enter viable U87MG cells, producing bright NIR fluorescence. In contrast, the fluorescence was much weaker in the presence of active caspase-3 in apoptotic cells. This decrement of intercellular fluorescence could be hindered by caspase-3 inhibitor (Z-VAD-fmk). Different from the fluorescence imaging results, the PA signal in apoptotic cells was approximately 3-fold higher than that of viable cells or Z-VAD-fmk pretreated apoptotic cells. Upon incubation with free ICG, no significant PA signal difference was observed between viable cells and apoptotic cells. Apoptotic cells incubated with control probe (**1-Ac**) also

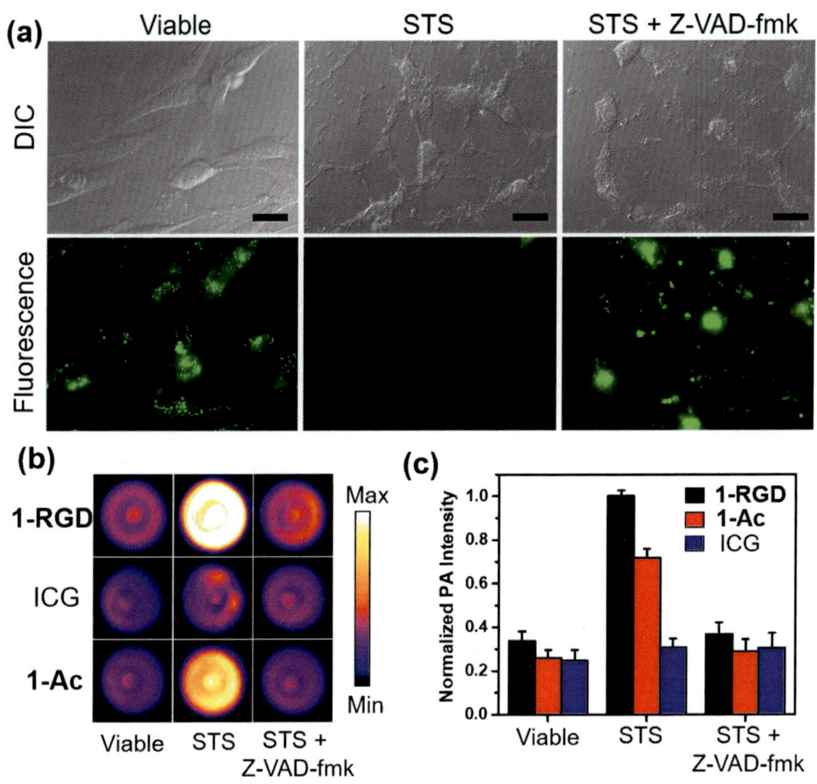

Fig. 9 Characterization of **1-RGD** in cells. (A) Fluorescence images of viable, STS-induced apoptotic or Z-VAD-fmk pretreated apoptotic U87MG cells after incubating with **1-RGD** at 37 °C for 24 h. Scale bars: 20 μm. (B) PA images and (C) quantification of the normalized PA intensities ($\lambda_{ex} = 780$ nm) of viable, STS-induced apoptotic or Z-VAD-fmk pretreated apoptotic U87MG cells after incubating with **1-RGD**, **1-Ac** or ICG for 24 h. Values are mean ± SD (n = 3). *Reproduced with permission from John Wiley and Sons (Wang Y., Hu X., Weng J., Li J., Fan Q., Zhang Y., et al. (2019). A Photoacoustic Probe for the Imaging of Tumor Apoptosis by Caspase-Mediated Macrocyclization and Self-Assembly.* Angewante Chemie International Edition, 58, *4886–4890. https://doi.org/10.1002/anie.201813748).*

showed enhanced PA signal relative to viable cells. However, the normalized PA intensity was lower than that of **1-RGD** (0.71 vs 1.0). These results suggested that active caspase-3 in apoptotic cells could activate **1-RGD** producing significantly enhanced PA signal. In addition, the targeting effect of c-RGD could enhance the active delivery of **1-RGD**, which further amplified the PA signals. Here we describe the step-by-step protocol for fluorescence imaging and photoacoustic imaging of apoptotic cells and viable cells.

Equipment

(1) Centrifuge tube (1.5 mL)
(2) Micropipettes and tips (1 mL, 200 μL, 10 μL, and 2.5 μL)
(3) PCR tube (200 μL)
(4) Cell culture dish (60 × 12 mm)
(5) 4-well glass bottom dish
(6) Centrifuge
(7) Olympus IX73 fluorescent inverted microscope
(8) Photoacoustic imaging system (Nexus128, Endra Life Sciences)

Reagents

(1) DMEM medium (containing 10% FBS and 1%penicillin/streptomysin)
(2) **1-RGD** stock solution (10 mM in DMSO)
(3) Staurosporine (STS) stock solution (1 mM in DMSO)
(4) *Z*-VAD-fmk stock solution (10 mM)
(5) 1 × PBS
(6) Trypsin

Procedure

4.5.1. Fluorescence imaging of viable or apoptotic U87MG cells incubated with **1-RGD**

1. Seed approximately 5×10^4 U87MG cells (in 500 μL DMEM medium) onto 4-well glass bottom dish
2. Incubate at 37 °C in the incubator until approach 70–80% confluence
3. Meanwhile, dilute the stock solution of STS (1 mM) in DMEM medium to generate a 2 μM solution. Dilute the stock solution of **1-RGD** (10 mM) in DMEM medium to generate a 10 μM solution. Dilute the stock solution of Z-VAD-fmk (10 mM) in DMEM medium to generate a 50 μM solution
4. For apoptotic group and inhibitory group, replace the medium with 500 μL of STS solution (2 μM, prepared in step 3). For viable group, replace the medium with 500 μL of fresh DMEM medium
5. Incubate for 4 h at 37 °C in the incubator
6. Remove the medium and wash each well with 1 × PBS for two times. NOTE: PBS should be added slowly to avoid floating the apoptotic cells
7. For apoptotic group and viable group, add **1-RGD** solution (2 μM, prepared in step 3) to each well and incubate for 24 h at

37 °C in the incubator. For inhibitory group, add Z-VAD-fmk solution (50 μM, prepared in step 3) and incubate for 30 min at 37 °C in the incubator. After 30 min incubation, replace the medium with fresh medium containing 50 μM Z-VAD-fmk and 10 μM **1-RGD**. Incubate for 24 h at 37 °C in the incubator

8. Remove the medium and wash each well with 1 × PBS for three times
9. Acquire fluorescence images of each well by using an Olympus IX73 fluorescent inverted microscope (Cy7 filter)
10. Use the same protocol to acquire fluorescence images of cells incubated with **1-Ac** or ICG

4.5.2. Photoacoustic imaging of viable or apoptotic U87MG cells incubated with **1-RGD**

1. Seed approximately 1×10^6 U87MG cells (in 2 mL DMEM medium) onto three 60 × 12 mm cell culture dish (Including viable group, apoptotic group, and inhibitory group)
2. Incubate at 37 °C in the incubator until approach 70–80% confluence
3. Treat the cells of each group according to steps 3–8 in 4.4.1
4. Add 500 μL of trypsin solution to each dish and incubate for 2 min at 37 °C in the incubator
5. Add 500 μL of DMEM culture medium and collect the cells in 1.5 mL centrifuge tube
6. After centrifuge at 1000 rpm or 4 min, add 200 μL of PBS to resuspend the cells. Transfer the cells into a 200 μL PCR tube to form cell pellets after centrifuge at 14000 rpm. Optional: Directly inject the cell suspension into the fine bore polythene tube to prepare PA sample is also practicable
7. Warm up the photoacoustic imaging system at least 30 min
8. Put the PCR tube vertically into the bowl of the photoacoustic imaging system. Fix the tube with a piece of medical transparent tape. Add deionized water to the bowl to completely cover the PCR tube
9. Acquire PA data of each sample (λ_{ex} = 780 nm). Quantify the average PA signals of each sample after reconstruction of PA images using the data analyzing software (OrsiriX Lite)
10. Use the same protocol to acquire photoacoustic images of cells incubated with **1-Ac** or ICG

5. In vivo application of 1-RGD

Having demonstrated that **1-RGD** has a good ability for PA imaging of caspase-3 activity in vitro, we then applied **1-RGD** for in vivo PA imaging. In vivo experiments should performed with institutional approval. The in vivo experiments that we carried out were approved by the Institutional Animal Care and Use Committee (IACUC) of Nanjing University.

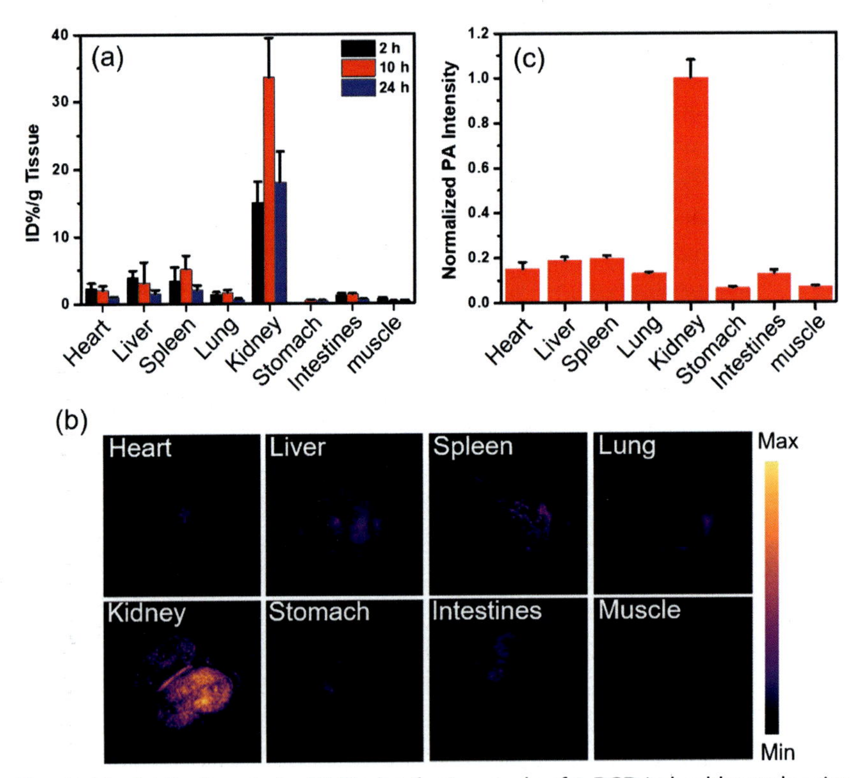

Fig. 10 Biodistribution study. (A) Biodistribution study of **1-RGD** in healthy nude mice after i.v. injection of 20 nmol of **1-RGD** at 2, 10 and 24 h. The amount of **1-RGD** was measured based on the ICG fluorescence intensity in collected organs after i.v. injection. (B) Ex vivo PA images and (C) normalized PA intensity of different organs resected from healthy nude mice at 10 h after i.v. injection of 20 nmol **1-RGD**. *Reproduced with permission from John Wiley and Sons (Wang Y., Hu X., Weng J., Li J., Fan Q., Zhang Y., et al. (2019). A Photoacoustic Probe for the Imaging of Tumor Apoptosis by Caspase-Mediated Macrocyclization and Self-Assembly.* Angewante Chemie International Edition, 58, *4886–4890. https://doi.org/10.1002/anie.201813748).*

5.1 Biodistribution study

We first investigated the biodistribution of **1-RGD** and confirmed the predominant renal clearance of **1-RGD** (Fig. 10). The protocol for biodistribution study is described as below.

Equipment

(1) Centrifuge tube (1.5 mL)
(2) Micropipettes and tips (1 mL, 200 μL, 10 μL, and 2.5 μL)
(3) 96-well black plate (transparent bottom)
(4) Microplate reader (SPARK 20 M, TECAN)
(5) Photoacoustic imaging system (Nexus128, Endra Life Sciences)

Reagents

(1) Saline (containing 5% castor oil)
(2) **1-RGD** stock solution (10 mM in DMSO)
(3) 1 × PBS

Procedure

1. Dilute the stock solution of **1-RGD** (10 mM) in saline (containing 5% castor oil) to generate a 100 μM solution
2. To nine healthy nude mice, inject 200 μL of diluted **1-RGD** solution through tail vein
3. Sacrifice three mice at 2, 10, and 24 h post injection, respectively
4. Resect heart, liver, spleen, lung, kidney, stomach, intestine, and muscle of each mice. Wash these organs with 1 × PBS
5. In the meantime, warm up the photoacoustic imaging system at least 30 min
6. Put the organ directly in the bowl of the photoacoustic imaging system and add deionized water to the bowl to completely cover the organ
7. Acquire PA data of each sample (λ_{ex} = 780 nm). Quantify the average PA signals of each sample after reconstruction of PA images using the data analyzing software (OrsiriX Lite)
8. Cut the organs into small pieces. Add the tissues into 1.5 mL centrifuge tubes
9. Add 1 × PBS (0.5 g tissue/mL PBS) to each tube and fully grind the tissue for 2 min by using a homogenate machine
10. Add 100 μL of each tissue homogenate to a 96-well plate and acquire the fluorescence intensity at 820 nm (λ_{ex} = 770 nm) by using a microplate reader

11. Prepare appropriate concentration of **1-RGD** solution to acquire the fluorescence intensity under the same conditions. Plot fluorescence intensity against **1-RGD** concentration to generate a standard curve

12. Use the standard curve to calculate the amount of **1-RGD**

13. Calculate the percent injected dose per gram tissue (ID%/g) (Fig. 10).

5.2 In vivo photoacoustic imaging

To evaluate the ability of **1-RGD** for PA imaging of caspase-3 activity in vivo, we inject three doses of doxorubicin ($5\,mg\,kg^{-1}$) through tail vein in nude mice bearing subcutaneous U87MG tumor to prepare chemotherapy model. The control group was prepared by injecting three doses of saline. **1-RGD**, **1-Ac** or ICG (20 nmol) was i.v. injected into mice 2 days after the final dose of DOX or saline.

We found that the PA intensity enhancement in DOX-treated tumors is 4.4-fold higher than that in saline-treated tumors after i.v. injection of **1-RGD** for 10 h. This enhancement further increased to 22-fold at 24 h. In contrast, negligible PA signal difference observed between DOX- and saline-treated tumors of mice injected with free ICG. The non-targeted control probe **1-Ac** could also be activated by caspase-3 in apoptotic tumors and produce higher PA intensity in DOX-treated tumors than that of saline-treated tumors. However, the maximum PA intensity enhancement in apoptotic tumors at 10 h is much lower as compared with **1-RGD** (as shown in Fig. 11). These results demonstrated that **1-RGD** was advantageous over **1-Ac** for PA imaging of tumor apoptosis in vivo. By analyzing the 3D-reconstructed PA image of entire DOX-treated tumor, we found that the PA signals appeared heterogeneously within tumor tissue. We hypothesized that the hotspots in tumor could be related to **1-RGD** activation and retention by active caspase-3 (see Fig. 12). Here we describe the detailed protocol for in vivo PA imaging of tumor apoptosis by applying **1-RGD**.

Equipment

(1) Centrifuge tube (1.5 mL)

(2) Micropipettes and tips (1 mL, 200 μL, 10 μL, and 2.5 μL)

(3) BD insulin syringe (1 mL)

(4) Photoacoustic imaging system (Nexus128, Endra Life Sciences)

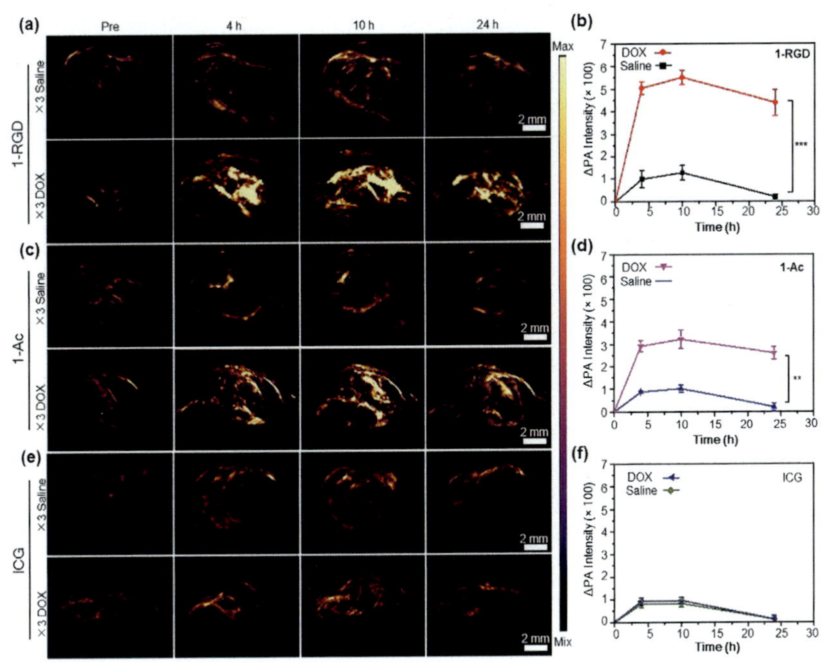

Fig. 11 PA imaging of tumor apoptosis. (A, C, E) Representative PA maximum imaging projection (MIP) images with coronal view of saline-treated (×3 Saline) or DOX-treated (×3 DOX) U87MG tumors in mice. PA images were acquired before (Pre) and 4, 10, 24 h after i.v. injection of 20 nmol (A) **1-RGD**, (B) **1-Ac** or (c) ICG. (B, D, F) Quantification of the PA intensity enhancement (ΔPA) in saline-treated (×3 Saline) or DOX-treated (×3 DOX) U87MG tumors following i.v. injection of 20 nmol (B) **1-RGD**, (D) **1-Ac** or (F) ICG. Values are mean ± SD (n = 3, ** $p < 0.01$, *** $p < 0.001$). *Reproduced with permission from John Wiley and Sons (Wang Y., Hu X., Weng J., Li J., Fan Q., Zhang Y., et al. (2019). A Photoacoustic Probe for the Imaging of Tumor Apoptosis by Caspase-Mediated Macrocyclization and Self-Assembly.* Angewante Chemie International Edition, 58, *4886–4890. https://doi.org/10.1002/anie.201813748).*

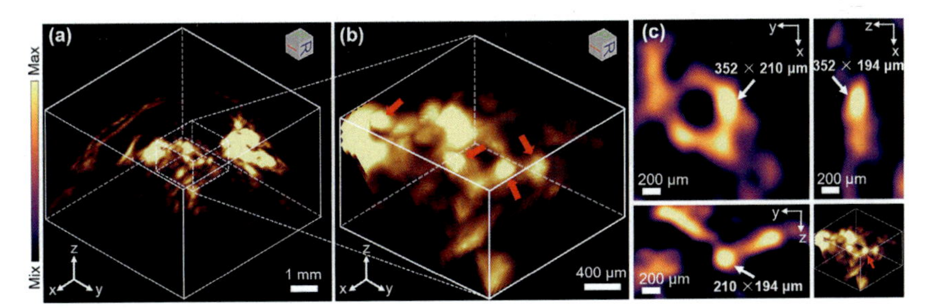

Fig. 12 (A) 3D Reconstruction PA image of tumor region in DOX-treated mice at 10 h after injection of **1-RGD** (20 nmol). (B) Enlarged 3D reconstruction PA image of the selected area as marked by the white box in (A). Red arrows show the views of representative individual apoptotic region (hotspot) within tumor tissues. (C) Enlarged 3D reconstruction PA images in a 3D-slice in tumor tissues. White arrows show the same apoptotic region in xy, xz and yz panels. The size of the apoptotic region was measured to be ~352 × 210 × 194 µm. *Reproduced with permission from John Wiley and Sons (Wang Y., Hu X., Weng J., Li J., Fan Q., Zhang Y., et al. (2019). A Photoacoustic Probe for the Imaging of Tumor Apoptosis by Caspase-Mediated Macrocyclization and Self-Assembly.* Angewante Chemie International Edition, 58, *4886–4890. https://doi.org/10.1002/anie.201813748).*

Reagents

(1) DMEM medium (containing 10% FBS and 1%penicillin/streptomysin)
(2) Matrigel
(3) Sterile Saline (containing 5% castor oil)
(4) **1-RGD** stock solution (10 mM in DMSO)
(5) **1-Ac** stock solution (10 mM in DMSO)
(6) ICG stock solution (10 mM in DMSO)
(7) 1 × PBS

Procedure

1. Prepare U87MG cells (cultured in 10 cm cell culture dish)
2. Thaw Matrigel in ice bath
3. Trypsinize the cells. After centrifuge at 1000 rpm for 4 min, add DMEM medium (1 mL per dish) to resuspend the cells. Count the cells and prepare a cell suspension in a 1:1 (v/v) mixture of DMEM medium and Matrigel (2×10^6 per 50 μL). NOTE: The whole process should be carried out in an ice bath to prevent the Matrigel from solidifying
4. Subcutaneously inject 50 μL of the U87MG suspension into the right flank of each mice using an insulin syringe
5. About 10–15 days after implantation or until the aspect of the tumor reached 0.7–0.9 cm, randomly divide all of the mice into two groups (×3 DOX group and ×3 Saline group)
6. For ×3 DOX group (chemotherapy group), inject 5 $mg\,kg^{-1}$ DOX (in sterile saline) through tail vein once every 4 days for a total of three times. For ×3 Saline group (control group), inject 200 μL of sterile saline through tail vein once every 4 days for a total of three times
7. Measure mouse body weight and tumor size (width and length) every other day during treatment. The approximate tumor volume can be calculated using the formula $(\text{length} \times \text{width}^2)/2$
8. 2 days after the final injection, divide the mice treated with DOX or saline into three groups (1-RGD group, 1-Ac group, and ICG group) respectively
9. For 1-RGD group, inject **1-RGD** (200 μL, 100 μM, dissolved in saline containing 5% castor oil) through tail vein. For 1-Ac group, inject **1-Ac** (200 μL, 100 μM, dissolved in saline containing 5% castor oil) through tail vein. For ICG group, inject ICG (200 μL, 100 μM, dissolved in saline containing 5% castor oil) through tail vein
10. Warm up the photoacoustic imaging system before acquiring PA data at least 30 min
11. Put the mice in a chamber, introduce isoflurane/oxygen gas into the chamber to anesthetize the mice. During the remaining experiment,

the isoflurane/oxygen gas should continuously piped through a nose cone to keep the mice anesthetized

12. Place the mice in the bowl of the photoacoustic imaging system with the tumor facing down. Add appropriate amount of deionized water to cover the whole tumor
13. Acquire PA data ($\lambda_{ex}=780$ nm). Quantify the average PA signals of each sample after reconstruction of PA images using the data analyzing software (OrsiriX Lite)

6. Summary

In summary, we detailed the synthesis, in vitro characterization, and in vivo application of the tumor targeting and caspase-3 activatable PA probe **1-RGD**, for noninvasive PA imaging of DOX-induced tumor apoptosis in living mice. **1-RGD** can undergo fast intramolecular macrocyclization and self-assembly into ICG-containing nanoparticles after interaction with active caspase-3, producing enhanced PA signals in DOX-treated tumors as compared with saline-treated tumors. By virtue of reconstructed high-resolution 3D PA images, **1-RGD** can detect the distribution of apoptotic regions within a whole tumor, allowing for noninvasive evaluation of tumor therapeutic efficacy at early stage in vivo. Such a enzyme-responsive in situ self-assembly approach can act as a promising strategy for the development of other activatable PA probes for in vivo imaging of enzyme activity with high sensitivity and spatial resolution.

References

Brindle, K. (2008). New approaches for imaging tumour responses to treatment. *Nature Reviews Cancer*, *8*, 94–107.

Chen, X., Zhen, X., & Pu, K. (2017). Nanoparticle regrowth enhances photoacoustic signals of semiconducting macromolecular probe for in vivo imaging. *Advanced Materials*, *29*, 1703693.

De la Zerda, A., Bodapati, S., Teed, R., May, S. Y., Tabakman, S. M., Liu, Z., et al. (2012). Family of enhanced photoacoustic imaging agents for high-sensitivity and multiplexing studies in living mice. *ACS Nano*, *6*, 4694–4701.

De La Zerda, A., Zavaleta, C., Keren, S., Vaithilingam, S., Bodapati, S., Liu, Z., et al. (2008). Carbon nanotubes as photoacoustic molecular imaging agents in living mice. *Nature Nanotechnology*, *3*, 557–562.

Edgington, L. E., Berger, A. B., Blum, G., Albrow, V. E., Paulick, M. G., Lineberry, N., et al. (2009). Noninvasive optical imaging of apoptosis by caspase-targeted activity-based probes. *Nature Medicine*, *15*, 967–973.

Hight, M. R., Cheung, Y.-Y., Nickels, M. L., Dawson, E. S., Zhao, P., Saleh, S., et al. (2014). A peptide-based positron emission tomography probe for in vivo detection of caspase activity in apoptotic cells. *Clinical Cancer Research*, *20*, 2126–2135.

Huang, R., Wang, X., Wang, D., Liu, F., Mei, B., Tang, A., et al. (2013). Multifunctional fluorescent probe for sequential detections of glutathione and caspase-3 in vitro and in cells. *Analytical Chemistry*, *85*, 6203–6207.

Kim, C., Favazza, C., & Wang, L. V. (2010). In vivo photoacoustic tomography of chemicals: High-resolution functional and molecular optical imaging at new depths. *Chemical Reviews*, *110*, 2756–2782.

Levi, J., Kothapalli, S. R., Ma, T.-J., Hartman, K., Khuri-Yakub, B. T., & Gambhir, S. S. (2010). Design, synthesis, and imaging of an activatable photoacoustic probe. *Journal of the American Chemical Society*, *132*, 11264–11269.

Li, H., Zhang, P., Smaga, L. P., Hoffman, R. A., Chan, J., et al. (2015). Photoacoustic probes for ratiometric imaging of copper(II). *Journal of the American Chemical Society*, *137*, 15628–15631.

Liang, G., Ronald, J., Chen, Y., Ye, D., Pandit, P., Ma, M. L., et al. (2011). Controlled self-assembling of gadolinium nanoparticles as smart molecular magnetic resonance imaging contrast agents. *Angewandte Chemie International Edition*, *50*, 6283–6286.

Maxwell, D., Chang, Q., Zhang, X., Barnett, E. M., & Piwnica-Worms, D. (2009). An improved cell-penetrating, caspase-activatable, near-infrared fluorescent peptide for apoptosis imaging. *Bioconjugate Cheminstry*, *20*, 702–709.

Miao, Q., & Pu, K. (2016). Emerging designs of activatable photoacoustic probes for molecular imaging. *Bioconjugate Chemistry*, *27*, 2808–2823.

Miao, Q., & Pu, K. (2018). Organic semiconducting agents for deep-tissue molecular imaging: Second near-infrared fluorescence, self-luminescence, and photoacoustics. *Advanced Materials*, *30*, 1801778.

Mizukami, S., Takikawa, R., Sugihara, F., Hori, Y., Tochio, H., Wälchli, M., et al. (2008). Paramagnetic relaxation-based ^{19}F MRI probe to detect protease activity. *Journal of the American Chemical Society*, *130*, 794–795.

Mizukami, S., Takikawa, R., Sugihara, F., Shirakawa, M., & Kikuchi, K. (2009). Dual-function probe to detect protease activity for fluorescence measurement and ^{19}F MRI. *Angewante Chemie International Edition*, *48*, 3641–3643.

Nejadnik, H., Ye, D., Lenkov, O. D., Donig, J. S., Martin, J. E., Castillo, R., et al. (2015). Magnetic resonance imaging of stem cell apoptosis in arthritic joints with a caspase activatable contrast agent. *ACS Nano*, *9*, 1150–1160.

Nguyen, Q.-D., Smith, G., Glaser, M., Perumal, M., Årstad, E., & Aboagye, E. O. (2009). Positron emission tomography imaging of drug-induced tumor apoptosis with a caspase-3/7 specific [^{18}F]-labeled isatin sulfonamide. *Proceedings of the National Academy of Sciences of the United States of America*, *106*, 16375–16380.

Ntziachristos, V. (2010). Going deeper than microscopy: The optical imaging frontier in biology. *Nature Methods*, *7*, 603–614.

Pop, C., & Salvesen, G. S. (2009). Human caspases: Activation, specificity, and regulation. *Journal of Biological Chemistry*, *284*, 21777–21781.

Pu, K., Chattopadhyay, N., & Rao, J. (2016). Recent advances of semiconducting polymer nanoparticles in in vivo molecular imaging. *Journal of Controlled Release*, *240*, 312–322.

Pu, K., Shuhendler, A. J., Jokerst, J. V., Mei, J., Gambhir, S. S., Bao, Z., et al. (2014). Semiconducting polymer nanoparticles as photoacoustic molecular imaging probes in living mice. *Nature Nanotechnology*, *9*, 233–239.

Ray, P., De, A., Patel, M., & Gambhir, S. S. (2008). Monitoring caspase-3 activation with a multimodality imaging sensor in living subjects. *Clinical Cancer Research*, *14*, 5801–5809.

Razgulin, A., Ma, N., & Rao, J. (2011). Strategies for in vivo imaging of enzyme activity: An overview and recent advances. *Chemical Society Reviews*, *40*, 4186–4216.

Reinhardt, C. J., & Chan, J. (2018). Development of photoacoustic probes for in vivo molecular imaging. *Biochemistry*, *57*, 194–199.

Shalini, S., Dorstyn, L., Dawar, S., & Kumar, S. (2014). Old, new and emerging functions of caspases. *Cell Death and Differentiation, 22*, 526–539.

Shen, B., Jeon, J., Palner, M., Ye, D., Shuhendler, A, Chin, F. T., et al. (2013). Positron emission tomography imaging of drug-induced tumor apoptosis with a caspase-triggered nanoaggregation probe. *Angewandte Chemie International Edition, 52*, 10511–10514.

Shi, H., Kwok, R. T. K., Liu, J., Xing, B., Tang, B. Z., & Liu, B. (2012). Real-time monitoring of cell apoptosis and drug screening using fluorescent light-up probe with aggregation-induced emission characteristics. *Journal of the American Chemical Society, 134*, 17972–17981.

Su, H., Chen, G., Gangadharmath, U., Gomez, L. F., Liang, Q., Mu, F., et al. (2013). Evaluation of [^{18}F]-CP18 as a PET imaging tracer for apoptosis. *Molecular Imaging and Biology, 15*, 739–747.

Sun, C., Wen, L., Zeng, J., Wang, Y., Sun, Q., Deng, L., et al. (2016). One-pot solventless preparation of PEGylated black phosphorus nanoparticles for photoacoustic imaging and photothermal therapy of cancer. *Biomaterials, 91*, 81–89.

Wang, L. V. (2009). Multiscale photoacoustic microscopy and computed tomography. *Nature Photonics, 3*, 503–509.

Wang, L. V., & Hu, S. (2012). Photoacoustic tomography: In vivo imaging from organelles to organs. *Science, 335*, 1458–1462.

Wang, H., Zhang, Q., Chu, X., Chen, T., Ge, J., & Yu, R. (2011). Graphene oxide–peptide conjugate as an intracellular protease sensor for caspase-3 activation imaging in live cells. *Angewante Chemie International Edition, 50*, 7065–7069.

Wu, C., Zhang, R., Du, W., Cheng, L., & Liang, G. (2018). Alkaline phosphatase-triggered self-assembly of near-infrared nanoparticles for the enhanced photoacoustic imaging of tumors. *Nano Letters, 18*, 7749–7754.

Yang, Q., Cai, S., Forrest, M. L., Cui, H., & Yang, X. (2011). In vivo photoacoustic imaging of chemotherapy-induced apoptosis in squamous cell carcinoma using a near-infrared caspase-9 probe. *Journal of Biomedical Optics, 16*, 116026.

Ye, D., Liang, G., Ma, M. L., & Rao, J. (2011). Controlling intracellular macrocyclization for the imaging of protease activity. *Angewante Chemie International Edition, 50*, 2275–2279.

Ye, D., Shuhendler, A. J., Cui, L., Tong, L., Tee, S. S., Tikhomirov, G., et al. (2014). Bioorthogonal cyclization-mediated in situ self-assembly of small-molecule probes for imaging caspase activity in vivo. *Nature Chemistry, 6*, 519–526.

Ye, D., Shuhendler, A. J., Pandit, P., Brewer, K. D., Tee, S. S., Cui, L., et al. (2014). Caspase-responsive smart gadolinium-based contrast agent for magnetic resonance imaging of drug-induced apoptosis. *Chemical Science, 5*, 3845–3852.

Yin, C., Zhen, X., Fan, Q., Huang, W., & Pu, K. (2017). Degradable semiconducting oligomer amphiphile for ratiometric photoacoustic imaging of hypochlorite. *ACS Nano, 11*, 4174–4182.

Yuan, Y., Sun, H., Ge, S., Wang, M., Zhao, H., Wang, L., et al. (2015). Controlled intracellular self-assembly and disassembly of ^{19}F nanoparticles for MR imaging of caspase 3/7 in zebrafish. *ACS Nano, 9*, 761–768.

Zhang, Y., Jeon, M., Rich, L. J., Hong, H., Geng, J., Zhang, Y., et al. (2014). Non-invasive multimodal functional imaging of the intestine with frozen micellar naphthalocyanines. *Nature Nanotechnology, 9*, 631–638.

Zhang, J., Smaga, L. P., Satyavolu, N. S. R., Chan, J., & Lu, Y. (2017). DNA aptamer-based activatable probes for photoacoustic imaging in living mice. *Journal of the American Chemical Society, 139*, 17225–17228.

CHAPTER THREE

Quantitatively visualizing the activity of MMP-2 enzyme *in vivo* using a ratiometric photoacoustic probe

Ling Yin[b], Anna Wang[a], Haibin Shi[a,*], and Mingyuan Gao[a,*]
[a]State Key Laboratory of Radiation Medicine and Protection, School for Radiological and Interdisciplinary Sciences (RAD-X) and Collaborative Innovation Center of Radiation Medicine of Jiangsu Higher Education Institutions, Soochow University, Suzhou, PR China
[b]Department of Chemistry and Chemical Engineering, Jining University, Qufu, PR China
*Corresponding authors: e-mail address: hbshi@suda.edu.cn; gaomy@iccas.ac.cn

Contents

Methods in Enzymology, Volume 657
ISSN 0076-6879
https://doi.org/10.1016/bs.mie.2021.06.035

Abstract

In this chapter, we describe the recent progress in development of small-molecule probes for quantitatively imaging of matrix metalloproteinase-2 (MMP-2) activity in living mice. We provide the detailed protocols for synthesis, characterization, and validation of a new multimodal probe QC with the near-infrared (NIR), single-photon emission computed tomography (SPECT), and photoacoustic (PA) imaging capabilities for accurate and quantitative detection of MMP-2 *in vivo*. We believe that this probe developed in our research group would offer a useful tool for precise evaluation of tumor metastasis as well as therapeutic efficacy.

1. Introduction

Tumor microenvironment (TME) is characterized by some special factors such as low extracellular pH, high level of reactive oxygen species, hypoxia, and over-expressed certain proteases in comparison with normal tissue. Therefore, TME has been recognized as reliable cancer signatures for clinical tumor diagnosis (Blum et al., 2015; Huang & Lovell, 2017; Joyce & Pollard, 2009; Sawyers, 2008; Shanmugam, Selvakumar, & Yeh, 2014; Turk, 2006). Matrix metalloproteinases (MMPs) as a family of zinc-dependent endopeptidases are normally inactive zymogens, and can be activated through the cleavage of the prodomains by extracellular enzymes (Sternlicht & Werb, 2001). MMPs play important role in the growth, invasion, metastasis and angiogenesis of malignant tumors (Black et al., 2015; Egeblad & Werb, 2002; Ferrari et al., 2019; Gallo et al., 2014; Lee, Xie, & Chen, 2010; Tang, Wu, Zhang, Zhang, & Jiang, 2016). For instance, MMP-2 was drawn great attention as a kind of reliable and valuable biomarker for cancer diagnosis and treatment (Dudani, Ibrahim, Kirkpatrick, Warren, & Bhatia, 2018; Kessenbrock, Plaks, & Werb, 2010; Winer, Adams, & Mignatti, 2018; Wisdom et al., 2018). Among them, MMP-2 is over-expressed in many types of tumors, and is highly relative to tumor invasion and metastasis (Egeblad & Werb, 2002; Wong et al., 2011). Hence, it is very meaningful to develop a smart probe that can specifically and sensitively detect MMP-2 *in vivo* for early diagnosis and treatment of cancers. Although many MMP-2-responsive fluorescent probes have been developed for detection of tumors, they are hard to be applied for real-time imaging of MMP-2 activity in living system due to the low spatial resolution and limited tissue penetration. Thus, developing novel imaging probes targeting to MMP-2 with high sensitivity and specificity is of great clinical significance for early diagnosis and evaluation of tumor therapeutic efficacy.

In this section, we introduce the design and synthesis of a novel MMP-2-responsive and fluorescence resonance energy transfer (FRET)-based probe QC in detail. Furthermore, the selection of fluorescent NIR dyes, peptide substrates and the quenching agents for construction of probe QC are discussed. Besides, the detailed methods and protocols for both *in vitro* and *in vivo* imaging of MMP-2 are also described in this chapter (Yin et al., 2019).

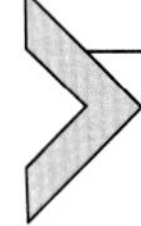

2. Design and construction of MMP-2-responsive and FRET-based probe

In this chapter, we report such an MMP-2-activatable multimodal imaging probe that can be used for quantitatively visualization of

MMP-2 activity *in vivo* through NIR, SPECT, and ratiometric photoacoustic imaging techniques. In general, a peptide-based probe QSY21-GPLGVRGY-Cy5.5 is developed by connecting the NIR fluorophore Cy5.5 and fluorescence quencher QSY21 together with a peptide substrate. The peptide sequence was chosen as GPLGVRGY that can be specifically cleaved between the G (Gly) and V (Val) residues by MMP-2 enzyme. Near-infrared emission (650–900 nm) is widely used for *in vivo* application due to its deep tissue penetration and less background interference from water, hemoglobin, and deoxyhemoglobin compared to visible light (Weissleder & Ntziachristos, 2003). Heptamethylene cyanines that are commonly composed of benzooxazole, benzothiazole, indolyl, 2-quinoline or 4-quinoline subclasses are chosen for NIR fluorophore (Frangioni, 2003). Among them, Cy5.5 has good water solubility and low toxicity for living system. The GPLGVRGY peptide can be labeled by Cy5.5 through the cycloaddition click reaction between alkyne and azide. Because QSY21 is a well-known effective quencher for Cy5.5 (Kabelác, Zimandl, Fessl, Chval, & Lankas, 2010), the fluorescence of Cy5.5 can be quenched efficiently based on the FRET mechanism when Cy5.5 and QSY21 are spaced in close proximity ($<$10 nm) (Johansson & Cook, 2003; Wu & Brand, 1994). However, the quenched fluorescence can be restored when the peptide linker is cut by MMP-2 enzyme, which can be used for detection of MMP-2.

Since probe QC is composed of a hydrophilic Cy5.5 and hydrophobic QSY21 groups, the whole probe is amphiphilic, which causes QC to self-assemble into uniform nanoparticles spontaneously in aqueous solution. The fluorescence measurements revealed that the QC particles initially had weak fluorescence, while strong fluorescence signals were determined after incubation with MMP-2, suggesting that the FRET-based fluorescence quenching is broken. Moreover, the linearly decreased PA signal at 680 nm was observed against the increasing concentrations of MMP-2, whereas the PA signal at 730 nm remained almost unchanged. Taking advantage of this ratiometric PA feature, probe QC can be served as a promising PA imaging contrast for quantitative detection of MMP-2 activity *in vivo* (Scheme 1).

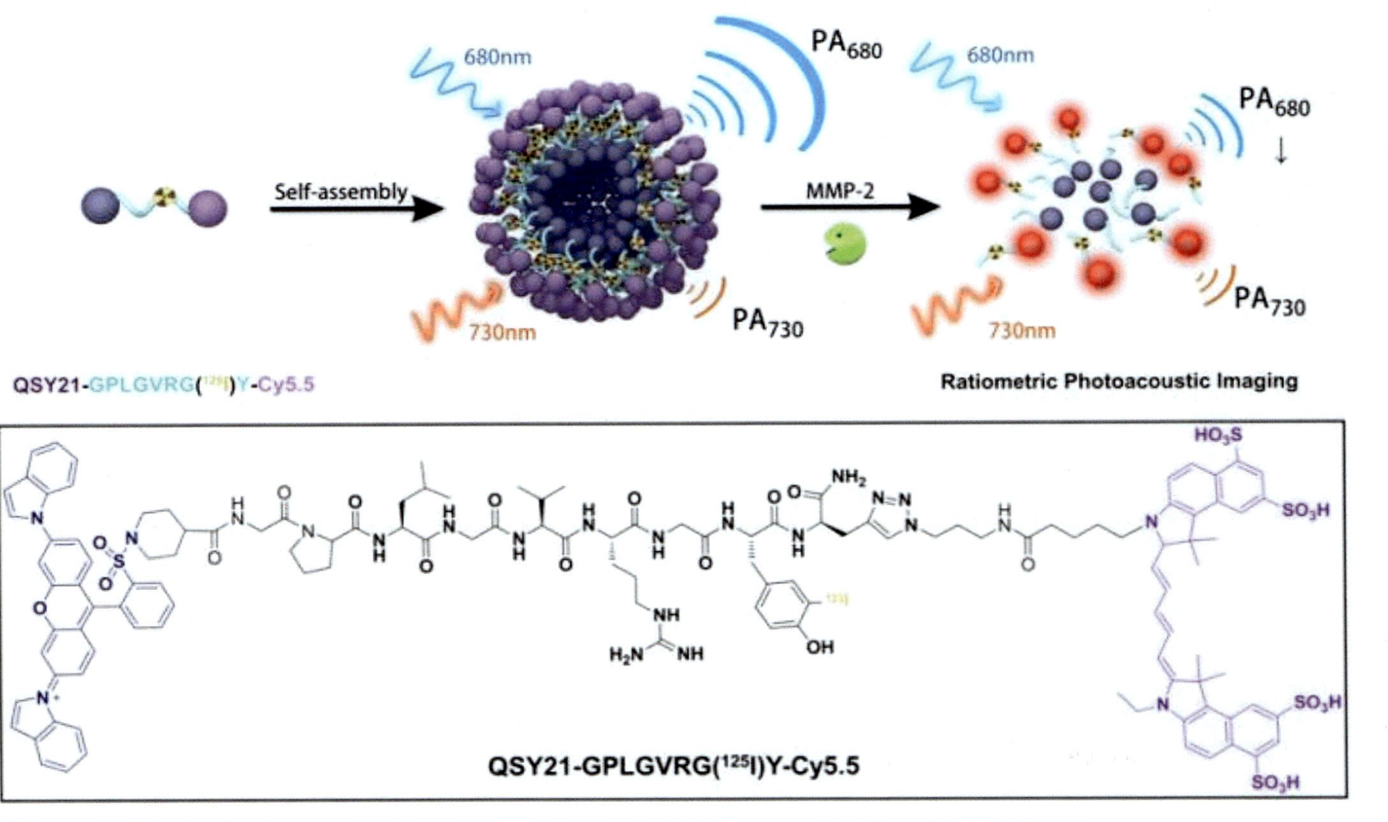

Scheme 1 A probe for noninvasive detection of MMP-2 activity through fluorescence/ratiometric PA imaging.

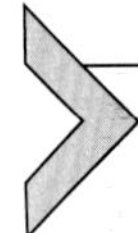

3. Synthesis of QSY21-GPLGVRGY-Cy5.5 (QC) and control probe QSY21-GLALGPGY-Cy5.5

Both probe QC and control can be synthesized in two steps according to the previously reported method (Yin et al., 2019). The starting materials include peptide substrates, NIR dye Cy5.5-N_3 and fluorescence quencher QSY21-NHS (QSY21 carboxylic acid, succinimidyl ester) (Scheme 2). All the solvents are obtained from commercial suppliers and used without further purification. The final products are purified by prep-HPLC and characterized by HR-MS.

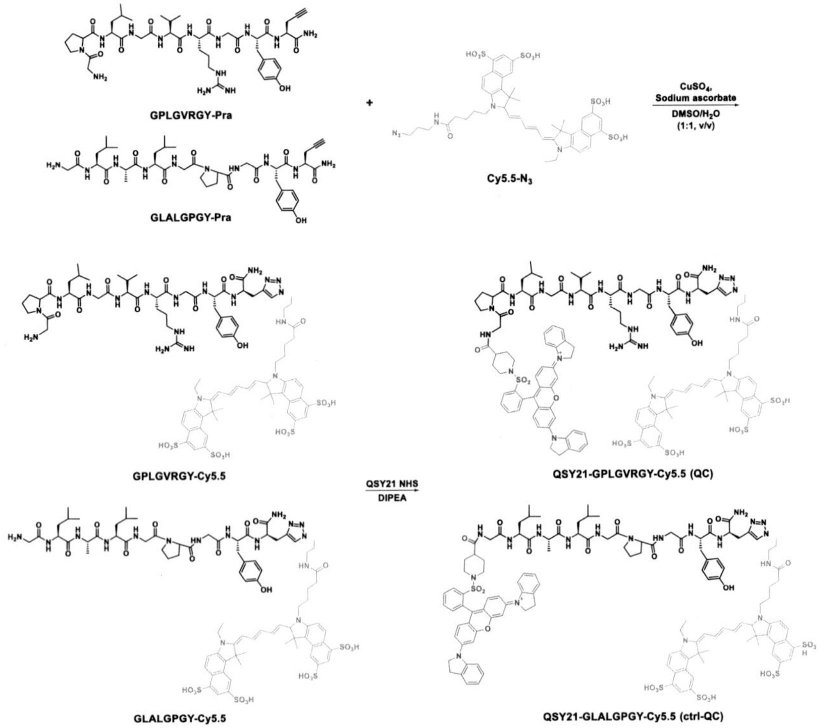

GPLGVRGY-Pra + Cy5.5-N_3 → ($CuSO_4$, Sodium ascorbate, DMSO/H_2O (1:1, v/v)) → GPLGVRGY-Cy5.5 → (QSY21 NHS, DIPEA) → QSY21-GPLGVRGY-Cy5.5 (QC)

GLALGPGY-Pra + Cy5.5-N_3 → GLALGPGY-Cy5.5 → QSY21-GLALGPGY-Cy5.5 (ctrl-QC)

Scheme 2 Synthetic routes of probe QSY21-GPLGVRGY-Cy5.5 (QC) and control probe QSY21-GLALGPGY-Cy5.5.

3.1 GPLGVRGY-Cy5.5

1. Add GPLGVRGY-Pra (5.01 mg, 0.0055 mmol, 1.1 equiv.), Cy5.5-N_3 (4.93 mg, 0.0050 mmol, 1.0 equiv), and 1 mL DMSO to foil-covered sample bottle.

2. Sodium ascorbate (1.19 mg, 0.006 mmol, 1.2 equiv.) dissolved in 0.5 mL milli-Q water is dropped into the 0.5 mL milli-Q water solution of $CuSO_4$ (0.48 mg, 0.003 mmol, 0.6 equiv.), and the mixed solution is then dropped into the former DMSO solution.
3. The reaction mixture is stirred in the dark at room temperature and monitored by HPLC until the reaction is completed in almost 3 h. HPLC running condition: a water phase (phase A) acidified with trifluoroacetic acid to 0.1% and second phase (phase B) acetonitrile acidified with 0.1% trifluoroacetic acid. The gradient used begin of 0–3 min, with elution of 1% B; 3–35 min linear gradient from 1% B up to 65% B at a flow rate of 1 mL min^{-1}.
4. The products are purified by HPLC. The solvent is removed from the separated solution with rotary evaporator, and the residual solvent is lyophilized with lyophilizer.
5. The pure products are characterized by HR-MS (Agilent 6230 Accurate-Mass Time-of-Flight Mass Spectrometer).

3.2 QSY21-GPLGVRGY-Cy5.5 (QC)

1. Add GPLGVRGY-Cy5.5 (1.14 mg, 0.0006 mmol, 1.2 equiv.), QSY21 NHS ester (0.41 mg, 0.0005 mmol, 1.0 equiv.), and 1 mL dried DMF to foil-covered sample bottle.
2. 20 μL of DIPEA is added into the former mixture solution.
3. The reaction mixture is stirred in the dark at room temperature and monitored by HPLC until the reaction is completed in about 5 h.
4. The products are purified by HPLC (see Fig. 1). The solvent is removed from the separated solution with rotary evaporator, and the residual solvent is lyophilized with lyophilizer.
5. The pure products are characterized by HR-MS.

3.3 GLALGPGY-Cy5.5

1. Add GLALGPGY-Pra (4.62 mg, 0.0055 mmol, 1.1 equiv.), Cy5.5-N_3 (4.93 mg, 0.0050 mmol, 1.0 equiv.), and 1 mL DMSO to foil-covered sample bottle.
2. Sodium ascorbate (1.19 mg, 0.006 mmol, 1.2 equiv.) dissolved in 0.5 mL milli-Q water is dropped into the 0.5 mL milli-Q water solution of $CuSO_4$ (0.48 mg, 0.003 mmol, 0.6 equiv.), and the mixed solution is then dropped into the former DMSO solution.
3. The reaction mixture is stirred in the dark at room temperature and monitored by HPLC until the reaction is completed in about 3 h.

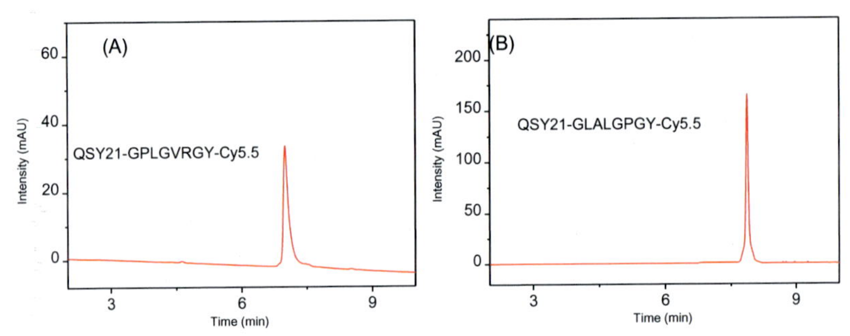

Fig. 1 The HPLC profiles of QSY21-GPLGVRGY-Cy5.5 (A) and QSY21-GLALGPGY-Cy5.5 (B). HPLC running condition: a water phase (phase A) acidified with trifluoracetic acid to 0.1% and second phase (phase B) acetonitrile acidified with 0.1% trifluoracetic acid. The gradient used begin of 0–3 min, with elution of 1% B; 3–35 min linear gradient from 1% B up to 65% B at a flow rate of 1 mL min^{-1}.

HPLC running condition: a water phase (phase A) acidified with trifluoracetic acid to 0.1% and second phase (phase B) acetonitrile acidified with 0.1% trifluoracetic acid. The gradient used begin of 0–3 min with elution of 1% B; 3–35 min linear gradient from 1% B up to 65% B at a flow rate of 1 mL min^{-1}.

4. The products are purified by HPLC. The solvent is removed from the separated solution with rotary evaporator, and the residual solvent is lyophilized with lyophilizer.
5. The pure products are characterized by HR-MS (Agilent 6230 Accurate-Mass Time-of-Flight Mass Spectrometer).

3.4 QSY21-GLALGPGY-Cy5.5 (QC control)

1. Add GLALGPGY-Cy5.5 (1.10 mg, 0.0006 mmol, 1.2 equiv.), QSY21 NHS ester (0.41 mg, 0.0005 mmol, 1.0 equiv.), and 1 mL dried DMF to foil-covered sample bottle.
2. 20 μL of DIPEA is added into the former solution.
3. The reaction mixture is stirred in the dark at room temperature and monitored by HPLC until the reaction is completed in about 5 h.
4. The products are purified by HPLC (see Fig. 1). The solvent is removed from the separated solution with rotary evaporator, and the residual solvent is lyophilized with lyophilizer.
5. The pure products are characterized by HR-MS.

4. *In vitro* characterization of QSY21-GPLGVRGY-Cy5.5 (QC) and control QSY21-GLALGPGY-Cy5.5

For the convenience of later use, both QC and control probes are dissolved into DMSO solution and quantified by HPLC to certain concentration using standard concentration of CY5.5-N_3 solution as reference.

4.1 Equipment

1—Microbalance
1—HPLC (Agilent 1260 high performance liquid chromatography)
3—HPLC sample bottle
3—Dram vials per compound (1 mL capacity)
3—Pipetting gun and heads (P10, P100, P1000)

4.2 Reagents

QSY21-GPLGVRGY-Cy5.5 (QC)
QSY21-GLALGPGY-Cy5.5 (Control)
Cy5.5-N_3
Organic solvent: DMSO, acetonitrile, trifluoracetic acid
Aqueous buffer: Milli-Q water

4.3 Procedure

1. According to the molecular weight of Cy5.5-N_3, the required sample mass and solvent volume of solution (10 mM) are calculated.
2. Label the dram vials as Cy5.5-N_3, QC, or Control.
3. Adjust the microbalance to zero and acquire the tare weight of "Cy5.5-N_3" dram vial.
4. Add Cy5.5-N_3 into the dram vial, and accurately weigh 4.93 mg Cy5.5-N_3.
5. 0.5 mL of DMSO is accurately measured with pipetting gun and added into the former vial to obtain 10 mM Cy5.5-N_3 solution.
6. QSY21-GPLGVRGY-Cy5.5 (QC) or QSY21-GLALGPGY-Cy5.5 (Control) is dissolved in DMSO to give a solution with concentration of 10 mM.
7. Take the same volume of Cy5.5-N_3, QSY21-GPLGVRGY-Cy5.5 (QC) and QSY21-GLALGPGY-Cy5.5 (Control) and detected by HPLC.

Choosing 10 mM Cy5.5-N_3 as a reference, the concentration of QSY21-GPLGVRGY-Cy5.5 (QC) and QSY21-GLALGPGY-Cy5.5 (Control) solutions are quantified according to the area of the maximum absorption peak of Cy5.5.

8. The calibrated QSY21-GPLGVRGY-Cy5.5 (QC), QSY21-GLALGPGY-Cy5.5 (Control) and Cy5.5-N_3 solutions are foil-covered and stored in −20 °C refrigerator away from light for later use.

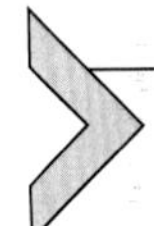

5. Hydrodynamic size profile and temporal evolution of the QC probe particles in PBS and DMEM

In order to evaluate the particle size and stability of the probes *in vivo*, it is necessary to study the size and stability of QC in PBS (0.01 M, pH = 7.4) and DMEM medium with 10% FBS *in vitro*.

5.1 Equipment

1—Particle size analyzer (Nano ZS90, Malvern)
1—Particle size analyzer cuvette
3—Pipetting gun and heads (P10, P100, P1000)
2—Eppendorf tubes (2.0 mL capacity)

5.2 Reagents

QSY21-GPLGVRGY-Cy5.5 (QC)
Organic solvents: DMSO
Aqueous buffer: PBS (0.01 M, pH = 7.4), DMEM medium with 10% FBS

5.3 Procedure

1. Calculate the volumes of QC (10 mM) solution and PBS needed for preparation of QC (40 μM, 2 mL) in PBS solution.
2. Accurately measure QC (10 mM) and PBS to make up 2 mL of QC (40 μM) in PBS solution
3. Clean the cuvette with alcohol and let it dry.
4. Warm up the particle size analyzer at least 30 min in advance of the experiments
5. Set up the experimental test parameters.
6. Add the QC (40 μM) into the cuvette and cap the cuvette to minimize solvent evaporation

7. According to the instruction of the instrument, open the cover of sample pool, and put the cuvette in sample pool followed by starting the measurement.
8. Repeat the measurement for three times.
9. Store the foil-covered solution in 4 °C refrigerator, measuring once a day for 30 days.
10. Calculate the volume of QC (10 mM) and DMEM needed to prepare 2 mL of QC (40 μM) DMEM solution.
11. Accurately measure QC (10 mM) and DMEM to make up 2 mL of QC (40 μM) DMEM solution.
12. Repeat steps 3–8 of QC (40 μM) in DMEM medium with 10% FBS.
13. Store the foil-covered solution in 4 °C refrigerator, measuring once a day for 7 days.

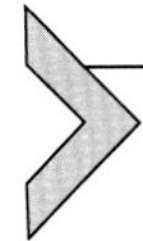

6. Photophysical characterization of QSY21-NHS, Cy5.5-N_3 and QC

UV–Vis absorption spectra of QSY21-NHS, Cy5.5-N_3 and QC in HEPES buffer or DMSO with different volume ratios are measured to give the suitable solvent condition and spectral absorption change.

6.1 Equipment

1—UV–Vis spectrometer (UV-3600, Shimadzu)
1—Quartz cuvette with corresponding cuvette cap
3—Pipetting gun and heads (P10, P100, P1000)
15—Eppendorf tubes (1.0 mL capacity)

6.2 Reagents

QSY21-NHS (10 mM)
Cy5.5-N_3 (10 mM)
QC (10 mM)
Organic solvents: DMSO
Aqueous buffer: HEPES buffer (20 mM, pH = 7.4)

6.3 Procedure

1. Calculate the volume of QSY21-NHS (10 mM), Cy5.5-N_3 (10 mM), QC (10 mM) and solvent (formed by HEPES buffer and DMSO with different volume ratios) to prepare the solutions as Table 1.

Table 1 Sample preparation of QSY21-NHS, Cy5.5-N_3 and QC.

		QSY21-NHS (1 mL)	Cy5.5-N_3 (1 mL)	QC (1 mL)
HEPES buffer/DMSO (Vol-%)	0:100	8 μM	8 μM	8 μM
	10:90	8 μM	8 μM	8 μM
	30:70	8 μM	8 μM	8 μM
	40:60	8 μM	8 μM	8 μM
	90:10	8 μM	8 μM	8 μM

2. Accurately measure the volume of QSY21-NHS (10 mM), Cy5.5-N_3 (10 mM), QC (10 mM) and solvent to prepare the solution as Table 1.
3. Clean the cuvette with alcohol and let it dry.
4. Warm up the spectrophotometer at least 15 min in advance of the experiment.
5. Set up the experiment to collect absorbance spectra from 500 to 900 nm at 0.5 nm increments.
6. Add a known volume of DMSO or HEPES buffer to the quartz cuvette (0.5–1 mL) and cap the cuvette to minimize solvent evaporation.
7. Blank the spectrophotometer using the baseline function.
8. Take a small amount of QSY21-NHS (8 μM) solution and wash the inner wall of the cuvette, pour it away, and then pour the QSY21-NHS (8 μM) solution into the cuvette for measurement.
9. Collect an absorbance spectrum.
10. Repeat steps 8–9 for each solution of the Table 1.
11. Export the data and process (see Fig. 2).

7. *In vitro* enzyme test

The *in vitro* reactivity of QC toward MMP-2 enzyme is verified by determining the fluorescence and photoacoustic change of the solution containing QC and MMP-2.

7.1 Equipment

1—Fluorescence spectrometer (FLS980, Edinburgh)
1—IVIS® spectrum *in vivo* imaging system (PerkinElmer)
1—Electron microscope (Tecnai G2 Spirit, FEI)

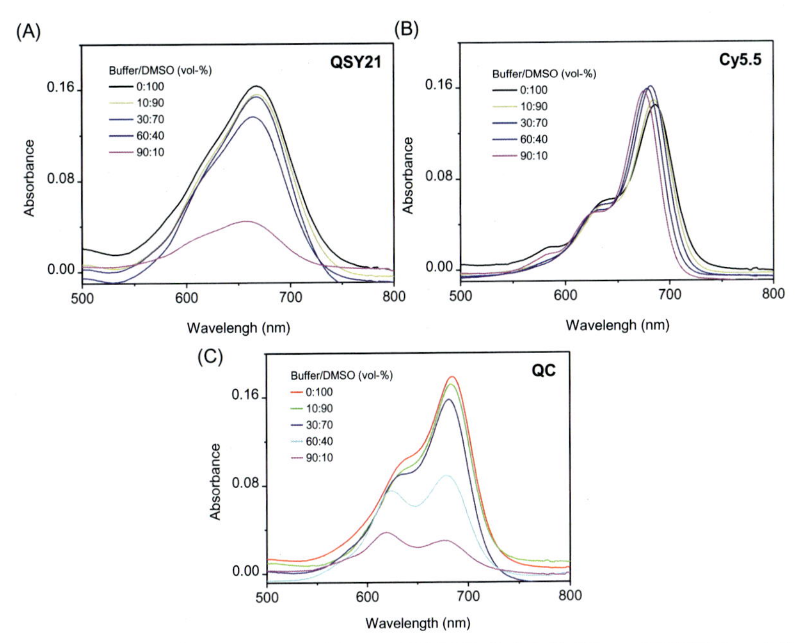

Fig. 2 UV–Vis absorption spectra of QSY21-NHS (A), Cy5.5-N_3 (B), and QC (C) in solution formed by HEPES buffer (pH = 7.4) and DMSO with different volume ratios (the concentration of QSY21-NHS, Cy5.5-N_3 and QC is of 8 μM).

1—Multispectral optoacoustic tomography scanner (MSOT, iThera medical)
1—Micro temperature-controlled oscillator (Thermo Mixer-Heat/Cool)
1—10 mm Quartz fluorescence cuvette with corresponding cuvette cap
3—Pipetting gun and heads (P10, P100, P1000)
60—Eppendorf tubes (0.5, 1.0 mL capacity)
2—polypropylene (PP) straw (0.4 × 20 cm)

7.2 Reagents

QC (10 mM)
QSY21-GLALGPGY-Cy5.5 (10 mM)
Matrix metalloproteinase-2 (MMP-2, 220 ng μL^{-1})
Inhibitor GM6001
Organic solvents: DMSO
Aqueous buffer: HEPES buffer (20 mM, pH = 7.4), PBS buffer (0.01 M, pH = 7.4)

7.3 Procedure

1. Calculate the volume of HEPES buffer and 220 ng μL^{-1} MMP-2 needed for preparation of 1 ng μL^{-1} MMP-2 (1 mL).
2. In ice bath, accurately measure the volume of HEPES buffer and 220 ng μL^{-1} MMP-2 to prepare 1 ng μL^{-1} MMP-2 in Eppendorf tube.
3. Label and store the tube in −20 °C refrigerator.
4. Calculate the volume of QC (10 mM) or QSY21-GLALGPGY-Cy5.5 (10 mM) and DMSO needed for preparation of QC (0.5 mM) or QSY21-GLALGPGY-Cy5.5 (1 mL, 0.5 mM).
5. Accurately measure the volume of QC (10 mM) or QSY21-GLALGPGY-Cy5.5 (10 mM) and DMSO needed for preparation of QC (0.5 mM) or QSY21-GLALGPGY-Cy5.5 (0.5 mM) in foil-covered Eppendorf tubes.
6. Label and store the foil-covered tubes in −20 °C refrigerator.
7. 1 mL of 100 μM inhibitor GM6001 solution is prepared with milli-Q water in Eppendorf tube, label and store the tube in 4 °C refrigerator.
8. Calculate the volume of QC (0.5 mM), HEPES buffer, and MMP-2 (1 ng μL^{-1}) needed for preparation of QC (4.0 μM) with different concentrations of MMP-2 (0, 5, 10, 20, 40, 80, 160 and 320 ng mL^{-1}).
9. Place nine Eppendorf tubes in the tube rack, label each tube with (0, 5, 10, 20, 40, 80, 160, 320 and 320 + inhibitor).
10. According to the calculation, accurately measure the volume of QC (0.5 mM), HEPES buffer and MMP-2 (1 ng μL^{-1}) to prepare QC (4.0 μM) with different concentrations of MMP-2 (0, 5, 10, 20, 40, 80, 160 and 320 ng mL^{-1}) in corresponding tubes.
11. GM6001 (100 μM) is treated with 320 ng mL^{-1} MMP-2 for 0.5 h before incubation with QC (4.0 μM).
12. Place all the tubes in 37 °C thermostatic oscillator for 2 h.
13. Turn on the mainframe power of fluorescence spectrometer, refrigerate the detector to −20 °C, turn on the xenon lamp power switch, and wait until the xenon lamp reads "ready to start."
14. Setup the experiment excitation wavelength (675 nm) and emission wavelength (694 nm) to collect the fluorescence spectra from 685 to 800 nm at 0.1 nm increments.
15. Clean the cuvette with alcohol and let it dry.
16. Take a small amount of blank sample to wash the inner wall of the cuvette, pour it away, and then pour the blank sample solution into the cuvette for measurement.

17. Collect and save the data.
18. Repeat steps 15–17 for each tubes of solutions.
19. Calculate the volume of QC (0.5 mM), QSY21-GLALGPGY-Cy5.5 (0.5 mM), HEPES buffer and 1 ng μL^{-1} MMP-2 needed for preparation of QC (4.0 μM) or QSY21-GLALGPGY-Cy5.5 (4.0 μM) with 320 ng mL^{-1} of MMP-2.
20. Place two Eppendorf tubes in the tube rack, label each tube with QC or control.
21. According to the calculation, accurately measure the volume of QC (0.5 mM), QSY21-GLALGPGY-Cy5.5 (0.5 mM), HEPES buffer and 1 ng μL^{-1} MMP-2 to prepare QC (4.0 μM) and QSY21-GLALGPGY-Cy5.5 (4.0 μM) with 320 ng mL^{-1} MMP-2 in corresponding tube.
22. Repeat steps 12–17 for each solution of the tubes.
23. After measuring the fluorescence intensity, the QC (4.0 μM) with different concentrations of MMP-2 (0, 5, 10, 20, 40, 80, 160 and 320 ng mL^{-1}) are successively transferred to the black cell culture plate.
24. Turn on the IVIS® system.
25. Select the Cy5.5 parameter from the operating system as the test condition.
26. Place the black cell culture plate containing the samples into the imaging chamber.
27. Click "start" to image.
28. Save the scanned images.
29. The solution containing QC (4.0 μM) and MMP-2 (0 and 320 ng mL^{-1}) are also transferred to HPLC bottles, and detected by HPLC.

To further study the specificity and selectivity of probe QC toward MMP-2, the probes are treated to different kinds of enzymes followed by fluorescence measurement.

1. Calculate and prepare different kinds of enzymes (1 ng μL^{-1}) including BamHI, BSA, Furin, RNase, MMP-9, and MMP-2.
2. Calculate the volume of QC (0.5 mM), HEPES buffer and different kinds of enzymes (1 ng μL^{-1}) needed for preparation of QC probe (4 μM) with different kinds of enzymes (320 ng mL^{-1}).
3. Place six Eppendorf tubes in the tube rack, label each tube with number 1–6.
4. According to the calculation, accurately measure the volume of QC (0.5 mM), HEPES buffer and different kinds of enzymes (1 ng μL^{-1}) needed to prepare of QC probe (4 μM) with different kinds of enzymes (320 ng mL^{-1}).
5. Repeat steps 12–17 for each tubes of solution.

8. Morphology study

1. Calculate the volume of Cy5.5-N_3 (10 mM), QSY21-NHS (10 mM), QC (10 mM), HEPES buffer, DMSO and MMP-2 (1 ng μL^{-1}) needed for preparation of Cy5.5-N_3 (8 μM), QSY21-NHS (8 μM), QC (8 μM) and QC (8 μM + 640 ng mL^{-1} of MMP-2) in 1 mL solution containing 90% HEPES buffer and 10% DMSO (Vol-%).
2. Place five PVC tubes in the tube rack, label each tube with number 1–5.
3. According to the calculation, accurately measure the volume of Cy5.5-N_3 (10 mM), QSY21-NHS (10 mM), QC (10 mM), HEPES buffer, DMSO and MMP-2 (1 ng μL^{-1}) needed for preparation of Cy5.5-N_3 (8 μM), QSY21-NHS (8 μM), QC (8 μM) and QC (8 μM + 640 ng mL^{-1} of MMP-2) in corresponding tube from 1 to 4.
4. Place the tubes in 37 °C thermostatic oscillator for 2 h.
5. The stocking QC (10 mM) solution in tube 5 is diluted with PBS buffer to provide QC (40 μM) solution.
6. Five samples are dropped onto the electron microscope net, and the solvent is evaporated at room temperature away upon light irradiation.
7. Open the transmission electron microscope program.
8. Check the working state of transmission electron microscope.
9. Load the number 1 sample into the sample rod.
10. Insert the sample rod.
11. Add filament current.
12. Start the operation, find the sample, focus, and take pictures.
13. End operation and return filament current.
14. Return the sample table to zero and take out the sample bar.
15. Remove the sample from the sample rod.
16. Repeat steps 9–15 for the other samples.
17. Burn data on a CD.

9. PA property study of probes

1. Cut 0.4 cm thick polypropylene (PP) straw into 5 cm long sections, label each section with 1, 2, 3 and 4.
2. The Cy5.5-N_3 (8 μM), QSY21-NHS (8 μM), QC (8 μM) and QC (8 μM + 640 ng mL^{-1} of MMP-2) samples are transferred to the four PP tubes and filled respectively. Both ends of each tube are sealed with sealing film and stored away from light.

3. Open the MSOT and launch the program, chamber is filled with water.
4. Set the measure wavelength to 680, 694, 730, 750, 780, 800 nm, and scan once every 0.3 cm interval.
5. Fix the PP tube of sample 1 on the test stand and adjust the direction of the PP tube.
6. Fine-tune the direction of sample tube through the program to find the best position and start the test.
7. Repeat steps 5–6 for the other samples.
8. Using the method above, measure PA of QC (0.25 μM) in the presence of different concentrations MMP-2 (0, 5, 10, 20, 40, 80, 160, 320 and 640 ng mL^{-1}).
9. The tomographic images are reconstructed using a linear model-based inversion. Size and resolution presets are 25 mm ROI (region of interest) with 100 μm resolution.
10. Photoacoustic signal intensities are measured by ROI analysis using the MSOT imaging system software package (see Fig. 3).

10. Cell culture

The cytotoxicity of QC toward 4T1 cells and NIH/3T3 cells is measured using MTT assay. Confocal microscopy images of 3T3 cells, and 4T1 cells incubated with the QC probes (4 μM) in the absence or presence of inhibitor GM6001 (100 μM).

10.1 Equipment

1—EnSpire® Multimode Plate Reader (PerkinElmer)
1—Fluorescence microscope (FV1200, Olympus)
2—Single cell culture plate
2—96-well cell culture plate
3—8-well laser confocal cell culture plate
3—Pipetting gun and heads (P10, P100, P1000)
36—Eppendorf tubes (0.5 mL, 1.0 mL capacity)

10.2 Reagents

QC (0.5 mM).
GM6001 (100 μM)
MTT assay
Hoechst 33342 stain (0.1 mg/mL)
RPMI 1640 medium

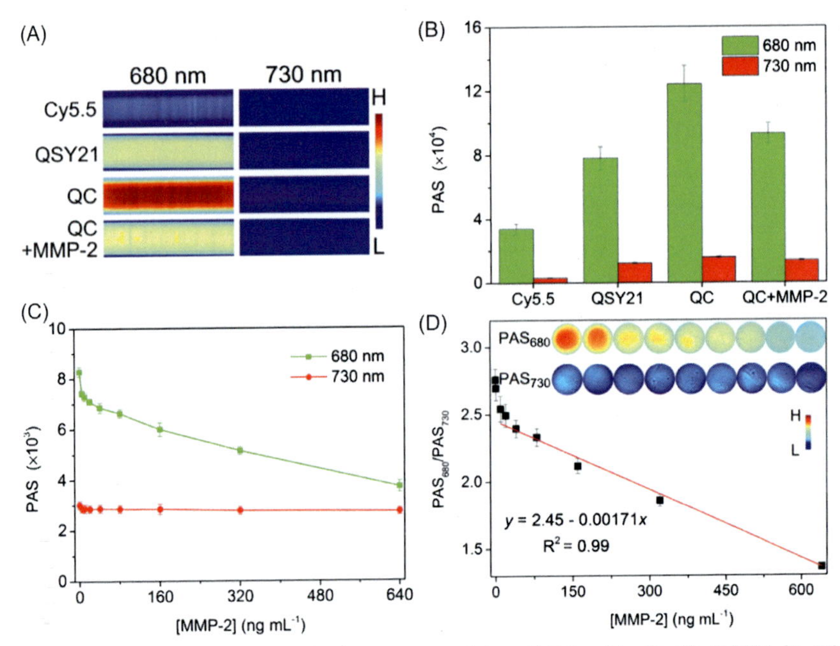

Fig. 3 Photoacoustic images of solutions containing Cy5.5 azide (8 μM), QSY21 (8 μM), QC (8 μM), or a mixture of QC (8 μM) and MMP-2 (640 ng mL^{-1}), recorded under illumination of different wavelength laser lights (680 nm and 730 nm) (A), photoacoustic signal (PAS) intensity recorded under different excitations (B), MMP-2 concentration dependent PAS intensity of QC probe (0.25 μM) recorded in the presence of different amounts of MMP-2 (C), and the correlation between the $\Delta PAS_{680}/\Delta PAS_{730}$ of the QC probe (0.25 μM) and the concentration of MMP-2 recorded at 37 °C after incubation for 2 h (Inset: PA images of the QC probe solutions containing different amounts of MMP-2) (D). The error bars represent standard deviations of three separate measurements.

DMEM medium
Fetal bovine serum
Penicillin streptomycin
Trypsin-EDTA (0.25%)
PBS buffer (0.01 M, pH = 7.4)

Cell: The murine breast carcinoma cell line 4T1, murine embryonic fibroblast cell line NIH/3T3

10.3 Procedure

10.3.1 Cytotoxicity study using MTT

1. Calculate the volume of QC (0.5 mM), RPMI 1640 medium (10% fetal bovine serum and 1% penicillin streptomycin) and DMEM medium

(10% fetal bovine serum and 1% penicillin streptomycin) needed for preparation of different concentrations of QC (0, 2, 4, 8, 16, 32 and 64 μM) with RPMI 1640 or DMEM medium.

2. According to calculation above, prepare different concentrations of QC (0, 2, 4, 8, 16, 32 and 64 μM) with RPMI 1640 or DMEM medium stored away from light.
3. Warm RPMI 1640 and DMEM medium in a 37 °C water bath.
4. 4T1 cells are cultured in RPMI 1640 medium (10% fetal bovine serum and 1% penicillin streptomycin) at 37 °C in a humidified atmosphere of 5% CO_2. The NIH/3T3 cells are cultured in DMEM medium (10% fetal bovine serum and 1% penicillin streptomycin) at 37 °C in a humidified atmosphere of 5% CO_2. The cells are cultured until 75% confluence is reached, and experiments are performed.
5. 4T1 and NIH/3T3 cells are planted at a density of 1×10^4 cells/well in 96-well cell culture plate at 37 °C in a humidified atmosphere of 5% CO_2.
6. After growing for 24 h, the cells are washed with PBS once and incubated with 100 μL different concentrations of QC (0, 2, 4, 8, 16, 32, and 64 μM).
7. 5 mg/mL MTT is prepared by diluting 50 mg/mL MTT solution with cell medium according to the volume ratio of 1:9, stored in foil-covered centrifuge tube.
8. After 24 h of incubation, the cells are washed with PBS once and incubated by 100 μL of MTT solution (5 mg/mL) in each well for 2 h away from light at 37 °C in a humidified atmosphere of 5% CO_2.
9. Remove the original medium, add 100 μL DMSO into each well and cover it with tin foil.
10. The absorbance is measured using EnSpire® Multimode Plate Reader at a wavelength of 490 nm.
11. Save and process the data (see Fig. 4).

11. Confocal

1. 4T1 and NIH/3T3 cells are planted at a density of 2×10^3 cells/well in 8-well laser confocal cell culture plate at 37 °C in a humidified atmosphere of 5% CO_2 respectively.
2. After growing for 24 h, 4 wells of 4T1 cells are washed with PBS once and incubated with 100 μL of inhibitor GM6001 (100 μM) each well for 30 min at 37 °C in a humidified atmosphere of 5% CO_2.

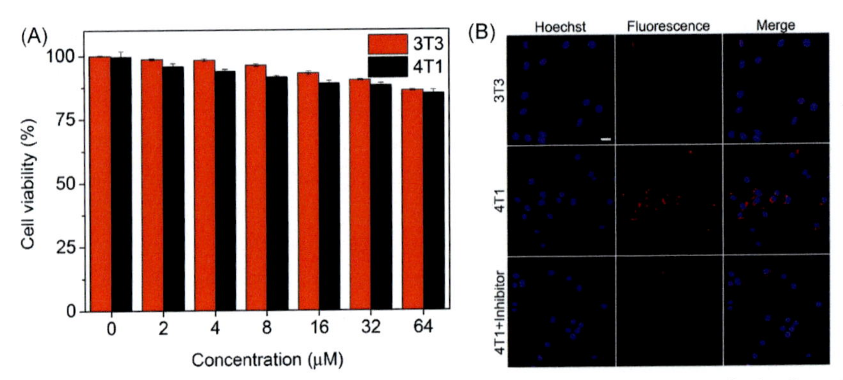

Fig. 4 MTT assays on the viability of murine embryonic fibroblast 3T3 cells (red) and murine breast carcinoma 4T1 cells (black) after incubated with the different concentrations of the QC probe, *i.e.*, 0, 2, 4, 8, 16, 32, and 64 μM (the error bars represent standard deviations of six parallel measurements) (A). Confocal microscopy images 3T3 cells (top), 4T1 cells (middle), and 4T1 cells incubated with the QC probes (4 μM) for 8 h in the absence (top, middle) or presence of 100 μM of the enzyme inhibitor GM6001 (bottom) (the embedded scale bar corresponds to 20 μm) (B).

3. Cell culture plates with 4T1 and 3T3 cells are washed with PBS once, then add 100 μL QC (4 μM) and incubate for 8 h at 37 °C in a humidified atmosphere of 5% CO_2.
4. 4T1 and 3T3 cells are washed with PBS once, then add 100 μL Hoechst 33342 staining solution and incubate for 15 min at 37 °C in a humidified atmosphere of 5% CO_2 away from light.
5. Wash 4T1 and 3T3 cells three times with PBS, and add 50 μL fresh medium each well.
6. Turn on the computer of confocal, mercury lamp power, microscope, stage controller, microscope touch panel, scanning head and laser of FV1200 laser scanning confocal microscope in sequence.
7. Open the application and select test parameters.
8. Place the cell culture plate on the oiled platform and adjust the position continuously for scanning and photographing.
9. Save and export the images, and process the images with the appropriate software.
10. Shut down in the reverse order as start (see Fig. 4).

12. Mice tumor model

Female BALB/c athymic nude mice with body weights of 18–20 g purchased from ChangZhou Cavensla Experimental Animal Technology Co. Ltd are housed under standard conditions (25 ± 2 °C/60% ± 10%

relative humidity) with 12h light/dark cycle. The tumors were grafted by subcutaneous inoculation of 1×10^6 4T1 cells in about 50μL PBS buffer (0.01M, pH=7.4) into the right front flank of each mouse. Fluorescence, SPECT and PAI studies are carried out when tumor size reached about 50–250mm^3. Mice used up in the experiment are sacrificed according to ethical requirements.

12.1 Equipment

1—IVIS® Spectrum *in vivo* imaging system (PerkinElmer)
1—Confocal Microscopy (Cellvizio, Mauna Kea)
1—Fluorescence microscope (FV1200, Olympus)
1—Multispectral Optoacoustic Tomography scanner (MSOT, iThera medical)
1—U-SPECT (micro-SPECT, MILabs BV)
3—sterile insulin Syringe
3—Pipetting gun and heads (P10, P100, P1000)
28—Eppendorf tubes (0.5mL, 1.0mL capacity)

12.2 Reagents

QC (0.5mM)
Control probe (0.5mM)
Inhibitor GM6001
PBS (0.01M, pH=7.4)
Ultrasonic coupling agent

12.3 Procedure

12.3.1 In vivo *fluorescence imaging*

1. 1.5mM inhibitor GM6001 solution is prepared with PBS, labeled, and stored in Eppendorf tube
2. Calculate and measure the volume of QC (0.5mM) and PBS needed for preparation of QC (4.0μM) solution, labeled and stored in Eppendorf tube.
3. Open the living confocal fluorescence microscope, and set up test parameters.
4. Turn on the IVIS® system.
5. Select the Cy5.5 parameter from the operating system as the test condition.
6. Three nude mice with tumor sizes (about 50mm^3) are selected and labeled as 1, 2, and 3, respectively.

7. The No. **3** mouse is taken out and intratumorally injected with inhibitor GM6001 (1.5 mM, 50 μL) after gas anesthesia (3% isoflurane mixed with oxygen gas, 0.5 L/min).
8. 0.5 h later, QC (60 μM, 200 μL) is injected into mice **2** and **3** through tail vein after gas anesthesia.
9. The three nude mice are scanned by living confocal fluorescence microscope and IVIS at different time points (0, 0.5, 1, 2, 24 and 48 h).
10. Repeat the previous procedures 6–8.
11. 24 h after QC injection through tail vein, the mice are sacrificed and removed out heart, liver, spleen, lung, kidney, tumor, then rinsed with PBS.
12. The removed tumor tissues and organs are placed on the black paper in the corresponding order and imaged with IVIS.
13. The fluorescence images are analyzed by vendor software to separate autofluorescence from chromophore signals through spectral unmixing algorithms (see Fig. 5).

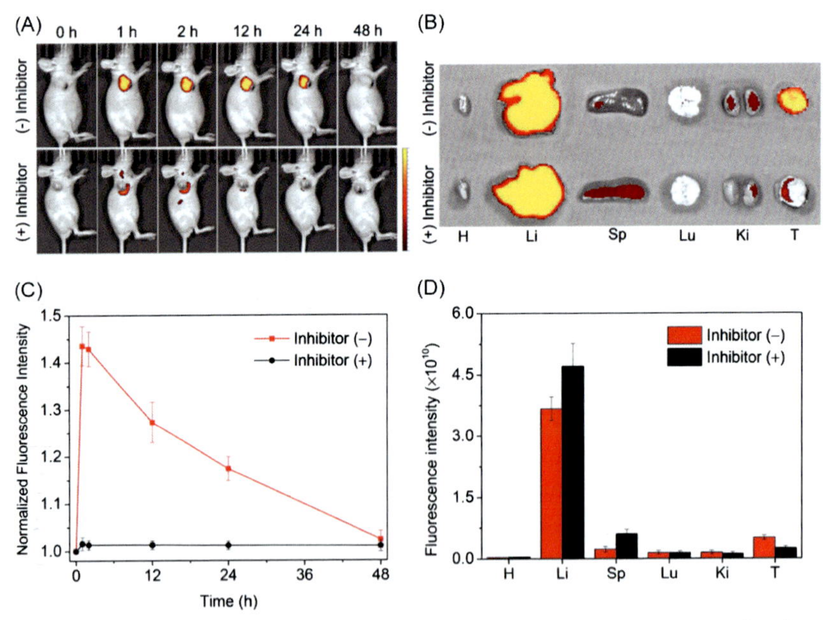

Fig. 5 *In vivo* fluorescence imaging of 4T1 tumor-bearing nude mice injected with the QC probe (60 μM, 200 μL) (top) or GM6001 (1.5 mM, 50 μL) 0.5 h posterior to the injection of QC probe (bottom) (A), *ex vivo* fluorescence images of major organs and tumors dissected from the mice 24 h post-injection of the probe (H: heart, Li: liver, Sp: spleen, Lu: lung, Ki: kidney, T: tumor) (B), normalized fluorescence signal intensity recorded from the tumor site of the mice shown in Fig. 5A (C), fluorescence intensities of major organs, and tumors presented in Fig. 5B (D).

14. The tumors of above are collected for cryosection.
15. The fluorescence of cryosection is observed with FV1200 laser scanning confocal microscope.

13. *In vivo* SPECT imaging

QC is radiolabeled with $Na^{125}I$ using the direct chloramine-T method. The ^{125}I-labeled probe is separated from free iodine-125 by C18 column (Scheme 3).

1. QC (40 μg) is dissolved in PBS buffer (0.01 M, 200 μL, pH = 7.4), labeled with $Na^{125}I$ (18.5 MBq, 500 μCi), and then chloramine-T (100 μg) in PBS is added.
2. The reaction mixture is incubated for 10 min with shaking at room temperature.
3. The crude reaction is then passed over a C18-SepPak (pretreated first with 95% EtOH and then PBS), washed with 4 mL PBS to remove free iodine-125, and then eluted into 1.0 mL of 95% EtOH in yield of 70–80%.
4. After that, the solution is heated to 40 °C for 6 h with shaking to evaporate EtOH, and controlled the volume of the solution to 50 μL.
5. For imaging, the probe is further diluted into 200 μL PBS.
6. 4T1 tumor-bearing nude mice are intravenously injected with probe (9.25 MBq, 250 μCi, 200 μL).
7. The whole-body SPECT imaging is performed under general anesthesia by inhalation of 3% isoflurane mixed with oxygen gas (0.5 L/min).
8. The mice are maintained in a prone position on a heated animal bed at 37 °C.
9. The SPECT images are acquired with a 60% energy window peaked at 30 keV, with an acquisition time 15 min per frame. The representative SPECT images are acquired at different time points (15, 60, 120, and 240 min).
10. After SPECT acquisition, the animal remained anesthetized, the images are obtained at an X-ray voltage of 55 kVp, and anode current of 615 μA in accurate mode using full angle, 3 frames averaging.
11. The projection data are reconstructed with U-SPECT.
12. After SPECT data reconstruction, the images are analyzed using PMOD software (PMOD™, version 3.6, PMOD Technologies Ltd., Zurich, Switzerland).

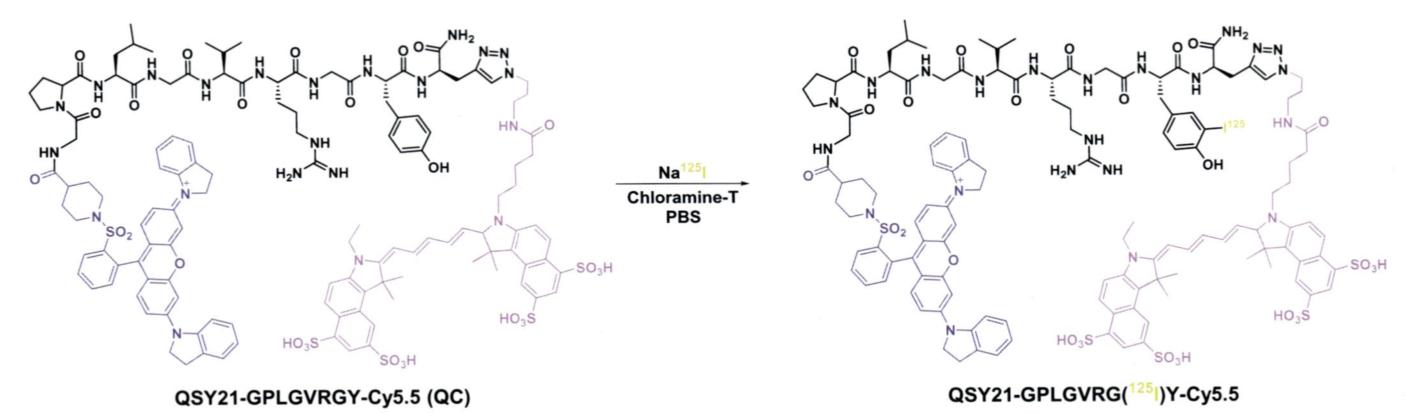

Scheme 3 Preparation and radioiodine labeling of probe QSY21-GPLGVRGY-Cy5.5.

14. *In vivo* photoacoustic imaging

1. Two nude mice with tumor (about 50 mm^3) are selected and marked as QC and Inhibitor, respectively.
2. The mouse is intratumorally injected with GM6001 (1.5 mM, 50 μL) after gas anesthesia.
3. 30 min later, the mouse is anesthetized by 1.5% isoflurane with oxygen gas (0.5 L/min) delivered *via* a nose cone, and placed into the water bath of MSOT to maintain its body temperature at 37 °C.
4. Fine-tune the direction of the mouse through the program to find the best position of tumor.
5. QC (60 μM, 200 μL) is injected into the mice through tail vein followed by imaging the tumors at different time points (0, 15, 30, 45, 60, 75, 90, 105 and 120 min) by Multispectral Optoacoustic Tomography scanner with excitation light at 680, 694, 730, 750, 780 and 800 nm.
6. The mouse injected with QC is imaged as the previous procedures 3–5 without inhibitor.
7. 4T1 tumor-bearing nude mice are divided into two groups according to the size of tumors (50, 90, 120, 160, and 230 mm^3) for QC group and control groups.
8. The mice in QC and control groups with different size of tumors are imaged in the same way, recorded at 2 h post injection of QC probe (60 μM, 200 μL) or control probe (60 μM, 200 μL).
9. After imaging reconstruction, the signal of probes at tumor site is measured by ROI analysis using the MSOT imaging system software package.
10. Tumor volume (V) is calculated as follows: V (mm^3) $= 1/2 \times$ length (mm) $\times$ width (mm)^2. The activity of MMP-2 in tumors with different sizes is calculated by normalizing the protease activity obtained by comparing PA signal of 680 nm to the one of 730 nm referring to the tumor volume (see Fig. 6).

15. *In vitro* determination of MMP-2 expression

15.1 Equipment

1—Bio-Rad
3—Sterile insulin syringe
3—Pipetting gun and heads (P10, P100, P1000)
28—Eppendorf tubes (0.5 mL, 1.0 mL capacity)

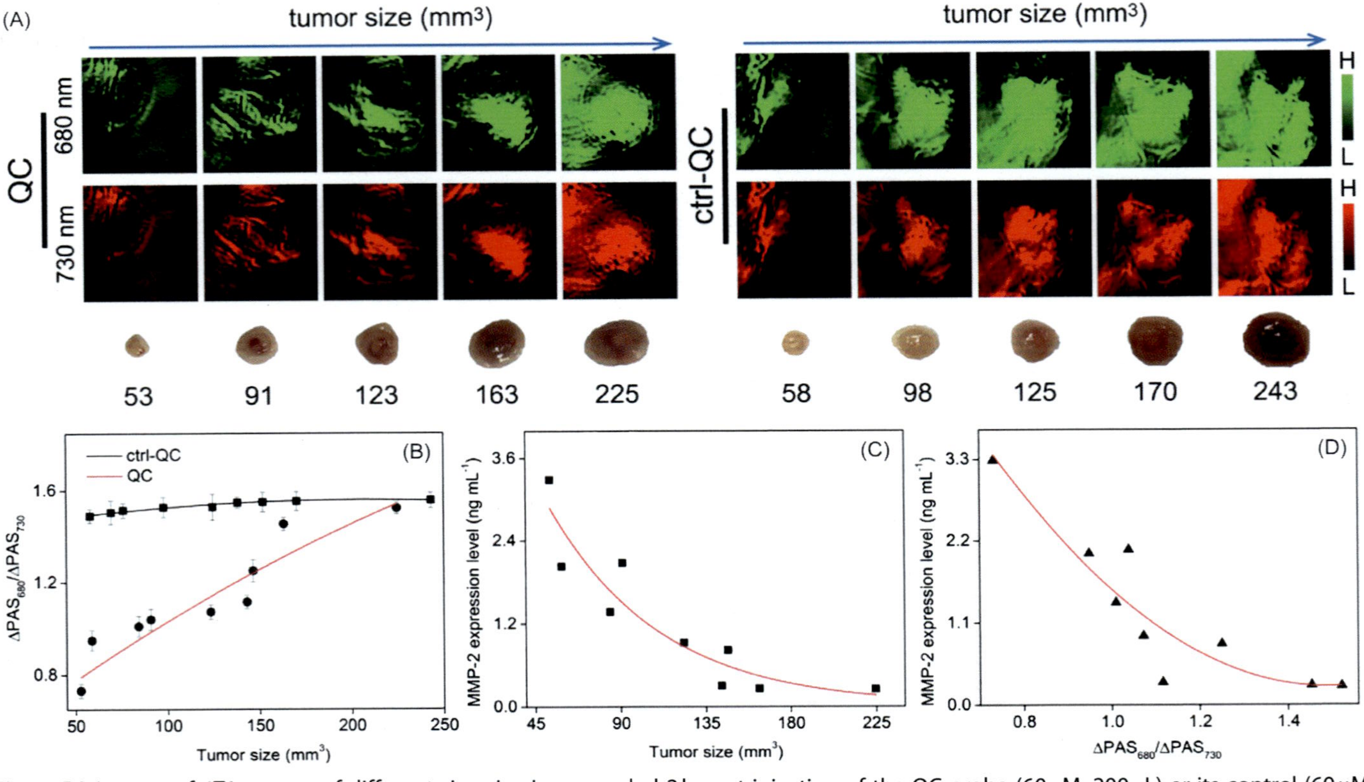

Fig. 6 PA images of 4T1 tumors of different sizes *in vivo*, recorded 2 h post injection of the QC probe (60 μM, 200 μL) or its control (60 μM, 200 μL) through 680 nm and 730 nm channels, together the photographs of the corresponding tumors harvested right after the PA imaging (A), the ratiometric signal $\Delta PAS_{680}/\Delta PAS_{730}$ against the tumor size (B), the tumor size-dependent MMP-2 expression level determined through conventional method (C) for noninvasively determining it through the $\Delta PAS_{680}/\Delta PAS_{730}$ signal (D).

15.2 Reagents

Bradford assay
PBST buffer (0.01 M)
PER Mammalian Protein Extraction Reagent (Thermo Fisher Scientific, Waltham, MA)
Pierce™ ECL Western Blotting Substrate (Thermo Fisher Scientific, Waltham, MA)

15.3 Procedure

The expression level of MMP-2 within tumorous tissues is determined through western blot method.

1. The tissues extracted from different sizes of tumors (50–250 mm^3) are frozen in liquid nitrogen and stored at −80 °C until use.
2. The tissues are grinded and homogenized with protein extraction reagent.
3. After centrifugation at 10,000 rpm for 10 min at 4 °C, the soluble fraction obtained is subjected to Bradford assay for measuring the concentration of the total proteins.
4. The anti-MMP2 antibody is immobilized in the well plate upon incubation overnight at 4 °C and rinsed with 1 × PBST buffer for three times.
5. The tissue lysate is then incubated in plate for 60 min at 37 °C and rinsed with 1 × PBST buffer.
6. Western blotting substrate is used to detect the chemiluminescence by Tanon 4200 automatic chemiluminescence image analysis system after exposure.
7. Use the Gel-Pro analyzer software to make gray analysis quantitatively.
8. By comparing with a MMP-2 standard substance, the MMP-2 expression level is determined and normalized according to the total protein concentration (see Fig. 6).

16. Concluding remarks

In summary, we developed an activatable fluorescence probe for optically detecting the tumor-associated protease MMP-2. We describe in detail of synthesis, properties, and the specificity of the probe toward MMP-2 enzyme *in vitro*, as well as fluorescence, SPECT, and photoacoustic imaging of tumors *in vivo*. We thus believe that this probe may provide a valuable tool for noninvasively investigating the malignant behavior of tumor-related proteases *in vivo*.

References

Black, K. C. L., Akers, W. J., Sudlow, G., Xu, B., Laforest, R., & Achilefu, S. (2015). Dual-radiolabeled nanoparticle SPECT probes for bioimaging. *Nanoscale*, 7, 440–444. https://doi.org/10.1039/c4nr05269b.

Blum, A. P., Kammeyer, J. K., Rush, A. M., Callmann, C. E., Hahn, M. E., & Gianneschi, N. C. (2015). Stimuli-responsive nanomaterials for biomedical applications. *Journal of the American Chemical Society*, *137*, 2140–2154. https://doi.org/10.1021/ja510147n.

Dudani, J. S., Ibrahim, M., Kirkpatrick, J., Warren, A. D., & Bhatia, S. N. (2018). Classification of prostate cancer using a protease activity nanosensor library. *Proceedings of the National Academy of Sciences of the United States of America*, *115*, 201805337. https://doi.org/10.1073/pnas.1805337115.

Egeblad, M., & Werb, Z. (2002). New functions for the matrix metalloproteinases in cancer progression. *Nature Reviews Cancer*, *2*, 161–174. https://doi.org/10.1038/nrc745.

Ferrari, R., Martin, G., Tagit, O., Guichard, A., Cambi, A., Voituriez, R., et al. (2019). MT1-MMP directs force-producing proteolytic contacts that drive tumor cell invasion. *Nature Communications*, *10*, 4886. https://doi.org/10.1038/s41467-019-12930-y.

Frangioni, J. V. (2003). In vivo near-infrared fluorescence imaging. *Current Opinion in Chemical Biology*, 7, 626–634. https://doi.org/10.1016/j.cbpa.2003.08.007.

Gallo, J., Kamaly, N., Lavdas, I., Stevens, E., Nguyen, Q. D., Wylezinska-Arridge, M., et al. (2014). CXCR4-targeted and MMP-responsive iron oxide nanoparticles for enhanced magnetic resonance imaging. *Angewandte Chemie International Edition*, *53*, 9550–9554. https://doi.org/10.1002/anie.201405442.

Huang, H., & Lovell, J. F. (2017). Advanced functional nanomaterials for theranostics. *Advanced Functional Materials*, *27*, 201603524. https://doi.org/10.1002/adfm.201603524.

Johansson, M. K., & Cook, R. M. (2003). Intramolecular dimers: A new design strategy for fluorescence-quenched probes. *Chemistry*, *9*, 3453. https://doi.org/10.1002/chem.2003 90237.

Joyce, J. A., & Pollard, J. W. (2009). Microenvironmental regulation of metastasis. *Nature Reviews Cancer*, *9*, 239–252. https://doi.org/10.1038/nrc2618.

Kabelác, M., Zimandl, F., Fessl, T., Chval, Z., & Lankas, F. (2010). A comparative study of the binding of QSY 21 and rhodamine 6G fluorescence probes to DNA: Structure and dynamics. *Physical Chemistry Chemical Physics*, *12*, 9677–9684. https://doi.org/10.1039/c004020g.

Kessenbrock, K., Plaks, V., & Werb, Z. (2010). Matrix metalloproteinases: Regulators of the tumor microenvironment. *Cell*, *141*, 52–67. https://doi.org/10.1016/j.cell.2010.03.015.

Lee, S., Xie, J., & Chen, X. (2010). Peptides and peptide hormones for molecular imaging and disease diagnosis. *Chemical Reviews*, *110*, 3087–3111. https://doi.org/10.1021/cr900361p.

Sawyers, C. L. (2008). The cancer biomarker problem. *Nature*, *452*, 548–552. https://doi.org/10.1038/nature06913.

Shanmugam, V., Selvakumar, S., & Yeh, C. S. (2014). Near-infrared light-responsive nanomaterials in cancer therapeutics. *Chemical Society Reviews*, *43*, 6254–6287. https://doi.org/10.1039/c4cs00011k.

Sternlicht, M. D., & Werb, Z. (2001). How matrix metalloproteinases regulate cell behavior. *Annual Review of Cell and Developmental Biology*, *17*, 463–516. https://doi.org/10.1146/annurev.cellbio.17.1.463.

Tang, Y., Wu, Z., Zhang, C. H., Zhang, X. L., & Jiang, J. H. (2016). Enzymatic activatable self-assembled peptide nanowire for targeted therapy and fluorescence imaging of tumors. *Chemical Communications*, *52*, 3631–3634. https://doi.org/10.1039/c5cc10591a.

Turk, B. (2006). Targeting proteases: Successes, failures and future prospects. *Nature Reviews Drug Discovery*, *5*, 785–799. https://doi.org/10.1038/nrd2092.

Weissleder, R., & Ntziachristos, V. (2003). Shedding light onto live molecular targets. *Nature Medicine*, *9*, 123–128. https://doi.org/10.1038/nm0103-123.

Winer, A., Adams, S., & Mignatti, P. (2018). Matrix metalloproteinase inhibitors in cancer therapy: Turning past failures into future successes. *Molecular Cancer Therapeutics*, *17*, 1147–1155. https://doi.org/10.1158/1535-7163.MCT-17-0646.

Wisdom, K. M., Adebowale, K., Chang, J., Lee, J. Y., Nam, S., Desai, R., et al. (2018). Matrix mechanical plasticity regulates cancer cell migration through confining microenvironments. *Nature Communications*, *9*, 4144. https://doi.org/10.1038/s41467-018-06641-z.

Wong, C., Stylianopoulos, T., Cui, J., Martin, J., Chauhan, V. P., Jiang, W., et al. (2011). Multistage nanoparticle delivery system for deep penetration into tumor tissue. *Proceedings of the National Academy of Sciences of the United States of America*, *108*, 2426–2431. https://doi.org/10.1073/pnas.1018382108.

Wu, P. G., & Brand, L. (1994). Resonance energy transfer: Methods and applications. *Analytical Biochemistry*, *218*, 1–13. https://doi.org/10.1006/abio.1994.1134.

Yin, L., Sun, H., Zhang, H., He, L., Qiu, L., Lin, J., et al. (2019). Quantitatively visualizing tumor-related protease activity in vivo using a ratiometric photoacoustic probe. *Journal of the American Chemical Society*, *141*, 3265–3273. https://doi.org/10.1021/jacs.8b13628.

CHAPTER FOUR

Aluminum naphthalocyanine conjugate as an MMP-2-activatable photoacoustic probe for *in vivo* tumor imaging

Koji Miki*, Naoto Imaizumi, Kohei Nogita, Masahiro Oe, Huiying Mu, Wenting Huo, and Kouichi Ohe*

Department of Energy and Hydrocarbon Chemistry, Graduate School of Engineering, Kyoto University, Kyoto, Japan

*Corresponding authors: e-mail address: kojimiki@scl.kyoto-u.ac.jp; ohe@scl.kyoto-u.ac.jp

Contents

Abstract

Matrix metalloproteinase-2 (MMP-2), which is one of MMPs family, is known as an extracellular gelatinase controlling cancer cell adhesion, growth, and metastasis. Because of the great interest in MMP-2 activity, the detailed protocols for evaluating MMP-2-responsive contrast agents, especially photoacoustic probes for *in vivo* use, are helpful for researchers in the field. We here describe the detailed synthetic procedure of MMP-2-activatable photoacoustic probe **AlNc-*pep*-PEG** consisting of aluminum naphthalocyanine, MMP-2-responsive peptide sequence, and poly(ethylene glycol), which has recently been developed in our research group. The detailed measurement protocol of photoacoustic signal intensity *in vitro* and *in vivo* by using in-house built photoacoustic signal measurement system and photoacoustic imaging apparatus are also summarized.

Methods in Enzymology, Volume 657
ISSN 0076-6879
https://doi.org/10.1016/bs.mie.2021.07.001

1. Introduction

Matrix metalloproteinases (MMPs, or matrix metallopeptidases) are a family of zinc-containing extracellular endopeptidases which can cleave specific peptide sequences in extracellular matrix proteins (Brinckerhoff & Matrisian, 2002; Fields, 2019; Yang & Rosenberg, 2015). Since the extracellular matrix plays an important role in adhesion, cell and tissue growth, and cell cycle regulation including apoptosis, many researchers focused their attention on the MMPs activity of cells, especially tumor cells. MMPs consists of more than 20 subtypes and are divided into subclasses, such as collagenases including MMP-1 and MMP-8, gelatinases including MMP-2 and MMP-9, stromelysins including MMP-3 and MMP-10, and others. Among them, matrix metalloprotease-2 (MMP-2) is known as an overexpressed proteinase in specific tumor tissues, which can control tumor metastasis and angiogenesis (Foda & Zucker, 2001; Mook, Frederiks, & Van Noorden, 2004). Therefore, the detection as well as monitoring of the MMP-2 activity is of great importance in the field of chemical probes. Although MMP-2 concentration in body fluids, which is correlated to cancer malignancy, can be assessed by several methods including zymographic detection (Ren et al., 2015; Eissa et al., 2007; Rocca, Pucci-Minafra, Marrazzo, Taormina, & Minafra, 2004), these methods are not suitable for evaluation and visualization of MMP-2-activity in the target tissues.

The role of MMP-2 in cancer progression is also important. Because MMP-2 can degrade type IV collagen which consist of the basement membrane of tissues, it is well known that MMP-2 tightly relates to the metastatic progression of cancers including the degradation of basement membranes. Based on these reasons, MMP-2-activatable fluorescence probes have been developed to evaluate the MMP-2 activity in cells as well as *in vivo* (Akers et al., 2012; Wu, Wang, Ding, & Xu, 2017; Zhang, Bresee, Fields, & Edwards, 2014). Although some of the turn-on fluorescence probes clearly visualize MMP-2 activity in cells, signals from deeper tissues are of much interest for better understanding and monitoring the MMP-2 activity. In the last two decades, some of MMP-2-responsive photoacoustic probes have been developed to apply them to *in vivo* visualization (Levi et al., 2010, 2013; Peng et al., 2019; Yin et al., 2019). However, all MMP-2-responsive photoacoustic probes so far reported consist of near-infrared (NIR) dyes with photobleaching disadvantages. Based on these background, we have recently developed the first example of MMP-2-activatable photoacoustic probe on the basis of photostable aluminum naphthalocyanine (AlNc).

We here summarize the probe design of MMP-2-activatable photoacoustic probe **AlNc-*pep*-PEG**, synthetic procedures, and validation of **AlNc-*pep*-PEG**. With regard to evaluation of photoacoustic signal intensity of probes, we here describe the in-house-built photoacoustic signal measurement system, consisting of a pulse laser, transducer, and oscilloscope. The evaluation of **AlNc-*pep*-PEG** *in vivo* by using photoacoustic tomographer is also described.

2. Selection of photoacoustic photosensitizer

For designing the proteinase-activatable photoacoustic probes, the increment of photoacoustic signal intensity after enzymatic reaction is required. Photosensitizing dyes bearing activatable properties can be divided in two categories: the enzymatic reaction (1) controls the quenched/emissive states of dyes or (2) induces the aggregation/dispersion of dyes. In the former approach, dyes are capped by the quenching groups which acts as a substrate of enzymes. After the enzymatic reaction, the removal of the quenching group leads to the generation of the emissive dye. Because of this simplicity, the former approach is frequently used for detection of enzymes; however, this strategy is difficult to be used for enzymes which cannot completely remove the substrate from the quenched dye. Since MMP-2 cleaves the middle of peptide sequences, such as PLGLAG (Jiang et al., 2004) and PLGVRG (Yin et al., 2019), the remaining peptide residue after MMP-2-mediated reaction usually prevents turning on the quenched dye. Therefore, we selected the latter approach; the MMP-2-mediated reaction induces the aggregation of dyes for enhancing the photoacoustic signal intensity.

2.1 Photoacoustic imaging

Photoacoustic (PA) imaging is one of the most powerful and reliable modalities to detect cancers by the irradiation of less-invasive NIR pulse laser (Wang & Hu, 2013; Weber, Beard, & Bohndiek, 2015). It is well accepted that NIR dyes which efficiently absorb photoenergy and generate PA waves through nonradiative decay are suitable as PA contrast agents (Gujrati, Mishra, & Ntziachristos, 2017; Wang & Zhang, 2021). For high-contrast PA imaging, the following three requirements are considered to be important for organic dyes to apply as PA contrast agents: (i) NIR non-emissive dyes with high molar extinction coefficients for strong PA wave generation (Cheng, Li, Tang, & Yoon, 2020), (ii) high photostability under continuous pulse laser irradiation for repeated diagnosis and biological assessment (Ji, Cheng, Yuan, Müllen, & Yin, 2019; Wang, Geng, Cong, Shen, & Yu, 2019), and (iii) activatable property to reduce false positive signal from unreacted probes in normal tissues or blood vessel (Cheng & Pu, 2020; Knox & Chan, 2018) should be involved.

2.2 Selection of near-infrared dye backbone

Based on the three requirements mentioned above, we selected naphthalocyanine (MNc) which is one of the photostable NIR dyes with a high ε value ($\sim 1 \times 10^5$ cm^{-1} M^{-1}) (Lu, Lim, Javitt, Heinmiller, & Prud'homme, 2017). Despite their properties being suitable as PA contrast agents, there are a limited number of water-soluble MNcs applying to PA imaging, probably because their low solubility in solvents has hampered their functionalization. Lovell and *co*-workers reported tin naphthalocyanine (SnNc) conjugated with poly(ethylene glycol) (PEG) and its application to *in vivo* PA imaging of blood vessels (Huang et al., 2016). Although their report supports the applicability of MNcs in *in vivo* PA imaging, the PA imaging of target tissues using MNcs with "always-on" property may provide low contrast images with background noise. **AlNc-*pep*-PEG** with "activatable" and "ratiometric" property has a great potential for sensitive detection of MMP-2 activity as well as high contrast PA imaging *in vivo*.

3. Design of MMP-2-responsive photoacoustic probe AlNc-*pep*-PEG

AlNc-*pep*-PEG consists of aluminum naphthalocyanine (AlNc), MMP-2-responsive peptide sequence, and water-soluble PEG chain (see Fig. 1). AlNc exhibits high molar extinction coefficient in *N*,*N*-dimethylformamide (DMF, $\sim 1.0 \times 10^5$ cm^{-1} M^{-1}) and water ($> 5 \times 10^4$ cm^{-1} M^{-1}) in the NIR region (680–780 nm). MMP-2-responsive peptide sequence PLGLAG having acetyl-capped cysteine at the *N* terminus can conjugate to PEG having a maleimide moiety under mild conditions. The standard condensation reaction of an amino substituent at the axial ligand of AlNc with free C terminus of peptide sequence is designed for efficient and mild conjugation.

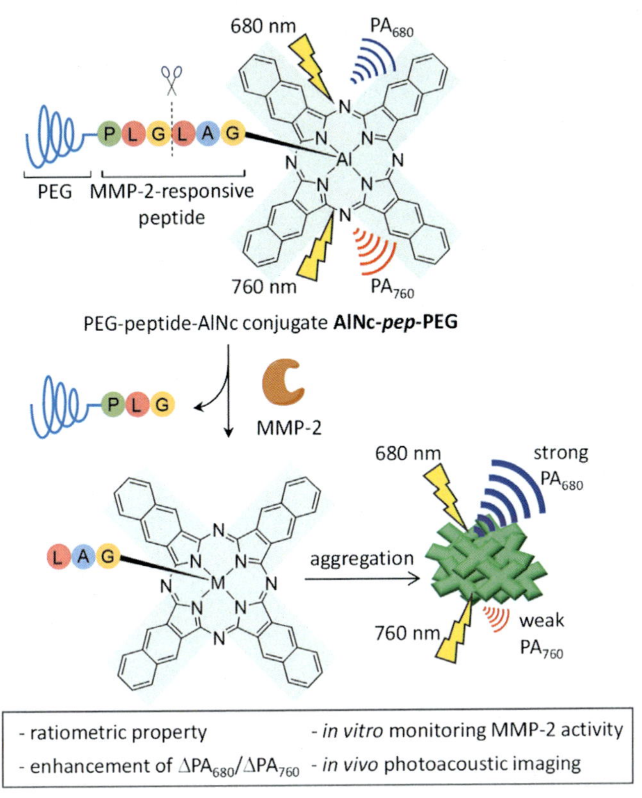

Fig. 1 MMP-2-triggered AlNc-aggregation for photoacoustic signal enhancement. P: proline, L: leucine, G: glycine, A: alanine. *Reproduced with permission from ACS: Miki, K., Imaizumi, N., Nogita, K., Oe, M., Mu, H., Huo, W., et al. (2021). MMP-2-activatable photoacoustic tumor imaging probes based on Al- and Si-naphthallocyanines.* Bioconjugate Chemistry. *doi: 10.1021/acs.bioconjchem.1c00266.*

4. Synthesis of AlNc-*pep*-PEG

AlNc-*pep*-PEG can be synthesized in two steps from AlNc **1** (Ford et al., 1992) bearing hydroxy group as an axial ligand (Scheme 1). **AlNc-PEG** without an MMP-2-responsive peptide is similarly synthesized by conjugating with poly(ethylene glycol) monomethyl ether (PEG, average molecular weight = 2000) derivative **HOOC-PEG** bearing a carboxy group. PEG-peptide conjugate is prepared from a peptide sequence CPLGLAG and PEG (average molecular weight = 2000) bearing a maleimide

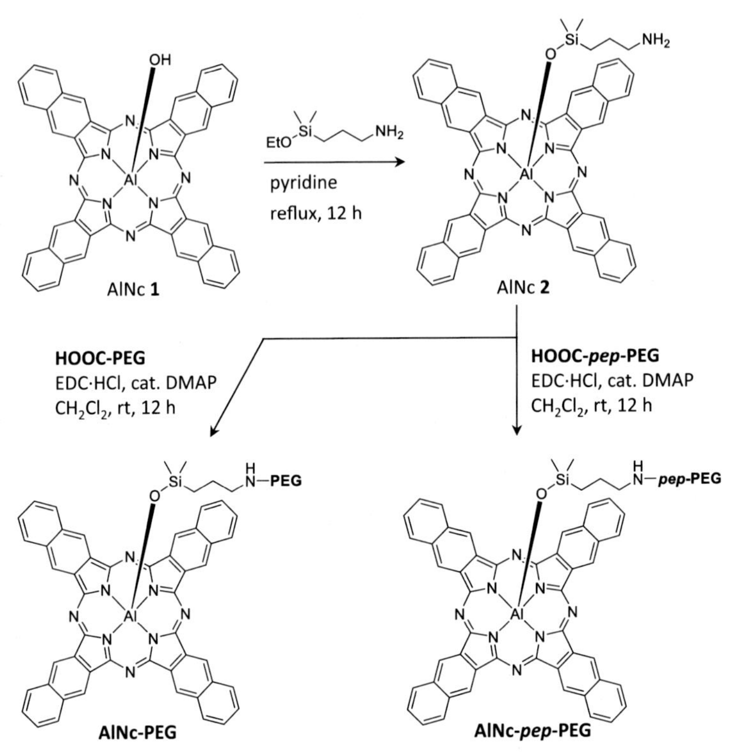

Scheme 1 Synthesis of **AlNc-*pep*-PEG** and **AlNc-PEG.**

moiety (Scheme 2). The peptide sequence CPLGLAG having acetyl cap at the *N* terminus is prepared by standard solid-phase synthesis using 2-chlorotrityl resin.

PEG derivative HOOC-PEG.

1. Preheated an oil bath at 60 °C
2. Add poly(ethylene glycol) derivative $H_2N(CH_2CH_2O)_{45}CH_3$ (1.0 g, 0.50 mmol, average molecular weight = 2000) to 25 mL flame-dried Schlenk flask equipped with a magnetic stirring bar under a gentle flow of nitrogen gas
3. Heat the reaction vessel in an oil bath for 8 h under reduced pressure
4. Add tetrahydrofuran dry (THF, 5 mL), dry triethylamine (Et_3N, 0.5 mL), and succinic anhydride (0.60 g, 0.60 mmol) to the reaction vessel at room temperature
5. Stir the mixture at room temperature for 24 h
6. Concentrate the reaction mixture by rotary evaporation
7. Wash the residue with ether (5 mL × 2).

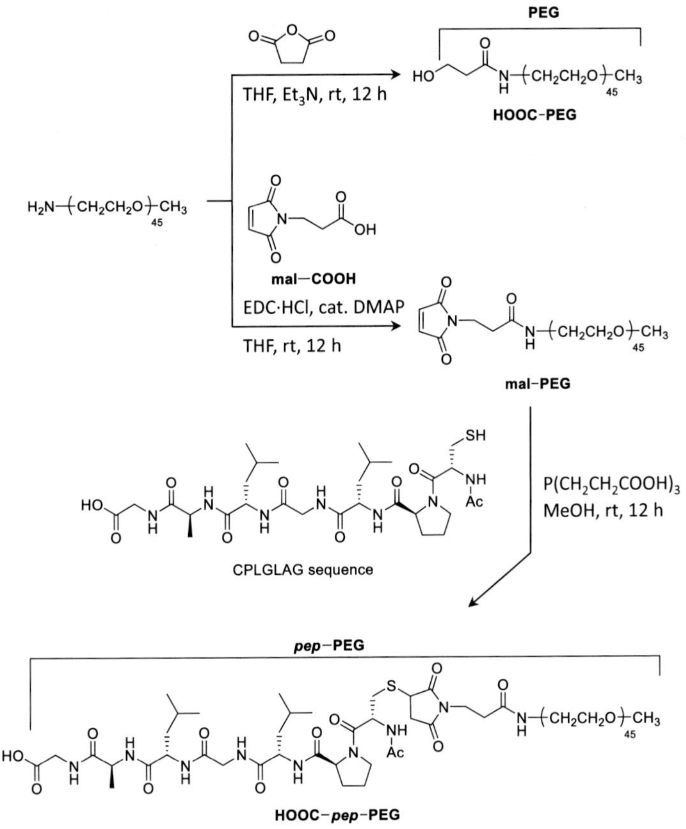

Scheme 2 Synthesis of **HOOC-PEG** and **HOOC-*pep*-PEG**.

8. Dry the crude solid under reduced pressure and use directly to the next step without further purification

PEG derivative mal-PEG

1. Add poly(ethylene glycol) derivative $H_2N(CH_2CH_2O)_{45}CH_3$ (1.0g, 0.50mmol, average molecular weight=2000), **mal-COOH** (0.17g, 1.0mol, Mantovani et al., 2005), 1-ethyl-(3-dimethylaminopropyl) carbodiimide hydrochloride (EDC·HCl, 0.19g, 1.0mol), 4-dimethylaminopyridine (DMAP, 6.0mg, 10mol%), and dry THF (10mL) to a 25mL flame-dried Schlenk flask equipped with a magnetic stirring bar and a rubber septum under a gentle flow of nitrogen gas
2. Stir the mixture at room temperature for 12h
3. Concentrate the reaction mixture by rotary evaporation
4. Dissolve the residue in THF (0.5mL) and add ether (10mL) to the mixture

5. Collect the precipitate and wash it with ether (5 mL × 2).
6. Dry the crude solid under reduced pressure and use directly to the next step without further purification

PEG derivative HOOC-*pep*-PEG

1. Add **mal-PEG** (62 mg, 30 μmol), CPLGLAG peptide sequence (20 mg, 30 μmol), tris(2-carboxyethyl)phosphine hydrochloride (9 mg, 30 μmol), and dry methanol (10 mL) to a 25 mL flame-dried Schlenk flask equipped with a magnetic stirring bar and a rubber septum under a gentle flow of nitrogen gas
2. Stir the mixture at room temperature for 12 h
3. Concentrate the reaction mixture by rotary evaporation
4. Dissolve the residue in H_2O (2 mL).
5. Dialyze the aqueous solution against H_2O for 1 day by using Spectra/Por 6 (MWCO = 3.5 K).
6. Lyophilize the resulting aqueous solution to afford **HOOC-*pep*-PEG** (72 mg, 26 μmol, 88%).

AlNc 2

1. Preheat an oil bath at 120 °C
2. Add AlNc **1** (30 mg, 40 μmol), (3-aminopropyl)dimethylethoxysilane (38 μL, 0.20 mmol), and pyridine (100 mL) to a 200 mL flame-dried two-necked round-bottom flask equipped with a Dimroth condenser, a magnetic stirring bar, and a glass stopcock under a gentle flow of nitrogen gas
3. Heat the reaction vessel in an oil bath for 12 h
4. Concentrate the reaction mixture by rotary evaporation
5. Dissolve the residue in dichloromethane (2 mL)
6. Wash the organic solution with an aqueous solution of copper sulfate (1 M, 5 mL).
7. Dry the organic layer over magnesium sulfate and concentrate it by rotary evaporation to obtain a crude product **2** (12 mg, 14 μmol, 36%). Use directly without further purification. NOTE: AlNc **2** gradually decomposes to form insoluble materials within 1 day, even stored as a solid state in a freezer

AlNc-PEG

1. Add **HOOC-PEG** (23 mg, 11 μmol), dry dichloromethane (CH_2Cl_2, 5 mL), EDC·HCl (9.0 mg, 44 μmol), and a solution of DMAP (0.3 mg, 2.5 μmol) in CH_2Cl_2 (0.1 mL) to a 50 mL flame-dried two-necked round-bottom flask equipped with a magnetic stirring bar and a rubber septum under a gentle flow of nitrogen gas and stir the mixture at room temperature for 1 h

2. Add AlNc **2** (10 mg, 11 μmol) and stir at room temperature for 12 h
3. Concentrate the reaction mixture by rotary evaporation
4. Dissolve the residue in H_2O (5 mL).
5. Centrifuge the supernatant and dialyze the aqueous solution against water for 1 d by using Spectra/Por 6 (MWCO = 25 K).
6. Lyophilize the resulting aqueous solution to afford **AlNc-PEG** (15 mg, 5.1 μmol, 46%) as a green solid. **AlNc-PEG** can be stored as a solid in a freezer (−20 °C) in dark for several months without noticeable decomposition

AlNc-*pep*-PEG

1. Add **HOOC-*pep*-PEG** (30 mg, 11 μmol), dry CH_2Cl_2 (5 mL), EDC·HCl (9.0 mg, 44 μmol), and a solution of DMAP (0.3 mg, 2.5 μmol) in CH_2Cl_2 (0.1 mL) to a 50 mL flame-dried two-necked round-bottom flask equipped with a magnetic stirring bar and a rubber septum under a gentle flow of nitrogen gas and stir the mixture at room temperature for 1 h
2. Add AlNc **2** (10 mg, 11 μmol) and stir at room temperature for 12 h
3. Concentrate the reaction mixture by rotary evaporation
4. Dissolve the residue in H_2O (5 mL).
5. Centrifuge the supernatant and dialyze the aqueous solution against water for 1 d by using Spectra/Por 6 (MWCO = 25 K).
6. Lyophilize the resulting aqueous solution to afford **AlNc-*pep*-PEG** (23 mg, 6.4 μmol, 58%) as a green solid. **AlNc-*pep*-PEG** can be stored as a solid in a freezer (−20 °C) in dark for several months without noticeable decomposition

5. *In vitro* characterization of AlNc-*pep*-PEG

Before cell experiments, it is necessary to assess the photophysical properties and MMP-2 responsiveness of **AlNc-*pep*-PEG**. We describe our characterization by UV–vis absorption measurement and photoacoustic signal measurement. We recommend using the 2 mM DMF solution as a stock solution for *in vitro* and *in vivo* experiments to expedite the experiments.

5.1 Photophysical characterization of AlNc-*pep*-PEG

The wavelength of absorption maximum and molar extinction coefficient at the wavelength is summarized in Table 1. This information allows the users to evaluate the concentration of **AlNc-*pep*-PEG** in sample solutions.

Table 1 Photophysical properties.

Compound	λ_{max} (nm, DMF)	ε (cm^{-1} M^{-1}, DMF)	λ_{max} (nm, water)	ε (cm^{-1} M^{-1}, water)
AlNc-*pep*-PEG	788	1.0×10^5	726	5.4×10^4

Equipment.
1–Microbalance
2–Vials per compounds (5 mL capacity)
1–UV–vis spectrometer
3–Quartz cuvettes (10 mm × 10 mm) with corresponding cuvette caps.
3–Micropipettes and tips (P10, P200, P1000).
Reagents
AlNc-*pep*-PEG (~5 mg)
DMF or water (MilliQ)
Procedure

1. Warm up the spectrometer at least 30 min before the experiment
2. Weigh a small amount of **AlNc-*pep*-PEG** (~4 mg) in a vial
3. Calculate the volume of solvent and add the appropriate amount of DMF or water (*ca.* 3.7 mg/0.5 mL) to prepare 2 mM solution
4. Cap the vial and shake vigorously to dissolve **AlNc-*pep*-PEG**. If necessary, sonicate the mixture until the solid is fully dissolved. NOTE: The 2 mM solution can be stored in refrigerator more than 1 week without decrement of absorbance
5. Add a DMF or aqueous solution of **AlNc-*pep*-PEG** (2 mM, 15 μL) and DMF or water (2985 μL) to a new vial and mix the solution by pipetting to prepare 10 μM solution
6. Add DMF or water (3 mL) in two cuvettes and conduct baseline correction
7. Transfer the 10 μM solution to a cuvette and measure the absorbance spectrum. Confirm the reproducibility by re-measuring the absorbance spectrum
8. Repeat the steps 2–7 to check the reproducibility. Identify the wavelength of maximum absorbance of **AlNc-*pep*-PEG** solution. Confirm the linear relationship between absorbance and concentration and calculate the molar extinction coefficient

5.2 Photoacoustic validation of AlNc-*pep*-PEG

To monitor the photoacoustic signal intensity before and after enzymatic reaction, it is necessary to detect the change of photoacoustic signal intensity. The photoacoustic signal intensity measurement system can be setup by using following equipment (see Fig. 2). The suitable pulse laser power range is 0.1–20 mJ/cm^{-2} for photoacoustic signal measurement (pulse width: 5–20 ns).

Equipment.

1–Nitrogen (N_2) laser (GL-3300, Photon Technology International)

1–Dye laser (GL-302, Photon Technology International)

1–Ultrasonic transducer (Panametrics-NDT V303, Olympus)

1–Preamplifier (Model 5682, Olympus)

1–Silicon biased detector (DET10A/M, Thorlabs)

1–Digital phosphor oscilloscope (TDS2012C, Tektronix)

1–Power meter (PM100USB, Thorlabs)

1–Sensor with broadband coating (ES111C, Thorlabs)

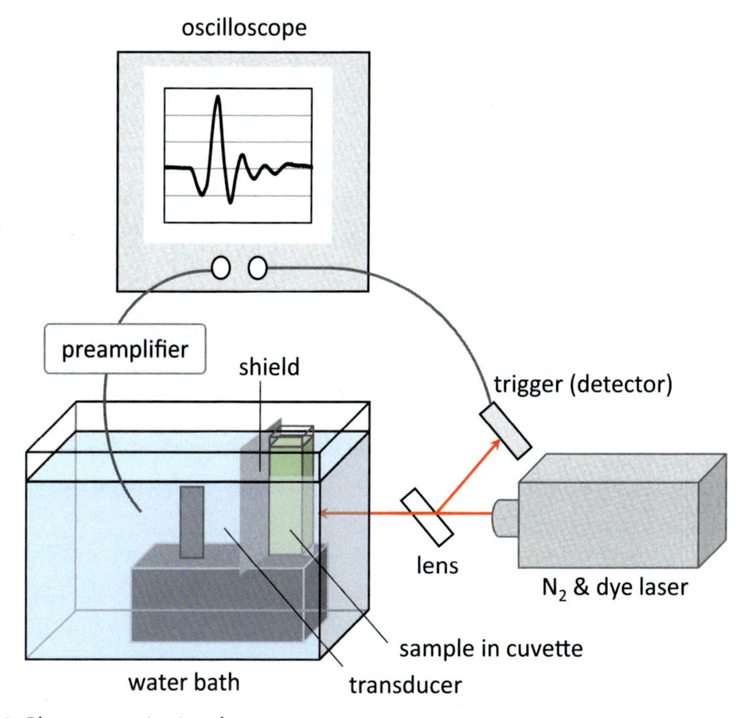

Fig. 2 Photoacoustic signal measurement system.

1–Lens (transmission >90%)
1–Aluminum foil (shield)
1–Data processor (computer)
1–Black-colored paper
3–Vials
4–Quartz cuvette (10 mm × 10 mm) with corresponding cuvette caps
2–Borosilicate cuvette (1 mm length × 10 mm width)
1–UV–vis spectrometer
3–Micropipettes and tips (P10, P200, P1000)
Reagents
DOTC iodide (3-ethyl-2-[7-(3-ethyl-2(3*H*)-benzoxazolylidene)-1,3,5-heptatrienyl]benzoxazolium iodide)
Oxazine 170 perchlorate (5,9-Bis(ethylamino)-10-methylbenzo[*a*]phenoxazin-7-ium perchlorate)
Rhodamine B or other visible dye
Dimethyl sulphoxide (DMSO) and ethanol (EtOH)
AlNc-*pep*-PEG (2 mM stock solution in DMF)
MilliQ water
Procedure

1. Warm up the UV–vis spectrometer at least 30 min before the experiment
2. Prepare solutions of DOTC iodide (1.5 mg in DMSO (3 mL)), oxazine 170 perchlorate (1.2 mg in EtOH (3 mL)), and rhodamine B (6 mg in EtOH (3 mL)) in vials. Cap the vial and shake vigorously to dissolve the dye. Transfer to quartz cuvettes
3. Add water in water bath
4. Set the quartz cuvette of rhodamine B solution in dye laser. Flow nitrogen gas into the laser system and turn on the N_2 laser. Set the pulse frequency to 6–10 Hz
5. Put a black paper in borosilicate cuvette and fill with water. Set this cuvette at the sample position. To centralize the laser, move the sample holder at the proper position. NOTE: another visible dye can be used for this purpose
6. Turn on the preamplifier, oscilloscope, and PC to detect the photoacoustic signal from black paper. NOTE: to avoid the direct irradiation of laser light to transducer, aluminum foil is placed between transducer and sample cuvette. When the distance between transducer and sample cuvette is 30 mm, the photoacoustic wave can be detected after 20–25 μs after silicon biased detector (trigger) detected the pulse laser light. If the signal is weak, connect one more preamplifier in series

7. Shield the N_2 laser light by using an aluminum foil and change to the quartz cuvette of DOTC iodide for 760 nm irradiation or oxazine 170 perchlorate for 680 nm irradiation. Measure the laser power between sample solution and a lens by using power meter and sensor. The laser power is an average value for 3 min measurement
8. Dilute a 2 mM stock DMF solution of **AlNc-*pep*-PEG** with MilliQ
9. Measure the UV–vis absorption spectrum of a sample solution of **AlNc-*pep*-PEG** (30 μM) to determine the absorbance at 680 nm and 760 nm
10. Transfer a sample solution to borosilicate cuvette and set it at the sample holder
11. Capture the average photoacoustic wave generated under irradiation of 128 pulses by the oscilloscope (see Fig. 3).
12. Check the laser power again and confirm the power was not changed. NOTE: if the laser power was decreased, change the dye solution to the new one and retry the measurement steps 7–11
13. Subtract the lowest value of signal from the largest value to calculate the photoacoustic signal intensity. Normalized photoacoustic signal intensity by absorbance and laser power. Repeat the steps 7–12 to check the reproducibility. NOTE: Since PA signal intensities are measured in NIR region, the cuvette materials (quartz, borosilicate, *etc.*) do not significantly affect the results. Polystyrene disposal cuvette can be used for this purpose

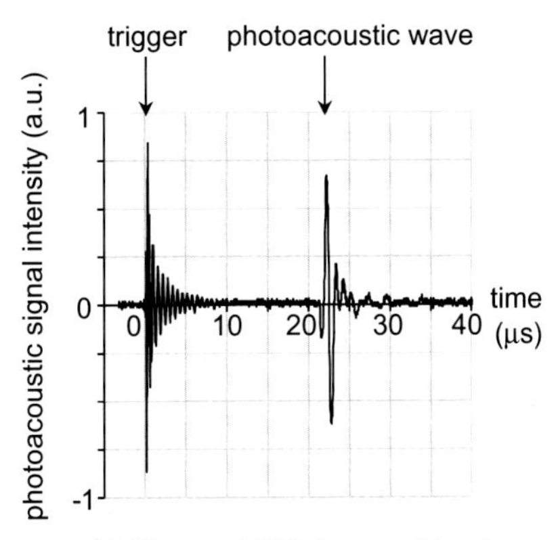

Fig. 3 Photoacoustic wave of **AlNc-*pep*-PEG** detected by the oscilloscope. *Reproduced with permission from ACS: Miki, K., Imaizumi, N., Nogita, K., Oe, M., Mu, H., Huo, W., et al. (2021). MMP-2-activatable photoacoustic tumor imaging probes based on Al- and Si-naphthallocyanines.* Bioconjugate Chemistry. *doi: 10.1021/acs.bioconjchem.1c00266.*

5.3 Enzyme-responsiveness of AlNc-*pep*-PEG

To monitor the MMP-2-responsiveness of **AlNc-*pep*-PEG**, we measure the photoacoustic signal intensity at 680 and 760 nm during incubation of **AlNc-*pep*-PEG** with MMP-2. To confirm the MMP-2-specific response, the photoacoustic signal intensity of **AlNc-*pep*-PEG** treated with other enzymes is similarly measured. To check the MMP-2-specific photoacoustic signal increment, the incubation of **AlNc-PEG** with MMP-2 is conducted.

Equipment

1–Photoacoustic signal measurement system

1–UV–vis spectrometer

1–Bench-top shaking water bath

3–Quartz cuvette (10 mm × 10 mm)

1–Borosilicate cuvette (1 mm length × 10 mm width)

3–Micropipettes and tips (P10, P200, P1000)

Reagents

AlNc-*pep*-PEG (2 mM solution in DMF)

AlNc-PEG (2 mM solution in DMF)

MilliQ water and phosphate buffered saline (pH 7.2)

Recombinant human MMP-2, or other recombinant human metalloproteinases

GM6001

Procedure

1. Warm up the UV–vis spectrometer at least 30 min before the experiment
2. Setup the photoacoustic signal measurement system according to the Section 5.2
3. Dissolve MMP-2 in Milli-Q water to prepare 5.0 μM solution. Dissolve other MMPs in Milli-Q water in the similar manner. NOTE: the temperature is controlled according to the instruction
4. Mix the stock DMF solution (7.5 μL, 2 mM) of **AlNc-*pep*-PEG** and phosphate buffered saline (2992 μL) in a vial. NOTE: Concentration of MMP-2: 2 nM; **AlNc-*pep*-PEG**: 5 μM. For inhibition experiment, the phosphate buffered saline of GM6001 (100 μM) was prepared and used for dilution in step 4
5. Add enzyme solution (3.0 μL) to the vial and mix gently by pipetting
6. Incubate the vial in bench-top shaking water bath
7. Transfer the reaction mixture in quartz cuvette and measure the absorption spectrum

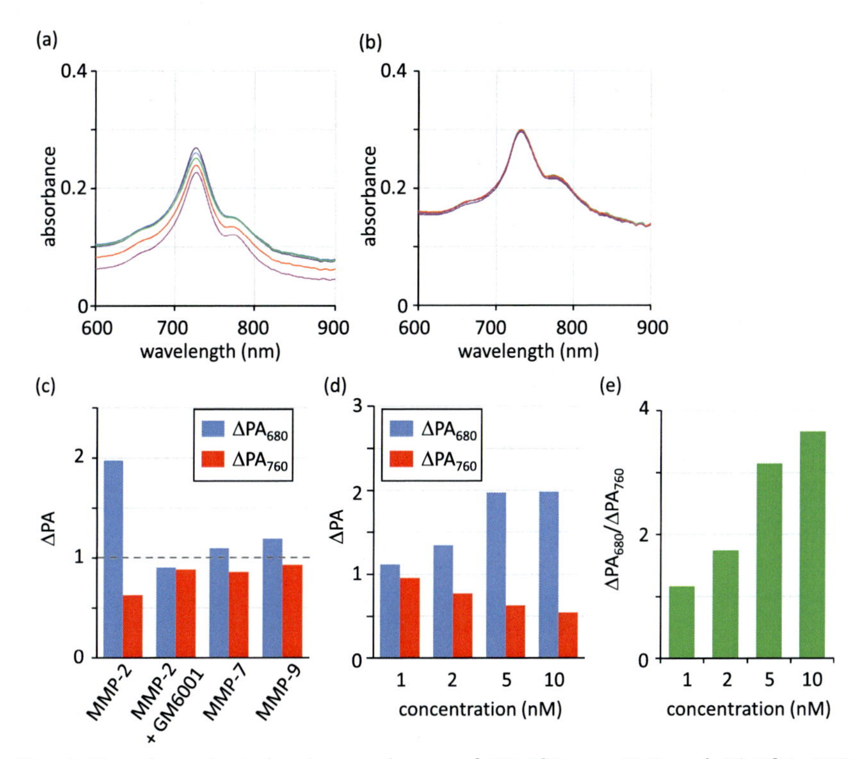

Fig. 4 Time-dependent absorbance change of (A) **AlNc-*pep*-PEG** and (B) **AlNc-PEG** (5.0 μM, pH 7.2) during incubation with MMP-2 (5.0 nM). Just after (purple), 15 min (blue), 30 min (green), 60 min (red), and 120 min (pink). (C) Photoacoustic signal intensity change (ΔPA) of **AlNc-*pep*-PEG** (30 μM, pH 7.2) after incubation with enzyme (0.32 μg/mL) for 2 h. Values correspond to the mean value of two independent experiments. GM6001: inhibitor of MMP-2. (D) Photoacoustic signal intensity change (ΔPA) of **AlNc-*pep*-PEG** and (E) ratio of photoacoustic intensity change ($\Delta PA_{680}/\Delta PA_{760}$) of **AlNc-*pep*-PEG** (30 μM, pH 7.2) after incubation with MMP-2 for 2 h. *Reproduced with permission from ACS: Miki, K., Imaizumi, N., Nogita, K., Oe, M., Mu, H., Huo, W., et al. (2021). MMP-2-activatable photoacoustic tumor imaging probes based on Al- and Si-naphthallocyanines.* Bioconjugate Chemistry. *doi: 10.1021/acs.bioconjchem.1c00266.*

8. Transfer the reaction mixture in borosilicate cuvette and measure photoacoustic signal intensity
9. Calculate the photoacoustic signal intensity change (ΔPA) and ratio of photoacoustic intensity change ($\Delta PA_{680}/\Delta PA_{760}$) (see Fig. 4). ΔPA is calculated by the following equation; $\Delta PA = PA_1 - PA_0$, when PA_1 and PA_0 are photoacoustic signal intensities after and before incubation, respectively
10. Repeat the steps 3–9 to check the reproducibility

5.4 MMP-2-responsiveness of AlNc-*pep*-PEG in live cells

To monitor the MMP-2-responsiveness of **AlNc-*pep*-PEG** in cells, we incubate **AlNc-*pep*-PEG** with MMP-2-overexpressed human sarcoma cell line HT-1080 and measure the photoacoustic signal intensity at 680 nm and 760 nm. The photoacoustic signal intensities from cells are summarized in Table 2.

Equipment.

1–Photoacoustic signal measurement system

1–12-well plate, coated for cell culture

1–Cell culturing clean bench

1–Incubator with 5% CO_2 atmosphere at 37 °C

1–Cell scraper, disposal

1–Borosilicate cuvette (1 mm length × 10 mm width).

3–Micropipettes and tips (P10, P200, P1000).

Reagents

HT-1080 cell

AlNc-*pep*-PEG (2 mM solution in DMF)

Milli-Q water and phosphate buffered saline (pH 7.2)

Eagle's minimum essential medium (EMEM), with 10% FBS and 1% penicillin and streptomycin

Phosphate buffered saline (pH 7.2)

Procedure

1. Seed HT-1080 in 12-well plate at a density of 5×10^4 cells/well
2. Incubate the cells for 24 in the incubator
3. Dilute a stock DMF solution of **AlNc-*pep*-PEG** (2 mM) with EMEM to prepare 30 μM solution
4. Setup the photoacoustic signal measurement system according to the Section 5.2
5. Wash the cells with EMEM (2 mL × 2) and add the EMEM solution of **AlNc-*pep*-PEG** (1 mL)
6. Incubate the cells for 2 or 4 h in the incubator

Table 2 Photoacoustic signal intensity change (ΔPA) and ratio of PA intensity change ($\Delta PA_{680}/\Delta PA_{760}$) of **AlNc-*pep*-PEG** during the incubation of HT-1080.

	2 h incubation			4 h incubation		
	ΔPA_{680}	ΔPA_{760}	$\Delta PA_{680}/\Delta PA_{760}$	ΔPA_{680}	ΔPA_{760}	$\Delta PA_{680}/\Delta PA_{760}$
AlNc-*pep*-PEG	1.9 ± 08	0.4 ± 0.1	4.3 ± 1.6	4.3 ± 1.2	0.8 ± 0.2	5.4 ± 1.0

7. Wash the cells with phosphate buffered saline (2 mL × 2) and detach cells by cell scraper
8. Suspend cells in phosphate buffered saline (0.2 mL).
9. Transfer the reaction mixture in borosilicate cuvette and measure photoacoustic signal intensity
10. Calculate the photoacoustic signal intensity change (ΔPA) and ratio of photoacoustic intensity change ($\Delta PA_{680}/\Delta PA_{760}$). ΔPA is calculated by the following equation; $\Delta PA = PA_1 - PA_0$, when PA_1 and PA_0 are photoacoustic signal intensities after and before incubation, respectively
11. Repeat the steps 1–10 to check the reproducibility

6. Characterization of AlNc-*pep*-PEG *in vivo*

To monitor the increment of photoacoustic signal intensity of **AlNc-*pep*-PEG** *in vivo*, we inject **AlNc-*pep*-PEG** to HT-1080-bearing mice and measure the photoacoustic intensity at the tumor site.

Equipment.

1–Photoacoustic imaging apparatus (Nexus 128, Endra Life Sciences)

1–Vaporizer equipped with isoflurane and oxygen for anesthesia

2–sterile syringes

Animals

1–Female BALB/c mice, 6–9 weeks of age

Reagents

AlNc-*pep*-PEG (2 mM solution in DMF)

AlNc-PEG (2 mM solution in DMF)

Sterile saline

Procedure

1. Warm up the photoacoustic imaging apparatus at least 30 min before the experiment
2. Prepare a 100 μM solution of **AlNc-*pep*-PEG** using 2 mM stock solution in DMF
3. Anesthetize the mouse in an isoflurane-oxygen chamber and keep the anesthetized state under a flow of isoflurane-oxygen atmosphere by using nose corn in the photoacoustic imaging apparatus
4. Inject the **AlNc-*pep*-PEG** solution (200 μL) intravenously using syringe
5. Place the mouse in the imaging apparatus and measure the photoacoustic signal intensity at 680, 760, and 850 nm (3 min measurement time in each wavelength).

6. Bring the mouse back to rearing cage
7. At the desired time point, repeat steps 3, 5, and 6 to measure photoacoustic signal intensity
8. Repeat steps 2–7 for **AlNc-PEG** using other mice
9. Subtract the photoacoustic signal intensity at 850 nm from those at 680 and 760 nm to omit the effect of photoacoustic signals from hemoglobin
10. Process the data by OsiriX according to the instruction. The region of interest (ellipse shape, 10 mm × 14 mm axes) is used
11. Calculate the photoacoustic signal intensity change (ΔPA_{vivo}) and ratio of photoacoustic intensity change ($\Delta PA_{680,vivo}/\Delta PA_{760,vivo}$) (see Fig. 5).

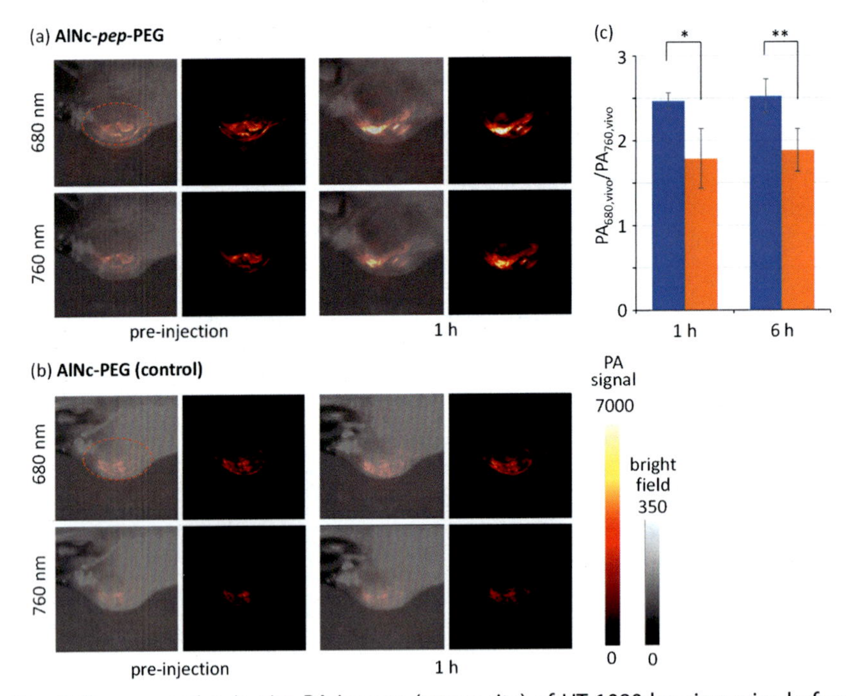

Fig. 5 Representative *in vivo* PA images (tumor site) of HT-1080-bearing mice before and 1 h after *i.v.* injection of (A) **AlNc-*pep*-PEG** and (B) **AlPc-PEG** (100 μM in saline, dose, 200 μL). PA signal image (right) and merged image of PA signal image with bright field image (left). Red dotted circle: tumor site. Photoirradiation wavelength: 680 and 760 nm. (C) PA signal intensity ratio ($PA_{680,vivo}/PA_{760,vivo}$) at the tumor site of HT-1080-bearing mice 1 and 6 h after *i.v.* injection of **AlNc-*pep*-PEG** (blue) and **AlNc-PEG** (orange). Averages and standard deviations ($n = 3$) are shown. Two-tailed Student's *t*-test: $^{*}P < 0.1$, $^{**}P < 0.05$. *Reproduced with permission from ACS: Miki, K., Imaizumi, N., Nogita, K., Oe, M., Mu, H., Huo, W., et al. (2021). MMP-2-activatable photoacoustic tumor imaging probes based on Al- and Si-naphthallocyanines.* Bioconjugate Chemistry. *doi: 10.1021/acs.bioconjchem.1c00266.*

ΔPA is calculated by the following equation; $\Delta PA_{vivo} = PA_{1,vivo} - PA_{0,vivo}$, when $PA_{1,vivo}$ and $PA_{0,vivo}$ are photoacoustic signal intensities hours after injection and before injection, respectively

12. Repeat the steps 1–11 to check the reproducibility

7. Concluding remarks

Here, we describe the synthesis and evaluation of **AlNc-*pep*-PEG**, which is an MMP-2-activatable photoacoustic probe for *in vivo* tumor imaging. The simple and accessible photoacoustic signal measurement system is also demonstrated, which is useful for facile evaluation of the photoacoustic signal generating ability of probes. We believe that the tools and methods described are helpful for researchers who are interested in enzyme-activatable photoacoustic probes.

Key resources table

Reagent or resource	Source	Identifier
Chemicals, peptides, and recombinant proteins		
Recombinant human MMP-2	PeproTech Inc., USA	420-02
Recombinant human MMP-7	BioLegend, USA	761302
Recombinant human MMP-9	RayBiotech, Inc.	230-30037-10
Experimental models: cell lines		
HT-1080	National Institutes of Biomedical Innovation, Health and Nutrition	JCRB9113

Note that not all areas will be used in every protocol.

References

Akers, W. J., Xu, B., Lee, H., Sudlow, G. P., Fields, G. B., Achilefu, S., et al. (2012). Detection of MMP-2 and MMP-9 activity *in vivo* with a triple-helical peptide optical probe. *Bioconjugate Chemistry*, *23*, 656–663. https://doi.org/10.1021/bc300027y.

Brinckerhoff, C. E., & Matrisian, L. M. (2002). Matrix metalloproteinases: A tail of a frog that became a prince. *Nature Reviews Molecular Cell Biology*, *3*, 207–214. https://doi.org/10.1038/nrm763.

Cheng, H.-B., Li, Y., Tang, B. Z., & Yoon, J. (2020). Assembly strategies of organic-based imaging agents for fluorescence and photoacoustic bioimaging applications. *Chemical Society Reviews*, *49*, 21–31. https://doi.org/10.1039/C9CS00326F.

Cheng, P., & Pu, K. (2020). Activatable phototheranostic materials for imaging-guided cancer therapy. *ACS Applied Materials & Interfaces*, *12*, 5286–5299. https://doi.org/10.1021/acsami.9b15064.

Eissa, S., Ali-Labib, R., Swellam, M., Bassiony, M., Tash, F., & El-Zayat, T. M. (2007). Noninvasive diagnosis of bladder cancer by detection of matrix metalloproteinases (MMP-2 and MMP-9) and their inhibitor (TIMP-2) in urine. *European Urology*, *52*, 1388–1397. https://doi.org/10.1016/j.eururo.2007.04.006.

Fields, G. B. (2019). Mechanism of action of novel drugs targeting angiogenesis-promoting matrix metalloproteinases. *Frontiers Immunology*, *10*, 1278. https://doi.org/10.3389/fimmu.2019.01278.

Foda, H. D., & Zucker, S. (2001). Matrix metalloproteinases in cancer invasion, metastasis and angiogenesis. *Drug Discovery Today*, *6*, 478–482. https://doi.org/10.1016/S1359-6446(01)01752-4.

Ford, W. E., Rodgers, M. A. J., Schechtman, L. A., Sounik, J. R., Rihter, B. D., & Kenney, M. E. (1992). Synthesis and photochemical properties of aluminum, gallium, silicon, and tin naphthalocyanines. *Inorganic Chemistry*, *31*, 3371–3377. https://doi.org/10.1021/ic00042a009.

Gujrati, V., Mishra, A., & Ntziachristos, V. (2017). Molecular imaging probes for multi-spectral optoacoustic tomography. *Chemical Communications*, *53*, 4653–4672. https://doi.org/10.1039/C6CC09421J.

Huang, H., Wang, D., Zhang, Y., Zhou, Y., Geng, J., Chitgupi, U., et al. (2016). Axial PEGylation of tin octabutoxy naphthalocyanine extends blood circulation for photoacoustic vascular imaging. *Bioconjugate Chemistry*, *27*, 1574–1578. https://doi.org/10.1021/acs.bioconjchem.6b00280.

Ji, C., Cheng, W., Yuan, Q., Müllen, K., & Yin, M. (2019). From dyestuff chemistry to cancer theranostics: The rise of rylenecarboximides. *Accounts of Chemical Research*, *52*, 2266–2277. https://doi.org/10.1021/acs.accounts.9b00221.

Jiang, T., Olson, E. S., Nguyen, Q. T., Roy, M., Jennings, P. A., & Tsien, R. Y. (2004). Tumor imaging by means of proteolytic activation of cell-penetrating peptides. *Proceedings of the National Academy of Sciences of the United States of America*, *101*, 17867–17872. https://doi.org/10.1073/pnas.0408191101.

Knox, H. J., & Chan, J. (2018). Acoustogenic probes: A new frontier in photoacoustic imaging. *Accounts of Chemical Research*, *51*, 2897–2905. https://doi.org/10.1021/acs.accounts.8b00351.

Levi, J., Kothapalli, S.-R., Bohndiek, S., Yoon, J.-K., Dragulescu-Andrasi, A., Nielsen, C., et al. (2013). Molecular photoacoustic imaging of follicular thyroid carcinoma. *Clinical Cancer Research*, *19*, 1494–1502. https://doi.org/10.1158/1078-0432.CCR-12-3061.

Levi, J., Kothapalli, S. R., Ma, T.-J., Hartman, K., Khuri-Yakub, B. T., & Gambhir, S. S. (2010). Design, synthesis, and imaging of an activatable photoacoustic probe. *Journal of the American Chemical Society*, *132*, 11264–11269. https://doi.org/10.1021/ja104000a.

Lu, H. D., Lim, T. L., Javitt, S., Heinmiller, A., & Prud'homme, R. K. (2017). Assembly of macrocycle dye derivatives into particles for fluorescence and photoacoustic applications. *ACS Combinatorial Science*, *19*, 397–406. https://doi.org/10.1021/acscombsci.7b00031.

Mantovani, G., Lecolley, F., Tao, L., Haddleton, D. M., Clerx, J., Cornelissen, J. J. L. M., et al. (2005). Design and synthesis of N-maleimido-functionalized hydrophilic polymers via copper-mediated living radical polymerization: A suitable alternative to PEGylation chemistry. *Journal of the American Chemical Society*, *127*, 2966–2973. https://doi.org/10.1021/ja0430999.

Mook, O. R. F., Frederiks, W. M., & Van Noorden, C. J. F. (2004). The role of gelatinases in colorectal cancer progression and metastasis. *Biochimica et Biophysica Acta*, *1705*, 69–89. https://doi.org/10.1016/j.bbcan.2004.09.006Get. rights and content.

Peng, J., Yang, Q., Xiao, Y., Shi, K., Liu, Q., Hao, Y., et al. (2019). Tumor microenvironment responsive drug-dye-peptide nanoassembly for enhanced tumor-targeting, penetration, and photo-chemo-immunotherapy. *Advanced Functional Materials*, *29*, 1900004. https://doi.org/10.1002/adfm.201900004.

Ren, F., Tang, R., Zhang, X., Madushi, W. M., Luo, D., Dang, Y., et al. (2015). Overexpression of MMP family members functions as prognostic biomarker for breast cancer patients: A systematic review and meta-analysis. *PLos One*, *10*, e0135544. https://doi.org/10.1371/journal.pone.0135544.

Rocca, G. L., Pucci-Minafra, I., Marrazzo, A., Taormina, P., & Minafra, S. (2004). Zymographic detection and clinical correlations of MMP-2 and MMP-9 in breast cancer sera. *British Journal of Cancer*, *90*, 1414–1421. https://doi.org/10.1038/sj.bjc.6601725.

Wang, X., Geng, Z., Cong, H., Shen, Y., & Yu, B. (2019). Organic semiconductors for photothermal therapy and photoacoustic imaging. *ChemBioChem*, *20*, 1628–1636. https://doi.org/10.1002/cbic.201800818.

Wang, L. V., & Hu, S. (2013). Photoacoustic tomography: In vivo imaging from organelles to organs. *Science*, *335*, 1458–1462. https://doi.org/10.1126/science.1216210.

Wang, S., & Zhang, X. (2021). Design strategies of photoacoustic molecular probes. *ChemBioChem*, *22*, 308–316. https://doi.org/10.1002/cbic.202000514.

Weber, J., Beard, P., & Bohndiek, S. E. (2015). Contrast agents for molecular photoacoustic imaging. *Nat Methods*, *13*, 639–650. https://doi.org/10.1038/nmeth.3929.

Wu, Y., Wang, A., Ding, X., & Xu, F.-J. (2017). Versatile functionalization of poly(methacrylic acid) brushes with series of Proteolytically cleavable peptides for highly sensitive protease assay. *ACS Applied Materials & Interfaces*, *9*, 127–135. https://doi.org/10.1021/acsami.6b12033.

Yang, Y., & Rosenberg, G. A. (2015). Matrix metalloproteinase as therapeutic targets for stroke. *Brain Research*, *1623*, 30–38. https://doi.org/10.1016/j.brainres.2015.04.024.

Yin, L., Sun, H., Zhang, H., He, L., Qiu, L., Lin, J., et al. (2019). Quantitatively visualising tumor-related protease activity *in vivo* using a ratiometric photoacoustic probe. *Journal of the American Chemical Society*, *141*, 3265–3273. https://doi.org/10.1021/jacs.8b13628.

Zhang, X., Bresee, J., Fields, G. B., & Edwards, W. B. (2014). Near-infrared triple-helical peptide with quenched fluorophores for optical imaging of MMP-2 and MMP-9 proteolytic activity in vivo. *Bioorganic Medicinal Chemistry Letters*, *24*, 3786–3790. https://doi.org/10.1016/j.bmcl.2014.06.072.

CHAPTER FIVE

Alkaline phosphatase-triggered self-assembly of near-infrared nanoparticles for the enhanced photoacoustic imaging of tumors

Chengfan Wu[a], Rui Zhang[b], Wei Du[a], Liang Cheng[b], and Gaolin Liang[a,c,*]

[a]Hefei National Laboratory of Physical Sciences at Microscale, Department of Chemistry, University of Science and Technology of China, Hefei, AH, China
[b]Institute of Functional Nano & Soft Materials (FUNSOM), Jiangsu Key Laboratory for Carbon-Based Functional Materials and Devices, Soochow University, Suzhou, JS, China
[c]State Key Laboratory of Bioelectronics, School of Biological Sciences and Medical Engineering, Southeast University, Nanjing, JS, China
*Corresponding author: e-mail address: gliang@ustc.edu.cn

Contents

Methods in Enzymology, Volume 657
ISSN 0076-6879
https://doi.org/10.1016/bs.mie.2021.06.028

Abstract

In this chapter, we discuss the need for the development of enzyme-activatable probes in the field of tumor-targeted photoacoustic (PA) imaging, then we give a brief description of the innovation of designing alkaline phosphatase (ALP)-activatable probes for PA imaging. After that, we provide detailed protocols for the syntheses and characterizations of a near-infrared photoacoustic imaging probe, **1P**, developed in our research group. With this tool, **1P** could form nanoparticles **1-NPs** under the catalysis of ALP and thus could be used to enhance PA imaging both in vitro and in vivo.

1. Introduction

Cancer is one of the leading causes of death worldwide and precise diagnosis of cancer is vitally important for improving survival of the patients (Moscow, Fojo, & Schilsky, 2018). Nowadays, various imaging techniques have been developed for the diagnosis of cancer, such as magnetic resonance imaging (MRI) (Dong et al., 2017), positron emission tomography (PET) (Ni et al., 2018), computed tomography (CT) (Lopci et al., 2018), single photon emission computed tomography (SPECT) (Chen et al., 2018), and optical imaging (OI) (Feng & Liu, 2018). In addition to these, photoacoustic (PA) imaging is a new imaging technique developed in recent years (Liu et al., 2015). Combining optical excitation with ultrasonic detection, PA imaging shows unique advantages in the diagnosis of superficial cancer with high spatial resolution and tissue-penetrating depth (Dragulescu-Andrasi, Kothapalli, Tikhomirov, Rao, & Gambhir, 2013; Kothapalli et al., 2012; Wang & Hu, 2012). A variety of PA contrast agents have been developed to enhance PA signal of tumors, such as carbon nanotubes (Kim, Galanzha, Shashkov, Moon, & Zharov, 2009), porphyrins (Huynh et al., 2015), fluorescent proteins (Filonov et al., 2012), metallic nanoparticles (Chen et al., 2014), near infrared fluorescent dyes (Ahn, Yao, Wang, & Belfield, 2012).

The tumor microenvironment is characterized by low pH, low oxygen and high expression of some enzymes (Wan, Chen, Shi, Li, & Ma, 2014). PA probes that specifically response to tumor microenvironment can be used for enhancing PA imaging of tumors. Therefore, a microenvironment-enzyme-triggered self-assembly strategy is developed for PA imaging of tumors.

Compared with nonenzyme-activatable PA probes, enzyme-activatable probes have higher specificity and can be used for early diagnosis of related diseases. For example, an ATG4B-sensitive nanoprobe for enhancing PA imaging and optimized autophagy-mediated chemotherapy of tumors in mice was reported by Wang et al. (Lin et al., 2017), a furin-activatable oligomerizable probe for PA imaging of furin-like activity in living subjects was developed by Gambhir et al. (Dragulescu-Andrasi et al., 2013).

Alkaline phosphatase (ALP) is widely distributed at different levels in human organs. It is usually distributed in bones and liver and less in placenta and kidneys (Freeman, Finder, Gill, & Willner, 2010; Li, Zhen, Wang, Li, & Huang, 2013). In addition, ALP is overexpressed by some malignant tumor cells, such as HepG2 (Zhan, Cai, He, Wang, & Yang, 2018), HeLa (Fishman et al., 1968), MESSA/Dx5 (Zheng, Chen, et al., 2016; Zheng, Tang, et al., 2016) and Saos-2 (Zhou, Du, & Xu, 2016), implying that it can be used as a universal biomarker for the detection of ALP-related cancers. ALP induced dephosphorylation produces more hydrophobic products and thus it is usually accompanied by the self-assembly of products (Feng, Wang, Chen, & Xu, 2017). Based on this principle, ALP-activatable probes have been widely developed for bioimaging applications. For example, Xu et al. have designed ALP-sensitive optical probes for cell membrane imaging (Wang, Feng, Del Signore, Rodal, & Xu, 2018) Yang et al. have developed ALP-activatable probes for organelle imaging (Wang et al., 2016). However, to the best of our knowledge, there has been no report of using an ALP-activatable probe for enhanced PA imaging of tumor so far. This challenge motivated our group's interest in developing ALP-activatable PA probes. The near-infrared photoacoustic imaging probe, **1P**, developed (Fig. 1) in our research group could be applied for highly sensitive diagnosis of superficial cancers overexpressing ALP in the near future.

In this chapter, we provide a brief primer on principles of PA tomography and outline a detailed guide to our group's development of **1P**, a molecule that can be dephosphorylated by ALP to form **1**, and then **1** self-assembles into nanoparticles **1-NPs** to induce the self-quenching of the NIR fluorescence but, in the meantime, enhances the PA signal for tumor imaging.

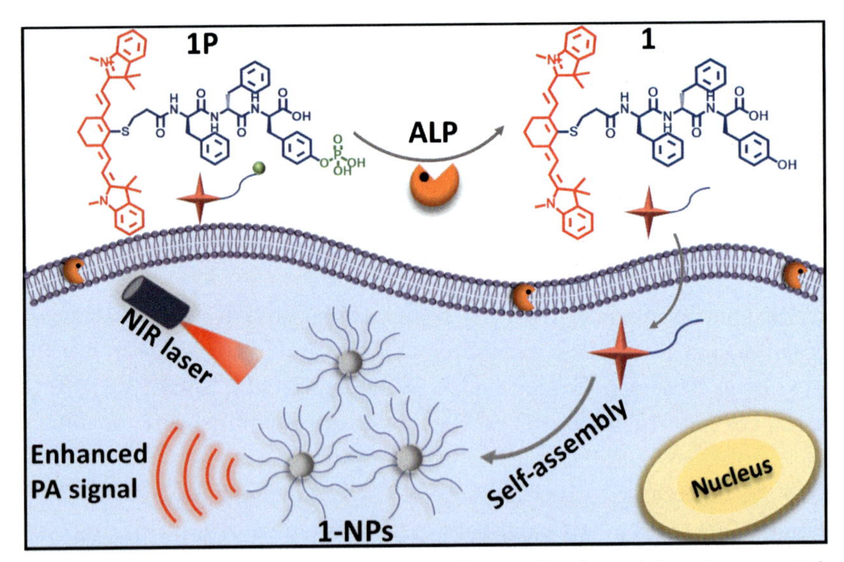

Fig. 1 Schematic illustration of ALP-triggered self-assembly of near-infrared nanoparticles for the enhanced PA imaging of tumors. *Reproduced with permission from ACS: Wu, C., Zhang, R., Du, W., C, L., & Liang, G. (2018) Alkaline phosphatase-triggered self-assembly of near-infrared nanoparticles for the enhanced photoacoustic imaging of tumors. Nano Letters, 18, 12, 7749–7754. https://pubs.acs.org/doi/abs/10.1021/acs.nanolett.8b03482.*

2. Design of 1P and 1

IR775-Phe-Phe-Tyr(H_2PO_3)-OH (**1P**) consists of a NIR dye IR775 and a well-studied Phe-Phe-Tyr(H_2PO_3)-OH motif. The NIR dye IR775 not only serves as an exogenous contrast agent for PA imaging but also increases the hydrophobicity of **1P** for self-assembly. The Phe-Phe-Tyr(H_2PO_3)-OH motif could be used for ALP-triggered self-assembly (Zheng, Chen, et al., 2016; Zheng, Tang, et al., 2016). In the tumor microenvironment, **1P** will be dephosphorylated by the secreted ALP to yield the more hydrophobic product **1**, which is taken up by tumor cells and self-assembles into **1-NPs**. Formation of **1-NPs** will induce the self-quenching of the NIR fluorescence but, in the meantime, enhances the PA signal for tumor imaging (see Fig. 1).

3. Syntheses of 1P and 1

1P and **1** are synthesized by the following detailed procedures (see Schemes 1–3). The method is simple and safe.

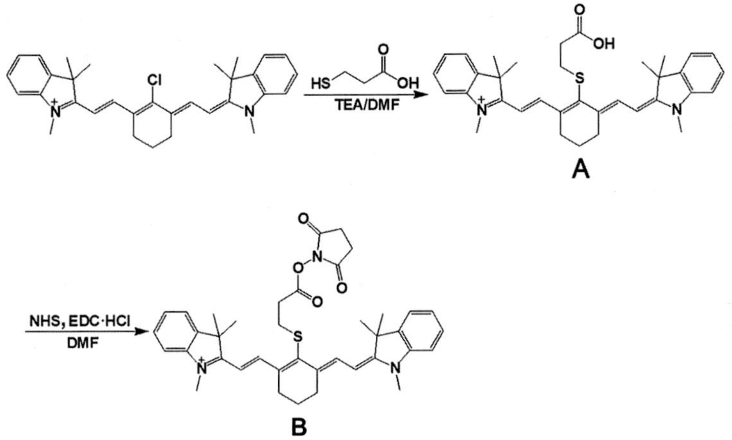

Scheme 1 Synthetic route to **A** and **B**. *Reproduced with permission from ACS: Wu, C., Zhang, R., Du, W., C, L., & Liang, G. (2018) Alkaline phosphatase-triggered self-assembly of near-infrared nanoparticles for the enhanced photoacoustic imaging of tumors. Nano Letters, 18, 12, 7749–7754. https://pubs.acs.org/doi/abs/10.1021/acs.nanolett.8b03482.*

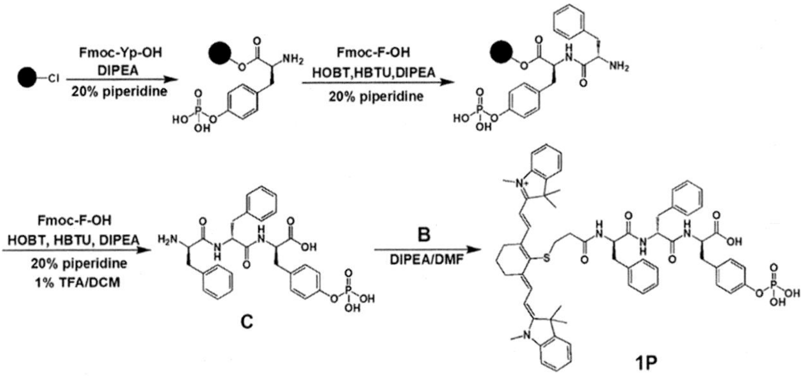

Scheme 2 Synthetic route to **1P**. *Reproduced with permission from ACS: Wu, C., Zhang, R., Du, W., C, L., & Liang, G. (2018) Alkaline phosphatase-triggered self-assembly of near-infrared nanoparticles for the enhanced photoacoustic imaging of tumors. Nano Letters, 18, 12, 7749–7754. https://pubs.acs.org/doi/abs/10.1021/acs.nanolett.8b03482.*

Synthesis of A

1. Well disperse IR775 chloride (200 mg, 0.39 mmol) in *N*,*N*-dimethylformamide (DMF, 2 mL) to form a green and homogeneous solution
2. Add the above solution to a 25 mL round bottom flask equipped with a stirring magneton and a rubber septum
3. Place the round bottom flask on a magnetic stirrer and stir it

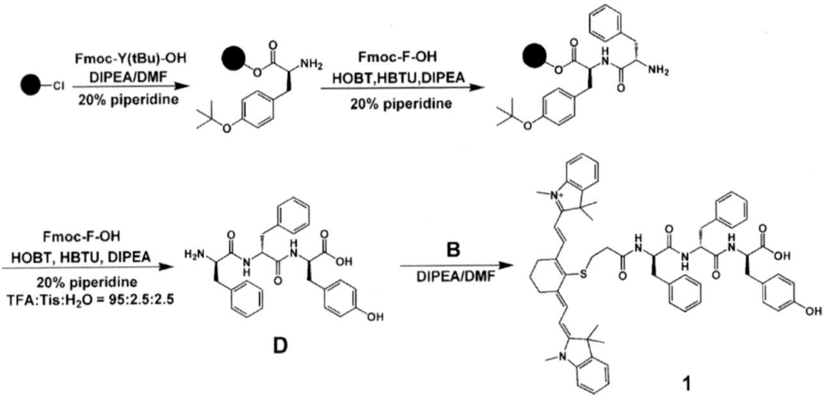

Fmoc-Y(tBu)-OH
DIPEA/DMF
20% piperidine
Fmoc-F-OH
HOBT,HBTU,DIPEA
20% piperidine
Fmoc-F-OH
HOBT, HBTU, DIPEA
20% piperidine
TFA:Tis:H_2O = 95:2.5:2.5
D
B
DIPEA/DMF
1

Scheme 3 Synthetic route to **1**. *Reproduced with permission from ACS: Wu, C., Zhang, R., Du, W., C, L., & Liang, G. (2018) Alkaline phosphatase-triggered self-assembly of near-infrared nanoparticles for the enhanced photoacoustic imaging of tumors. Nano Letters, 18, 12, 7749–7754. https://pubs.acs.org/doi/abs/10.1021/acs.nanolett.8b03482.*

4. Add 3-mercapto propionic acid (80 μL, 0.88 mmol) and triethylamine (TEA, 170 μL, 1.23 mmol) in turn into the above mixture
5. Further stir the mixture overnight at room temperature to yield the crude product **A**
6. Purify the crude product by high-performance chromatography (HPLC) using water-acetonitrile added with 0.1% TFA as the eluent to get the high purity product **A** (120 mg, about 56% purity by HPLC).

Synthesis of B

1. Well disperse compound **A** (50 mg, 0.1 mmol) in 1 mL DMF
2. Add the solution to a 25 mL round bottom flask equipped with a stirring magnet on and a rubber septum
3. Place the round bottom flask on a magnetic stirrer and stir it
4. Disperse N-hydroxysuccinimide (NHS, 41.6 mg, 0.36 mmol) in 100 μL DMF and 1-ethyl-(3-(dimethylamino)propyl)-carbodiimide hydrochloride (EDC·HCl, 52 mg, 0.27 mmol) in 200 μL DMF
5. Add NHS and EDC to the stirring solution above
6. Further stir the solution overnight to yield IR775 NHS ester **B** (about 95% purity by HPLC).

Synthesis of C

Solid phase peptide synthesis (SPPS) is used to prepare the peptide Phe-Phe-Tyr(H_2PO_3)-OH (**C**) and the peptide Phe-Phe-Tyr(tBu)-OH (**D**) and the detailed procedures are as follows.

1. Amino acids used in the experiment are weighed in advance and placed in 10 mL centrifuge tubes. Seal the centrifuge tubes with sealing film and make some holes in it. The amino acids are placed in the lyophilizer overnight
2. Well disperse 0.3 g (0.33 mmol) resin in 2 mL anhydrous DMF, add it to ox horn tube and shake it on a shaker for 30–40 min. This step activates the resin and removes impurities. After resin activation is complete, extrude the liquid from the cored end of the ox horn tube with the washing ear ball and wash it with DMF for at least 3 times
3. Dissolve Fmoc-Tyr(H_2PO_3)-OH 0.1962 g (0.66 mmol) in 2 mL anhydrous DMF. Add it to a 5 mL centrifuge tube. Add 115 μL (0.66 mmol) DIEPA in it and homogenize the mixture by ultrasound. Then add the solution to ox horn tube and shake it on the shaker for about 8 h
4. Extrude the liquid from the ox horn tube with the washing ear ball and wash it with 2 mL anhydrous DMF for 3 times to remove the unreacted amino acids
5. Add 26 μL methanol and 2 mL DMF to ox horn tube. Shake it on the shaker for about 30 min. This procedure seals off the unreacted chlorine active sites on the resin
6. Repeat step 4
7. Add the deprotection reagent (20% piperidine in DMF) to remove Fmoc protection group and shake it on a shaker for 15 min. The operation is repeated for three times. Then repeat step 4
8. Take a little bit of the resin obtained from the above reaction and add it into a 1.5 mL centrifuge tube for Kaiser test. One drop of Kaiser reagent A, B and C is respectively added to the centrifuge tube. The centrifuge tube is then placed in a metal thermostatic bath at 105 °C. After 1 min, the tube is taken out and the color is observed. If it's dark blue, continue the steps below; If it's yellow, continue step 7
9. Fmoc-Phe-OH 0.1635 g (0.55 mmol), anhydrous 1-hydroxybenzotrizole (HOBT, 0.07432 g) and O-Benzotriazole-N,N,N′,N′-tetraMethyl-uroniuM-hexafluorophosphate (HBTU, 0.2086 g) are dissolved in 2 mL anhydrous DMF. Add the solution to a 5 mL centrifuge tube. Then add 96 μL (0.33 mmol) DIEPA. Add the mixture to ox horn tube and shake it on a shaker for 8 h
10. Repeat step 4, then repeat step 8. If it's yellow, continue the next step. If it is blue, return to step 9
11. Repeat steps 7–10 again

12. Wash the resin with isopropanol for 3 times, then wash it with hexane for 3 times, extrude the liquid from the ox horn tube with the washing ear ball. Prepare dichloromethane containing 1% trifluoroacetic acid (300 μL TFA, 30 mL DCM). Add the solution to the ox horn tube and the oligopeptide is cut from the resin. The crude product is obtained by collecting the liquid extruded from the ox horn tube. Repeat this procedure until the color of resin turns dark brown. The washed solution is collected. TFA, DCM and other organic solvents are removed by rotary evaporator. Add about 35 mL ether to precipitate the oligopeptide. Then disperse the oligopeptide in the ether by ultrasound. Transfer the ether containing the oligopeptide to a 50 mL centrifuge tube. Then take another 50 mL centrifuge tube and add deionized water in it. The quality between the two centrifuge tubes is not more than 0.1 g. The two centrifuge tubes are put into a refrigerated centrifuge, the temperature of the refrigerated centrifuge is adjusted to 0 °C and the speed is adjusted to 4000 r/min. Then the two tubes are centrifuged for 20 min. After centrifugation, remove the upper ether from the tubes, then lyophilize the oligopeptides in the lyophilizer
13. Purify the crude product by HPLC using water-acetonitrile added with 0.1% TFA as the eluent to get the high purity end product **C**

Synthesis of 1P

1. Well disperse compound **C** (50 mg, 0.1 mmol) in 300 μL DMF
2. Add the solution and N, N-Diisopropylethylamine (DIPEA, 110 μL) into the reaction mixture containing compound **B** obtained from the previous step
3. Further stir the above solution overnight at room temperature to get the crude product **1P**
4. Purify the crude product **1P** by HPLC using water-acetonitrile added with 0.1% TFA as the eluent to get the high purity end product **1P** (19 mg, about 19% purity by HPLC).

Synthesis of D

The synthetic route of the peptide Phe-Phe-Tyr(tBu)-OH (**D**) is similar to that of Phe-Phe-Tyr(H_2PO_3)-OH (**C**) by using solid phase synthetic peptide (SPPS). The other steps are the same except that the two steps are different. One is that in step 3, Fmoc-Tyr(H_2PO_3)-OH is replaced with Fmoc-Tyr(tBu)-OH; The second is that in the step 12, 95% trifluoroacetic acid is used to cut the oligopeptide from the resin.

Synthesis of 1

1. Well disperse compound **D** (63.6 mg, 0.14 mmol) in DMF
2. Add the solution and N, N-Diisopropylethylamine (DIPEA, 56 μL) into the reaction mixture containing compound **B** obtained from the previous step
3. Further stir the above solution overnight at room temperature to get the crude product **1**
4. Purify the crude product **1** by HPLC using water-acetonitrile added with 0.1% TFA as the eluent to get the high purity end product **1** (9 mg, about 8% purity by HPLC).

4. In vitro characterizations of 1P and 1

After successfully synthesizing **1P** and **1**, it is necessary to (i) assess the photophysical properties of **1P** and its dephosphorylation product **1**, (ii) demonstrate that **1P** could be dephosphorylated under the catalysis of ALP and then self-assemble to form nanoparticles (**1-NPs**), (iii) assess the photoacoustic properties of **1P** before and after incubation with ALP, and (iv) **1P** and **1** have high stability and biocompatibility. The equipment used in the experiment and the detailed procedures are as follows.

Equipment

Microbalance
1.5 mL centrifuge tube
1 mL centrifuge tube
Rack for storing centrifuge tubes
Pipetting gun (20–200 μL, 200–1000 μL)
−20 °C refrigerator

Reagents

Purified compounds (**1P** and **1**)
Dimethylsulfoxid (DMSO)

Procedure

1. Label the 25 mL round bottom flask containing the compound **1P** or **1**
2. Place several 1.5 mL centrifuge tubes in the tube rack
3. Set the balance to zero and use the subtraction method to acquire the approximate weight of the compound **1P** or **1**
4. Thaw DMSO in advance, take out 5 mL and place it in a centrifuge tube

5. Calculate the volume of DMSO required for 50 mM solution based on the molecular weight of the compound **1P** or **1**
6. Add DMSO to the 25 mL round bottomed flask at room temperature
7. Use ultrasound to fully dissolve **1P** or **1**
8. Transfer the above liquid to a 1.5 mL centrifuge tube, seal the centrifugal tube with sealing film and wrap it with foil
9. Use Beer-Lambert Law to determine the concentration of **1P** or **1**. Method of the measurement of the extinction coefficient is detailedly described in Section 4.1
10. Store the prepared stock solution in the refrigerator at −20 °C, and all the solutions required for the following experiments are taken from the stock solution

Before any experiments that require compound of interest (**1P** or **1**), warm the stock solution of **1P** and **1** to room temperature. Then prepare solutions at the concentration required for the following experiments. If the stock solution needs to be reconfigured later, the concentration of it need to be determined and calibrated using the same method.

4.1 Determination of molar extinction coefficient of 1P and 1

This experiment is used to determine the maximum absorption wavelength and molar extinction coefficient of **1P** and **1**. This information is crucial for the accurate determination of compound concentration in the subsequent experiments. Below, the measurement of molar extinction coefficient is described in detail.

Equipment

Microbalance
1.5 mL centrifuge tube
Rack for storing centrifuge tubes
A Pekin-Elmer lambda 25 UV–vis spectrophotometer
Quartz cuvettes (100 μL) for UV–vis absorption spectra determination
Pipetting gun (0.5–10 μL, 10–100 μL, 20–200 μL, 200–1000 μL)

Reagents

The raw material (IR775)
Purified compounds (**1P** and **1**)
Organic solvents: methanol (CH_3OH), DMSO

Procedure

1. Warm up the Pekin-Elmer lambda 25 UV–vis spectrophotometer at least 15 min before the experiment

2. Set experimental parameters, including scanning wavelength (from 300 to 900 nm), scanning speed, scanning interval, etc.
3. Prepare a stock solution of IR775 following the method mentioned at the beginning of this section
4. Add 100 μL CH_3OH to the quartz cuvette and cap the cuvette to minimize solvent evaporation
5. Baseline scanning of CH_3OH in the quartz cuvette from 200 to 900 nm with 0.5 nm increment and make it as a blank control
6. Use CH_3OH to dilute the stock solution of IR775 to 10, 100, 1000 times, etc. Add each solution to the corresponding 1.5 mL centrifuge tube
7. Mix thoroughly by pipetting up and down 10 times
8. Collect an absorbance spectrum from 200 to 900 nm, at 0.5 nm increments
9. At least collect five scans in which the peak absorbance value at 775 nm (the maximum absorption wavelength) is between 0.2 and 0.8 a.u
10. Continue to dilute the above solution if the peak absorbance value exceeds this range
11. Generate a linear regression for absorbance vs. concentration and use the Beer-Lambert Law to determine the extinction coefficient of IR775. Then we use the molar extinction coefficient of IR775 to approximate it as the molar extinction coefficient of **1P** and **1**
12. Prepare the stock solution of **1P** and **1** following the method mentioned at the beginning of this section
13. Repeat steps 4–10
14. Determine the maximum absorption wavelength of the selected compound (**1P** or **1**), and use Beer-Lambert Law to determine the concentration of the stock solution of **1P** and **1**

4.2 Dephosphorylation experiment of 1P

This experiment fully demonstrates the conversion of **1P** to more hydrophobic compound **1** under the catalysis of ALP in vitro (see Fig. 2). The dephosphorylation experiment is the basis of the formation of **1-NPs** and the subsequent photophysical and photoacoustic experiments.

Equipment

Microbalance
1.5 mL centrifuge tube
10 mL centrifuge tube

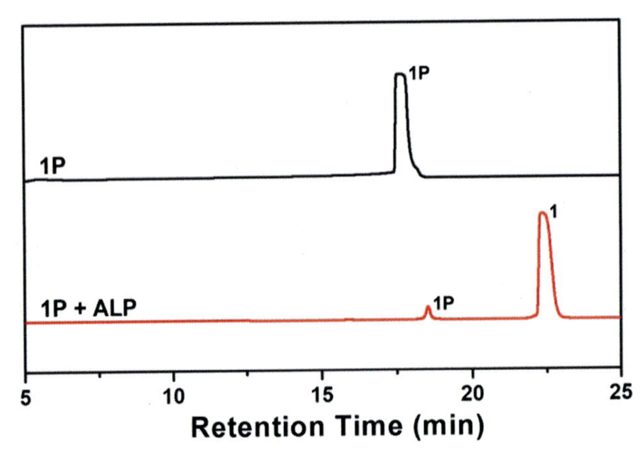

Fig. 2 HPLC traces of **1P** at 0.125 wt% in PB (0.2 M, pH 7.4) (black) and **1P** at 0.125 wt% in PB (0.2 M, pH 7.4) treated with 200 U/mL ALP at 37 °C for 12 h (red). Wavelength for detection: 775 nm. *Reproduced with permission from ACS: Wu, C., Zhang, R., Du, W., C, L., & Liang, G. (2018) Alkaline phosphatase-triggered self-assembly of near-infrared nanoparticles for the enhanced photoacoustic imaging of tumors. Nano Letters, 18, 12, 7749–7754. https://pubs.acs.org/doi/abs/10.1021/acs.nanolett.8b03482.*

Rack for storing centrifuge tubes

Pipetting gun (0.5–10 μL, 10–100 μL, 20–200 μL, 200–1000 μL)

Thermostatic water bath with adjustable temperature

An Agilent 1200 HPLC system equipped with a G1322A pump and in-line diode array UV detector using an Agilent Zorbax 300SB-C18 RP column with CH_3CN (0.1% of trifluoroacetic acid (TFA)) and water (0.1% of TFA) as the eluent

Reagents

0.2 M sodium dihydrogen phosphate

0.2 M disodium hydrogen phosphate

Ultrapure water (18.2 MΩ•cm)

Purified **1P**

ALP (M ≈ 56 kDa, obtained from BaoMan Inc. (Shanghai, China))

Procedure

1. Weigh 2 mg of **1P** using the microbalance and put it in a 1.5 mL centrifuge tube
2. Prepare an aqueous solution of 0.2 M sodium dihydrogen phosphate (A) and an aqueous solution of 0.2 M disodium hydrogen phosphate (B)
3. Add 1.9 mL A and 8.1 mL B to a 10 mL centrifuge tube to prepare 0.2 M PB (pH 7.4)
4. Add 200 μL PB buffer to the 1.5 mL centrifuge tube

5. Vortex, ultrasound to make **1P** fully dissolved
6. Add 40 units ALP to the above solution
7. Incubate the solution in a thermostatic water bath at 37 °C for 12 h
8. Take 10 μL of the above the incubated solution and inject it into HPLC for analysis

4.3 Measurement of the kinetic parameters of 1P for ALP dephosphorylation

Enzyme reaction kinetics mainly explores the reaction rate catalyzed by enzymes and various factors affecting the reaction rate. These factors include enzyme concentration, substrate concentration, pH, temperature, activators and inhibitors, etc. The purpose of this experiment is to measure the kinetic parameters of **1P** for ALP dephosphorylation and how fast could **1P** be nearly completely converted to **1**.

Equipment

1.5 mL centrifuge tube
Rack for storing centrifuge tubes
Pipetting gun (0.5–10 μL, 10–100 μL, 20–200 μL, 200–1000 μL)
Thermostatic water bath with adjustable temperature
An Agilent 1200 HPLC system equipped with a G1322A pump and in-line diode array UV detector using an Agilent Zorbax 300SB-C18 RP column with CH_3CN (0.1% of trifluoroacetic acid (TFA)) and water (0.1% of TFA) as the eluent

Reagents

Purified **1P**
ALP
Aqueous buffer: phosphate buffer (PB, 0.2 M, pH 7.4)

Procedure

1. Prepare a series of **1P** solutions of different concentrations in 20 mM PB buffer (7.5, 15, 30, 60, 90, 120, and 150 μM). At least 100 μL should be prepared for each concentration
2. For each concentration, 100 μL is taken and analyzed by HPLC in order from low to high concentration
3. Calculate the HPLC peak area corresponding to each concentration
4. Plotting the standard curve of HPLC peak areas vs. the amount of **1P**
5. Prepare a series of **1P** solutions of different concentrations in 20 mM PB buffer (7.5, 40, 60, 250, 500, 750, and 1250 μM. At least 100 μL should be prepared for each concentration

6. Add 600 nM ALP to each of **1P** solutions of different concentrations above and incubate it at 37 °C for 20 min
7. After 20 min, take out the solution and inactivate the enzyme by ultrasound to terminate the reaction
8. Inject each of the above solution into HPLC for analysis
9. According to the remaining HPLC peak area, use the first order rate equation to calculate the initial velocities of ALP dephosphorylation for each concentration
10. Plot the initial velocities against the initial concentrations of **1P** and fit it to the Michaelis-Menten model
11. Obtain the k_{cat}/K_M value of **1P** for ALP dephosphorylation using Lineweaver-Burk analysis

4.4 Photophysical characterizations of 1P and 1

UV–vis spectra and fluorescence spectra can be used to directly study the assembly behavior of the NIR dye in solution. When the NIR dye self-assemble in solution, the maximum absorption peak of it will red shift and broaden, what's more, the corresponding fluorescence intensity of it will decrease obviously. This experiment is used to determine the critical aggregation concentration (CAC) of **1P** and **1**. It is also vital for selecting suitable concentration of **1P** for following experiments.

4.4.1 UV–vis spectra of 1P and 1

Equipment

1.5 mL centrifuge tube
Rack for storing centrifuge tubes
A Pekin-Elmer lambda 25 UV–vis spectrophotometer
Quartz cuvettes (100 μL) for UV–vis absorption spectra determination
Pipetting gun (0.5–10 μL, 10–100 μL, 20–200 μL, 200–1000 μL)

Reagents

Purified compounds (**1P** and **1**)
Aqueous buffer: phosphate buffer (PB, 0.2 M, pH 7.4)

Procedure

1. Warm up the Pekin-Elmer lambda 25 UV–vis spectrophotometer at least 15 min before the experiment
2. Set experimental parameters, including scanning wavelength (from 200 to 900 nm), scanning speed, scanning interval, etc.
3. Add 100 μL PB to the quartz cuvette and cap the cuvette to minimize solvent evaporation

4. Baseline scanning of PB in the quartz cuvette from 200 to 900 nm with 0.5 nm increment and make it as a blank control
5. Prepare a series of **1P** solutions of different concentrations in 20 mM PB buffer (10, 15, 20, 25, and 30 μM). At least 100 μL should be prepared for each concentration
6. Mix each solution thoroughly by pipetting up and down 10 times
7. Measure the UV–vis spectra of **1P** from low to high concentrations. Collect an absorbance spectrum from 200 to 900 nm, at 0.5 nm increments

4.4.2 Fluorescence spectra of 1P and 1

Equipment

1.5 mL centrifuge tube
Rack for storing centrifuge tubes
A Hitachi FL-4600 fluorescence spectrophotometer (Hitachi High-Techonologies Corporation, Japan)
Thermostatic water bath with adjustable temperature
Quartz cuvettes (50 μL) for fluorescence spectra determination
Pipetting gun (0.5–10 μL, 10–100 μL, 20–200 μL, 200–1000 μL)

Reagents

Purified compounds (**1P** and **1**)
ALP
Aqueous buffer: phosphate buffer (PB, 0.2 M, pH 7.4)

Procedure

1. Warm up the Hitachi FL-4600 fluorescence spectrophotometer (Hitachi High-Techonologies Corporation, Japan) at least 15 min before the experiment
2. Set experimental parameters, including excitation wavelength (775 nm), emission wavelength (from 795 to 900 nm), slit width, voltage, scanning speed, scanning interval, etc.
3. Prepare a series of **1P** solutions of different concentrations in 20 mM PB buffer (0.98, 1.95, 3.91, 7.81, 15.63, 31.25, 62.5, and 125 μM). At least 50 μL should be prepared for each concentration
4. Mix each solution thoroughly by pipetting up and down 10 times
5. Measure the fluorescence spectra of **1P** from low to high concentrations. Collect an fluorescence emission spectrum from 780 to 900 nm, at 0.5 nm increments

6. Plot fluorescence intensity against the concentration of **1P**. Obtain two linear fitting curves and the intersection point is the critical aggregation concentration (CAC) of **1P**
7. For testing the CAC of **1**, incubate **1P** with ALP (200 U/mL) at 37 °C for 12 h in PB (pH 7.4, 20 mM) and dilute the incubation mixture to obtain serial solutions (0.98, 1.95, 3.91, 7.81, 15.63, 31.25, 62.5, and 125 μM). At least 100 μL should be prepared for each concentration
8. Repeat steps 4–6

4.5 Photophysical characterizations of 1P after incubation with ALP

We use UV–vis spectra and fluorescence spectra to study the self-assembly of **1P** after incubation with ALP. After dephosphorylation by ALP, **1** self-assembles into nanoparticles (i.e., **1-NPs**), then the maximum absorption peak of it will red shift and broaden, the corresponding fluorescence intensity at the maximum emission peak will decrease obviously.

4.5.1 UV–vis spectra of 1P after incubation with ALP

Equipment

1.5 mL centrifuge tube
Rack for storing centrifuge tubes
A Pekin-Elmer lambda 25 UV–vis spectrophotometer
Quartz cuvettes (100 μL) for UV–vis absorption spectra determination
Pipetting gun (0.5–10 μL, 10–100 μL, 20–200 μL, 200–1000 μL)

Reagents

Purified compounds (**1P** and **1**)
ALP
Aqueous buffer: phosphate buffer (PB, 0.2 M, pH 7.4)

Procedure

1. Warm up the Pekin-Elmer lambda 25 UV–vis spectrophotometer at least 15 min before the experiment
2. Set experimental parameters, including scanning wavelength (from 200 to 900 nm), scanning speed, scanning interval, etc.
3. Turn on the constant temperature mode of the UV–vis spectrophotometer and set it to 37 °C
4. Add 100 μL PB to the quartz cuvette and cap the cuvette to minimize solvent evaporation
5. Baseline scanning of PB in the quartz cuvette from 200 to 900 nm with 0.5 nm increment and make it as a blank control

6. Prepare **1P** solution at 15 μM. At least 100 μL should be prepared
7. Mix thoroughly by pipetting up and down 10 times
8. Measure UV–vis spectra of 15 μM **1P** at 0 min
9. Defrost the ALP in advance from −20 °C to 4 °C, then add ALP to the above solution, and immediately transfer it to the quartz cuvette and cap the cuvette to minimize solvent evaporation
10. Measure time course UV–vis spectra of 15 μM **1P** after incubation with 200 U/mL ALP at 37 °C (0, 30, 60, 90, 180, 240, and 300 min). Collect the corresponding absorbance spectrum from 200 to 900 nm, at 0.5 nm increments (see Fig. 3A).

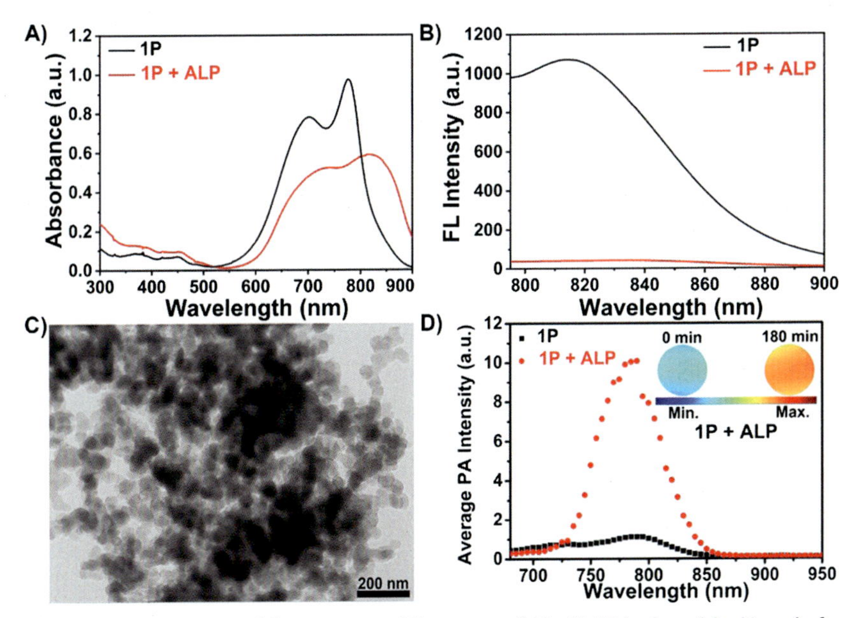

Fig. 3 The UV–vis (A) and fluorescence (B) spectra of 15 μM **1P** before (black) and after (red) incubation with 200 U/mL ALP at 37 °C for 3 h in 0.2 M PB (pH 7.4). Excitation: 775 nm. (C) TEM image of 15 μM **1P** after 3 h incubation with 200 U/mL ALP. (D) PA spectra of 15 μM **1P** before (black) and after (red) 3 h incubation with 200 U/mL ALP. Inset: PA phantom images of above incubation mixture at an excitation wavelength of 795 nm. *Reproduced with permission from ACS: Wu, C., Zhang, R., Du, W., C, L., & Liang, G. (2018) Alkaline phosphatase-triggered self-assembly of near-infrared nanoparticles for the enhanced photoacoustic imaging of tumors. Nano Letters, 18, 12, 7749–7754. https://pubs.acs.org/doi/abs/10.1021/acs.nanolett.8b03482.*

4.5.2 Fluorescence spectra of 1P after incubation with ALP

Equipment

1.5 mL centrifuge tube
Rack for storing centrifuge tubes
A Hitachi FL-4600 fluorescence spectrophotometer (Hitachi High-Techonologies Corporation, Japan)
Thermostatic water bath with adjustable temperature
Quartz cuvettes (50 μL) for fluorescence spectra determination
Pipetting gun (0.5–10 μL, 10–100 μL, 20–200 μL, 200–1000 μL)

Reagents

Purified compounds (**1P** and **1**)
ALP
Aqueous buffer: phosphate buffer (PB, 0.2 M, pH 7.4)

Procedure

1. Warm up the Hitachi FL-4600 fluorescence spectrophotometer (Hitachi High-Techonologies Corporation, Japan) at least 15 min before the experiment
2. Set experimental parameters, including excitation wavelength (775 nm), emission wavelength (from 795 to 900 nm), slit width, voltage, scanning speed, scanning interval, etc.
3. Turn on the constant temperature mode of the fluorescence spectrophotometer and set it to 37 °C
4. Prepare **1P** solution at 15 μM. At least 100 μL should be prepared
5. Mix thoroughly by pipetting up and down 10 times
6. Measure fluorescence spectra of 15 μM **1P** at 0 min
7. Add ALP (200 U/mL) to the above solution and immediately transfer it to to the quartz cuvette and cap the cuvette to minimize solvent evaporation
8. Measure time course fluorescence spectra of 15 μM **1P** after incubation with 200 U/mL ALP at 37 °C (from 0 to 255 min). Collect corresponding fluorescence emission spectrum from 795 to 900 nm, at 0.5 nm increments (see Fig. 3B).

4.6 Characterization of 1-NPs by transmission electron microscopy (TEM)

The purpose of this experiment is to directly observe the morphology of the assemblies after incubation **1P** with ALP. The TEM image clearly showed the formation of nanoparticles (i.e., **1-NPs**) with an average diameter of

39.4 ± 5.0 nm in the mixture after 3 h of incubation. In comparison, in the absence of ALP, only amorphous structures are seen in the TEM image of the **1P** solution.

Equipment

Pipetting gun for taking liquid (0.5–10 μL, 10–100 μL, 20–200 μL, 200–1000 μL)
Thermostatic water bath with adjustable temperature
Carbon-coated copper grids
Filter paper
Lyophilizer
A JEM-2100F field emission transmission electron microscope operated at an acceleration voltage of 200 kV

Reagents

Purified compounds (**1P** and **1**)
ALP
Aqueous buffer: phosphate buffer (PB, 0.2 M, pH 7.4)

Procedure

1. Prepare **1P** solution at 15 μM. At least 100 μL should be prepared
2. Mix thoroughly by pipetting up and down 10 times
3. Add ALP to the above solution, and immediately transfer it to the 1.5 mL centrifuge tube
4. Incubate the solution in a thermostatic water bath at 37 °C for 3 h
5. Carbon-coated copper grids are glow-discharged to increase their hydrophilicity before use
6. Place copper grids on filter paper
7. Gently immerse the carbon-coated side of the grid in the solution for 1 s
8. For the control group, gently immerse the carbon-coated side of the grid in the solution of 15 μM **1P** in PB (0.2 M, pH 7.4) for 1 s
9. Freeze-dry the carbon-coated copper grids overnight
10. Take TEM images of them (see Fig. 3C).

4.7 Application of 1P to ex vivo phantom PA signal enhancement

According to the principle of energy conservation, the fluorescence signal generated by the radiation transition process decreases, and the ultrasonic response generated by the thermoelastic expansion of the tissue structure enhances. The formation of **1-NPs** effectively quenches the fluorescence signal obviously. Here, we use this experiment to verify whether the PA signal in vitro is enhanced.

Equipment

1.5 mL centrifuge tube
Rack for storing centrifuge tubes
Pipetting gun (0.5–10 μL, 10–100 μL, 20–200 μL, 200–1000 μL)
Thermostatic water bath with adjustable temperature
FEP tubes
Syringe (1 mL)
A Vevo 2100 LAZR system (VisualSonics Inc.)

Reagents

Purified **1P**
ALP
Coupling agent
Aqueous buffer: phosphate buffer (PB, 0.2 M, pH 7.4)

Procedure

1. Warm up the Vevo 2100 LAZR system (VisualSonics Inc.) for at least 30 min prior to starting the experiment
2. Prepare **1P** solution at 15 μM. At least 100 μL should be prepared
3. Mix thoroughly by pipetting up and down 10 times
4. Add ALP to the above solution, and immediately transfer it to the 1.5 mL centrifuge tube
5. Incubate the solution in a thermostatic water bath at 37 °C for 3 h
6. Fix an FEP tube to the support required for the PA signal measurement
7. Using a 1 mL syringe, carefully transfer 200 μL of above solution into an FEP tube
8. Seal the tube on end by folding the edge of the tube upon itself and securing the crease with an FEP cap
9. Add the coupling agent needed for the PA experiment to the surface of the FEP tube
10. Coat the ultrasonic probe with the the coupling agent
11. Move the ultrasonic probe to make it full contact with the surface of the FEP tube
12. Acquire PA measurements with an an excitation wavelength of 795 nm and emission wavelength from 680 to 950 nm
13. For the control group, use a 1 mL syringe to carefully transfer 200 μL **1P** solution into an FEP tube
14. Repeat steps 8–12
15. Obtain corresponding PA phantom images at an excitation wavelength of 795 nm (see Fig. 3D).

4.8 Detection of ALP by PA signal enhancement

PA imaging indicates that PA signal of **1P** phantom increases by 6.4-folds after 3 h incubation with ALP, so we could use the PA spectra of **1P** to establish an assay for ALP detection. The limit of detection (LOD) of ALP of this assay is 1.00 U/mL (S/N=3).

Equipment

1.5 mL centrifuge tube
Rack for storing centrifuge tubes
Pipetting gun (0.5–10 μL, 10–100 μL, 20–200 μL, 200–1000 μL)
Thermostatic water bath with adjustable temperature
FEP tubes
Syringe (1 mL)
A Vevo 2100 LAZR system (VisualSonics Inc.)

Reagents

Purified **1P**
ALP
Coupling agent
Aqueous buffer: phosphate buffer (PB, 0.2 M, pH 7.4)

Procedure

1. Warm up the Vevo 2100 LAZR system (VisualSonics Inc.) for at least 30 min prior to starting the experiment. This ensures adequate heating of the optical system
2. Prepare five **1P** solutions at 15 μM. At least 100 μL of each should be prepared
3. Mix thoroughly by pipetting up and down 10 times
4. Add different concentrations of ALP (12.5, 25, 50, 100 and 200 U/mL) to each of the above solutions
5. Incubate the solutions in a thermostatic water bath at 37 °C for 3 h
6. Fix five FEP tubes to the support required for the experiment
7. Using a 1 mL syringe, carefully transfer 200 μL of each above solution successively into an FEP tube
8. Seal the tube on end by folding the edge of the tube upon itself and securing the crease with an FEP cap
9. Add the coupling agent needed for the PA experiment to the surface of the FEP tube
10. Coat the ultrasonic probe with the the coupling agent
11. Move the ultrasonic probe to make it full contact with the surface of the FEP tube

12. Acquire PA measurements with an an excitation wavelength of 795 nm and emission wavelength from 680 to 950 nm
13. Plot average PA intensity at 795 nm of 15 μM **1P** against different concentrations of ALP
14. Obtain the limit of detection (LOD) of ALP of this assay

4.9 Stability tests of 1P and 1

Cell endosomes and lysosomes and tumor microenvironments are characterized with low pH value, so it is important to study stability of **1P** and **1** under low pH values before applying **1P** for PA imaging of tumor. In this experiment, we incubate **1P** and **1** in PB buffer at different pH values simulating the tumor microenvironment for 6 h and then analyze their stability by HPLC. HPLC analysis clearly show that 15 μM of both **1P** and **1** are quite stable at pH 3–7 and 37 °C for 6 h.

Equipment

1.5 mL centrifuge tube
Rack for storing centrifuge tubes
pH paper
Pipetting gun (0.5–10 μL, 10–100 μL, 20–200 μL, 200–1000 μL)
Thermostatic water bath with adjustable temperature
An Agilent 1200 HPLC system equipped with a G1322A pump and in-line diode array UV detector using an Agilent Zorbax 300SB-C18 RP column with CH_3CN (0.1% of trifluoroacetic acid (TFA)) and water (0.1% of TFA) as the eluent

Reagents

Purified **1P**
M/L HCl
Aqueous buffer: phosphate buffer (PB, 0.2 M, pH 7.4)

Procedure

1. Prepare five **1P** solutions at 15 μM. At least 400 μL of each solution should be prepared
2. Use 0.1 M/L HCl and pH paper to adjust above **1P** solutions to different pH values (pH 3–7).
3. Incubate the above solutions in a thermostatic water bath at 37 °C for 6 h
4. Inject each of the above incubated solutions into HPLC for analysis
5. For testing the stability of **1**, prepare five **1** solutions at 15 μM. At least 400 μL of each solution should be prepared
6. Repeat steps 2–4 (see Fig. 4).

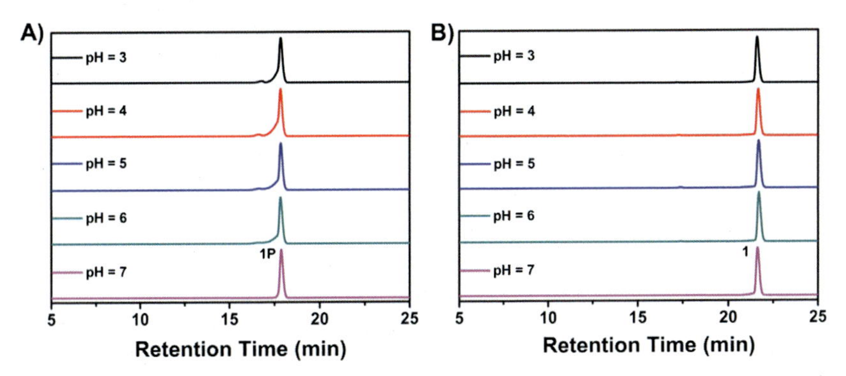

Fig. 4 (A) HPLC traces of 15 μM **1P** in 0.2 M PB buffer at different pH values and 37 °C for 6 h. Detection wavelength: 775 nm. (B) HPLC traces of 15 μM **1** in 0.2 M PB buffer at different pH values and 37 °C for 6 h. Detection wavelength: 775 nm. *Reproduced with permission from ACS: Wu, C., Zhang, R., Du, W., C, L., & Liang, G. (2018) Alkaline phosphatase-triggered self-assembly of near-infrared nanoparticles for the enhanced photoacoustic imaging of tumors. Nano Letters, 18, 12, 7749–7754. https://pubs.acs.org/doi/abs/10.1021/acs.nanolett.8b03482.*

4.10 Cellular uptake experiments

The cell uptake experiments are conducted mainly for proving that: (i) **1P** is indeed dephosphorylated by the secreted ALP to yield the more hydrophobic product IR775-Phe-Phe-Tyr-OH (**1**), which is then uptaken by tumor cells; ii) the concentration of **1** in the tumor cell far exceeds the CAC of it, which indicates that **1** indeed forms nanoparticles inside the cell.

Equipment

6 cm sterile cell culture dish

Incubator suitable for mammalian cell culture, with 5% CO_2 atmosphere

1.5 mL sterile centrifuge tube

Low speed centrifuge

Pipetting gun (0.5–10 μL, 10–100 μL, 20–200 μL, 200–1000 μL)

An Agilent 1200 HPLC system equipped with a G1322A pump and in-line diode array UV detector using an Agilent Zorbax 300SB-C18 RP column with CH_3CN (0.1% of trifluoroacetic acid (TFA)) and water (0.1% of TFA) as the eluent.

Reagents

HeLa cells

Dulbecco's modified Eagle's medium (DMEM, Hycolon) supplemented with 10% fetal bovine serum

Purified **1P** and **1**

DMSO
Trypsin
Aqueous buffer: phosphate buffer (PB, 0.2 M, pH 7.4), phosphate buffer saline (PBS, 10 mM, pH 7.4)

Procedure

1. HeLa cells are routinely cultured in preheated Dulbecco's modified Eagle's medium (DMEM, Hycolon) supplemented with 10% fetal bovine serum at 37 °C, 5% CO2, and humid atmosphere
2. Seed each 1.5 mL sterile centrifuge tube with HeLa cells (48.5×10^4 cells in each tube).
3. Dilute **1P** stock solution in serum-free DMEM to prepare solutions of 7.5, 15 and 30 μM. At least 600 μL of each solution should be prepared
4. Incubated HeLa cells with above culture medium at 37 °C for 0.5, 1, and 2 h
5. Centrifuge at 1000 rpm for 5 min to allow the cells to concentrate in the bottom of the tube
6. Transfer the supernatants to new 1.5 mL sterile centrifuge tubes
7. Use 100 μL dimethylsulfoxide (DMSO) to break cell membrane and fully dissolve **1P** and **1** inside the cell
8. The cell lysate, together with the supernatants, is separately injected into a HPLC system for analysis
9. Obtain HPLC traces of the culture media and the cell lysates of HeLa cells
10. Prepare a series of **1** solutions of different concentrations in 20 mM PB buffer (3.75, 7.5, 15, 30, 60, 90, and 120 μM). At least 100 μL should be prepared for each concentration
11. For each concentration, 100 μL is taken and analyzed by HPLC in order from low to high concentration
12. Calculate the HPLC peak area corresponding to each concentration
13. Plotting the standard curve of HPLC peak areas vs. the amount of **1**
14. Use the following formulas to calculate cell uptake of the dephosphation product **1**:

$$\text{Intracellular concentration} = n/(\text{cell number} \times 4 \times 10^{-9}\ \text{cm}^3)$$

$$n = (\text{HPLC Peak Area} + 2700.61)/10{,}491.41$$

An average size for the common HeLa cells used in cell culture is 15–20 μm in diameter for a suspended cell (volume $4/3\pi r^3 = 4 \times 10^{-9}\ \text{cm}^3$).

4.11 Biocompatibility test of 1P

Because **1P** is designed to be used for PA imaging of tumors, it is important to confirm that **1P** have high biocompatibility. Here, the 3-(4,5-dimethylthiazol-2-yl)-2,5 diphenyl-tetrazolium bromide (MTT) assay is conducted. The results show that more than 74.4% of HeLa cells survive in **1P** at incubation compound concentration up to 60 μM and time up to 12 h, suggesting that **1P** has very low cytotoxicity when applied for PA imaging.

Equipment

6 cm sterile cell culture dish
Incubator suitable for mammalian cell culture, with 5% CO_2 atmosphere
96-well plate, clear-bottomed, coated for cell culture
Pipetting gun (0.5–10 μL, 10–100 μL, 20–200 μL, 200–1000 μL)
1.5 mL sterile centrifuge tube
15 mL sterile centrifuge tube
Thermostatic water bath
ELISA reader (VARIOSKAN FLASH)

Reagents

HeLa cells
Dulbecco's modified Eagle's medium (DMEM, Hycolon) supplemented with 10% fetal bovine serum
Purified **1P** and **1**
Trypsin
Aqueous buffer: phosphate buffer (PB, 0.2 M, pH 7.4), phosphate buffer saline (PBS, 10 mM, pH 7.4)
MTT
DMSO

Procedure

1. HeLa cells are routinely cultured in preheated Dulbecco's modified Eagle's medium (DMEM, Hycolon) supplemented with 10% fetal bovine serum at 37 °C, 5% CO_2, and humid atmosphere
2. Cells growing in log phase are seeded into 96-well cell-culture plate at 2.5×10^3/well
3. The cells are incubated at 37 °C under 5% CO_2 overnight until they are attached to the surface of the 6 cm sterile cell culture dish
4. Dilute the stock solution of **1P** in DMEM to generate solutions of 7.5, 15, 30 and 60 μM in media

5. Add 100 μL of the appropriate **1P** solution to each well such that the cells will be incubated with 0, 7.5, 15, 30, or 60 μM of **1P**
6. Incubate the cells for 4, 8, or 12 h at 37 °C under 5% CO_2
7. A solution of 5 mg/mL MTT dissolved in PBS is added to each well of the 96-well plate (10 μL/well).
8. DMSO is added to dissolve the formazan after 4 h (100 μL/well).
9. Shake the 96-well plate in a thermostatic water bath at low speed for 10 min
10. Using an ELISA reader (VARIOSKAN FLASH) to detect its optical absorption at 490 nm. Each of the experiments is performed at least 3 times
11. The viability of cell growth is calculated using the following formula: viability (%) = (mean of absorbance value of treatment group/mean of absorbance value of control) × 100 (see Fig. 5).

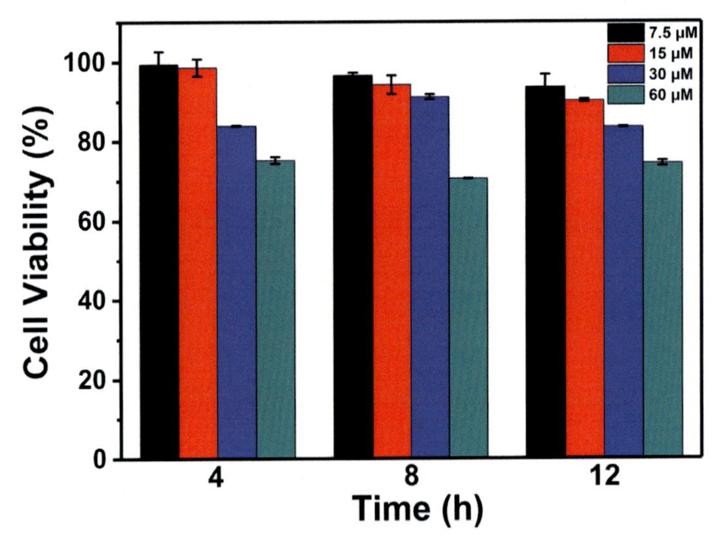

Fig. 5 Cell viability of HeLa cells incubated with **1P** at different concentrations for 4, 8, or 12 h. Each error bar represents the standard deviation of three independent experiments. *Reproduced with permission from ACS: Wu, C., Zhang, R., Du, W., C, L., & Liang, G. (2018) Alkaline phosphatase-triggered self-assembly of near-infrared nanoparticles for the enhanced photoacoustic imaging of tumors. Nano Letters, 18, 12, 7749–7754. https://pubs.acs.org/doi/abs/10.1021/acs.nanolett.8b03482.*

5. In vivo characterizations of 1P

We have validated ALP-triggered self-assembly and ex vivo phantom PA signal enhancement of **1P**. In the following experiments, we validate ALP-triggered self-assembly of **1-NPs** could enhance PA imaging in vivo. We first conduct PA imaging of ALP overexpressing tumors at different stages with high specificity. Then we quantitatively analyze ALP activity of tumors at different stages.

Female nude mice are purchased from Nanjing Peng Sheng Biological Technology Co Ltd. and used under protocols approved by Soochow University Laboratory Animal Center. The HeLa tumor models are generated by subcutaneous injection of 1×10^6 cells in 60 μL PBS into the right backside of nude mice. The mice are used for treatment when the tumor sizes are within 5–6 mm and 7–10 mm in diameter.

5.1 Application of 1P to enhanced PA imaging of tumor (5–10 mm in diameter)

This experiment demonstrates that **1P** could be applied for enhanced PA imaging in vivo with high specificity. Until the tumor sizes are within 5–10 mm in diameter, the nude mice are randomly divided into four groups ($n = 3$ for each group). **1P** is injected into each of those tumor-bearing nude mice through tumor direct injection. For control groups, each mouse is intravenously (i.v.) injected with *L*-phenylalanine (one type of ALP inhibitor) followed by the tumor-direct injection of **1P** 30 min later. PA imaging of each mouse is acquired by a Vevo 2100 LAZR system for 24 h. The results show that **1P** could be used for PA imaging of HeLa tumors at different stages with different ALP expressions. Quantitative analysis indicates that, compared to that in the control group, PA contrast in the experimental group enhances 2.3-folds at 4 h after **1P** injection.

Equipment

A Vevo 2100 LAZR system (VisualSonics Inc.)
Vaporizer equipped with isoflurane and oxygen (for anesthesia)
Sterile insulin syringes

Cells and animals

HeLa cells
Female nude BALB/c mice, 5 weeks old.

Reagents

Dulbecco's modified Eagle's medium (DMEM, Hycolon) supplemented with 10% fetal bovine serum
Serum-free DMEM
Trypsin
Matrigel
Purified **1P** and **1**
L-phenylalanine (L-Phe, one type of ALP inhibitor)
Aqueous buffer: phosphate buffer saline (PBS, 10mM, pH 7.4)

Procedure

1. Culture HeLa cells using standard conditions
2. Two hours before implantation, transfer the Matrigel to a wet ice bath to thaw. Keep cold until just prior to mixing with the cell suspension
3. Trypsinize and count cells. Prepare a cell suspension in a 3:1 mixture of serum-free medium and Matrigel
4. Implant 50 μL of the HeLa suspension ($3*10^6$ cells) into the subcutaneous space of each flank using an insulin syringe
5. Until the tumor sizes are within 5–10 mm in diameter, randomly divide the nude mice into four groups ($n=3$ for each group). Two groups with tumors of 5–6 mm and the other two with tumors of 7–10 mm
6. A total of 50 μL of **1P** at 0.5 μmol/kg is injected into each of those tumor-bearing nude mice through tumor-direct injection
7. For corresponding control groups, each mouse is intravenously (i.v.) injected with 20 μmol/kg L-phenylalanine (one type of ALP inhibitor) followed by the tumor-direct injection of **1P** 30 min later
8. Warm up the Vevo 2100 LAZR system (VisualSonics Inc.) for at least 30 min prior to starting the experiment. This ensures adequate heating of the optical system
9. Mice are anesthetized and fixed on the support
10. Add the coupling agent needed for the PA experiment to the surface of tumors of mice
11. Coat the ultrasonic probe with the the coupling agent
12. Move the ultrasonic probe to make it full contact with the surface of tumors
13. Acquire PA imaging of each mouse with an an excitation wavelength of 790 nm by a Vevo 2100 LAZR system for 24 h
14. Obtain in vivo real-time PA spectra from 770 to 860 nm extracted from the tumors in living mice from 0 to 24 h post-injection (see Fig. 6).

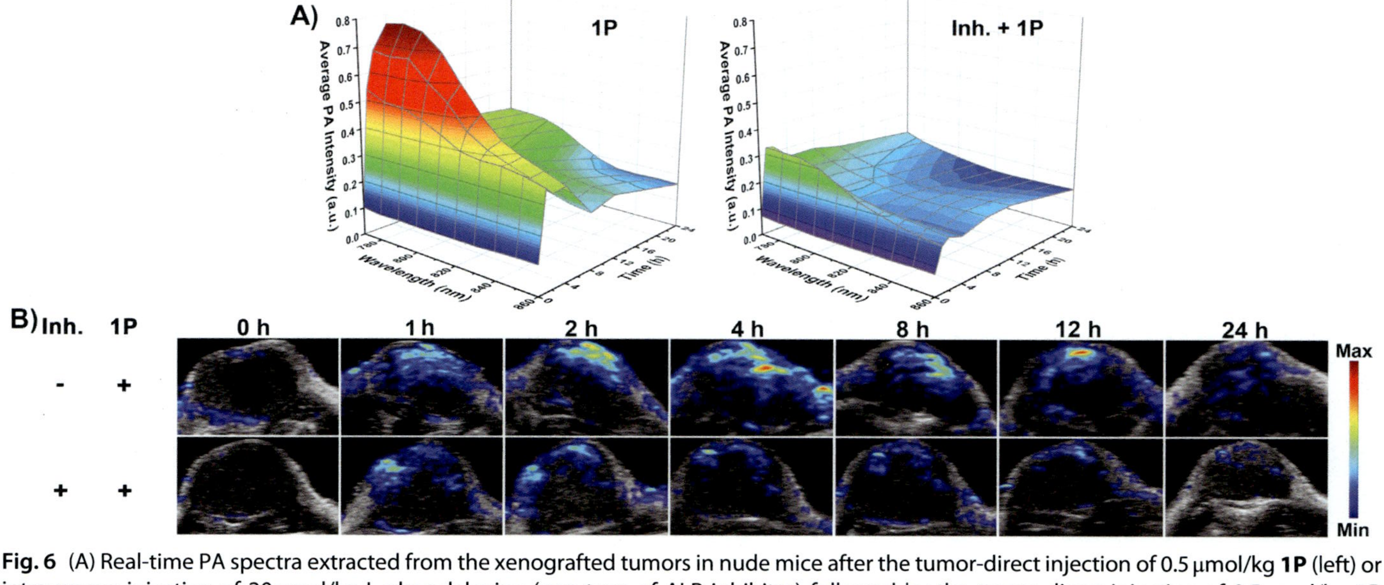

Fig. 6 (A) Real-time PA spectra extracted from the xenografted tumors in nude mice after the tumor-direct injection of 0.5 μmol/kg **1P** (left) or intravenous injection of 20 μmol/kg L-phenylalanine (one type of ALP inhibitor) followed by the tumor direct injection of 0.5 μmol/kg **1P** 30 min later (right). (B) Time-course PA images of tumors in transverse sections at 790 nm after injections of 0.5 μmol/kg **1P** with (bottom row) or without (top row) the pretreatment of an ALP inhibitor at 20 μmol/kg. *Reproduced with permission from ACS: Wu, C., Zhang, R., Du, W., C, L., & Liang, G. (2018) Alkaline phosphatase-triggered self-assembly of near-infrared nanoparticles for the enhanced photoacoustic imaging of tumors. Nano Letters, 18, 12, 7749–7754. https://pubs.acs.org/doi/abs/10.1021/acs.nanolett.8b03482.*

5.2 Application of 1P to PA imaging of muscles

When applying **1P** for PA imaging in vivo, it is necessary to validate its selectivity. Then we conduct in vivo PA imaging on normal tissues (muscles in this work). The results show that the PA spectra of **1P** injected muscles have similar patterns to but even lower signal intensity than those of control tumors, suggesting that **1P** is suitable for PA imaging of solid tumor.

Equipment

A Vevo 2100 LAZR system (VisualSonics Inc.)
Vaporizer equipped with isoflurane and oxygen (for anesthesia)
Insulin syringes

Animals

Female nude BALB/c mice, 5 weeks old

Reagents

Purified **1P**
Aqueous buffer: phosphate buffer saline (PBS, 10 mM, pH 7.4)

Procedure

1. Warm up the Vevo 2100 LAZR system (VisualSonics Inc.) for at least 30 min prior to starting the experiment. This ensures adequate heating of the optical system
2. Inject a total of 50 μL of **1P** at 0.5 μmol/kg into the thigh muscles of each nude mouse
3. Mice are anesthetized and fixed on the support
4. Add the coupling agent needed for the PA experiment to the surface of the thigh muscles of each nude mouse
5. Coat the ultrasonic probe with the coupling agent
6. Move the ultrasonic probe to make it full contact with the surface of muscles
7. Acquire PA imaging of each mouse with an excitation wavelength of 790 nm by a Vevo 2100 LAZR system for 24 h
8. Obtain PA spectra from 770 to 860 nm extracted from the muscles and PA images of the muscles at 790 nm and different times

5.3 Ex vivo PA imaging of different organs dissected from mice

Here, to prove that **1P** could be used for PA imaging of HeLa tumor with high specificity, ex vivo PA imaging of different organs (bone, kidney, lung, intestine, liver, heart, muscle, skin, stomach, spleen, brain, and tumor in this work) are conducted at 4 h after **1P** injection.

Equipment

A Vevo 2100 LAZR system (VisualSonics Inc.)
Carbon dioxide cylinder
Dissecting tools for mice

Cells and animals.

Female nude BALB/c mice, 5 weeks old.

Reagents

Purified **1P**
Aqueous buffer: phosphate buffer saline (PBS, 10 mM, pH 7.4)

Procedure

1. Inject a total of 50 μL of **1P** at 0.5 μmol/kg into the tumor of each nude mouse
2. Warm up the Vevo 2100 LAZR system (VisualSonics Inc.) for at least 30 min prior to starting the experiment. This ensures adequate heating of the optical system
3. Four hours after **1P** injection, the mice are euthanized and dissected. PA imaging of different organs (bone, kidney, lung, intestine, liver, heart, muscle, skin, stomach, spleen, brain, and tumor in this work) are conducted
4. Obtain PA spectra from 770 to 860 nm extracted from different organs and PA images of different organs at 790 nm and different times

5.4 Assays of ALP in tumor homogenates

The ALP expression in HeLa tumors at different stages is different. In order to test the sensitivity of **1P** to PA imaging of tumors at different stages, we detect ALP activity in tumors at different stages. The results show that ALP activity in large tumors ($650.0 \pm 6.3\,U\,L^{-1}$) is about 2.1-fold that in small tumors ($313.3 \pm 7.1\,U\,L^{-1}$).

Equipment

Incubator suitable for mammalian cell culture, with 5% CO_2 atmosphere
ELISA reader (VARIOSKAN FLASH)
Dissecting tools for mice
High speed centrifuge

Animals

Female nude BALB/c mice, 5 weeks old.

Reagents

Purified **1P**

Aqueous buffer: phosphate buffer saline (PBS, 10 mM, pH 7.4)

ALP detection kits (obtained from Boster Biological Technology Co. ltd.)

Procedure

1. Until the tumor sizes are within 5–10 mm in diameter, weigh each mouse and measure its tumor size using calipers. Calculate the tumor volume using the formula $V = (w^2 * l)/2$, where w is the tumor width and l is the tumor length
2. The mice are divided into two groups according to tumor size, one group with tumors of 5–6 mm and the other one with tumors of 7–10 mm
3. The tumors of the two groups of mice are dissected separately
4. Collect tumor homogenates, lyse them, and centrifuge them at 14,800 rpm for 4 min
5. Use PBS to dilute the supernatants (300 μL) 10 times
6. Determine ALP activities of tumor supernatants by commercially available kits. The higher optical density (O.D.) value represents the higher activity of ALP

6. Concluding remarks

Herein, we describe the synthesis and validation of **1P**, a NIR contrast agent for PA imaging ALP activity in vitro and in tumor. We envision that our strategy could be applied for highly sensitive diagnosis of superficial cancers overexpressing ALP in the near future. Moreover, here we provide a thorough and systematic approach for validating **1P** could be used for enhancing PA imaging both in vitro and in vivo, and we hope that this study could provide guidance for researchers interested in probing enzyme-activatable probes for PA imaging.

References

Ahn, H. Y., Yao, S., Wang, X., & Belfield, K. D. (2012). Near-infrared emitting squaraine dyes with high 2PA cross-sections for multiphoton fluorescence imaging. *ACS Applied Materials & Interfaces*, *4*, 2847–2854.

Chen, M., Guo, Z., Chen, Q., Wei, J., Li, J., Shi, C., et al. (2018). Pd nanosheets with their surface coordinated by radioactive iodide as a highperformance theranostic nanoagent for orthotopic hepatocellular carcinoma imaging and cancer therapy. *Chemical Science*, *9*, 4268–4274.

Chen, M., Tang, S., Guo, Z., Wang, X., Mo, S., Huang, X., et al. (2014). Core-shell Pd@au nanoplates as theranostic agents for in-vivo photoacoustic imaging, CT imaging, and photothermal therapy. *Advanced Materials*, *26*, 8210–8216.

Dong, L., Qian, J., Hai, Z., Xu, J., Du, W., Zhong, K., et al. (2017). Alkaline phosphatase-instructed self-assembly of gadolinium nanofibers for enhanced T_2-weighted magnetic resonance imaging of tumor. *Analytical Chemistry*, *89*, 6922–6925.

Dragulescu-Andrasi, A., Kothapalli, S. R., Tikhomirov, G. A., Rao, J., & Gambhir, S. S. (2013). Activatable oligomerizable imaging agents for photoacoustic imaging of furin-like activity in living subjects. *Journal of American Chemistry Society*, *135*, 11015–11022.

Feng, G., & Liu, B. (2018). Aggregation-induced emission (AIE) dots: Emerging theranostic nanolights. *Accounts of Chemical Research*, *51*, 1404–1414.

Feng, Z., Wang, H., Chen, X., & Xu, B. (2017). Self-assembling ability determines the activity of enzyme-instructed self-assembly for inhibiting cancer cells. *Journal of the American Chemical Society*, *139*, 15377–15384.

Filonov, G. S., Krumholz, A., Xia, J., Yao, J., Wang, L. V., & Verkhusha, V. V. (2012). Deep-tissue photoacoustic tomography of a genetically encoded near-infrared fluorescent probe. *Angewandte Chemie, International Edition*, *51*, 1448–1451.

Fishman, W. H., Inglis, N. R., Green, S., Anstiss, C. L., Gosh, N. K., Reif, A. E., et al. (1968). Immunology and biochemistry of regan isoenzyme of alkaline phosphatase in human cancer. *Nature*, *219*, 697–699.

Freeman, R., Finder, T., Gill, R., & Willner, I. (2010). Probing protein kinase (CK2) and alkaline phosphatase with CdSe/ZnS quantum dots. *Nano Letters*, *10*, 2192–2196.

Huynh, E., Leung, B. Y. C., Helfield, B. L., Shakiba, M., Gandier, J. A., Jin, C., et al. (2015). In situ conversion of porphyrin microbubbles to nanoparticles for multimodality imaging. *Nature Nanotechnology*, *10*, 325–332.

Kim, J. W., Galanzha, E. I., Shashkov, E. V., Moon, H. M., & Zharov, V. P. (2009). Golden carbon nanotubes as multimodal photoacoustic and photothermal high-contrast molecular agents. *Nature Nanotechnology*, *4*, 688–694.

Kothapalli, S. R., Ma, T.-J., Vaithilingam, S., Oralkan, O., Khuri-Yakub, B. T., & Gambhir, S. S. (2012). Deep tissue photoacoustic imaging using a miniaturized 2-D capacitive micromachined ultrasonic transducer array. *IEEE Transactions on Biomedical Engineering*, *59*, 1199–1204.

Li, C., Zhen, S., Wang, J., Li, Y., & Huang, C. (2013). A gold nanoparticles-based colorimetric assay for alkaline phosphatase detection with tunable dynamic range. *Biosensors & Bioelectronics*, *43*, 366–371.

Lin, Y., Wang, Y., Qiao, S., An, H., Wang, J., Ma, Y., et al. (2017). "In vivo self-assembled" nanoprobes for optimizing autophagy-mediated chemotherapy. *Biomaterials*, *141*, 199–209.

Liu, Z., Wang, B., Ma, Z., Zhou, Y., Du, L., & Li, M. (2015). Fluorogenic probe for the human ether-a-go-go-related gene potassium channel imaging. *Analytical Chemistry*, *87*, 2550–2554.

Lopci, E., Saita, A., Lazzeri, M., Lughezzani, G., Colombo, P., Buffi, N. M., et al. (2018). ^{68}Ga-PSMA positron emission tomography/computerized tomography for primary diagnosis of prostate cancer in men with contraindications to or negative multiparametric magnetic resonance imaging: A prospective observational study. *Journal of Urology*, *200*, 95–103.

Moscow, J. A., Fojo, T., & Schilsky, R. L. (2018). The evidence framework for precision cancer medicine. *Nature Reviews Clinical Oncology*, *15*, 183–192.

Ni, D., Jiang, D., Im, H. J., Valdovinos, H. F., Yu, B., Goel, S., et al. (2018). Radiolabeled polyoxometalate clusters: Kidney dysfunction evaluation and tumor diagnosis by positron emission tomography imaging. *Biomaterials*, *171*, 144–152.

Wan, Q., Chen, S., Shi, W., Li, L., & Ma, H. (2014). Lysosomal pH rise during heat shock monitored by a lysosome targeting near-infrared ratiometric fluorescent probe. *Angewandte Chemie, International Edition*, *53*, 10916–10920.

Wang, H., Feng, Z., Del Signore, S. J., Rodal, A. A., & Xu, B. (2018). Active probes for imaging membrane dynamics of live cells with high spatial and temporal resolution over extended time scales and areas. *Journal of the American Chemical Society*, *140*, 3505–3509.

Wang, H., Feng, Z., Wang, Y., Zhou, R., Yang, Z., & Xu, B. (2016). Integrating enzymatic self-assembly and mitochondria targeting for selectively killing cancer cells without acquired drug resistance. *Journal of the American Chemical Society*, *138*, 16046–16055.

Wang, L. V., & Hu, S. (2012). Photoacoustic tomography: In vivo imaging from organelles to organs. *Science*, *335*, 1458–1462.

Zhan, J., Cai, Y., He, S., Wang, L., & Yang, Z. (2018). Tandem molecular self-assembly in liver cancer cells. *Angewandte Chemie, International Edition*, *57*, 1813–1816.

Zheng, Z., Chen, P., Xie, M., Wu, C., Luo, Y., Wang, W., et al. (2016). Cell environment-differentiated self-assembly of nanofibers. *Journal of the American Chemical Society*, *138*, 11128–11131.

Zheng, Z., Tang, A., Guan, Y., Chen, L., Wang, F., Chen, P., et al. (2016). Nanocomputed tomography imaging of bacterial alkaline phosphatase activity with an iodinated hydrogelator. *Analytical Chemistry*, *88*, 11982–11985.

Zhou, J., Du, X., & Xu, B. (2016). Regulating the rate of molecular self-assembly for targeting cancer cells. *Angewandte Chemie, International Edition*, *55*, 5770–5775.

CHAPTER SIX

A TME-activated *in situ* nanogenerator for magnetic resonance/fluorescence/photoacoustic imaging

Shuyu Xu[a], Xiaoxiao Shi[a], Chengchao Chu[a,b,c,*], and Gang Liu[a,b,*]

[a]State Key Laboratory of Molecular Vaccinology and Molecular Diagnostics, Center for Molecular Imaging and Translational Medicine, School of Public Health, Xiamen University, Xiamen, China
[b]State Key Laboratory of Cellular Stress Biology, Innovation Center for Cell Biology, School of Life Sciences, Xiamen University, Xiamen, China
[c]Amoy Hopeful Biotechnology Co., Ltd., Xiamen, China
*Corresponding authors: e-mail address: chuchengchao@xmu.edu.cn; gangliu.cmitm@xmu.edu.cn

Contents

Abstract

With the rapid development of biomedical imaging, non-invasive imaging method particularly has been widely used in clinical diagnosis. Different imaging methods have their own advantages, such as higher resolution of optical imaging, deeper penetration of acoustic imaging, high resolution photoacoustic imaging (PAI), and multi-parameter

Methods in Enzymology, Volume 657
ISSN 0076-6879
https://doi.org/10.1016/bs.mie.2021.06.029

guidance of magnetic resonance imaging (MRI). Recent years, multimodal MRI, fluorescence imaging, PAI and others have been verified to play an important role in the field of molecular imaging and provided detail information for accurate *in vivo* diagnosis. Therefore, the design of multimodal probe that can integrate the above advantages to carry out combined imaging will be more accurate to assist diagnosis and treatment. While tumor microenvironment (TME) is highly critical in validating and optimizing current therapeutic strategies. Herein, we highlight the TME-triggered supramolecular system as an *in situ* nano-generator for multimodal imaging-guided treatment. The experimental protocols on the PAI/fluorescence imaging/MRI-based diagnosis are described in this chapter.

1. Introduction

Photoacoustic imaging (PAI) is a non-invasive biomedical imaging modality, which generates ultrasonic wave by irradiating the material with pulsed laser and reconstructs the image of light energy absorption distribution in the tissue. Interestingly, PAI breaks the limitation of low resolution caused by tissue scattering for traditional optical imaging technology, and shows high-contrast and sensitivity in deep tissue penetration. Considering its unique advantages, PAI plays a vital role in modern medical imaging (Chu et al., 2020), and has been widely applied in the diagnosis of tumors (Lin et al., 2019), vasculature, brain disease (Wang et al., 2019) and arthritis (Jo et al., 2018). PA probes can be simply divided into endogenous and exogenous forms. Endogenous molecular probes, including hemoglobin and melanin, can achieve high-resolution imaging of blood vessels and veins and accurate positioning of melanoma. However, due to the characteristics of endogenous probe, such as distribution *in vivo* and weak signal in deep foci, its application scope is limited. Otherwise, effective exogenous photoacoustic contrast agent can further enhance the contrast and resolution of visualization, result in high sensitivity of deep tumor diagnosis (Chan, Dodani, & Chang, 2012; Guo et al., 2017; Jia et al., 2018; Li, Lu, Zhao, Lei, & Zhang, 2018).

To date, a wide variety of PA probes have been exploited, including metallic nanomaterials (gold nanocrystals, Fe_3O_4, Pd nanosheets, *etc.*), carbon-based nanomaterials, organic small molecules (porphyrin, melanin, cyanine-based dye, *etc.*) and polymer nanoparticles (Fu, Zhu, Song, Yang, & Chen, 2019; García-Álvarez et al., 2020; Shi et al., 2020; Zhang et al., 2020). Our group has also made a series of explorations in the field of PA probes. Lin et al. (2019) highlighted a Fe^{3+}-mediated supramolecular assembly of ICG for cancer sonotheranostics. They confirmed the Fe^{3+}/ICG@MB seems to outperform free ICG and Fe^{3+}/ICG in PAI. As well as, Liu et al. (2018) prepared a manganese-doped polydopamine nanoparticles by a one-pot reaction of $KMnO_4$ and dopamine. The obtained MnEMNPs was applied in T1–T2 MRI/PAI for tumor detection and treatment. Furthermore, the MnEMNPs was a biodegradable nanoparticle in both tumor tissue and normal tissue, which could reduce its long-time *in vivo* toxicity.

Although PAI showed many advantages in diagnosis, the application of most PA probes is still restricted by undesirable off-target biodistribution, *in vivo* instability, poor biocompatibility and so on. Recently, multimodal imaging that combined use of several imaging could provide more comprehensive and accurate information of diseases. Fluorescence imaging is a method based on the absorption and scattering of light by substances and has the characteristics of short imaging time and high imaging sensitivity. Meanwhile, magnetic resonance imaging (MRI) uses signals generated by the resonance of atomic nuclei in a magnetic field. In addition, MRI has high resolution, which could provide rich diagnostic information with multiple parameters. Thus, the combine utilization of MRI, fluorescence imaging and PAI could conduct multi-scale imaging study and provided detail information for accurate diagnosis (Peet, Arvanitis, & Leach, 2012; Yue et al., 2019).

Compared to most normal tissue cells, critical cellular metabolism pathways are altered in cancer cells, and the tumor microenvironment (TME) was effective diagnostic targets regardless the tumor types. The design of TME-responsive multimodal imaging probe is of great significance in the tumor diagnosis and treatment. Here we highlight a novel approach to construct MnO_2/DVDMS nanotheranostic probe, which depends TME-triggered *in situ* supramolecular assembly, as well as can be monitored by activated MRI/fluorescence imaging/PAI signals (Chu et al., 2017).

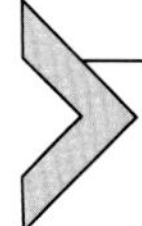

2. Materials

2.1 Reagents

Sinoporphyrin sodium (DVDMS, $C_{68}H_{66}N_8O_9Na_4$, MW: 1230.265) (Jiangxi Qinglong Group Co., Ltd.); Tetramethylammonium hydroxide (TMA·OH) (Sigma-Aldrich); $MnCl_2·4H_2O$ (Sigma-Aldrich); L-glutathione (GSH) (Sigma-Aldrich); H_2O_2 (Xiamen Luyin Reagent Glass Instrument

Co., Ltd); Reactive oxygen species Kit (DCFH-DA) (Sangon Biotech Co., Ltd); GSH assay kit (Sangon Biotech Co., Ltd); H_2O_2 Kit (Sangon Biotech Co., Ltd).

2.2 Equipment

Dynamic light scattering (DLS) and Zeta potential instrument (HORIBA).
Transmission electron microscope (TEM) (Tecnai G2 Spirit).
Atomic force microscopy (AFM) (Multimode).
Photoacoustic imaging (PAI) (Vevo LAZR-X).
Scanning electron microscope (SEM) (GeminiSEM 500).
X-ray energy spectrum (XPS) (XFORD INTRUMENTS AN).
Magnetic resonance imaging (MRI) (Bruker).

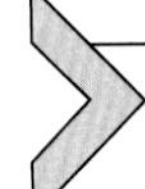

3. Methods

3.1 Preparation of Mn/DVDMS nanotheranostics (nanoDVD)

Manganese ion binds to the carboxyl groups on the backbone of DVDMS by interaction. Briefly, 1 mL Mn^{2+} solution was reacted with 1 mL DVDMS solution for 3 h at room temperature. NanoDVD formed could be visualized by TEM. UV–Vis absorption and XPS scan subsequently confirmed the successful Mn^{2+} modification in nanotheranostics.

3.2 Preparation of MnO_2/DVDMS

MnO_2 was synthesized according to the protocol reported previously (Zhao et al., 2014). Physisorption occurred as electrostatic interaction between MnO_2 and DVDMS. DVDMS (1 mg/mL) and MnO_2 (1 mg/mL) were stirred together for 2 h at room temperature. After 6 h of incubation, DVDMS were adsorbed on MnO_2 surface. Then 30 kDa MWCO Amicon filters were used to remove unreacted DVDMS.

3.3 *In situ* Mn/DVDMS nanoassemblies

The reaction of DVDMS and Mn^{2+} in tumor cells was studied by cell fluorescence imaging and TEM. First, the cells were incubated with 10 μg/mL of DVDMS for 9 h, then washed to remove the DVDMS which was not internalized by the cells. Subsequently, Mn^{2+} solution dispersed in the culture medium was added to co culture for 3 h. After the reaction, the remaining Mn^{2+} was washed away with PBS. For TEM, the cells were collected by cell scraper, then fixed with 4% paraformaldehyde solution and further encapsulated and sliced.

3.4 GSH/H_2O_2 (pH 5.5) microenvironmental responsiveness

MnO_2 could react with the GSH or H_2O_2 in acidic environment. In this study, 10 mL MnO_2/DVDMS solution (100 μg/mL) reacted with 1 mL H_2O_2 solution (20 mM, PH 5.5) and 1 mL GSH solution (2 mg/mL), respectively, for 15 min, and then the products were collected by 30 kDa MWCO Amicon filters. The products were characterized by TEM, XPS and zeta potential.

3.5 PAI study

3.5.1 *In vitro PA study of Mn/DVDMS and MnO_2/DVDMS*

The PAI performances of MnO_2/DVDMS and Mn/DVDMS after reaction in GSH and H_2O_2/H^+ solution was studied, respectively. Comparing the photoacoustic imaging and photoacoustic intensity of DVDMS, Mn/DVDMS, MnO_2, MnO_2/DVDMS at different concentrations.

3.5.2 *In vivo PAI of MnO_2/DVDMS*

In addition, the author explored the performance of MnO_2/DVDMS in tumor. *In situ* injection of 25 μL of DVDMS, MnO_2 and MnO_2/DVDMS solution (MnO_2 1.25 mg/kg, DVDMS 1.18 mg/kg). PAI was performed at the excitation wavelength of 690 nm at 1, 3, 6, 12 and 24 h post drug injection.

3.6 *In vivo* fluorescence and MR images

The accumulation and signal changes of MnO_2/DVDMS in tumor-bearing mice were further observed by fluorescence imaging and MR imaging. *In situ* injection of 25 μL of MnO_2/DVDMS solution (MnO_2 1.25 mg/kg, DVDMS 1.18 mg/kg). Fluorescence images was performed at 1, 3, 6, 12, and 24 h after injection, and T1-weighted MRI images was performed at 3, 6, and 24 h after injection.

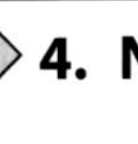

4. Notes

1. The photoacoustic data of this experiment were acquisition of Vevo LAZR-X photoacoustic imaging system.

2. 690 nm was selected as the excitation wavelength in the experiment, and the wavelength and other imaging parameters remained unchanged.
3. Different concentrations of GSH and H_2O_2 (pH 5.5) were added into MnO_2/DVDMS. The structural changes of the nanomaterials were verified by TEM, XPS, and the color changes of the solution, which confirmed the occurrence of the secondary assembly of the materials.
4. Comparing the photothermal properties of MnO_2/DVDMS and DVDMS, it was found that the effect of MnO_2 nanosheets leads to the photothermal enhancement. After GSH and H_2O_2/H^+ reaction, the temperature can be increased to 55 °C and 53 °C, respectively, which could be used in the photothermal removal of tumor.
5. MnO_2 can produce O_2 after reacting with H_2O_2/H^+, and oxygen could enhance the photodynamic effect. In order to verify the improvement of anaerobic environment, the expression of HIF-1 α was detected after adding MnO_2/DVDMS.

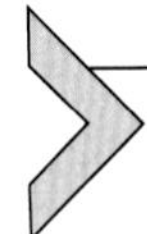

5. Results

5.1 MnO_2/DVDMS

The successful modification of DVDMS was verified by UV absorption peak at 630 nm and FT-IR (Fig. 1). After the modification of DVDMS, the nanostructures could exist stably in PBS, DMEM, and FBS.

5.2 MnO_2/DVDMS properties

The performance of MnO_2/DVDMS in the presence of GSH or H_2O_2 (pH 5.5) was studied (Fig. 2). The results showed that the material changed from nanosheets to spherical structure. This phenomenon could be reasonably explained by the decomposition of MnO_2 nanosheets in GSH or H_2O_2 (pH 5.5) solution to form Mn^{2+} and release of DVDMS. The resulting DVDMS and Mn^{2+} were further constructed to form Mn/DVDMS. The particle sizes of MnO_2 nanosheets, MnO_2/DVDMS, MnO_2/DVDMS + GSH and MnO_2/DVDMS + H_2O_2/H^+ were 128.4, 91.2, 197.2, and 122.4 nm, respectively.

For the material performance, first, the T1 weighted MRI imaging performance of the material was discussed. After the interaction of MnO_2/DVDMS with GSH, R1 increased to 34.7 times of the original, and after the interaction with H_2O_2/H^+, R1 increased to 32.5 times of the original. Then, the fluorescence imaging properties of the materials were investigated.

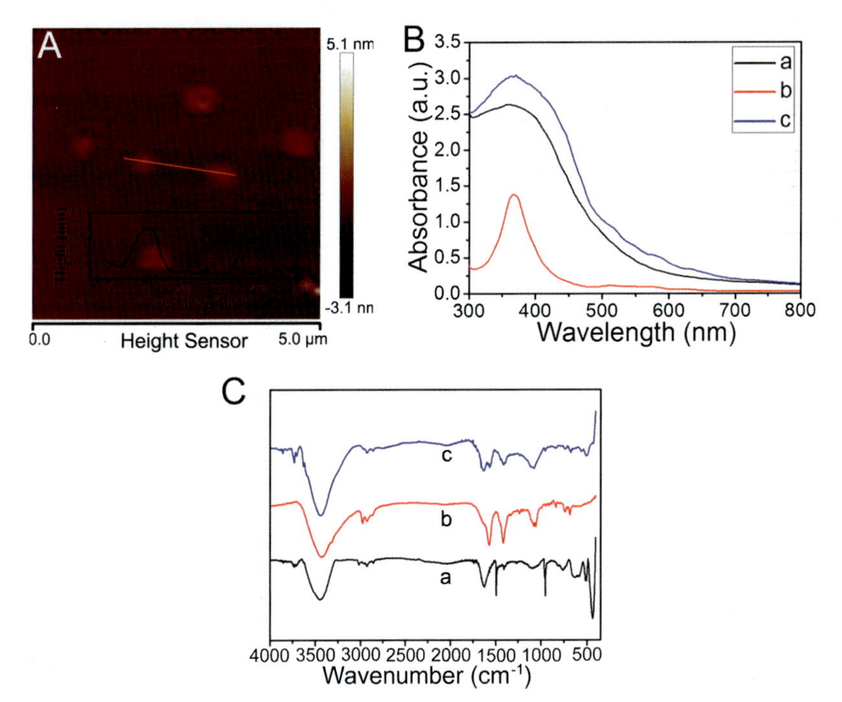

Fig. 1 (A) AFM of the MnO_2 nanosheet; (B) UV–vis absorption of the MnO_2 (a), DVDMS (b) and MnO_2/DVDMS (c); (C) FT-IR of the MnO_2 (a), DVDMS (b) and MnO_2/DVDMS (c). *From Chu, C., Lin, H., Liu, H., Wang, X., Wang, J., & Zhang, P., et al. (2017). Tumor microenvironment-triggered supramolecular system as an in situ nanotheranostic generator for cancer phototherapy.* Advanced Materials, 29, *1605928.*

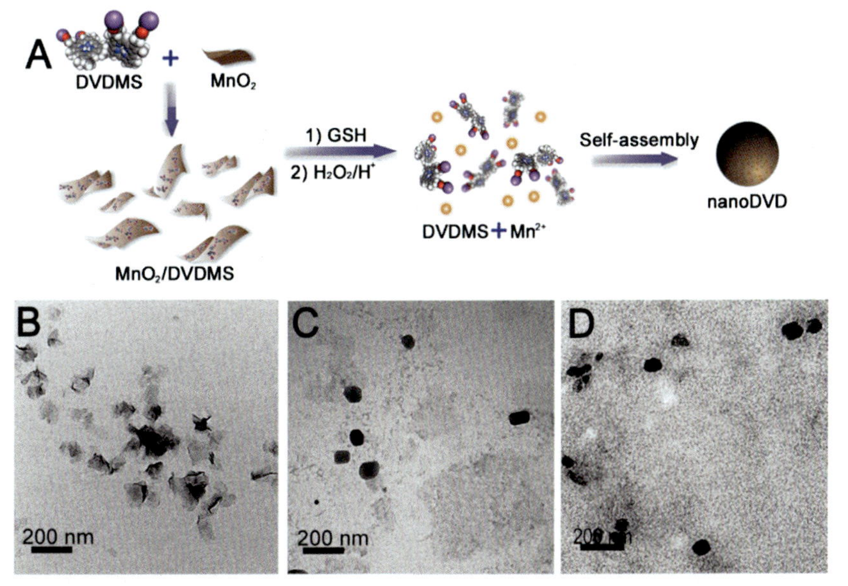

Fig. 2 (A) Schematic illustration of the MnO2/DVDMS reacted in GSH solution or H_2O_2 (pH 5.5). TEM of (B) MnO_2 nanosheet, (C) MnO_2/DVDMS+GSH, and (D) MnO_2/DVDMS +H_2O_2/H^+. *From Chu, C., Lin, H., Liu, H., Wang, X., Wang, J., & Zhang, P., et al. (2017). Tumor microenvironment-triggered supramolecular system as an in situ nanotheranostic generator for cancer phototherapy.* Advanced Materials, 29, *160592.*

With the modification of MnO_2 nanosheets, the fluorescence was quenched to only 13.8%. After the reaction of GSH with H_2O_2/H^+, MnO_2/DVDMS was assembled into Mn/DVDMS, and the fluorescence partially recovered. However, compared with pure DVDMS, the fluorescence still decreased significantly (Fig. 3). Therefore, this material could be used as MRI and fluorescence contrast agent to guide tumor treatment. In addition, due to the obvious photothermal properties of Mn/DVDMS after agglomeration, the material has strong photothermal treatment ability in tumor area. The imaging properties *in vivo* will be introduced in Section 5.3.

The photoacoustic performance of DVDMS and Mn/DVDMS was studied. From the research results, it is believed that the photoacoustic performance of Mn/DVDMS is better than Free DVDMS. While the

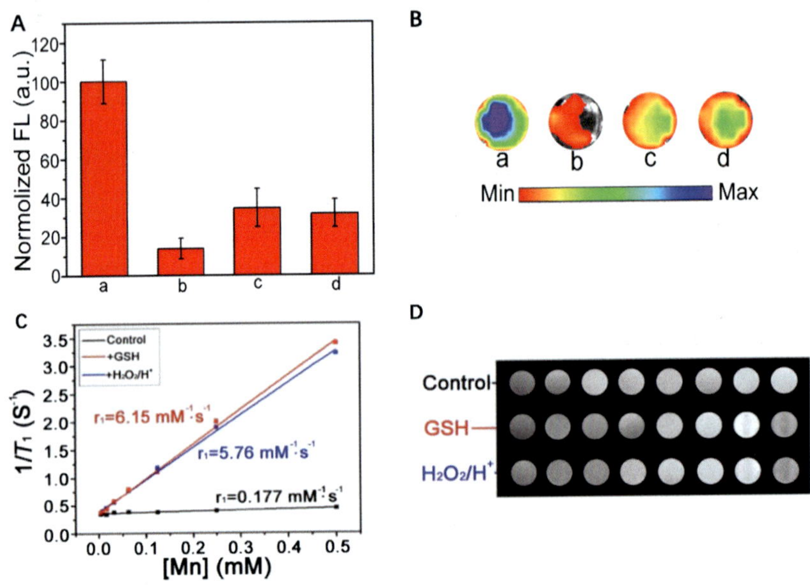

Fig. 3 ROI intensity (A) and fluorescence images (B) of DVDMS (100 μg/mL) (a), MnO_2/DVDMS (b), MnO_2/DVDMS + GSH (c) and MnO_2/DVDMS + H_2O_2/H^+ (d). (C) MR imaging study of the MnO_2/DVDMS: 1/T1 *versus* Mn^{2+} concentration for the MnO_2/DVDMS solution (black line), the MnO_2/DVDMS + GSH solution (red line), and MnO_2/DVDMS + H_2O_2/H^+ (blue line). (D) T1-weighted MRI phantom images. *From Chu, C., Lin, H., Liu, H., Wang, X., Wang, J., & Zhang, P., et al. (2017). Tumor microenvironment-triggered supramolecular system as an in situ nanotheranostic generator for cancer phototherapy.* Advanced Materials, 29, *160592.*

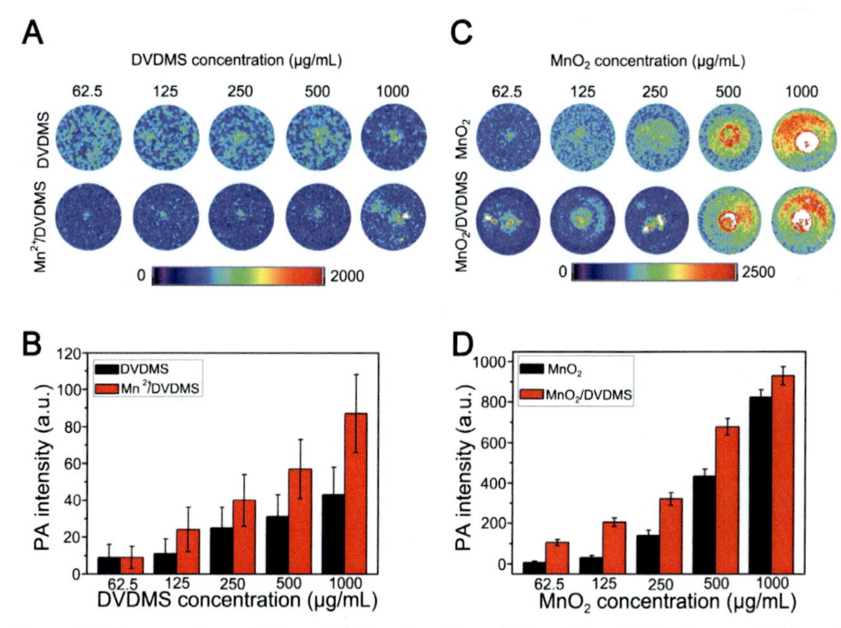

Fig. 4 PA images (A and C) and ROI value (B and D) of free DVDMS, nanoDVD and MnO_2/DVDMS of different concentrations (690 nm). *From Chu, C., Lin, H., Liu, H., Wang, X., Wang, J., & Zhang, P., et al. (2017). Tumor microenvironment-triggered supramolecular system as an in situ nanotheranostic generator for cancer phototherapy.* Advanced Materials, 29, *160592.*

photoacoustic performance of MnO_2 and MnO_2/DVDMS is significantly improved, as high as 10 times. That is to say, when MnO_2/DVDMS with strong photoacoustic signal is decomposed and assembled into Mn/DVDMS, the photoacoustic signal changes significantly (Fig. 4). This unique property can also be applied to the monitoring of biological reactions.

5.3 Multimodal images *in vivo*

In order to study the characteristics of DVDMS, MnO_2 and MnO_2/DVDMS in tumor, we injected DVDMS, MnO_2 and MnO_2/DVDMS into MCF-7 tumor bearing mice and monitored them by photoacoustic imaging. The experimental results showed that the change process of photoacoustic signal from strong to weak can be observed, which is consistent with the decomposition process of material from MnO_2/DVDMS to Mn/DVDMS in response to tumor microenvironment. *In vivo* experiments further demonstrated the responsiveness of the material and its ability to guide treatment.

Meanwhile, the fluorescence properties of tumor-bearing mice after intratumoral injection of MnO_2/DVDMS were investigated. The fluorescence intensity increased rapidly in 0–3 h and continued to 12 h and remained stable in 12–24 h (Fig. 5). Subsequently, T1-weighted MRI signals of mice injected with MnO_2/DVDMS were investigated. The results showed that T1 intensity gradually decreased, which could be attributed to the release of Mn^{2+} and enrichment in the tumor environment after the secondary assembly of Mn/DVDMs.

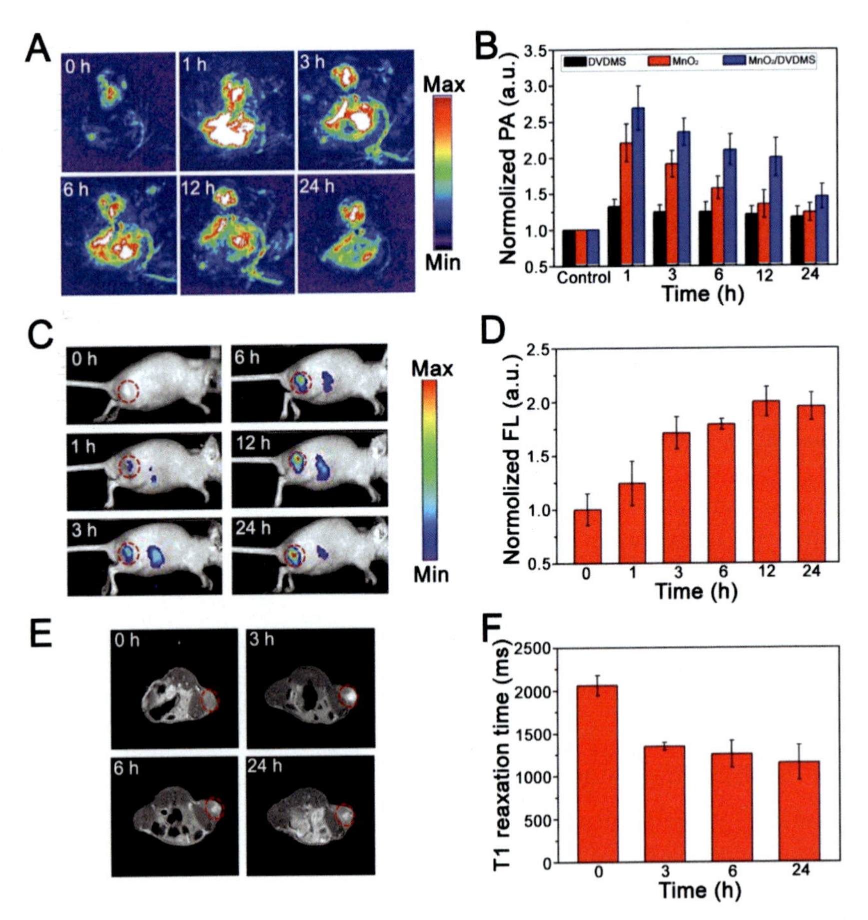

Fig. 5 (A) PA images of tumor before and after intratumoral injection of the DVDMS. (B) Region of interest (ROI) analysis of the PA signal. (C) Fluorescence images and (D) fluorescence signal intensity of the tumor. (E) T1-weighted MR images and (F) T1 relaxation time before and after injection of MnO_2/DVDMS. *From Chu, C., Lin, H., Liu, H., Wang, X., Wang, J., & Zhang, P., et al. (2017). Tumor microenvironment-triggered supramolecular system as an in situ nanotheranostic generator for cancer phototherapy.* Advanced Materials, 29, *160592.*

6. Summary

A nanomaterial MnO_2/DVDMS with the function of tumor microenvironment response was designed. MnO_2 is an effective carrier of DVDMS, and can produce Mn^{2+}, DVDMS and O_2 through the action of GSH and H_2O_2/H^+ in tumor microenvironment. Furthermore, Mn^{2+} assisted self-assembly of DVDMS occurs, resulting in the secondary assembly of Mn/DVDMS. This kind of nanomaterial has good performance of photoacoustic imaging, fluorescence imaging, and MRI. The changes of imaging signals with microenvironmental responsiveness provide an important reference for the diagnosis and treatment of tumors.

Acknowledgment

The authors acknowledge the support from Natural Science Foundation of China (NSFC) (Grant Nos. 81925019 and U1705281), the Major State Basic Research Development Program of China (Grant No. 2017YFA0205201), the Fundamental Research Funds for the Central Universities (20720190088 and 20720200019), Natural Science Foundation of Fujian Province of China (Grant No. 2018 J05144), the Program for New Century Excellent Talents in University of China (NCET-13-0502).

References

Chan, J., Dodani, S. C., & Chang, C. J. (2012). Reaction-based small-molecule fluorescent probes for chemoselective bioimaging. *Nature Chemistry*, *4*, 973–984.

Chu, C., Lin, H., Liu, H., Wang, X., Wang, J., Zhang, P., et al. (2017). Tumor microenvironment-triggered supramolecular system as an in situ nanotheranostic generator for cancer phototherapy. *Advanced Materials*, *29*, 1605928.

Chu, C., Yu, J., Ren, E., Ou, S., Zhang, Y., Wu, Y., et al. (2020). Multimodal photoacoustic imaging-guided regression of corneal neovascularization: A non-invasive and safe strategy. *Advancement of Science*, 7, 2000346.

Fu, Q., Zhu, R., Song, J., Yang, H., & Chen, X. (2019). Photoacoustic imaging: Contrast agents and their biomedical applications. *Advanced Materials*, *31*, e1805875.

García-Álvarez, R., Chen, L., Nedilko, A., Sánchez-Iglesias, A., Rix, A., Lederle, W., et al. (2020). Optimizing the geometry of photoacoustically active gold nanoparticles for biomedical imaging. *ACS Photonics*, 7, 646–652.

Guo, Z., Zhu, S., Yong, Y., Zhang, X., Dong, X., Du, J., et al. (2017). Synthesis of BSA-coated BiOI@Bi2 S3 semiconductor heterojunction nanoparticles and their applications for radio/photodynamic/photothermal synergistic therapy of tumor. *Advanced Materials*, *29*, 1704136.

Jia, Q., Ge, J., Liu, W., Zheng, X., Chen, S., Wen, Y., et al. (2018). A magnetofluorescent carbon dot assembly as an acidic H2 O2-driven oxygenerator to regulate tumor hypoxia for simultaneous bimodal imaging and enhanced photodynamic therapy. *Advanced Materials*, *30*, e1706090.

Jo, J., Tian, C., Xu, G., Sarazin, J., Schiopu, E., Gandikota, G., et al. (2018). Photoacoustic tomography for human musculoskeletal imaging and inflammatory arthritis detection. *Photoacoustics*, *12*, 82–89.

Li, B., Lu, L., Zhao, M., Lei, Z., & Zhang, F. (2018). An efficient 1064 nm NIR-II excitation fluorescent molecular dye for deep-tissue high-resolution dynamic bioimaging. *Angewandte Chemie (International Ed. in English)*, *57*, 7483–7487.

Lin, H., Li, S., Wang, J., Chu, C., Zhang, Y., Pang, X., et al. (2019). A single-step multi-level supramolecular system for cancer sonotheranostics. *Nanoscale Horizons*, *4*, 190–195.

Liu, H., Chu, C., Liu, Y., Pang, X., Wu, Y., Zhou, Z., et al. (2018). Novel intrapolymerization doped manganese-eumelanin coordination nanocomposites with ultrahigh relaxivity and their application in tumor theranostics. *Advancement of Science*, *5*, 1800032.

Peet, A., Arvanitis, T., & Leach, M. (2012). Functional imaging in adult and paediatric brain tumours. *Nature Reviews. Clinical Oncology*, *9*, 700–711.

Shi, X., Zhang, Y., Tian, Y., Xu, S., Ren, E., Bai, S., et al. (2020). Multi-responsive bottlebrush-like unimolecules self-assembled nano-riceball for synergistic sono-chemotherapy. *Small Methods*, *2021*, 2000416.

Wang, S., Sheng, Y., Yang, Z., Hu, D., Long, X., Feng, G., et al. (2019). Activatable small-molecule photoacoustic probes that cross the blood-brain barrier for visualization of copper(ii) in mice with Alzheimer's disease. *Angewandte Chemie International Edition*, *58*, 12415–12419.

Yue, Y., Huo, F., Cheng, F., Zhu, X., Mafireyi, T., Strongin, R., et al. (2019). Functional synthetic probes for selective targeting and multi-analyte detection and imaging. *Chemical Society Reviews*, *48*(15), 4155–4177.

Zhang, Y., Wang, X., Chu, C., Zhou, Z., Chen, B., Pang, X., et al. (2020). Genetically engineered magnetic nanocages for cancer magneto-catalytic theranostics. *Nature Communications*, *11*, 5421.

Zhao, Z., Fan, H., Zhou, G., Bai, H., Liang, H., Wang, R., et al. (2014). Activatable fluorescence/MRI bimodal platform for tumor cell imaging via MnO_2 nanosheet-aptamer nanoprobe. *Journal of the American Chemical Society*, *136*, 11220–11223.

Near-infrared II photoacoustic probes for nitric oxide sensing

Melissa Y. Lucero, Amanda K. East, and Jefferson Chan*

Department of Chemistry and Beckman Institute for Advanced Science and Technology, University of Illinois at Urbana – Champaign, Urbana, IL, United States

*Corresponding author: e-mail address: jeffchan@illinois.edu

Contents

Abstract

In this chapter, we introduce a two-phase tuning approach for developing highly sensitive photoacoustic probes for imaging nitric oxide (NO) in the near-infrared (NIR)-II window. Due to the synthetically challenging nature of current NIR-II dye platforms, our two-phase tuning approach circumvents this issue by first allowing one to

Methods in Enzymology, Volume 657
ISSN 0076-6879
https://doi.org/10.1016/bs.mie.2021.06.032

tune the reactivity using a synthetically accessible dye. We have used a physical organic workflow to understand the reaction kinetics and identify the most reactive sensing component. The selected reactive trigger is then introduced to phase two where it is appended to a range of well-established NIR-II dyes. This strategy is used to select the ideal photoacoustic probe for NIR-II imaging *in vivo*. Here, we have detailed procedures for synthesis, *in vitro* studies, and *in vivo* imaging.

1. Introduction

Photoacoustic (PA) imaging is a reliable *in vivo* imaging modality that utilizes light to generate an ultrasound readout. Since sound scatters significantly less than light does in tissue, PA imaging provides excellent spatial resolution *in vivo* (Wang & Yao, 2016). Indeed, label-free PA imaging has been employed in clinical trials (e.g., breast cancer). Only PA contrast agents have reached clinical trials; however, major efforts have been made toward the development of activatable PA probes which enable the detection of specific disease biomarkers (i.e., metals (Li, Zhang, Smaga, Hoffman, & Chan, 2015; Roberts et al., 2018; Wang et al., 2019a; Zhang et al., 2020), thiols (Chen et al., 2019; Lucero & Chan, 2020), enzymes (Cheng et al., 2020; Wang et al., 2019b; Wu et al., 2018; Yin et al., 2019), reactive oxygen and nitrogen species (Ikeno et al., 2019; Reinhardt, Xu, & Chan, 2020; Reinhardt, Zhou, Jorgensen, Partipilo, & Chan, 2018)). For instance, we have previously reported activatable PA probes for NO, namely APNO-5 and SR-APNO-3. However, these probes and many other PA probes in the literature are limited to imaging in the NIR-I region (650–900 nm). This region faces significant interference from endogenous chromophores such as hemoglobin, oxy-hemoglobin, lipids, and melanin. To overcome this challenge, recent work has been geared toward shifting the absorbance of PA probes into the NIR-II region (900–1100 nm) where the absorbance from biological chromophores is significantly attenuated (Sun et al., 2020; Upputuri & Pramanik, 2019; Zhang, Ning, Zeng, & Pu, 2021).

Here we have introduced a two-phase tuning strategy for the development of activatable probes specifically for the detection of NO. NO is a short-lived reactive nitrogen species involved in a variety of biological and pathological roles such as signaling, inflammation, and cancer (Thomas et al., 2008). Therefore, to achieve a good response *in vivo* we sought to first tune the reactivity before appending the NO-responsive trigger to the more synthetically challenging NIR-II dye platform.

In this chapter, we have described a general strategy for developing optimally tuned activatable PA probes for NIR-II imaging. We have provided detailed procedures to synthesize cyanine-based activatable PA probes, construct Hammett and Brønsted plots, and perform *in vitro* and *in vivo* PA imaging. Importantly, we demonstrate the enhanced PA signal provided by NIR-II imaging over NIR-I imaging.

2. Strategy for optimizing nitric oxide response

We have taken a physical organic approach to tune the reactivity with NO. Since NO has a short half-life, typically in the millisecond to microsecond range, we needed a highly reactive probe to achieve NO detection *in vivo*. To accomplish this, we tuned the reactivity of the *N*-nitrosation reaction by modifying the *para*-position of the aniline. By changing the electronics at this position, we can systematically enhance the rate of the reaction.

2.1 Design of photoacoustic probe

In general, the design of an activatable NIR-II PA probe must meet three criteria. First, the absorption of the turnover probe must be in the NIR-II window for deep tissue penetration. Second, the extinction coefficient must be greater than 10^4 M^{-1} cm^{-1} to generate a strong PA signal. Third, it must be biocompatible with low toxicity, good stability, high selectivity and reactivity.

In our design, we used a NIR-I cyanine dye (Cy7-Cl) for tuning NO-responsiveness due to its synthetic accessibility and suitable photophysical properties. A panel was synthesized with varying electron donating groups to establish Hammett and Brønsted plots (Fig. 1).

2.2 Synthesis of phase one probes

All probes in this panel can be synthesized in one step from Cy7-Cl. Two general methods were utilized to synthesize the panel for phase one tuning

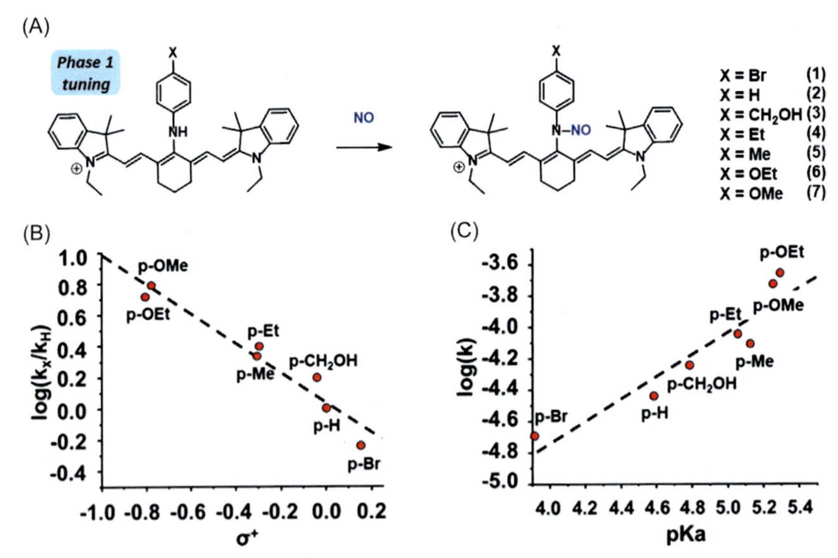

Fig. 1 (A) Schematic showing the reaction between the various NIR-I APNOs (compounds 1–7) and NO to form an *N*-nitroso product. (B) Hammett plot for the *N*-nitrosation reaction between para-substituted APNOs and NO (introduced as MAHMA-NONOate) at 25 °C. Dashed line represents the best linear fit. $R^2 = 0.94$. (C) Brønsted plots indicating the linear relationship between log(k) and the pK_a value of the conjugate acid form of each aniline. Dashed line represents the best linear fit. $R^2 = 0.87$. *Reprinted (adapted) with permission from Lucero, M. Y., East, A. K., Reinhardt, C. J., Sedgwick, A. C., Su, S., Lee, M. C. & Chan, J. (2021). Development of NIR-II photoacoustic probes tailored for deep-tissue sensing of nitric oxide.* Journal of the American Chemical Society, *143, 7196–7202. Copyright (2021) American Chemical Society.*

(Scheme 1). The following synthetic procedures require understanding of standard organic chemistry techniques and basic synthetic laboratory equipment. All running solvents used for purification are defined as percent volume to volume.

2.2.1 Procedure

2.2.1.1 General method A

1. Add Cy7-Cl (1 equiv.) and the corresponding para-substituted aniline (5 equiv.) into a 50 mL round-bottom flask equipped with a magnetic stir bar and rubber septum.
2. Add anhydrous DMSO dried over 4 Å molecular sieves into the flask to dissolve the combined solids.
3. Treat the reaction mixture with triethylamine dried over 4 Å molecular sieves (5 equiv.).

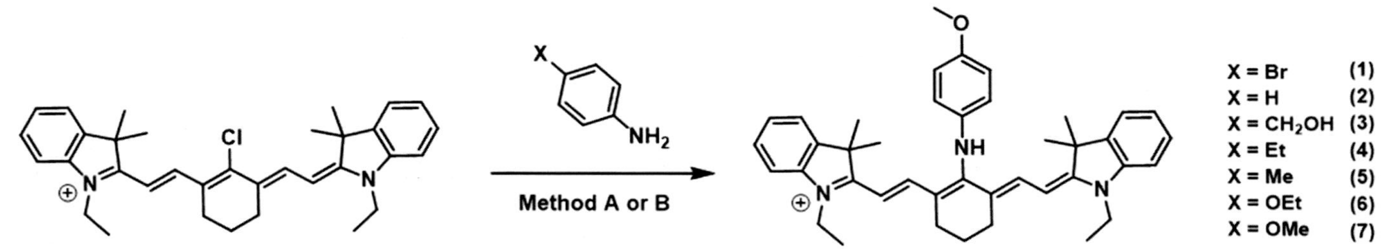

Method A or B

X = Br (1)
X = H (2)
X = CH_2OH (3)
X = Et (4)
X = Me (5)
X = OEt (6)
X = OMe (7)

Scheme 1 Synthetic scheme for compounds 1–7.

4. Place the reaction flask into an oil bath.
5. Heat the reaction to 55 °C for 8 h before cooling to room temperature.
6. A high flow of N_2 gas was used to remove DMSO overnight.
7. Alternatively, DMSO can be removed by diluting the reaction mixture with ethyl acetate or CH_2Cl_2 and performing brine washes (3 ×). The organic layer can then be dried over sodium sulfate, filtered and concentrated.
8. Resuspend the residue in CH_2Cl_2 and wash with 1 M HCl.
9. Then wash the organic layer with a saturated aqueous KI solution.
10. Dry the organic layer over Na_2SO_4 and concentrate via rotary evaporation.
11. Purify the resultant residue via silica gel chromatography and isolate a shiny purple-blue solid (5% MeOH/CH_2Cl_2 and 10%/45%/45% MeOH/Toluene/CH_2Cl_2). Note: Allow the column to slow drip and do not apply air pressure. This ensures better separation from unreacted Cy7-Cl.

2.2.1.2 General method B

1. Add Cy7-Cl (1 equiv.) and the corresponding para-substituted aniline (3 equiv.) into a 50 mL round-bottom flask equipped with a magnetic stir bar and rubber septum.
2. Add anhydrous DMF dried over 4 Å molecular sieves into the flask to dissolve the combined solids.
3. Treat the reaction mixture with triethylamine dried over 4 Å molecular sieves (5 equiv.).
4. Place the reaction flask into an oil bath.
5. Heat the reaction to 55 °C for 8 h before cooling to room temperature.
6. Remove DMF by diluting the reaction mixture with brine. Transfer to a separatory funnel, and extract with CH_2Cl_2 (3 ×).
7. Wash the combined organic layers with and equal volume of brine (1 ×), 1 M HCl (1 ×), and finally a saturated aqueous KI solution (1 ×).
8. Dry the organic layer over Na_2SO_4 and concentrate via rotary evaporation.
9. Purify the resultant residue via silica gel chromatography (5% MeOH/ CH_2Cl_2) and isolate a shiny blue solid. Note: Allow the column to slow drip and do not apply air pressure. This ensures better separation from unreacted starting material.

Refer to Table 1. for exact amounts and measurements.

Table 1 Amounts for synthesis of compounds 1–7.

Compound	Method	Cy7-Cl	*p*-x-aniline	Triethylamine	Solvent	Product yield
1	A	500 mg, 0.80 mmol	673 mg, 4.0 mmol	545 μL, 4.0 mmol	13 mL	8.2 mg, 0.011 mmol, 1.4%
2	A	500 mg, 0.80 mmol	684 mg, 4.0 mmol	545 μL, 4.0 mmol	13 mL	54.3 mg, 0.078 mmol, 10.0%
3	B	500 mg, 0.80 mmol	360.7 mg, 2.3 mmol	327 μL, 2.3 mmol	8 mL	67.3 mg, 0.112 mmol, 11.5%
4	B	500 mg, 0.80 mmol	284.4 mg, 2.3 mmol	327 μL, 2.3 mmol	8 mL	110.2 mg, 0.152 mmol, 19.4%
5	B	500 mg, 0.80 mmol	360.7 mg, 2.3 mmol	408 μL, 2.3 mmol	8 mL	66.4 mg, 0.094 mmol, 12.0%
6	B	500 mg, 0.80 mmol	322.0 mg, 2.3 mmol	327 μL, 2.3 mmol	8 mL	182.5 mg, 0.247 mmol, 31.5%
7 (APNO-780)	B	500 mg, 0.80 mmol	289.1 mg, 2.3 mmol	408 μL, 2.3 mmol	8 mL	101.9 mg, 0.140 mmol, 18.0%

2.3 Photophysical characterization of phase one probes

Each probe was evaluated *in vitro* to obtain their absorbance maxima, extinction coefficient, and relative rate constants (Table 2).

2.3.1 Equipment

1—UV–Vis spectrophotometer
1—Glass cuvette with corresponding cuvette cap
4—Micropipettes and tips (P2.5, P10, P200, P1000)

2.3.2 Reagents

3 aliquots each of compounds 1, 2, 3, 4, 5, 6 and 7
Degassed 10 mM KOH solution
Organic solvents: DMSO
Aqueous buffers: 50 mM HEPES containing 0.01% CrEL (pH 7.2) and 20 mM ethanolic potassium phosphate buffer (pH 7.4)

2.3.3 Procedure for determining absorbance maxima

1. Pipet 999 μL of 20 mM ethanolic potassium phosphate buffer into a 1.5 mL cuvette.
2. Insert the cuvette into the UV–Vis spectrophotometer.
3. Perform a baseline correction.
4. Prepare a 2 mM probe stock solution in DMSO.
5. Pipet 1 μL of the probe stock solution into the cuvette. Mix the solution thoroughly by pipetting up and down 5–10 times. Then, perform a scan from 250 to 900 nm.

Table 2 Summary of photophysical properties of compounds 1–7 in ethanolic potassium phosphate buffer (pH 7.4).

Compound	APNO λ_{abs} (nm)	tAPNO λ_{abs} (nm)
1	763	800
2	758	808
3	744	804
4	746	808
5	744	806
6	727	802
7 (APNO-780)	727	807

6. Repeat steps 1, 2, and 5 three times to obtain spectra in triplicate.
7. Find the maximum wavelength using the software tool or by exporting the data into Microsoft Excel then sorting the data from largest to smallest.

2.3.4 Procedure for determining extinction coefficients

1. Pipet 1000 μL of 20 mM ethanolic potassium phosphate buffer into a 1.5 mL cuvette.
2. Insert the cuvette into the UV–Vis spectrophotometer.
3. Perform a baseline correction.
4. Prepare known concentrations of the probes in DMSO.
5. Pipet 1 μL of the probe stock solution into the cuvette. Mix the solution thoroughly by pipetting up and down 5–10 times. Then, perform a scan from 250 to 900 nm.
6. Continue to perform step 5 until at least six scans are performed. Note: the absorbance should range from 0.05 to 1.5 to obtain an appropriate linear change in absorbance.
7. Perform steps 1, 2, 5, and 6 three times to obtain experimental triplicates.
8. Use the Beer-Lambert law to calculate the extinction coefficient for each probe.

2.3.5 Procedure for determining relative rate constants for Hammett and Brønsted plots

1. Pipet 995 μL of 50 mM HEPES containing 0.01% CrEL buffer into a 1.5 mL cuvette.
2. Insert the cuvette into the UV–Vis spectrophotometer.
3. Perform a baseline correction.
4. Prepare a 2 mM probe stock solution of the probes in DMSO.
5. Pipet 1 μL of the probe stock solution into the cuvette. Mix the solution thoroughly by pipetting up and down 5–10 times.
6. Perform a scan from 250 to 900 nm.
7. Prepare a 50 mM stock solution of MAHMA-NONOate in degassed 10 mM KOH. Keep the MAHMA-NONOate stock solution on ice.
8. Pipet 4 μL of the MAHMA-NONOate stock solution into the cuvette at 25 °C.
9. Perform a scan from 250 to 900 nm every 30 s for 10 min and a final endpoint scan at 30 min to obtain the absorbance maximum of the fully turned over probe.
10. Repeat steps 5, 6, 8, and 9 two more times for each compound to obtain spectra in triplicate.

11. Plot the change in the absorbance of the turned over probe at its absorbance maximum over time to obtain the relative rates of reaction for each probe.
12. For the Hammett plot, generate a linear regression of the logarithmic value of relative rate for the compound of interest normalized to the non-substituted aniline (k_x/k_H) derivative vs σ_p^+. The Brønsted plot can be generated by plotting the logarithmic value of the relative rate vs the pK_a of the anilinium acid.

3. Selection of nitric oxide responsive trigger

The *p*-methoxyaniline trigger was selected as the optimal probe for NO sensing. It exhibited the highest relative rate constant and could fully turnover with as low as 1 μM of MAHMA-NONOate in 20 mM ethanolic potassium phosphate buffer (pH 7.4). Although *p*-ethoxyaniline had a nearly identical response, the *p*-methoxy is a smaller group which will not add unnecessary bulk to the NIR-II versions.

4. Design of phase two probes

After identifying compound 7 (also known as APNO-780) as the most sensitive toward NO, we designed five NIR-II probes derived from IR-26, IR-1061, IR-1048, Et-1080, and Flav7 (Fig. 2). IR-26, IR-1061, and IR-1048 are commercially available dyes that have previously been used for NIR-II optical imaging. In addition, Et-1080 (Feng et al., 2019) and Flav7 (Cosco et al., 2017) have successfully been employed for *in vivo* imaging. Here, we sought to identify the optimal NIR-II dye platform for NO sensing based on the following criteria: extinction coefficient, absorbance shift, NO response, and biocompatibility.

5. Synthesis of phase two probes

Procedures to synthesize phase two probes require an understanding of standard organic chemistry techniques and access to standard laboratory equipment.

5.1 Compound 8

1. Weigh out 40 mg of IR-26 (0.053 mmol, 1 equiv.) and add to a 20 mL scintillation vial.
2. Dissolve IR-26 by adding 1 mL of DMF.

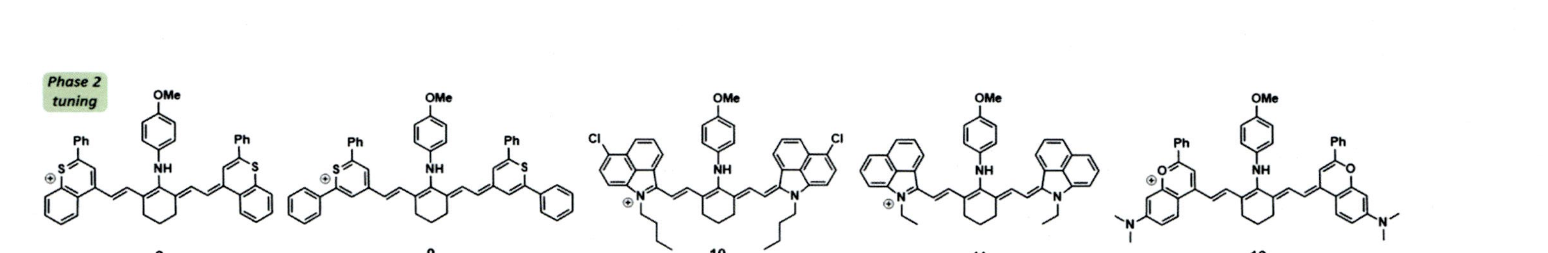

Fig. 2 Panel of NO-responsive probes for NIR-II PA imaging. *Reprinted (adapted) with permission from Lucero, M. Y., East, A. K., Reinhardt, C. J., Sedgwick, A. C., Su, S., Lee, M. C. & Chan, J. (2021). Development of NIR-II photoacoustic probes tailored for deep-tissue sensing of nitric oxide.* Journal of the American Chemical Society, *143, 7196–7202. Copyright (2021) American Chemical Society.*

3. Weigh out 14 mg of *p*-anisidine (0.11 mmol, 2 equiv.) and add to the reaction mixture.
4. Stir the reaction mixture at room temperature for 0.5 h.
5. Remove DMF via rotary evaporation.
6. Dissolve the resulting residue in CH_2Cl_2 (25 mL).
7. Wash the organic solution with an aqueous potassium iodide solution (saturated, 3 × 25 mL).
8. Dry the organic layer over sodium sulfate and concentrate by rotary evaporation.
9. Perform a trituration using cold acetone to obtain a shiny red solid (15 mg, 0.021 mmol, 38%).

5.2 Compound 9

1. Weigh out 40 mg of IR-1061 (0.053 mmol, 1 equiv.) and add to a 20 mL scintillation vial.
2. Dissolve IR-1061 by adding 1 mL of DMF.
3. Weigh out 13 mg of *p*-anisidine (0.11 mmol, 2 equiv.) and add to the reaction mixture.
4. Stir the reaction mixture at room temperature for 1 h.
5. Remove DMF via rotary evaporation.
6. Dissolve the resulting residue in CH_2Cl_2 (25 mL).
7. Wash the organic solution with an aqueous potassium iodide solution (saturated, 3 × 25 mL).
8. Dry the organic layer over sodium sulfate and concentrate by rotary evaporation.
9. Purify the residue via silica gel chromatography (5% MeOH/CH_2Cl_2).
10. Perform a trituration using diethyl ether to obtain a shiny red solid (22 mg, 0.029 mmol, 55%).

5.3 Compound 10 (APNO-1080)

1. Weigh out 80 mg of IR-1048 (0.11 mmol, 1 equiv.) and add to a 20 mL scintillation vial.
2. Dissolve IR-1048 by adding 1 mL of DMF.
3. Weigh out 27 mg of *p*-anisidine (0.10 mmol, 0.9 equiv.) and add to the reaction mixture.
4. Stir the reaction mixture at room temperature for 4 h.
5. Dilute the reaction mixture with CH_2Cl_2 (50 mL).
6. Pour the reaction mixture into a separatory funnel.

7. Wash the organic layer with saturated NaCl (50 mL).
8. Wash the organic layer further with saturated NaCl (6 × 25 mL).
9. Dry the organic layer over sodium sulfate and concentrate by rotary evaporation.
10. Purify the residue via silica gel chromatography (5% MeOH/CH_2Cl_2).
11. Perform a trituration using diethyl ether to obtain a dark solid (21 mg, 0.029 mmol, 25%).

5.4 Compound 11

1. Weigh out 50 mg of Et-1080 (0.08 mmol, 1 equiv.) and add to a 20 mL scintillation vial.
2. Dissolve Et-1080 by adding 2 mL of DMF.
3. Weigh out 100 mg of *p*-anisidine (0.81 mmol, 10 equiv.) and add to the reaction mixture.
4. Stir the reaction mixture at 40 °C for 24 h.
5. Remove DMF via rotary evaporation.
6. Dissolve the resulting residue in CH_2Cl_2 (25 mL).
7. Wash the organic solution with an aqueous potassium iodide solution (saturated, 3 × 25 mL).
8. Dry the organic layer over sodium sulfate and concentrate by rotary evaporation.
9. Purify the residue via silica gel chromatography (5% MeOH/CH_2Cl_2).
10. Perform a trituration using diethyl ether to obtain a dark solid (18 mg, 0.027 mmol, 34%).

5.5 Compound 12

1. Weigh out 50 mg of Flav7 (0.075 mmol, 1 equiv.) and add to a 20 mL scintillation vial.
2. Dissolve Flav7 by adding 2 mL of DMF.
3. Weigh out 76 mg of *p*-anisidine (0.62 mmol, 8.2 equiv.) and add to the reaction mixture.
4. Stir the reaction mixture at 40 °C for 24 h.
5. Partition the reaction mixture with CH_2Cl_2 (50 mL) and saturated NaCl (50 mL).
6. Pour the reaction mixture into a separatory funnel and isolate the organic layer.
7. Wash the organic layer with a saturated solution of NaCl (6 × 25 mL).

8. Dry the organic layer over sodium sulfate and concentrate by rotary evaporation.
9. Purify the residue via silica gel chromatography (5% MeOH/CH_2Cl_2).
10. Perform a trituration using diethyl ether to obtain a dark red solid (10 mg, 0.013 mmol, 18%).

6. Photophysical characterization of phase two probes

The phase two probes were characterized via UV–Vis spectroscopy as a proxy for PAI. We were primarily interested in whether the NIR-II versions would maintain the same reactivity as APNO-780. In addition, we needed a NIR-II PA probe that had a largest absorbance shift for optimal resolution. To achieve this, the NIR-II probe must be reasonably soluble under physiological conditions.

Each probe was evaluated *in vitro* to obtain their absorbance shift and NO response.

6.1 Equipment

1—UV–Vis spectrophotometer
1—Glass cuvette with corresponding cuvette cap
4—Micropipettes and tips (P2.5, P10, P200, P1000)

6.2 Reagents

3 aliquots each of compounds 8, 9, 10, 11, 12
Degassed 10 mM KOH solution
MAHMA-NONOate
Organic solvents: DMSO
Aqueous buffers: 20 mM ethanolic potassium phosphate buffer (pH 7.4)

6.3 Procedures

1. Pipet 989 μL of 20 mM ethanolic potassium phosphate buffer into a 1.5 mL cuvette.
2. Insert the cuvette into the UV–Vis spectrophotometer.
3. Perform a baseline correction.
4. Prepare a 5 mM probe stock solution in DMSO.
5. Pipet 1 μL of the probe stock solution into the cuvette and perform a scan from 850 to 1150 nm.
6. Prepare a 50 mM stock solution of MAHMA-NONOate in degassed 10 mM KOH. Keep the MAHMA-NONOate stock solution on ice.

Table 3 Summary of photophysical properties of compounds 8–12 in ethanolic potassium phosphate buffer (pH 7.4).

Compound	APNO λ_{abs} (nm)	tAPNO λ_{abs} (nm)
8	Insoluble	Insoluble
9	816	1078
10 (APNO-1080)	874	1080
11	878	1043
12	790	No turnover

7. Pipet 10 μL of the MAHMA-NONOate stock solution into the cuvette and perform a scan from 850 to 1150 nm.
8. Repeat steps 1–3, 5, and 7 two more times to obtain spectra in triplicate.
9. Find the maximum wavelengths using the software tool or by exporting the data into Microsoft Excel then sorting the data from largest to smallest (Table 3).

7. Selection of APNO-1080 and *in vitro* characterization

Based on the data in Table 3., we determined that compound 10 (APNO-1080) was the ideal probe for *in vivo* imaging of NO, due to its relatively good solubility, large NIR-II absorbance shift, and rapid reactivity. Furthermore, we investigated its biocompatibility and deep tissue PA imaging capability.

7.1 Biocompatibility evaluation of APNO-1080

To determine whether APNO-1080 would be suitable for *in vivo* imaging we evaluated its cytotoxicity via an MTT assay and potential off-target reactivity with a selection of biologically relevant analytes (Fig. 3).

7.1.1 Equipment

1—Spectra Max M2 plate reader
1—24-well plate
1—96-well plate
1—Cell culture incubator
1—Biosafety cabinet
4—Micropipettes and tips (P2.5, P10, P200, P1000)

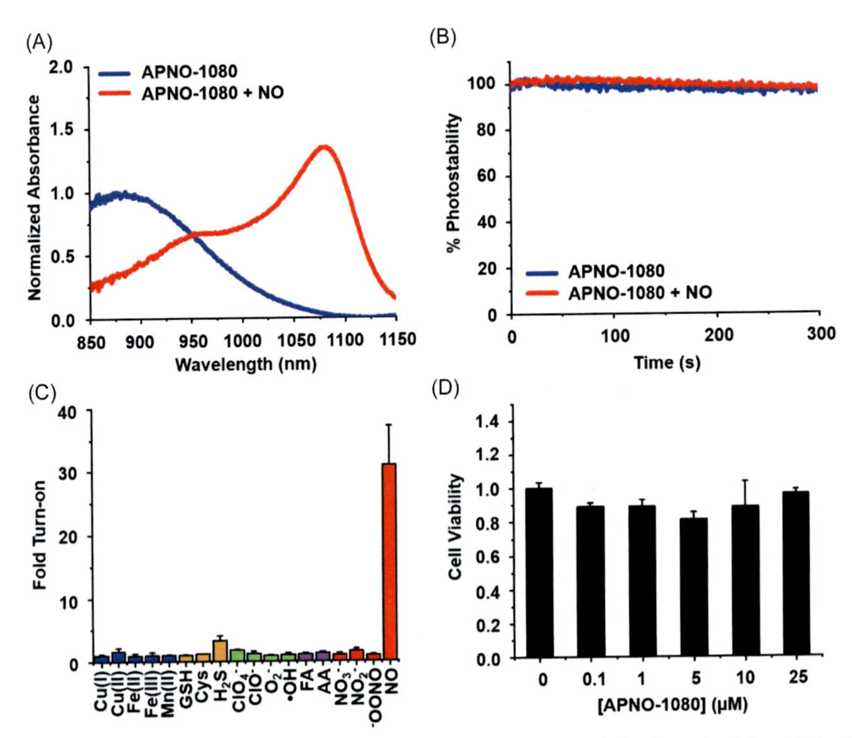

Fig. 3 (A) Normalized absorbance spectra of APNO-1080 with (red) and without (blue) NO treatment. (B) Photostability assay in which APNO-1080 or the *N*-nitrosated product (50 μM) is continuously irradiated with a pulsed laser at their respective λ_{abs} for 300 s. (C) Reaction of APNO-1080 (5 μM) with biologically relevant metal ions (1 mM), GSH (1 mM), Cys (500 μM), H_2S (100 μM), reactive oxygen species (1 mM), reactive carbonyl species (1 mM), reactive nitrogen species (1 mM), $ONOO^-$ (50 μM), and NO (100 μM) after 1 h incubation. (D) MTT viability assay in A549 cells after 24 h incubation at 37 °C. *Reprinted (adapted) with permission from Lucero, M. Y., East, A. K., Reinhardt, C. J., Sedgwick, A. C., Su, S., Lee, M. C. & Chan, J. (2021). Development of NIR-II photoacoustic probes tailored for deep-tissue sensing of nitric oxide.* Journal of the American Chemical Society, *143, 7196–7202. Copyright (2021) American Chemical Society.*

7.1.2 Reagents

1 aliquot of compound 10 (APNO-1080)

Trypsin

MTT regent, 3-(4,5-dimethylthiazol-2-yl)-2,5-diphenyl tetrazolium bromide

A549 cells

Organic solvents: DMSO

Phosphate buffer solution (PBS)

Serum-free Dulbecco's Modified Eagle Media
Ham's F-12K media containing 10% fetal bovine serum

7.1.3 Procedure for MTT viability assay

1. Trypsinize cells from a T-75 cell culture flask.
2. Resuspend cells with Ham's F-12K media (10% FBS) in a 1:10 dilution.
3. Count cells in the suspension using an automated cell counter.
4. Dilute cells to obtain a 14 mL suspension of cells with a concentration of 100,000 cells/mL.
5. Plate cells into a 24-well cell culture plate by pipetting 500 μL of the cell suspension into each well.
6. Incubate the cells for 24 h under standard conditions (5% CO_2).
7. Aspirate the media from each well.
8. Pipet 500 μL of APNO-1080 into the plate as shown in Scheme 2 with the following concentrations: 0, 0.1, 1, 5, 10, 25 μM.
9. Incubate the cells for 24 h.
10. To prepare the MTT solution, weigh out 5 mg of MTT and dissolve in 1 mL of PBS.
11. Further dilute the MTT solution to a total volume of 20 mL using PBS.
12. After the 24 h incubation, replace the media with 500 μL of the MTT solution.
13. Incubate the cells for 1 h under standard cell culture conditions.
14. Lyse the cells using 500 μL of DMSO.

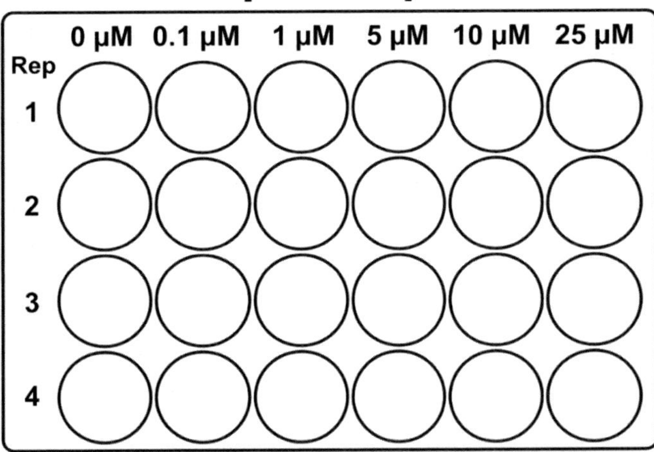

Scheme 2 MTT assay schematic in a 24-well plate.

15. Transfer the lysed cells to a 96-well plate.
16. Measure the absorbance at 555 nm of each well using a plate reader.
17. To calculate the relative cell viability, use the following equation:

$$\text{Viability} = \text{Abs}_{555} \text{ of x [APNO-1080]} / \text{Abs}_{555} \text{ of } 0\,\mu\text{M APNO-1080}$$

7.2 Comparison of APNO-1080 and APNO-780

For *in vitro* PA studies, we used tissue phantoms which mimic light scattering in tissues. Here, we describe how to prepare the tissue phantoms and perform PA imaging. We compared the PA intensity of APNO-1080 and APNO-780 after treating with NO in tissue phantoms with 0.5 and 3 cm thickness.

7.2.1 Equipment

1—Multispectral optoacoustic tomographer (MSOT inVision 128, iTheraMedical)
1—Microwave
1—Plastic mold for 1 cm diameter phantom
1—Plastic mold for 6 cm diameter phantom
8—3 mm plastic straws
6—1.5 mL Eppendorf tubes
1—Hot glue pen
4—Micropipettes and tips (P2.5, P10, P200, P1000)

7.2.2 Reagents

1 aliquot of compound 10 (APNO-1080) and compound 7 (APNO-780)
Agarose LE
2% milk
MAHMA-NONOate
Degassed 10 mM KOH solution
100 mL Deionized water
10 L Deuterated water
Organic solvents: DMSO
Aqueous buffers: 20 mM ethanolic potassium phosphate buffer (pH 7.4)

7.2.3 Procedure for tissue phantom imaging of APNO-780 and APNO-1080 via PA

1. Prepare the tissue phantom suspension by suspending 750 mg of agarose LE in 50 mL of deionized water.
2. Heat the suspension for 1 min in a microwave until the suspension becomes a viscous, translucent gel.

3. Pipet 1 mL of 2% milk into the gel and mix before it solidifies.
4. Pour the gel into a plastic mold (1 cm diameter).
5. Insert a straw into the mold as a placeholder for subsequent imaging.
6. Cool the gel at room temperature until it solidifies.
7. Repeat steps 1–6 for the second mold (6 cm diameter).
8. Carefully remove the straw and tissue phantom from its mold.
9. Store the phantoms in DI water to prevent the phantoms from drying out.
10. Next, turn on the MSOT inVision 128 and choose the D_2O setting.
11. Set the water temperature to 25 °C.
12. Pipet 999 μL of 20 mM ethanolic potassium phosphate buffer into an Eppendorf tube.
13. Prepare a 5 mM probe stock solution in DMSO.
14. Pipet 1 μL of the probe stock solution into the Eppendorf tube.
15. Prepare a 50 mM stock solution of MAHMA-NONOate in degassed 10 mM KOH. Keep the MAHMA-NONOate stock solution on ice.
16. Pipet 10 μL of the MAHMA-NONOate stock solution into the Eppendorf tube.
17. Close one end of a plastic straw using hot glue.
18. Inject approximately 200 μL of the reaction solution into a plastic straw and close the other end with hot glue.
19. Insert the straw into the tissue phantom (1 cm diameter).
20. Place the phantom into the holder of the MSOT inVision 128.
21. Irradiate the sample using the maximum absorbance wavelength of the corresponding turnover product.
22. Process the data after reconstruction by selecting equal ROIs, and quantify the mean PA intensity from each ROI, using the MSOT software.
23. Repeat steps 12–21 three times to obtain the data in triplicate.
24. Repeat steps 12–22 for the second probe and the second phantom (6 cm diameter) (Fig. 4).

8. *In vivo* characterization of APNO-1080

Once the biocompatibility evaluation and *in vitro* characterization were completed, APNO-1080 was used for *in vivo* imaging. To perform *in vivo* experiments, training in ethics and practice of handling animals and institutional approvals are required. All the *in vivo* experiments are performed upon approval from the Institutional Animal Care and Use Committee of the University of Illinois at Urbana – Champaign, following

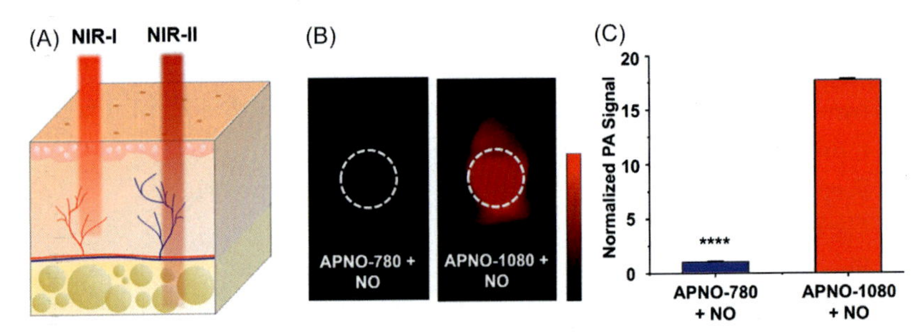

Fig. 4 (A) Schematic illustrating that NIR-II light can penetrate deeper into tissue compared to NIR-I light. (B) Representative PA images of APNO-780 and APNO-1080 (10 μM) treated with NO and overlaid with a 3 cm thick tissue imaging phantom. (C) Quantified data from (B). Error bars = SD ($n = 3$) Statistical analysis was performed using a two-tailed Student's t-test ($\alpha = 0.05$). ****: $p < 0.0001$. *Reprinted (adapted) with permission from Lucero, M. Y., East, A. K., Reinhardt, C. J., Sedgwick, A. C., Su, S., Lee, M. C. & Chan, J. (2021). Development of NIR-II photoacoustic probes tailored for deep-tissue sensing of nitric oxide.* Journal of the American Chemical Society, *143, 7196–7202. Copyright (2021) American Chemical Society.*

the principles outlined by the American Physiological Society on research animal use. For the 4T1-Luc breast cancer model study, BALB/c mice between 6 and 8 weeks old are acquired from a breeding colony. For the A549-Luc2 heterotopic lung cancer model, Nu/J mice were acquired from The Jackson Laboratory.

8.1 Application of APNO-1080 in a breast cancer model and a liver metastatic model of lung cancer

Next, we evaluated whether APNO-1080 could detect NO in an orthotopic murine model of breast cancer. The generation of a 4T1-Luc breast cancer model was based on published protocols (Yadav et al., 2021). To showcase the performance of APNO-1080, we imaged lung cancer in a heterotopic model. This model was used as a proxy for liver metastasis of lung cancer. It is important to note that PA imaging of the liver using probes is difficult to achieve due to the overwhelming background signal caused by the blood. We overcome this barrier by using NIR-II light, which also permits deeper light penetration. The heterotopic lung cancer model was established using published protocols (Lucero & Chan, 2020).

8.1.1 Equipment

1—Multispectral optoacoustic tomographer (MSOT inVision 128, iTheraMedical)

1—Vaporizer equipped with isoflurane and oxygen (for anesthesia)

21—Sterile Eppendorf tubes
21—Sterile insulin syringes
1—Electric shaver, for hair removal
1—Bottle depilatory cream, for hair removal

8.1.2 Animals

6—BALB/c mice
6—Orthotopic 4T1-Luc breast cancer models
3—Nu/J mice
6—Heterotopic A549-Luc2 lung cancer models

8.1.3 Reagents

APNO-1080
Proparacaine
Sterile saline containing 10% DMSO by volume

8.1.4 Procedure

1. Turn on the MSOT inVision 128 and choose the D_2O setting.
2. Set the water temperature to 34 °C.
3. Use isoflurane/oxygen chamber to anesthetize the mouse and to keep it anesthetized, place it under a stream of isoflurane through a nose cone for the remainder of the experiment.
4. Shave and remove residual hair from both flanks of the mouse using depilatory cream and washing with warm damp gauze. Note: Nu/J mice do not have hair and do not need to be shaved.
5. Position the animal such that the tumor is faced down in the animal holder.
6. Place the animal in the imaging chamber and allow the temperature of the mouse and water chamber to equilibrate for 10–15 min.
7. Perform a background PA scan of the tumor using the following wavelengths: 700, 730, 760, 800, 850, 874, 900, 905, 930, 1000, 1065, 1079, 1150 nm. 10 frames were recorded at every imaging wavelength. These wavelengths were selected to allow subsequent spectral unmixing.
8. Administer a drop of proparacaine to the right eye of the mouse.
9. Formulate 50 μM APNO-1080 in the saline/DMSO solution under sterile conditions and inject 100 μL retro-orbitally.
10. Acquire a PA scan of the tumor after 30 min using the same imaging conditions in steps 6–7.
11. Repeat steps 3–10 for each cancer model.

12. Process the data after reconstruction of three-dimensional images by selecting equal ROIs, and quantify the mean PA signal from each ROI, using the MSOT software.
13. Calculate the PA fold turn-on by using the ratio of the PA intensity after administering APNO-1080 and pre-injection (Fig. 5).

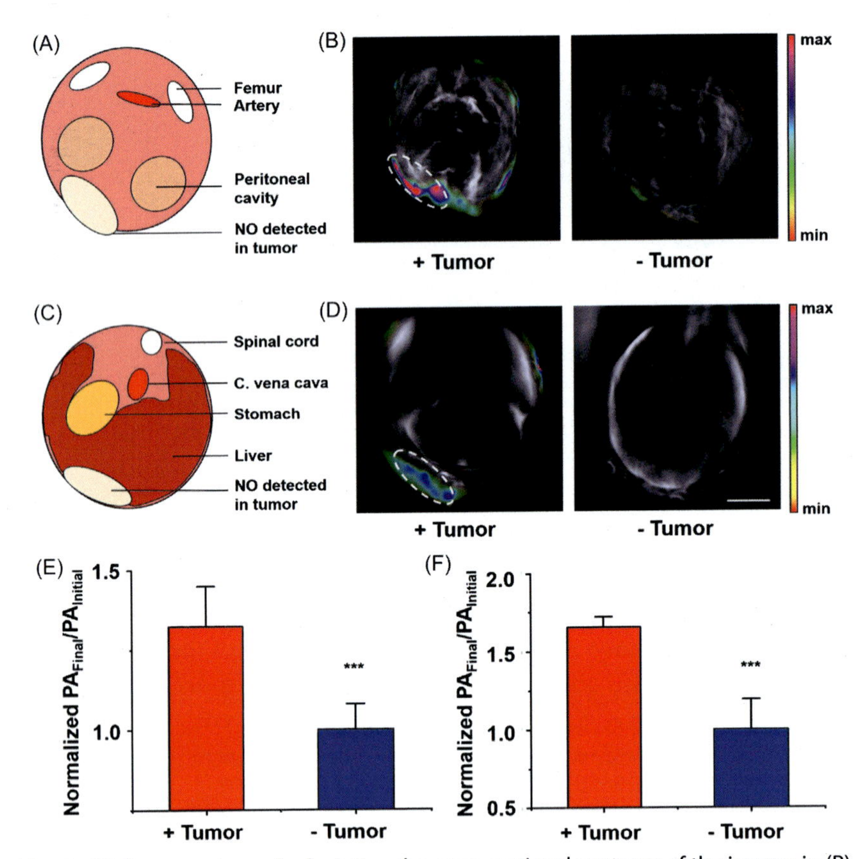

Fig. 5 (A) Cartoon schematic depicting the cross-sectional anatomy of the images in (B). (B) Representative PA images of a 4T1-Luc tumor, as well as the tumor-less control after treatment with APNO-1080 (50 μM). Scale bar = 5 mm. (C) Cartoon schematic depicting the cross-sectional anatomy of images in (D). (D) Representative PA images of a heterotopic A549-Luc2 tumor, as well as the tumor-less control after treatment with APNO-1080. Scale bar = 5 mm. (E) Normalized PA fold turn-on in 4T1-Luc tumors after treatment with APNO-1080. Error bars = SD ($n = 6$). (F) Normalized PA fold turn-on in A549 tumors ($n = 6$) and the tumor-less control ($n = 3$) after treatment with APNO-1080. Error bars = SD. Statistical analysis was performed using a two-tailed Student's t-test ($\alpha = 0.05$). ***: $p < 0.001$. *Reprinted (adapted) with permission from Lucero, M. Y., East, A. K., Reinhardt, C. J., Sedgwick, A. C., Su, S., Lee, M. C. & Chan, J. (2021). Development of NIR-II photoacoustic probes tailored for deep-tissue sensing of nitric oxide.* Journal of the American Chemical Society, *143, 7196–7202. Copyright (2021) American Chemical Society.*

9. Concluding remarks

With this work, we have introduced a two-phase tuning approach to achieve optimally sensitive PA probes for NO in the NIR-II region. This method can be applied toward the development of other NIR-II probes for *in vivo* imaging of important biological analytes. We have provided a facile method to tune the reactivity of a probe using readily accessible NIR dyes. In addition, we found the ideal NIR-II dye platform for our purpose by screening through the most commonly used NIR-II dyes. Importantly, this work showcases the enhanced sensitivity of NIR-II imaging compared to a NIR-I analog.

References

Chen, Z., et al. (2019). An optical/photoacoustic dual-modality probe: Ratiometric in/ex vivo imaging for stimulated H_2S upregulation in mice. *Journal of the American Chemical Society*, *141*, 17973–17977.

Cheng, P., et al. (2020). Fluoro-photoacoustic polymeric renal reporter for real-time dual imaging of acute kidney injury. *Advanced Materials*, *32*, 1908530.

Cosco, E. D., et al. (2017). Flavylium polymethine fluorophores for near- and shortwave infrared imaging. *Angewandte Chemie*, *129*, 13306–13309.

Feng, W., et al. (2019). Lighting up NIR-II fluorescence in vivo: An activable probe for noninvasive hydroxyl radical imaging. *Analytical Chemistry*, *91*, 1.

Ikeno, T., et al. (2019). Design and synthesis of an activatable photoacoustic probe for hypochlorous acid. *Analytical Chemistry*, *91*, 9086–9092.

Li, H., Zhang, P., Smaga, L. P., Hoffman, R. A., & Chan, J. (2015). Photoacoustic probes for Ratiometric imaging of copper(II). *Journal of the American Chemical Society*, *137*, 15628–15631.

Lucero, M. Y., & Chan, J. (2020). Towards personalized medicine: Photoacoustic imaging enables companion diagnosis and targeted treatment of lung cancer. *ChemRxiv*. Preprint https://doi.org/10.26434/chemrxiv.11888214.v2.

Reinhardt, C. J., Xu, R., & Chan, J. (2020). Nitric oxide imaging in cancer enabled by steric relaxation of a photoacoustic probe platform. *Chemical Science*, *11*, 1587–1592.

Reinhardt, C. J., Zhou, E. Y., Jorgensen, M. D., Partipilo, G., & Chan, J. (2018). A ratiometric acoustogenic probe for in vivo imaging of endogenous nitric oxide. *Journal of the American Chemical Society*, *140*, 1011–1018.

Roberts, S., et al. (2018). Calcium sensor for photoacoustic imaging. *Journal of the American Chemical Society*, *140*, 2718–2721.

Sun, A., et al. (2020). Evaluation of visible NIR-I and NIR-II light penetration for photoacoustic imaging in rat organs. *Optics Express*, *28*, 9002.

Thomas, D. D., et al. (2008). The chemical biology of nitric oxide: Implications in cellular signaling. *Free Radical Biology and Medicine*, *45*, 18–31.

Upputuri, P. K., & Pramanik, M. (2019). Photoacoustic imaging in the second near-infrared window: A review. *Journal of Biomedical Optics*, *24*, 1.

Wang, L. V., & Yao, J. (2016). A practical guide to photoacoustic tomography in the life sciences. *Nature Methods*, *13*, 627–638.

Wang, S., et al. (2019a). Activatable small-molecule photoacoustic probes that cross the blood–brain barrier for visualization of copper(II) in mice with Alzheimer's disease. *Angewandte Chemie*, *131*, 12545–12549.

Wang, Y., et al. (2019b). A photoacoustic probe for the imaging of tumor apoptosis by caspase-mediated macrocyclization and self-assembly. *Angewandte Chemie, 131*, 4940–4944.
Wu, Y., et al. (2018). Activatable probes for diagnosing and positioning liver injury and metastatic tumors by multispectral optoacoustic tomography. *Nature Communications, 9*, 3983.
Yadav, A. K., et al. (2021). NIR bioluminescence probe enables discovery of diet-induced modulation of the tumor microenvironment via nitric oxide. *ChemRxiv*. Preprint. https://doi.org/10.26434/chemrxiv.14176835.v1.
Yin, L., et al. (2019). Quantitatively visualizing tumor-related protease activity in vivo using a ratiometric photoacoustic probe. *Journal of the American Chemical Society, 141*, 3265–3273.
Zhang, J., Ning, L., Zeng, Z., & Pu, K. (2021). Development of second near-infrared photoacoustic imaging agents. *Trends in Chemistry, 3*, 305–317.
Zhang, C., et al. (2020). Design and synthesis of a ratiometric photoacoustic probe for in situ imaging of zinc ions in deep tissue in vivo. *Analytical Chemistry, 92*, 6382–6390.

CHAPTER EIGHT

Monitoring tumor growth with a novel NIR-II photoacoustic probe

Menglei Zha[a], Jen-Shyang Ni[b], Yaxi Li[a], and Kai Li[a,*]
[a]Shenzhen Key Laboratory of Smart Healthcare Engineering, Department of Biomedical Engineering, Southern University of Science and Technology (SUSTech), Shenzhen, China
[b]Department of Chemical and Materials Engineering, Photo-sensitive Material Advanced Research and Technology Center (Photo-SMART), National Kaohsiung University of Science and Technology, Kaohsiung, China
*Corresponding author: e-mail address: lik@sustech.edu.cn

Contents

Abstract

In this chapter, we designed and synthesized a series of thiadiazoloquinoxaline (TQ)-based semiconducting polymers (SPs) with a broad absorption covering from NIR-I to NIR-II regions. Theoretical calculation suggests that the BTD-TQE with ester-substituted TQ-acceptor shows a large dihedral angle and narrow adiabatic energy as well as low radiative decay, resulting in higher reorganization energy for efficient photoinduced

Methods in Enzymology, Volume 657
ISSN 0076-6879
https://doi.org/10.1016/bs.mie.2021.06.033

nonradiative decay (PNRD). As a result, the obtained BDT-TQE SP-cored nanoparticles, a NIR-II PA probe, exhibit a highest NIR-II photothermal conversion efficiency (61.6%) and achieved PA tracking of *in situ* hepatic tumor growth for 20 days. Herein, we propose a strategy to construct an effective NIR-II photoacoustic reagent through the enhanced PNRD effect of twisted intramolecular charge transfer (TICT), thereby extending the application of NIR-II PA reagents in *in vivo* bioimaging.

1. Introduction

Cancer is one of the leading causes of death in the world and the incidence rate continuously rises (Bray et al., 2018). In the past few years, millions of people died of cancer each year. Besides, in 2020, the efficiency of diagnosis and treatment of cancer are hampered by the Coronavirus Disease 2019 (COVID-19) pandemic. For example, due to the closure of medical institutions, the diagnosis and treatment are significantly delayed, which ultimately leads to an increase in the death rate of cancer patients (Nogueira, Yabroff, & Bernstein, 2020). Therefore, no one will question the importance of investing in the understanding of cancer progress, to benefit the prevention and treatment of the diseases. To address this challenge, real-time and/or long-term monitoring of tumor growth is one of the effective methods to prevent, diagnose, and treat cancer (Liu, Li, & Liu, 2015; Ni, Li, Yue, Liu, & Li, 2020). Specifically, advanced imaging techniques can be used to track the labeled tumor cells to monitor the disease progression, evaluate the anti-cancer efficacy of treatment, unveil the critical steps involved in metastasis, and provide visible targets for anti-metastasis drug development (Kim, Lemaster, Chen, Li, & Jokerst, 2017; Rosenholm et al., 2016). Thus, taking advantage of the imaging enhancement to understand cancer progression, we have better ways to achieve the final goal of beating cancer.

Recently, photoacoustic (PA) imaging has been an emerging modality among varied imaging techniques (Li & Pu, 2019; Mallidi, Luke, & Emelianov, 2011; Xu & Wang, 2006). It has immense potential for detecting ultrasonic signals with laser excitation and can act as a portable standalone modality for imaging blood vessels and other regions of interest with or without contrast agents (Lyu, Li, & Pu, 2019; Wu, Zhang, Du, Cheng, & Liang, 2018; Zhou et al., 2018). Fortunately, it retains the advantages of low light scattering and dissipation by tissues from ultrasound (US) for relatively deep tissue penetration, with a high spatial/temporal resolution inherited from optical imaging (Attia et al., 2019; Mantri & Jokerst, 2020; Zhang et al., 2017). The contrast agent is a key factor in PA imaging. However, so far, the strategic guidance for constructing high-efficient organic PA agents is still very rare. Thus, we designed and synthesized a new series of SPs, BDT-TQP, BDT-TQT, and BDT-TQE, with varied substituting groups in the acceptor units. After detailed theoretical calculation and experimental validation, we hereby offer a molecular guideline to design advanced PA contrast agents through photoinduced nonradiative decay (PNRD) enhancement by twisted intramolecular charge transfer (TICT) within skeletons of SPs. Thanks to this strategy, the yielded BDT-TQE NPs have optimal photothermal conversion and PA efficiency, which assures the application in real-time monitoring of tumor growth *in vivo* (Zha et al., 2020).

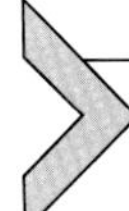

2. Before you begin

2.1 The cell line

1. 4T1 cells

   4T1 cell is a murine mammary carcinoma cell line, and the cells grow fast and are easy to form tumors *in vivo*.
2. HepG2 cells

   HepG2 cell is a human hepatoma cell line, which is prone to form tumors in the liver of BALB/c nude mice.

2.2 The buffers

1. Modified RPMI 1640 media
   - **(a)** Roswell park memorial institute (RPMI) 1640 medium (Gibco)
   - **(b)** 10% Fetal bovine serum (FBS) (Gibco)
   - **(c)** 1% 100 × penicillin-streptomycin solution (Biosharp)
2. Modified DMEM media
   - **(a)** Dulbecco's modified eagle medium (DMEM) medium (Gibco)
   - **(b)** 10% Fetal bovine serum (FBS) (Gibco)
   - **(c)** 1% 100 × penicillin-streptomycin solution (Biosharp)

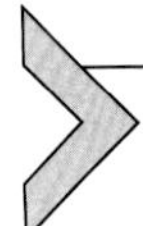

3. Key resources table

Note that not all areas will be used in every protocol.

Reagent or resource	Source	Identifier
Antibodies		
Bacterial and virus strains		
Biological samples		
Chemicals, peptides, and recombinant proteins		
$Pd_2(dba)_3$	ITC	T2184
Tri-*o*-tolylphosphine	Sigma-Aldrich	704,903
Toluene (PhMe)	Sigma-Aldrich	89,680
Methanol	J&K	949,300
Acetone	Sigma-Aldrich	90,872
Chloroform ($CHCl_3$)	Sigma-Aldrich	48,520-U
DSPE-PEG_{2000}	Nanocs	PG1-DS-2k
DSPE-PEG_{2000}-Mal	Nanocs	PG2-DSML-2k
Dichloromethane (CH_2Cl_2)	Sigma-Aldrich	02575
N,*N*-dimethylformamide (DMF)	Sigma-Aldrich	72,438
Tetrahydrofuran (THF)	Sigma-Aldrich	78,445
Cysteine-modified TAT peptides	GL Biochem (Shanghai, China)	296,229

—cont'd

Reagent or resource	Source	Identifier
Critical commercial assays		
Cell counting kit-8 (CCK-8) assay	Beyotime Biotechnology	C0038
Deposited data		
Experimental models: Cell lines		
4T1 cells	Xiehe Cell Bank of the Chinese Academy of Medical Sciences (Beijing, People's Republic of China)	
HepG2 cells		
Experimental models: Organisms/strains		
BALB/c nude mice	Beijing Vital River Laboratory Animal Technology Co., Ltd.	
Oligonucleotides		
Recombinant DNA		
Software and algorithms		
Origin		
Gaussian 16		
Other		

4. Design and synthesis of BDT-TQ

Through adjusting the structures of donor and acceptor units, the absorption of SPs can be readily regulated to NIR-I (750–950 nm) and even NIR-II (1000–1700 nm) regions, for satisfactory imaging performance in practice (Ni et al., 2018). In general, the good photoacoustic performance of SPs always requires the promoted nonradiative decay pathway upon laser irradiation. To achieve this goal, it remains an urgent demand to explore new strategies with elucidated mechanism for optimized molecular structures. To address this issue, we designed a series of SP molecules and proposed a designing principle with validation from theoretical calculations. According to the results of the calculation, the molecules of this series of SPs (see Scheme 1), BDT-TQP, BDT-TQT, and BDT-TQE, are synthesized.

4.1 Theoretical calculations of BDT-TQ

This new series of SPs consist of electron-donating unit, 4,8-bis((2-ethylhexyl)oxy)benzo[1,2-*b*:4,5-*b*′]dithiophene (BDT), and electron-deficient group, [1,2,5]thiadiazolo[3,4-*g*]quinoxaline (TQ) derivatives. To validify the influence of electron-withdrawing ability of acceptor on the final optical properties, the TQ is substituted by varied groups, including alkoxy phenyl, alkyl thienyl, and ester, to yield TQP, TQT, and TQE (Dallos, Beckmann, Brunklaus, & Baumgarten, 2011; Graham et al., 2014; Hasegawa et al., 2017; Zhang, Fu, Zhang, & Xie, 2010). The molecular geometries of simplified repeating units (s-BDT-TQP, s-BDT-TQT, and

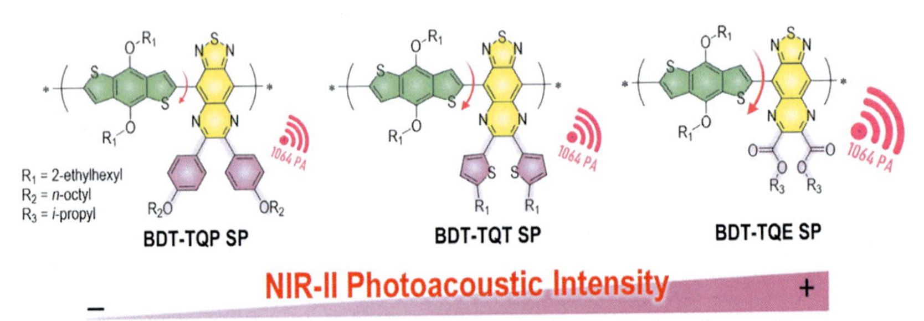

Scheme 1 Structures of semiconducting polymer-based photoacoustic agents. *Reproduced with permission from Zha, M., Lin, X., Ni, J. S., Li, Y., Zhang, Y., Zhang, X., et al. (2020). An ester-substituted semiconducting polymer with efficient nonradiative decay enhances NIR-II photoacoustic performance for monitoring of tumor growth.* Angewandte Chemie International Edition, *59(51), 23268–23276. https://doi.org/10.1002/anie.202010228.*

s-BDT-TQE) for SPs in the ground S_0 and excited S_1 states were calculated and optimized by Gaussian 16 with a time-dependent density functional theory (TD-DFT) at the level of B3LYP/6-31G*. The theoretical analysis is helpful to understand the nonradiative property of these units. The electron densities of highest occupied molecular orbital (HOMO) and lowest unoccupied molecular orbital (LUMO) for these simplified repeating units in S_0 and S_1 states are shown in Fig. 1. The results indicate that s-BDT-TQE displays more electron-hole separation in the S_1 state and larger reorganization energy between S_0 and S_1 geometries. Reorganization energy plays a vital part in the nonradiative decay pathway of excited molecule, which can promote the photoacoustic performance of organic probes. Thus, to gain insight into the nonradiative process of simplified repeating units, their vibration frequency data of the optimized S_0 and S_1 geometries were further analyzed by MOMAP software. These results indicate that the excited s-BDT-TQE nonradiatively releases energy back to the ground state through more molecular motions *via* structural reorganization than the others.

4.2 Synthesis of BDT-TQ semiconducting polymers

These SPs are synthesized *via* Stille coupling reaction, and the synthetic process is shown in Scheme 2.

4.2.1 BDT-TQP SP

1. Dissolve (4,8-bis((2-ethylhexyl)oxy)benzo[1,2-*b*:4,5-*b′*]dithiophene-2,6-diyl)bis-(trimethylstannane) (1 eq.), 4,9-dibromo-6,7-bis(4-(*n*-octyloxy)-phenyl)-[1,2,5]thiadiazolo[3,4-*g*]quinoxaline (1 eq.), $Pd_2(dba)_3$ (0.1 eq.), and tri-*o*-tolylphosphine (0.4 eq.) in absolute toluene (100 mM), and then remove the oxygen. Stir at 110 °C for 2 days.
2. Cool the mixture to room temperature, and then pour into methanol.
3. Filter and wrap the produced precipitate in filter paper.
4. Put the wrapped precipitate into a Soxhlet extractor, add 300 mL of methanol, and heat it at 110 °C for 12 h.
5. Stop heating and cool to room temperature.
6. Remove the methanol and then add 300 mL of acetone into the Soxhlet extractor, and heat at 110 °C for 12 h.
7. Stop heating and cool to room temperature.
8. Remove the acetone and then add 300 mL of chloroform into the Soxhlet extractor, and heat at 110 °C for 6 h.
9. Stop heating and cool to room temperature.
10. Dry the mixture liquid under vacuum, and obtain the deep-purple solid, 56% yield.

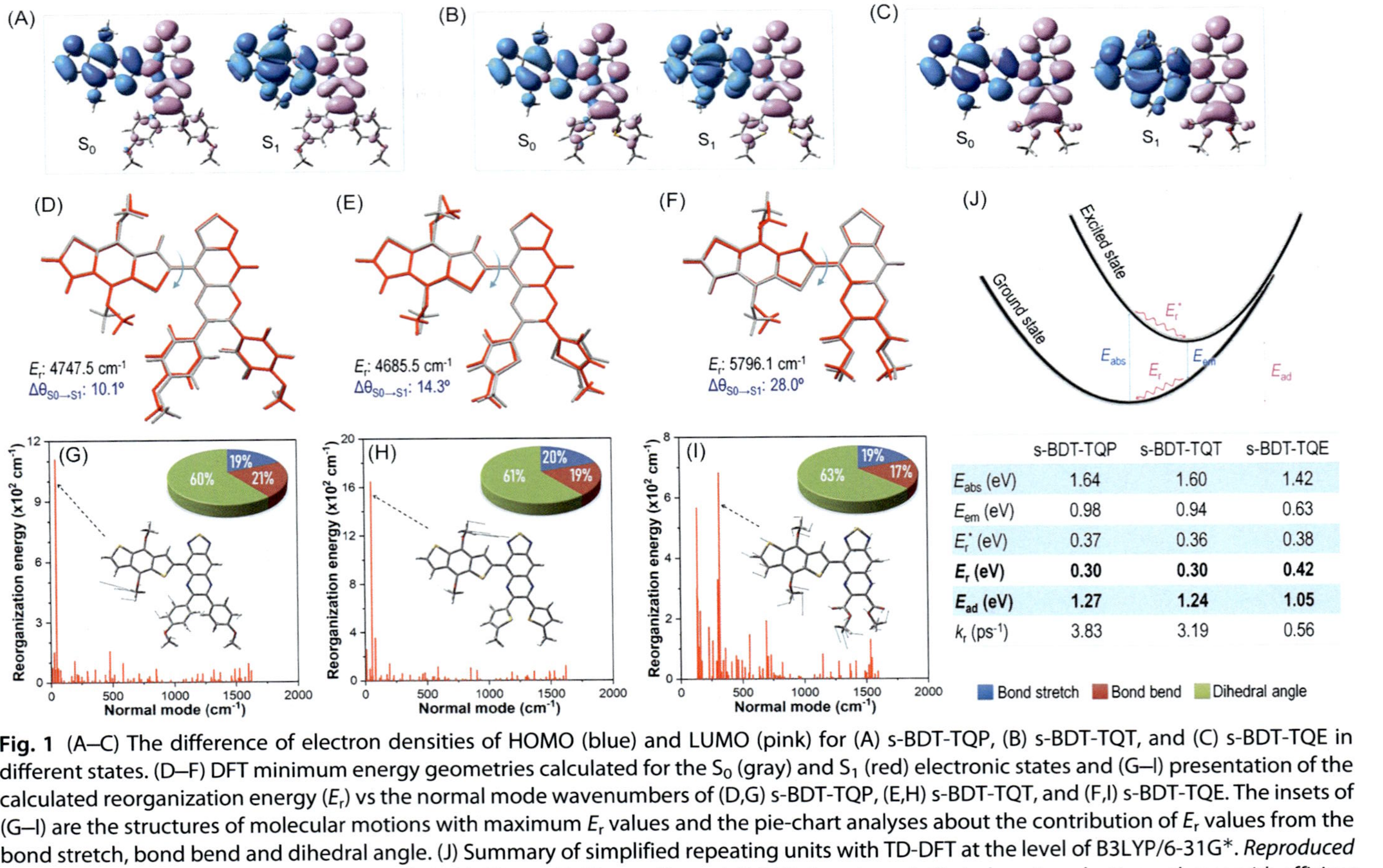

	s-BDT-TQP	s-BDT-TQT	s-BDT-TQE
E_{abs} (eV)	1.64	1.60	1.42
E_{em} (eV)	0.98	0.94	0.63
E_r^* (eV)	0.37	0.36	0.38
E_r (eV)	**0.30**	**0.30**	**0.42**
E_{ad} (eV)	**1.27**	**1.24**	**1.05**
k_r (ps^{-1})	3.83	3.19	0.56

Fig. 1 (A–C) The difference of electron densities of HOMO (blue) and LUMO (pink) for (A) s-BDT-TQP, (B) s-BDT-TQT, and (C) s-BDT-TQE in different states. (D–F) DFT minimum energy geometries calculated for the S_0 (gray) and S_1 (red) electronic states and (G–I) presentation of the calculated reorganization energy (E_r) vs the normal mode wavenumbers of (D,G) s-BDT-TQP, (E,H) s-BDT-TQT, and (F,I) s-BDT-TQE. The insets of (G–I) are the structures of molecular motions with maximum E_r values and the pie-chart analyses about the contribution of E_r values from the bond stretch, bond bend and dihedral angle. (J) Summary of simplified repeating units with TD-DFT at the level of B3LYP/6-31G*. *Reproduced with permission from Zha, M., Lin, X., Ni, J. S., Li, Y., Zhang, Y., Zhang, X., et al. (2020). An ester-substituted semiconducting polymer with efficient nonradiative decay enhances NIR-II photoacoustic performance for monitoring of tumor growth.* Angewandte Chemie International Edition, *59(51), 23268–23276. https://doi.org/10.1002/anie.202010228.*

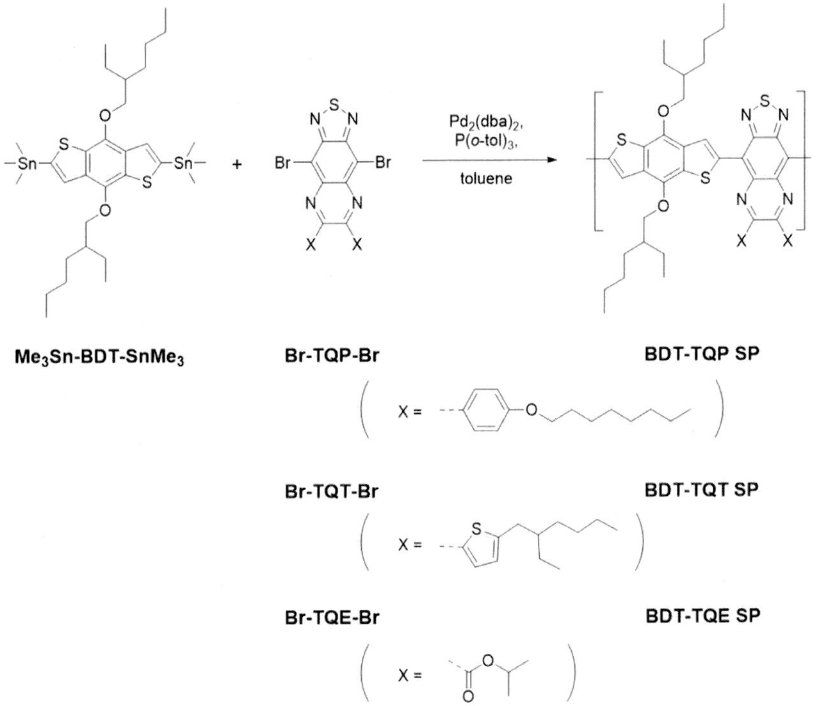

Me_3Sn-BDT-$SnMe_3$ Br-TQP-Br BDT-TQP SP

Br-TQT-Br BDT-TQT SP

Br-TQE-Br BDT-TQE SP

Scheme 2 Synthetic routes of BDT-TQ semiconducting polymers (SPs). *Reproduced with permission from Zha, M., Lin, X., Ni, J. S., Li, Y., Zhang, Y., Zhang, X., et al. (2020). An ester-substituted semiconducting polymer with efficient nonradiative decay enhances NIR-II photoacoustic performance for monitoring of tumor growth.* Angewandte Chemie International Edition, *59(51), 23268–23276. https://doi.org/10.1002/anie.202010228.*

4.2.2 BDT-TQT SP

1. Dissolve (4,8-bis((2-ethylhexyl)oxy)benzo[1,2-*b*:4,5-*b*′]dithiophene-2,6-diyl)bis-(trimethylstannane) (1 eq.), 44,9-dibromo-6,7-bis(5-(2-ehtylhexyl)thiophen-2-yl)-[1,2,5]thiadiazolo[3,4-*g*]quinoxaline (1 eq.), $Pd_2(dba)_3$ (0.1 eq.), and tri-*o*-tolylphosphine (0.4 eq.) in absolute toluene (100 mM), and then remove the oxygen. Stir at 110 °C for 2 days.
2. Cool the mixture to room temperature, and then pour into methanol.
3. Filter and then wrap the produced precipitate in filter paper.
4. Put the wrapped precipitate into a Soxhlet extractor, add 300 mL of methanol, and heat it at 110 °C for 12 h.
5. Stop heating and cool to room temperature.
6. Remove the methanol and then add 300 mL of acetone into the Soxhlet extractor, and heat at 110 °C for 12 h.
7. Stop heating and cool to room temperature.

8. Remove the acetone and then add 300 mL of chloroform into the Soxhlet extractor, and heat at 110 °C for 6 h.
9. Stop heating and cool to room temperature.
10. Dry the mixture liquid under vacuum, and obtain the deep-green solid, 58% yield.

4.2.3 BDT-TQE SP

1. Dissolve (4,8-bis((2-ethylhexyl)oxy)benzo[1,2-*b*:4,5-*b′*]dithiophene-2,6-diyl)bis-(trimethylstannane) (1 eq.), diisopropyl-4,9-dibromo [1,2,5]thiadiazolo[3,4-*g*]quinoxaline-6,7-dicar-boxylate (1 eq.), $Pd_2(dba)_3$ (0.1 eq.), and tri-*o*-tolylphosphine (0.4 eq.) are dissolved in absolute toluene (100 mM), and then remove the oxygen. Stir at 110 °C for 2 days.
2. Cool the mixture to room temperature, and then pour into methanol.
3. Filter and then wrap the produced precipitate in filter paper.
4. Put the wrapped precipitate into a Soxhlet extractor, add 300 mL of methanol, and heat it at 110 °C for 12 h.
5. Stop heating and cool to room temperature.
6. Remove the methanol and then add 300 mL of acetone into the Soxhlet extractor, and heat at 110 °C for 12 h.
7. Stop heating and cool to room temperature.
8. Remove the acetone and then add 300 mL of chloroform into the Soxhlet extractor, and heat at 110 °C for 6 h.
9. Stop heating and cool to room temperature.
10. Dry the mixture liquid under vacuum, and obtain the deep-purple solid, 66% yield.

4.3 Synthesis of BDT-TQ semiconducting polymer-based nanoparticles

The BDT-TQ semiconducting polymers (BDT-TQP, BDT-TQT, BDT-TQE) are soluble in common organic solvents, such as chloroform (CLF), tetrahydrofuran (THF), and dimethylformamide (DMF). However, good water solubility is a main precondition to facilitate their application in biological environment. To promote their stability and biocompatibility, these SPs were encapsulated with an amphiphilic polymer matrix (DSPE-PEG_{2000}) to form the water-dispersible photoacoustic nanoparticles (NPs) through nanoprecipitation.

4.3.1 BDT-TQ semiconducting polymer-based nanoparticles

4.3.1.1 Equipment

1—Microbalance

1—VCX150, Sonics

3—50 mL centrifuge tubes
3—1.5 mL microcentrifuge tubes
3—Dialysis tubing (molecular weight cut-off 10 kDa)
3—Amicon Ultra-16 Centrifugal Filters

4.3.1.2 Reagents

BDT-TQP, BDT-TQT, BDT-TQE SPs
DSPE-PEG_{2000}
MilliQ water
THF

4.3.1.3 Procedure

1. Dissolve BDT-TQP SPs (1 mg) and DSPE-PEG_{2000} (2 mg) in THF (1 mL) using an ultrasonic bath.
2. Quickly inject the liquid mixture into ultrapure water (9 mL).
3. Sonicate the mixture at 20% output for 2 min with a probe sonicator.
4. Transfer the nanoparticles solution to dialysis tubing against MilliQ water at room temperature for 24 h. Change the MilliQ water frequently.
5. Concentrate the solution to 1 mL using Amicon Ultra-16 Centrifugal Filters.
6. Store the obtained solution at 4 °C for future use.
7. Dissolve BDT-TQT SPs (1 mg) and DSPE-PEG_{2000} (2 mg) in THF (1 mL) using an ultrasonic bath, repeat steps 2–6.
8. Dissolve BDT-TQE SPs (1 mg) and DSPE-PEG_{2000} (2 mg) in THF (1 mL) using an ultrasonic bath, repeat steps 2–6.

5. Photophysical characterization of BDT-TQ *in vitro*

After synthesizing the BDT-TQ SPs and their nanoparticles, the next step is to choose the one with optimal photothermal conversion efficiency and optoacoustic performance to fabricate the cell tracer. The optical properties, photothermal performance, and photoacoustic imaging capability are the main factors to be taken into account. The biocompatibility and cell tracking performance of selected SP nanoparticles will be further studied.

5.1 Optical properties of BDT-TQ

The UV–vis-NIR absorption and photoluminescence (PL) spectra of BDT-TQP, BDT-TQT, and BDT-TQE are the main methods used to evaluate the photophysical properties. We can understand the maximum absorption wavelength, emission wavelength, and the potential of photoacoustic imaging from these spectral results. We further plotted the change of Stokes shift to Δf value for the SPs according to the Lippert-Mataga relationship, in order to learn more about the TICT feature of SPs. In order to gain insight into the photophysical property of SPs in the aggregation state, their fluorescent properties are further investigated in the mixtures of THF/water with different water fractions. These results can verify the conclusion of the previous theoretical calculation. The solutions of SP molecules and the corresponding nanoparticles are analyzed.

5.1.1 The UV–vis-NIR absorption of BDT-TQ SPs

5.1.1.1 Equipment

1—Shimadzu UV-2600 spectrometer
2—Quartz cuvettes with caps
3—1.5 mL microcentrifuge tubes

5.1.1.2 Reagents

BDT-TQP, BDT-TQT, BDT-TQE SPs
THF

5.1.1.3 Procedure

1. Warm up the spectrophotometer for 20 min before the experiment.
2. Dissolve the BDT-TQ SPs with THF, respectively.
3. Setup the experiment to collect absorbance spectra from 400 to 1400 nm, at 0.5 nm increments. Add the equal volume of THF to the quartz cuvettes and cap the cuvette to minimize solvent evaporation. Blank the spectrophotometer using the baseline function.
4. Gently remove the THF in a quartz cuvette, add equal volume of BDT-TQP SP solution and cap the cuvette to minimize solvent evaporation.
5. Collect an absorption spectrum when the peak absorbance value falls in the range of 0.1 and 1.0 au.
6. Discard the BDT-TQP SP solution and wash the cuvette with THF for several times to completely remove the BDT-TQP SP residue.
7. Add equal volume of BDT-TQT SP solution, cap the cuvette to minimize solvent evaporation and repeat step 5.
8. Discard the BDT-TQT SP solution and wash the cuvette with THF for several times to completely remove the BDT-TQT SP residue.
9. Add equal volume of BDT-TQE SP solution, cap the cuvette to minimize solvent evaporation and repeat step 5.
10. Discard the BDT-TQE SP solution and wash the cuvette with THF for several times to completely remove the BDT-TQE SP residue.

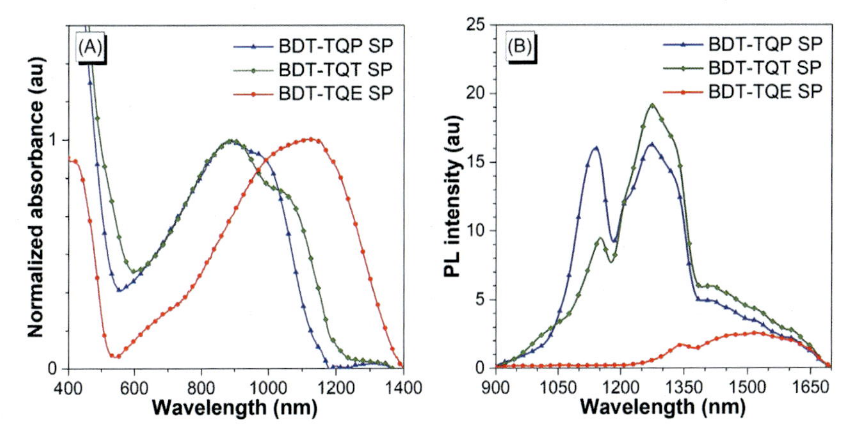

Fig. 2 The photophysical properties of BDT-TQ SPs. (A) Normalized UV–vis-NIR absorption and (B) photoluminescence spectra of SPs in THF. *Reproduced with permission from Zha, M., Lin, X., Ni, J. S., Li, Y., Zhang, Y., Zhang, X., et al. (2020). An ester-substituted semiconducting polymer with efficient nonradiative decay enhances NIR-II photoacoustic performance for monitoring of tumor growth.* Angewandte Chemie International Edition, *59(51), 23268–23276. https://doi.org/10.1002/anie.202010228.*

11. Discard the THF and wash the cuvettes with ethanol for several times, followed by washing with water repeatedly.
12. Identify the wavelengths of absorption maxima for the BDT-TQ SPs (see Fig. 2A).

5.1.2 The photoluminescence spectra of BDT-TQ SPs

5.1.2.1 Equipment

1—Hitachi F-4600 fluorescence spectrometer
1—Quartz cuvette with a cap
3—1.5 mL microcentrifuge tubes

5.1.2.2 Reagents

BDT-TQP, BDT-TQT, BDT-TQE SPs
THF

5.1.2.3 Procedure

1. Warm up the Hitachi F-4600 fluorescence spectrometer for few minutes before the experiment.
2. Dissolve the BDT-TQ SPs in THF, respectively.
3. Setup the experiment to collect photoluminescence spectra from 900 to 1700 nm.
4. Add a known volume of BDT-TQP SP solution to a quartz cuvette and cap the cuvette to minimize solvent evaporation.

5. Collect the photoluminescence spectra of solution upon excitation by the 808 nm laser.
6. Discard the BDT-TQP SP solution and wash the cuvette with THF for several times to completely remove the BDT-TQP SP residue.
7. Add equal volume of BDT-TQT SP solution to a quartz cuvette, cap the cuvette, and repeat step 5.
8. Discard the BDT-TQT SP solution and wash the cuvette with THF for several times to completely remove the BDT-TQT SP residue.
9. Add equal volume of BDT-TQE SP solution to a quartz cuvette, cap the cuvette, and repeat step 5.
10. Discard the BDT-TQE SP solution and wash the cuvette with THF for several times to completely remove the BDT-TQE SP residue.
11. Identify the wavelengths of emission maxima for the BDT-TQ SPs (see Fig. 2B).

5.1.3 The UV–vis-NIR absorption of BDT-TQ NPs

5.1.3.1 Equipment

1—Shimadzu UV-2600 spectrometer
2—Quartz cuvettes
3—1.5 mL microcentrifuge tubes

5.1.3.2 Reagents

BDT-TQP, BDT-TQT, BDT-TQE NPs (1 mg/mL)
MilliQ water

5.1.3.3 Procedure

1. Warm up the spectrophotometer for 20 min before the experiment.
2. Setup the experiment to collect absorbance spectra from 400 to 1400 nm, at 0.5 nm increments.
3. Add the equal volume of water to the quartz cuvettes, and then blank the spectrophotometer using the baseline function.
4. Meanwhile, dilute the BDT-TQP, BDT-TQT, BDT-TQE NP solutions to 100 μg/mL with water.
5. Remove the water in a quartz cuvette and add equal volume of BDT-TQP NP solution.
6. Collect an absorption spectrum when the peak absorbance value falls in the range of 0.1 and 1.0 au.
7. Discard the BDT-TQP NP solution and wash the cuvette with water for several times to completely remove the BDT-TQP NPs residue.

8. Add equal volume of BDT-TQT NP solution and repeat step 6.
9. Discard the BDT-TQT NP solution and wash the cuvette with water for several times to completely remove the BDT-TQT NPs residue.
10. Add equal volume of BDT-TQE NP solution and repeat step 6.
11. Discard the BDT-TQE NP solution and wash the cuvette with water for several times to completely remove the BDT-TQE NPs residue.
12. Identify the wavelengths of absorption maxima for the BDT-TQ NPs (see Fig. 3A).

5.1.4 The photoluminescence spectra absorption of BDT-TQ NPs

5.1.4.1 Equipment

1—Hitachi F-4600 fluorescence spectrometer
1—Quartz cuvette
3—1.5 mL microcentrifuge tubes

5.1.4.2 Reagents

BDT-TQP, BDT-TQT, BDT-TQE NPs (1 mg/mL)
MilliQ water

5.1.4.3 Procedure

1. Warm up the Hitachi F-4600 fluorescence spectrometer for few minutes before the experiment.

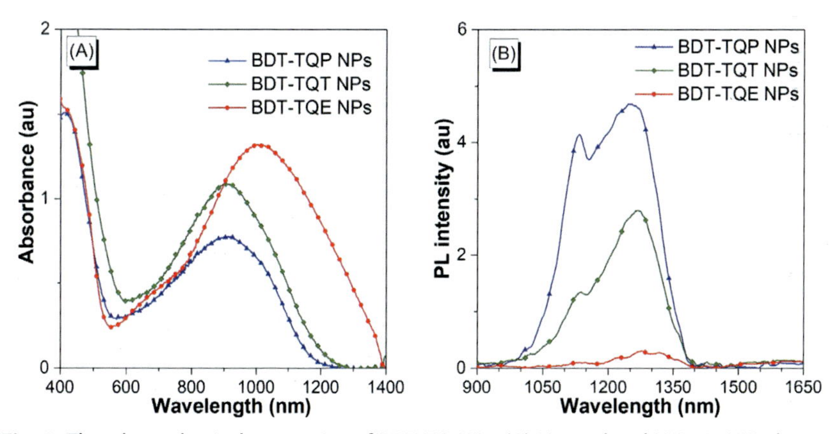

Fig. 3 The photophysical properties of BDT-TQ NPs. (A) Normalized UV–vis-NIR absorption and (B) photoluminescence spectra of NPs in water. *Reproduced with permission from Zha, M., Lin, X., Ni, J. S., Li, Y., Zhang, Y., Zhang, X., et al. (2020). An ester-substituted semiconducting polymer with efficient nonradiative decay enhances NIR-II photoacoustic performance for monitoring of tumor growth.* Angewandte Chemie International Edition, *59(51), 23268–23276. https://doi.org/10.1002/anie.202010228.*

2. Dilute the BDT-TQP, BDT-TQT, BDT-TQE NP solutions to 100 μg/mL with water.
3. Setup the experiment to collect photoluminescence spectra from 900 to 1700 nm.
4. Add a known volume of BDT-TQP NPs solution to a quartz cuvette.
5. Collect the photoluminescence spectra of solution upon excitation by the 808 nm laser.
6. Discard the BDT-TQP NP solution and wash the cuvette with water for several times to completely remove the BDT-TQP NPs residue.
7. Add equal volume of BDT-TQT NP solution to a quartz cuvette and repeat step 5.
8. Discard the BDT-TQT NP solution and wash the cuvette with water for several times to completely remove the BDT-TQT NPs residue.
9. Add equal volume of BDT-TQE NP solution to a quartz cuvette and repeat step 5.
10. Discard the BDT-TQE NP solution and wash the cuvette with water for several times to completely remove the BDT-TQE NPs residue.
11. Identify the wavelength of maximum emission for the BDT-TQ NPs (see Fig. 3B).

5.1.5 *The plot of Stokes shift against Δf value of solution for BDT-TQ SPs*

5.1.5.1 Equipment

1—Shimadzu UV-2600 spectrometer
1—Hitachi F-4600 fluorescence spectrometer
2—Quartz cuvette with a cap
30—1.5 mL microcentrifuge tubes

5.1.5.2 Reagents

BDT-TQP, BDT-TQT, BDT-TQE SPs
Toluene (PhMe)
Dichloromethane (CH_2Cl_2)
THF
N,*N*-dimethylformamide (DMF)
Chloroform ($CHCl_3$)

5.1.5.3 Procedure

1. Warm up the spectrophotometer for 20 min before the experiment.
2. Dissolve the BDT-TQ SPs with different organic solvents indicated above, respectively.

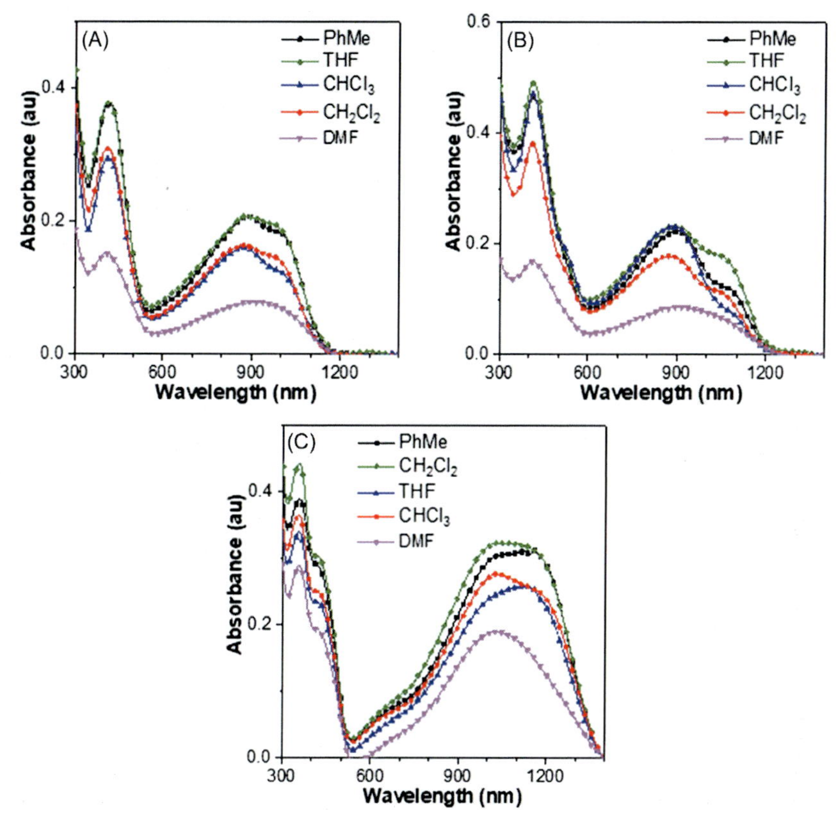

Fig. 4 Normalized UV–vis-NIR absorption of (A) BDT-TQP, (B) BDT-TQT, and (C) BDT-TQE SPs in the different solvents, toluene (PhMe), dichloromethane (CH_2Cl_2), tetrahydrofuran (THF), chloroform ($CHCl_3$), and *N,N*-dimethylformamide (DMF).

3. Repeat the steps 3–10 in the produce in "The UV-vis-NIR absorption of BDT-TQ SPs." *Note*: Wash the cuvette with the related solvents.
4. Identify the wavelengths of absorption maxima for the BDT-TQ SPs in different organic solvents indicated above (see Fig. 4).
5. Warm up the Hitachi F-4600 fluorescence spectrometer for few minutes before the experiment.
6. Meanwhile, dissolve the BDT-TQ SPs with different organic solvents indicated above, respectively.
7. Setup the experiment to collect photoluminescence spectra from 900 to 1700 nm.
8. Repeat the steps 4–10 in the produce in "The photoluminescence spectra of BDT-TQ SPs." *Note*: Wash the cuvette with the related solvents.
9. Record the photoluminescence spectra of BDT-TQ SPs in different solvents (see Fig. 5).

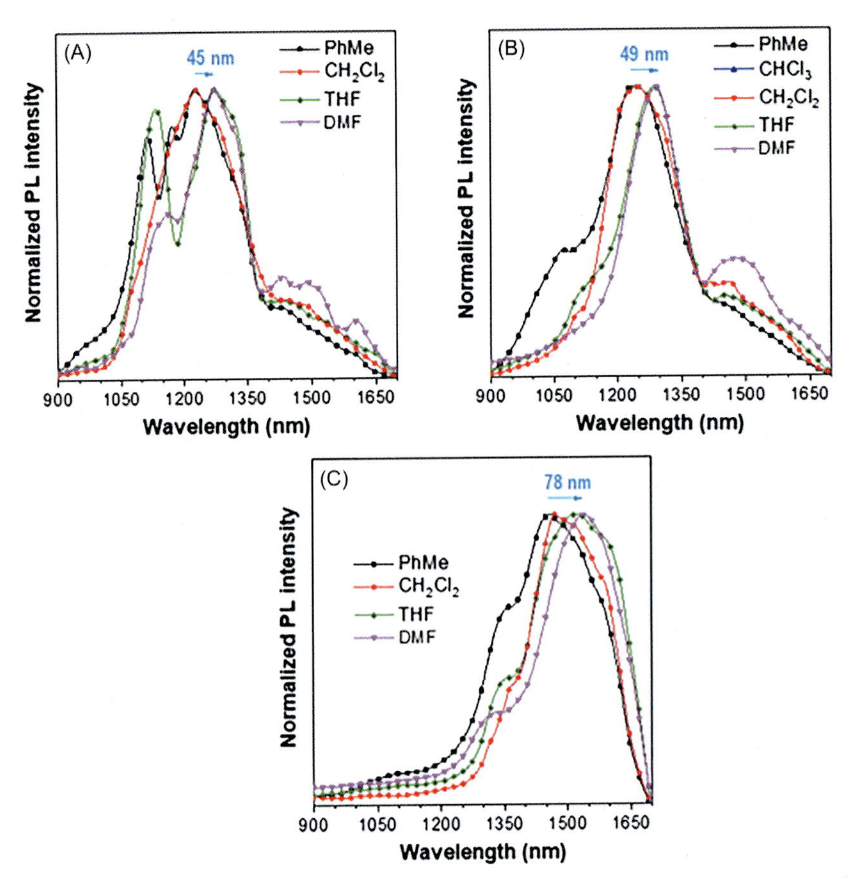

Fig. 5 Normalized photoluminescence spectra of (A) BDT-TQP, (B) BDT-TQT, and (C) BDT-TQE SPs in the different solvents, toluene (PhMe), dichloromethane (CH_2Cl_2), tetrahydrofuran (THF), and *N,N*-dimethylformamide (DMF).

Tip: Due to BDT-TQ SPs needed is very little, it is difficult to weigh the samples. BDT-TQ SP can be dissolved with THF to form a high concentration solution. Then add the solution into microcentrifuge tubes, open the lid to allow the solvent to evaporate. After the THF was completely evaporated, dissolve the BDT-TQ SP with different organic solvents, respectively.

The change of Stokes shift against Δf value of the solvent for SPs (see Fig. 6) are calculated according to the Lippert-Mataga equation:

$$\sigma_a - \sigma_f \cong \frac{2}{hc} \frac{(\mu^* - \mu)^2}{a^3} \times \Delta f + \text{constant}$$

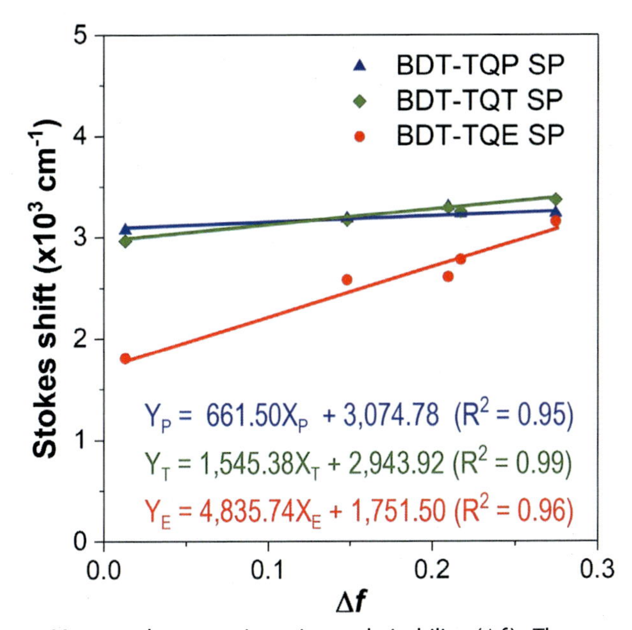

Fig. 6 Lippert-Mataga plots vs orientation polarizability (Δf). The organic solvents include toluene (PhMe), dichloromethane (CH_2Cl_2), tetrahydrofuran (THF), and *N,N*-dimethylformamide (DMF). *Reproduced with permission from Zha, M., Lin, X., Ni, J. S., Li, Y., Zhang, Y., Zhang, X., et al. (2020). An ester-substituted semiconducting polymer with efficient nonradiative decay enhances NIR-II photoacoustic performance for monitoring of tumor growth.* Angewandte Chemie International Edition, *59(51), 23268–23276. https://doi.org/10.1002/anie.202010228.*

where $\sigma_a - \sigma_f$ is the energy difference between absorption and emission maxima (i.e., Stokes shift), μ and μ^* are the dipole moments of molecule in S_0 and S_1 states, respectively, a is the cavity radius in which the material resides, h is Planck's constant, c is light speed, and Δf is the solvent orientation polarizability calculated from the following equation:

$$\Delta f = \frac{\varepsilon + 1}{2\varepsilon + 1} - \frac{n^2 - 1}{2n^2 + 1}$$

where ε and n are the dielectric constant and refractive index of solvent, respectively.

5.1.6 The photophysical property of BDT-TQ SPs in the aggregation state

We dissolve the BDT-TQ SPs in THF as a good solvent. To facilitate the formation of SP aggregates, we add the water in the THF solution because the SPs are insoluble in water. The more water added to the THF solution, the more obvious the aggregation phenomenon of SPs. The change of relative emission intensity (I/I_0) vs the water fraction (f_w) can reflect the photophysical property of BDT-TQ SPs in the aggregation state.

5.1.6.1 Equipment

1—Hitachi F-4600 fluorescence spectrometer
1—Quartz cuvette with a cap
33—1.5 mL microcentrifuge tubes

5.1.6.2 Reagents

BDT-TQP, BDT-TQT, BDT-TQE SPs
THF
MilliQ water

5.1.6.3 Procedure

1. Warm up the Hitachi F-4600 fluorescence spectrometer for few minutes before the experiment.
2. Dissolve the BDT-TQ SPs in mixtures of THF/water with different f_w, respectively. The water fractions include 0%, 10%, 20%, 30%, 40%, 50%, 60%, 70%, 80%, 90%.
3. Setup the experiment to collect photoluminescence spectra from 900 to 1700 nm.
4. Add a known volume of BDT-TQP SP with gradually increased f_w to a quartz cuvette respectively and cap the cuvette to minimize solvent evaporation. Collect the photoluminescence spectra after sequentially exciting the solutions with an 808 nm laser. *Note*: Clean and dry the quartz cuvette before changing a sample.
5. Discard the BDT-TQP SP aggregate solution and wash the cuvette to completely remove the BDT-TQP SP residue.
6. Add a known volume of BDT-TQT SP with gradually increased f_w to a quartz cuvette respectively and cap the cuvette to minimize solvent evaporation. Collect the photoluminescence spectra after sequentially exciting the solutions with an 808 nm laser.
7. Discard the BDT-TQT SP aggregate solution and wash the cuvette to completely remove the BDT-TQT SP residue.
8. Add a known volume of BDT-TQE SP with gradually increased f_w to a quartz cuvette respectively and cap the cuvette to minimize solvent

evaporation. Collect the photoluminescence spectra after sequentially exciting the solutions with an 808 nm laser.

9. Discard the BDT-TQE SP aggregate solution and wash the cuvette to completely remove the BDT-TQT SP residue.

10. Record the photoluminescence spectra of BDT-TQ SP aggregates with different water fractions (see Fig. 7).

Tip: Due to the poor solubility of BDT-TQ SP in water, it needs to be dissolved in THF to form a high concentration solution. Add different volumes

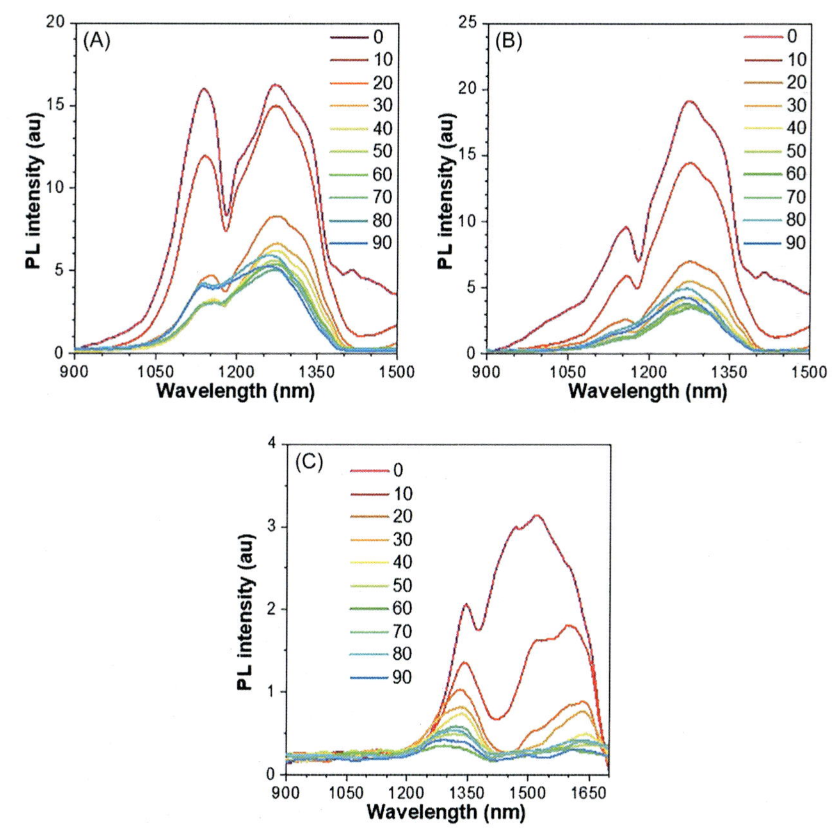

Fig. 7 Photoluminescence spectra of (A) BDT-TQP, (B) BDT-TQT, and (C) BDT-TQE SPs in THF with the different water fractions (vol%). *Reproduced with permission from Zha, M., Lin, X., Ni, J. S., Li, Y., Zhang, Y., Zhang, X., et al. (2020). An ester-substituted semiconducting polymer with efficient nonradiative decay enhances NIR-II photoacoustic performance for monitoring of tumor growth.* Angewandte Chemie International Edition, *59(51), 23268–23276. https://doi.org/10.1002/anie.202010228.*

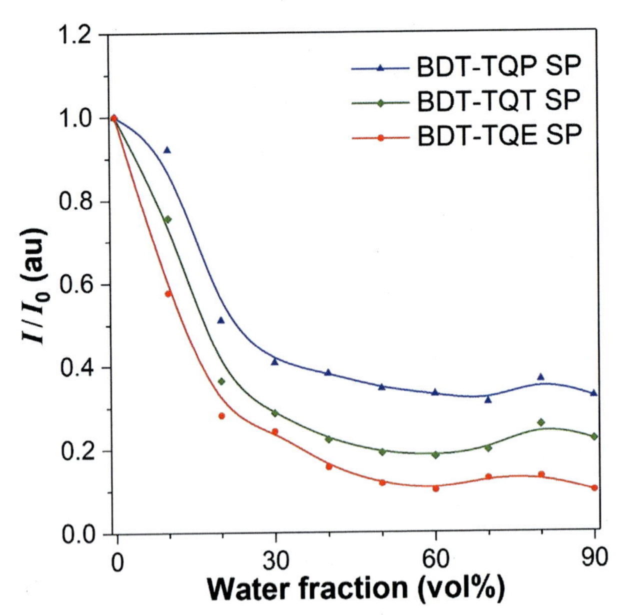

Fig. 8 The change of relative emission intensity (I/I_0) vs the water fraction (f_w) for BDT-TQ SPs. The SPs are dissolved in THF, followed by addition of water. *Reproduced with permission from Zha, M., Lin, X., Ni, J. S., Li, Y., Zhang, Y., Zhang, X., et al. (2020). An ester-substituted semiconducting polymer with efficient nonradiative decay enhances NIR-II photoacoustic performance for monitoring of tumor growth.* Angewandte Chemie International Edition, *59(51), 23268–23276. https://doi.org/10.1002/anie.202010228.*

of water into microcentrifuge tubes. Before starting the experiments, add the high concentration solution into the tubes to form the solutions with different f_W.

Record the maximum emission intensity (I) of BDT-TQ SP aggregates in different water fractions. The emission intensity of the SPs in THF solution is set as I_0, and draw change of relative emission intensity (I/I_0) vs the water fraction (see Fig. 8).

5.2 Photothermal performance of BDT-TQ nanoparticles

Among the photoluminescence spectra of all three types of NPs, the PL intensity for BDT-TQE NPs is extremely weaker due to strong TICT-induced fluorescence attenuation. This means that the laser-irradiated BDT-TQE NPs would readily convert absorbed photo-energy into non-radiative heat-energy. Herein, the photothermal performance of BDT-TQ NPs is detected to verify this hypothesis.

5.2.1 *Photothermal conversion of BDT-TQ NPs*

5.2.1.1 Equipment

1—FLIR E6 Thermal Camera
1—1060 nm laser (Changchun Laser Technology Co., Ltd)
1—Power density meter
1—96-well plate
2—Chemistry laboratory iron stands
1—Timer

5.2.1.2 Reagents

BDT-TQP, BDT-TQT, BDT-TQE NPs (1 mg/mL)
1 × PBS

5.2.1.3 Procedure

1. Fix the 1060 nm laser and FLIR E6 Thermal Camera on the holders, respectively.
2. Adjust the height of laser to ensure that the spot is as large as one well of 96-well plate.
3. Adjust the power density of spot to 1 W/cm^2 with a power density meter.
4. Dilute the BDT-TQP, BDT-TQT, BDT-TQE NP solutions to 100 μg/mL with 1 × PBS.
5. Add 200 μL of BDT-TQP NP solution into a well of the 96-well plate.
6. Turn on the thermal camera and laser. *Note*: The camera and laser should be turned on at the same time.
7. Turn off the laser after 10 min and keep the thermal camera working.
8. Turn off the thermal camera after another 10 min.
9. Add 200 μL of BDT-TQT and BDT-TQE NP solutions into other wells of the 96-well plate and repeat steps 6–8, respectively.
10. The time-dependent temperature changes for BDT-TQ NP solutions are shown in Fig. 9.

5.2.2 *Photothermal stability of BDT-TQE NPs*

5.2.2.1 Equipment

1—FLIR E6 Thermal Camera
1—1060 nm laser (Changchun Laser Technology Co., Ltd)
1—Power density meter
1—96-well plate
2—Chemistry laboratory iron stands
1—Timer

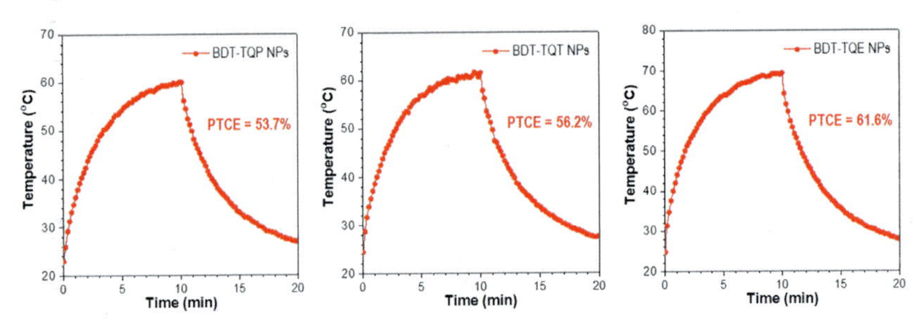

Fig. 9 The time-dependent temperature changes of BDT-TQ NPs in 1 × PBS (100 μg/mL) under the 1060 nm laser irradiation (1 W/cm^2). *Reproduced with permission from Zha, M., Lin, X., Ni, J. S., Li, Y., Zhang, Y., Zhang, X., et al. (2020). An ester-substituted semiconducting polymer with efficient nonradiative decay enhances NIR-II photoacoustic performance for monitoring of tumor growth.* Angewandte Chemie International Edition, *59(51), 23268–23276. https://doi.org/10.1002/anie.202010228.*

5.2.2.2 Reagents

BDT-TQE NPs (1 mg/mL)
1 × PBS

5.2.2.3 Procedure

1. Fix the 1060 nm laser and FLIR E6 Thermal Camera on the holders, respectively.
2. Adjust the height of laser to ensure that the spot is as large as one well of 96-well plate.
3. Adjust the power density of spot to 1 W/cm^2 with a power density meter.
4. Dilute the BDT-TQE NP solution to 100 μg/mL 1 × PBS.
5. Add 200 μL of BDT-TQE NP solution into a well of the 96-well plate.
6. Turn on the thermal camera and laser. *Note*: The camera and laser should be turned on at the same time.
7. Turn off the laser after 5 min to and turn on the laser after another 10 min.
8. Repeat the step 7 for four times and record the time-dependent temperature changes (see Fig. 10).

5.3 Photoacoustic imaging capability

The photothermal performance of BDT-TQ NPs denotes that these nanoparticles may hold great potential in PA imaging upon pulse laser irradiation. Herein, the photoacoustic imaging capability of nanoparticles is

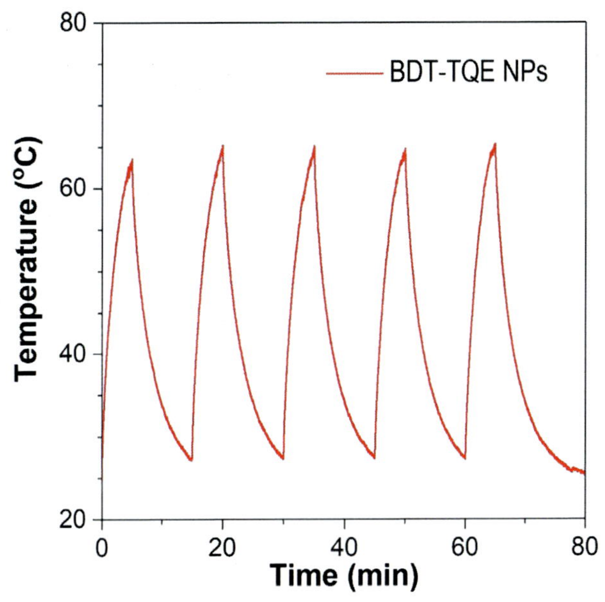

Fig. 10 The temperature changes of BDT-TQE NPs in 1 × PBS (100 μg/mL) over five laser-on/off cycles under the 1060 nm laser irradiation (1 W/cm^2). *Reproduced with permission from Zha, M., Lin, X., Ni, J. S., Li, Y., Zhang, Y., Zhang, X., et al. (2020). An ester-substituted semiconducting polymer with efficient nonradiative decay enhances NIR-II photoacoustic performance for monitoring of tumor growth.* Angewandte Chemie International Edition, *59(51), 23268–23276. https://doi.org/10.1002/anie.202010228.*

tested before they are used in biological applications. The PA signal intensity and photostability are the main factors that affect photoacoustic imaging capability of contrast agents. A custom-made acoustic-resolution photoacoustic computed tomography (PACT) system (Scheme 3) is designed for the evaluation the performance of BDT-TQ NP in PA imaging.

The PACT consists the Nd:YAG laser system, linear array transducer, and ultrasound system. The Nd:YAG laser system has an optical parametric oscillator (OPO) unit, and it can emit a 6–9 ns width laser pulse with pulse repetition rate of 20 Hz. The wavelength of the laser system is tuned between 400 and 2000 nm. To ensure that the surface of the animal model is uniformly illuminated, the fiber bundle is used to deliver the output light beam. The linear array transducer is used to capture the ultrasound (US) and PA (US/PA) data. After that, the data is digitalized *via* a research ultrasound system. The external trigger from the laser system is sent to synthesize the above data acquisition parts. Finally, the back-projection method

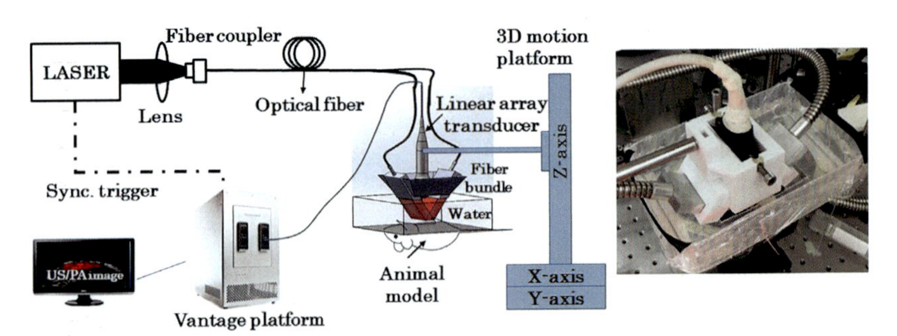

Scheme 3 A custom-made acoustic-resolution photoacoustic computed tomography (PACT) system with the dual-modality ultrasound and photoacoustic imaging. *Reproduced with permission from Zha, M., Lin, X., Ni, J. S., Li, Y., Zhang, Y., Zhang, X., et al. (2020). An ester-substituted semiconducting polymer with efficient nonradiative decay enhances NIR-II photoacoustic performance for monitoring of tumor growth.* Angewandte Chemie International Edition, *59(51), 23268–23276. https://doi.org/10.1002/anie.202010228.*

restores the coregistered US/PA images. In order to ensure safety, the laser energy should be lower than the American National Standards Institute (ANSI) safety limit. Therefore, in this experiment, the laser energy is set to $10\,mJ/cm^2$ at wavelengths of 808 and 1064 nm, which is lower than ANSI safety limit.

5.3.1 Equipment

1—Nd:YAG laser system (Quanta-Ray INDI-40-20, Spectra Physics, USA)
1—Linear array transducer (L11-4v, Verasonics, USA)
1—Research ultrasound system (Vantage 256, Verasonics, USA)
5—Capillaries

5.3.2 Reagents

BDT-TQP, BDT-TQT, BDT-TQE NPs (1 mg/mL)
$1\times$ PBS

5.3.3 Procedure

1. Dilute the BDT-TQ NP solutions to 0.25 mg/mL with $1\times$ PBS.
2. Draw the BDT-TQ NP solutions with capillaries, respectively.
3. Put the capillaries with liquid in the detection position, respectively.
4. Record the PA spectra of BDT-TQ NPs (see Fig. 11A) by using the PACT system, respectively. The laser energy set for PACT imaging is $2\,mJ/cm^2$.

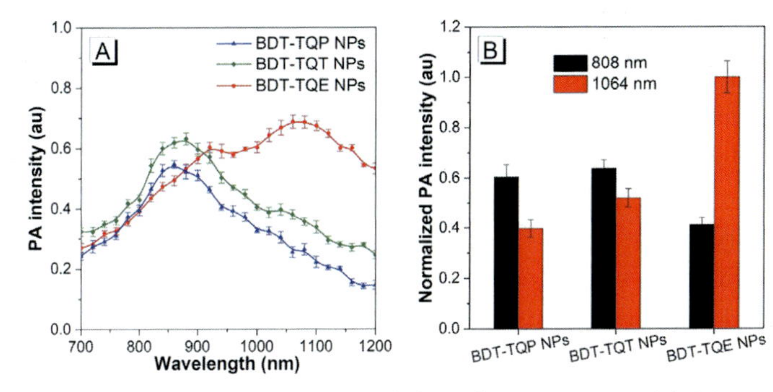

Fig. 11 The photoacoustic imaging capability of BDT-TQ NPs. (A) PA spectra and (B) their relative PA intensity at NIR-I (808 nm) or NIR-II (1064 nm) wavelengths of BDT-TQ NPs (0.25 mg/mL). *Reproduced with permission from Zha, M., Lin, X., Ni, J. S., Li, Y., Zhang, Y., Zhang, X., et al. (2020). An ester-substituted semiconducting polymer with efficient nonradiative decay enhances NIR-II photoacoustic performance for monitoring of tumor growth.* Angewandte Chemie International Edition, *59(51), 23268–23276. https://doi.org/10.1002/anie.202010228.*

5. Relative PA intensity at NIR-I (808 nm) or NIR-II (1064 nm) wavelengths of BDT-TQ NPs is shown in Fig. 11B according the recorded PA spectra.

5.4 Biocompatibility and cell tracking performance *in vitro*

Due to the strongest PA signal intensity and excellent photostability, BDT-TQE NPs are chosen for tracking cells. First of all, the excellent cell tracker must not have obvious biological toxicity. The biocompatibility of BDT-TQE NPs is first investigated through cell viability evaluation with the cell counting kit-8 (CCK-8) assay. The cell tracking performance is studied after confirming that BDT-TQE NPs have negligible influence on the cell proliferation. To improve the cell labeling efficiency for long-term tracking study, cell-penetrating peptides (Tat) are subsequently used to modify the surface of BDT-TQE NPs to afford BDT-TQE-Tat NPs. The photostability of modified BDT-TQE-Tat NPs is also detected in order to evaluate the long-term tracking capability.

5.4.1 Biocompatibility of BDT-TQE NPs

5.4.1.1 Equipment

1—Microplate reader
1—Microscope
2—96-well plates

16—1.5 mL microcentrifuge tubes
4—15 mL centrifuge tubes
1—Incubator suitable for mammalian cell culture, with 5% CO_2 atmosphere

5.4.1.2 Reagents

BDT-TQE NPs (1 mg/mL)
RPMI 1640 medium supplemented with 10% FBS and 1% penicillin/streptomycin
DMEM medium supplemented with 10% FBS and 1% penicillin/streptomycin
CCK-8 reagent
1 × PBS

5.4.1.3 Cells

4T1 cells
HepG2 cells

5.4.1.4 Procedure

1. Warm the modified RPMI 1640 and DMEM media in a 37 °C water bath, respectively.
2. Seed 4T1 cells supplemented by 100 μL of RPMI 1640 (5000 cells/well) and HepG2 cells supplemented by 100 μL of DMEM media (5000 cells/well) in 96 well plates, respectively.
3. Incubate the plates for 12 h at 37 °C, or until the cells approach 60% confluence in each well.
4. Dilute the BDT-TQE NP solution with modified DMEM or RPMI 1640 to afford sample solutions with 1, 5, 10, 20, 50, 100, 200 μg/mL of NPs. The NP concentration of modified DMEM or RPMI 1640 without BDT-TQE NPs is 0 μg/mL.
5. Remove the medium from each well, then add fresh 100 μL of BDT-TQE NP solutions at varied concentrations as above. Set six wells for each sample at every concentration.
6. Incubate the plates for another 24 h at 37 °C.
7. Dilute CCK-8 reagent to 10% with DMEM or RPMI 1640 medium.
8. Remove the medium from each well. Prepare six wells for each concentration, where add CCK-8 containing medium to three wells and pure complete medium to another three wells. *Note*: Avoid light when using CCK-8 solution.

9. Incubate the plates for 4h at 37°C.

10. Detect the absorbance of each well at 450nm with a microplate reader. The relative cell viability is calculated from the following equation to eliminate the background caused by interference from cells and NPs:

$$\text{Cell viability } (\%) = \frac{A - B}{A' - B'}$$

A: the absorbance of NPs-treated cells with CCK-8 solution at 450nm.

B: the absorbance of NPs-treated cells without CCK-8 solution at 450nm.

A′: the absorbance of untreated healthy cells with CCK-8 solution at 450nm.

B′: the absorbance of untreated healthy cells without CCK-8 solution at 450nm.

The results of cell viability are shown in Fig. 12.

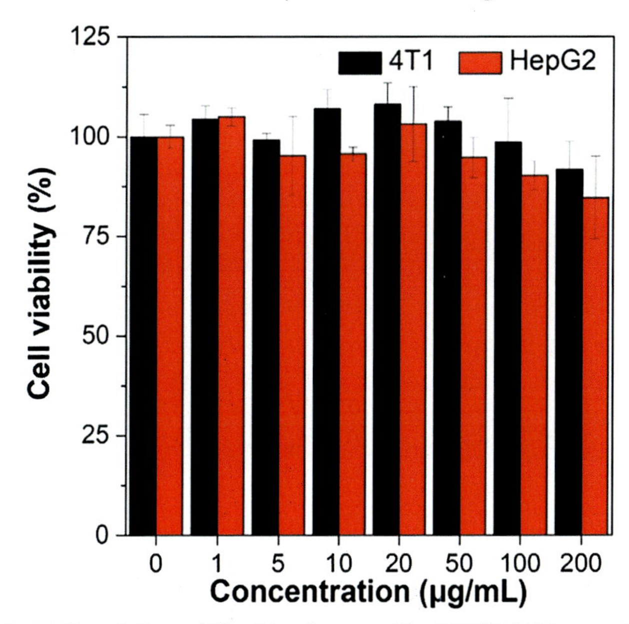

Fig. 12 Cell viability of 4T1 and HepG2 cells treated by BDT-TQE NPs at varied concentrations for 24h. *Reproduced with permission from Zha, M., Lin, X., Ni, J. S., Li, Y., Zhang, Y., Zhang, X., et al. (2020). An ester-substituted semiconducting polymer with efficient non-radiative decay enhances NIR-II photoacoustic performance for monitoring of tumor growth.* Angewandte Chemie International Edition, *59(51), 23268–23276. https://doi.org/10.1002/anie.202010228.*

5.4.2 Tat-functionalized BDT-TQE nanoparticles

5.4.2.1 Equipment

1—Microbalance
1—VCX150, Sonics
2—50 mL centrifuge tubes
1—1.5 mL microcentrifuge tube
2—Dialysis tubing (molecular weight cut-off 10 kDa)
2—Amicon Ultra-16 Centrifugal Filters

5.4.2.2 Reagents

BDT-TQE SPs
DSPE-PEG$_{2000}$
Cysteine-modified TAT peptides (cysteine located at the C-terminus, RKKRRQRRR-Cys)
10 × PBS
MilliQ water
THF

5.4.2.3 Procedure

1. Dissolve BDT-TQE (1 mg), DSPE-PEG$_{2000}$-MAL (1 mg), and DSPE-PEG$_{2000}$ (1 mg) in THF (1 mL) using an ultrasonic bath.
2. Quickly inject the liquid mixture into ultrapure water (9 mL).
3. Sonicate the mixture at 20% output for 2 min with a probe sonicator.
4. Transfer the nanoparticles solution to dialysis tubing against MilliQ water at room temperature for 24 h. Change the MilliQ water frequently.
5. Concentrate the solution to 5 mL using Amicon Ultra-16 Centrifugal Filters.
6. Add 0.56 mL of 10 × PBS to the concentrated solution.
7. Add 7.5 μL of Tat-peptide (dissolved in DMSO, 0.1 M) to the NPs in 1 × PBS. Stir at room temperature overnight.
8. Transfer the solution to dialysis tubing against MilliQ water at room temperature for 24 h. Change the MilliQ water frequently.

9. Concentrate the solution to 1 mL using Amicon Ultra-16 Centrifugal Filters.
10. Stored the obtained solution at 4 °C for future use.

5.4.3 Photostability of BDT-TQE-Tat NPs

5.4.3.1 Equipment

1—PACT system
5—Capillaries

5.4.3.2 Reagents

BDT-TQP, BDT-TQT, BDT-TQE NPs (1 mg/mL)
1 × PBS

5.4.3.3 Procedure

1. Dilute the BDT-TQE-Tat NP solution to 0.25 mg/mL with 1 × PBS.
2. Draw the BDT-TQE-Tat NP solution (0.25 mg/mL) with capillaries.
3. Put the capillary with BDT-TQE-Tat NPs in the detection position.
4. Expose the capillary under 808 or 1064 nm laser-irradiations for 8000 laser pulses (6.7 min), respectively. The laser energy set for PACT imaging is 2 mJ/cm^2 (see Fig. 13).

5.4.4 Cell tracking performance in vitro

5.4.4.1 Equipment

1—PACT system
1—Microscope
4—6-well plates
12—1.5 mL microcentrifuge tubes
24—Centrifuge tubes
14—Capillaries
1—Incubator suitable for mammalian cell culture, with 5% CO_2 atmosphere

5.4.4.2 Reagents

BDT-TQE-Tat NPs (1 mg/mL)
RPMI 1640 medium supplemented with 10% FBS and 1% penicillin/streptomycin

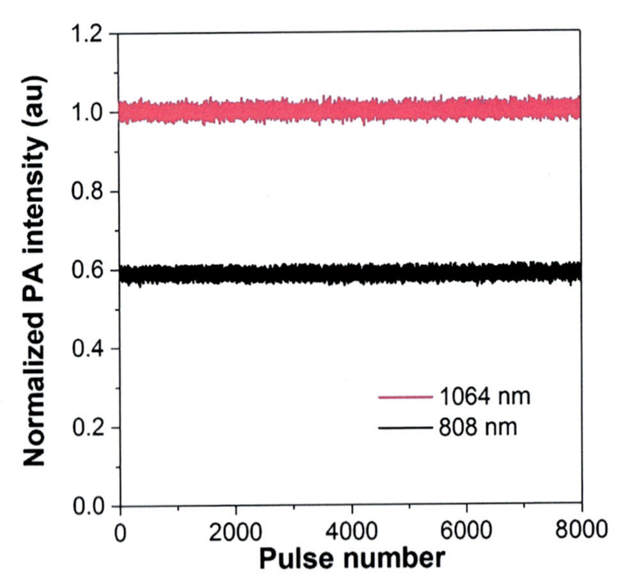

Fig. 13 The photostability of BDT-TQE-Tat NPs under 808 or 1064 nm laser-irradiations for 6.7 min. *Reproduced with permission from Zha, M., Lin, X., Ni, J. S., Li, Y., Zhang, Y., Zhang, X., et al. (2020). An ester-substituted semiconducting polymer with efficient non-radiative decay enhances NIR-II photoacoustic performance for monitoring of tumor growth.* Angewandte Chemie International Edition, *59(51), 23268–23276. https://doi.org/10.1002/anie.202010228.*

DMEM medium supplemented with 10% FBS and 1% penicillin/streptomycin
Paraformaldehyde (PFA, 4%)
Trypsin
1 × PBS
Deionized water (DIW)

5.4.4.3 Cells

4T1 cells
HepG2 cells

5.4.4.4 Procedure

1. Warm the modified RPMI 1640 and DMEM media in a 37 °C water bath, respectively.

2. Seed 4T1 cells supplemented by 2 mL of RPMI 1640 (60,000 cells/well), and HepG2 cells supplemented by 2 mL of DMEM (60,000 cells/well) in each well of the 6-well plates, respectively.
3. Incubate the plates for 12 h at 37 °C, or until the cells approach 80% confluence in each well.
4. Remove the media and wash the wells with 1 × PBS.
5. Add 1.8 mL of media and 200 μL of BDT-TQE-Tat NPs (1 mg/mL) in each well.
6. Incubate the plates for 24 h at 37 °C.
7. Remove the media and wash the wells with 1 × PBS. Add 500 μL of trypsin solution in each well.
8. Add 1 mL of relevant complete medium in each well after the cells start to detach.
9. Collect the cells in mixed liquid, respectively. Divide each cell suspension equally into two centrifuge tubes and centrifuge (1000 rpm, 5 min).
10. Discard the supernatant in centrifuge tubes. Add 400 μL of PFA in the two tubes (4T1 cells and HepG2 cells) to resuspend cells and add the suspension in two microcentrifuge tubes. The microcentrifuge tubes are stored at 4 °C.
11. Add 2 mL of relevant medium in another two tubes with live 4T1 cells and HepG2 cells to resuspend cells and add the suspension in two wells.
12. Incubate the plates for 24 h at 37 °C.
13. Repeat the steps 7–12, collect the cells after 1–5 days, respectively. Set the first batch of collected cells to day 0.
14. Centrifuge the total microcentrifuge tubes, wash and resuspend the cells in 1 × PBS buffer.
15. Add 1 × PBS to prepare the cell suspensions with the same cell density and draw the cell suspension with capillaries.
16. Put the capillary with suspension in the detection position.
17. Image the capillaries *via* the PACT system using both the 808 and 1064 nm lasers for excitation (5 mJ/cm^2).
18. Calculate the normalized PA signal intensities under 808 or 1064 nm of laser-irradiations and their relative intensities (I_{1064}/I_{808}) for 4T1 and HepG2 cells (see Fig. 14). *Note*: Do not store the labeled cells for a long time

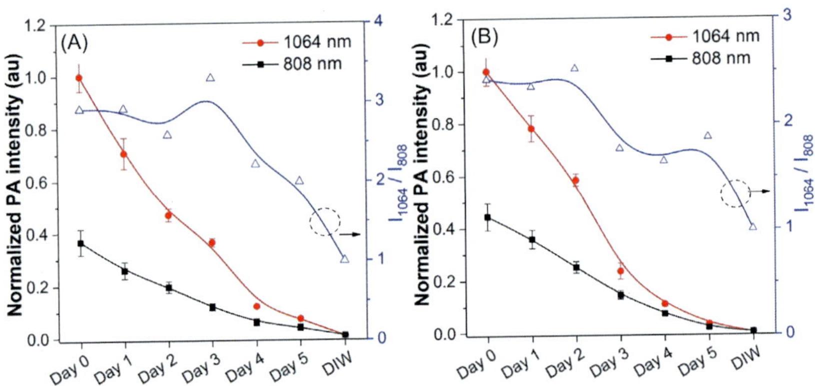

Fig. 14 Cell tracking performance of BDT-TQE-Tat NPs. Normalized PA signal intensities under 808 or 1064 nm of laser-irradiations and their relative intensities (I_{1064}/I_{808}) for (A) 4T1 and (B) HepG2 cells after incubation with BDT-TQE-Tat NPs at 37 °C for 24 h, followed by subculture for different days. *Reproduced with permission from Zha, M., Lin, X., Ni, J. S., Li, Y., Zhang, Y., Zhang, X., et al. (2020). An ester-substituted semiconducting polymer with efficient nonradiative decay enhances NIR-II photoacoustic performance for monitoring of tumor growth.* Angewandte Chemie International Edition, *59(51), 23268–23276. https://doi.org/10.1002/anie.202010228.*

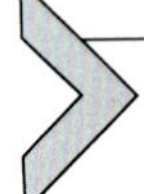

6. Characterization of BDT-TQE-Tat nanoparticles *in vivo*

The above results of cell tracking assay indicate that BDT-TQE-Tat NPs have great potential for long-term cell tracing studies with high PA intensity in the NIR-II window. Herein, BDT-TQE-Tat NPs are further explored in the use of *in vivo* biological applications as the cell tracer. The labeling and imaging of cells are important in various scenarios, holding the potential to understand more details in the biological progress and treatment of various diseases including cancer. Subsequently, the ability of BDT-TQE-Tat NPs for real-time *in vivo* monitoring of subcutaneous and *in situ* hepatic tumor growth is evaluated.

6.1 Monitoring of subcutaneous tumor growth

The reason of using subcutaneous 4T1 tumor is that the tumor model is relatively easy to establish. And the successful real-time *in vivo* monitoring of subcutaneous tumor is the precondition to applying the BDT-TQE-Tat NPs in more complicated cancer models. First of all, the 4T1 cells need to be labeled using BDT-TQE-Tat NPs. Then the labeled cells are subcutaneously injected into the back of mice. Next, the tumor growth is recorded using the PACT system serially at different times.

6.1.1 4T1 cells labeling

6.1.1.1 Equipment

1—Microscope
2—Cell culture dishes
2—1.5 mL microcentrifuge tubes
2—Centrifuge tubes
1—Incubator suitable for mammalian cell culture, with 5% CO_2 atmosphere

6.1.1.2 Reagents

BDT-TQE-Tat NPs (1 mg/mL)
RPMI 1640 medium supplemented with 10% FBS and 1% penicillin/streptomycin
1 × PBS
Trypsin

6.1.1.3 Cells

4T1 cells

6.1.1.4 Procedure

1. Culture 4T1 cells under standard conditions until the cells approach 80% confluence in culture dishes.
2. Warm the modified RPMI 1640 media in a 37 °C water bath.
3. Remove the media and wash the dish with 1 × PBS.
4. Add cell culture medium containing BDT-TQE-Tat NPs (50 μg/mL) in one dish.
5. Incubate the plates for 24 h at 37 °C.
6. Before implantation, transfer Matrigel to an ice bath to thaw. Store the Matrigel with ice box at 4 °C before mixing with the cell suspension.
7. Remove the media and wash the dish with 1 × PBS.
8. Detach the cells with Trypsin.
9. Collect the cells into two centrifuge tubes and centrifuge.
10. Wash the cells with 1 × PBS for three times and count cells.
11. Prepare the unlabeled and labeled cell suspensions with a 1:1 mixture of 1 × PBS and Matrigel, respectively.

6.1.2 PA imaging of subcutaneous tumor growth

6.1.2.1 Equipment

1—PACT system
1—Animal anesthesia machine
6—Insulin syringes

6.1.2.2 Reagents

Isoflurane

6.1.2.3 Cells and animals

4T1 cells
6—Female BALB/c nude mice, 4–6 weeks old

6.1.2.4 Procedure

1. Anesthetize the mice and inject subcutaneously 100 μL of the 4T1 suspension (2×10^5 cells) on their back using insulin syringes. Three mice are implanted with 4T1 cells labeled by BDT-TQE-Tat NPs, and the other three mice are implanted with the unlabeled 4T1 cells.
2. Put the mice in the detection position and image the mice with the PACT system under 808 or 1064 nm laser excitation with the same energy density of 10 mJ/cm^2. While recording the PA signal, the mice are under anesthesia. *Note*: Pay attention to the state of mice under anesthesia to prevent mice from dying during the test.
3. Set the post-implantation imaging to day 0, and then record the tumor growth on 2, 4, 6, 8, 10, 12, 15 day by US/PA imaging, respectively (see Fig. 15).

6.2 Monitoring of *in situ* hepatic tumor growth

The results of subcutaneous tumor growth show that the PA signals remain detectable even after 15 days, when the cells have grown into solid tumors with certain sizes. By contrast, the unlabeled subcutaneous tumors showed a negligible difference in PA signals over the observation period. Among them, the obvious and serious background noise of PA signals appear under irradiation of 808 nm laser, originating from the self-absorption of blood. Thus, compared to the unlabeled tumor, the NPs-labeled signal is enhanced by several folds. At 1064 nm excitation, the signal is enhanced up to 26.44

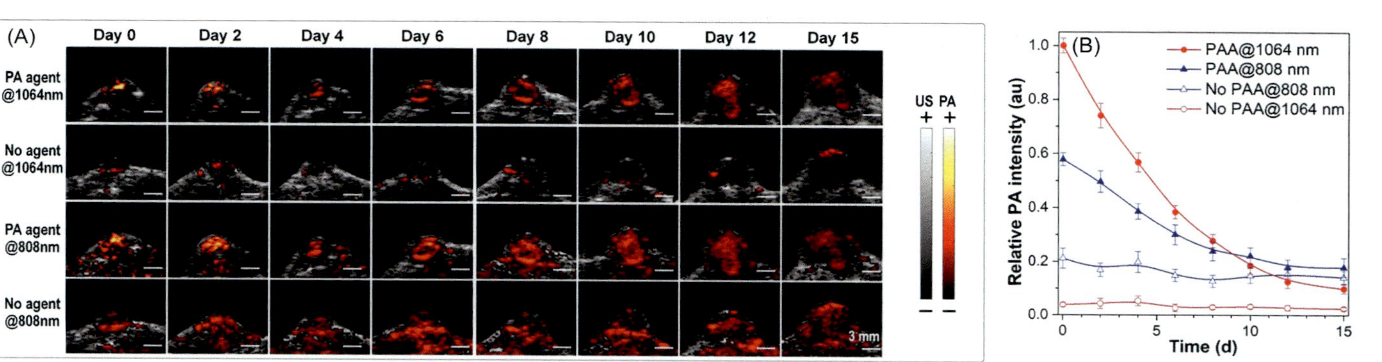

Fig. 15 Monitoring of the 4T1 subcutaneous tumor growth with time-dependent PA imaging based on PA agent (BDT-TQE-Tat NPs). (A) The merged ultrasound (US, gray scale) and PA (red-yellow scale) images of continuous *in vivo* subcutaneous tumor transplanted with 4T1 cells labeled or unlabeled by PA agents. Images are acquired under NIR-I (808 nm) or NIR-II laser (1064 nm) excitation with an identical energy density of 10 mJ/cm^2. (B) The relative changes of average PA signal intensity in the subcutaneous tumor regions. Scale bar: 3 mm. *Reproduced with permission from Zha, M., Lin, X., Ni, J. S., Li, Y., Zhang, Y., Zhang, X., et al. (2020). An ester-substituted semiconducting polymer with efficient nonradiative decay enhances NIR-II photoacoustic performance for monitoring of tumor growth.* Angewandte Chemie International Edition, *59(51), 23268–23276. https://doi.org/10.1002/anie.202010228.*

and 4.07 times for day 0 and day 15. Interestingly, the PA images suggest a $\sim$5.10 ± 1.63 mm of imaging depth with a contrast-to-noise ratio (CNR) of 22.93 ± 0.36 dB as well as the perfect contour of tumor progression in the NIR-II window. Due to the satisfactory results, we subsequently apply the BDT-TQE-Tat NPs in the monitoring *in situ* hepatic tumor growth to extend their biological applications in a more complicated biological environment.

6.2.1 HepG2 cells labeling

6.2.1.1 Equipment

1—Microscope
2—Cell culture dishes
2—1.5 mL microcentrifuge tubes
2—Centrifuge tubes
1—Incubator suitable for mammalian cell culture, with 5% CO_2 atmosphere

6.2.1.2 Reagents

BDT-TQE NPs (1 mg/mL)
DMEM medium supplemented with 10% FBS and 1% penicillin/streptomycin
1 × PBS
Trypsin

6.2.1.3 Cells

HepG2 cells

6.2.1.4 Procedure

1. Culture HepG2 cells using standard conditions until the cells approach 80% confluence in culture dishes.
2. Warm the modified DMEM media in a 37 °C water bath.
3. Remove the media and wash the dish with 1 × PBS.
4. Add cell culture medium containing BDT-TQE-Tat NPs (50 μg/mL) in one dish.
5. Incubate the plates for 24 h at 37 °C.
6. Before implantation, transfer Matrigel to an ice bath to thaw. Store the Matrigel with ice box at 4 °C before mixing with the cell suspension.
7. Remove the media and wash the dish with 1 × PBS.

8. Detach cells with Trypsin.
9. Collect the cells into two centrifuge tubes and centrifuge.
10. Wash the cells with $1 \times$ PBS for three times and count cells.
11. Prepare the unlabeled and labeled cell suspensions with a 1:1 mixture of $1 \times$ PBS and Matrigel, respectively.

6.2.2 *PA imaging of* in situ *hepatic tumor*

6.2.2.1 Equipment

1—PACT system
1—Animal anesthesia machine
6—Insulin syringes
1—Mice surgical instrument

6.2.2.2 Reagents

Isoflurane

6.2.2.3 Cells and animals

HepG2 cells
6—Female BALB/c nude mice, 4–6 weeks old

6.2.2.4 Procedure

1. Anesthetize the mice and cut the abdominal skin to expose the livers.
2. Inject 100 μL of the HepG2 suspension (1×10^6 cells) into the liver using insulin syringes. Three mice are implanted with HepG2 cells labeled by BDT-TQE-Tat NPs, and the other three mice are implanted with the unlabeled HepG2 cells.
3. Suture the abdominal skin after injection.
4. Put the mice in the detection position with supine position and image the mice with the PACT system under 808 or 1064 nm laser excitation with the same energy density of 10 mJ/cm^2. While recording the PA signal, the mice are under anesthesia. *Note*: Pay attention to the state of mice under anesthesia to prevent mice from dying during the test.
5. Set the post-implantation PA imaging to day 0, and then record the tumor growth on 2, 4, 6, 8, 10, 12, 16, 20 day by US/PA imaging, respectively (see Fig. 16).

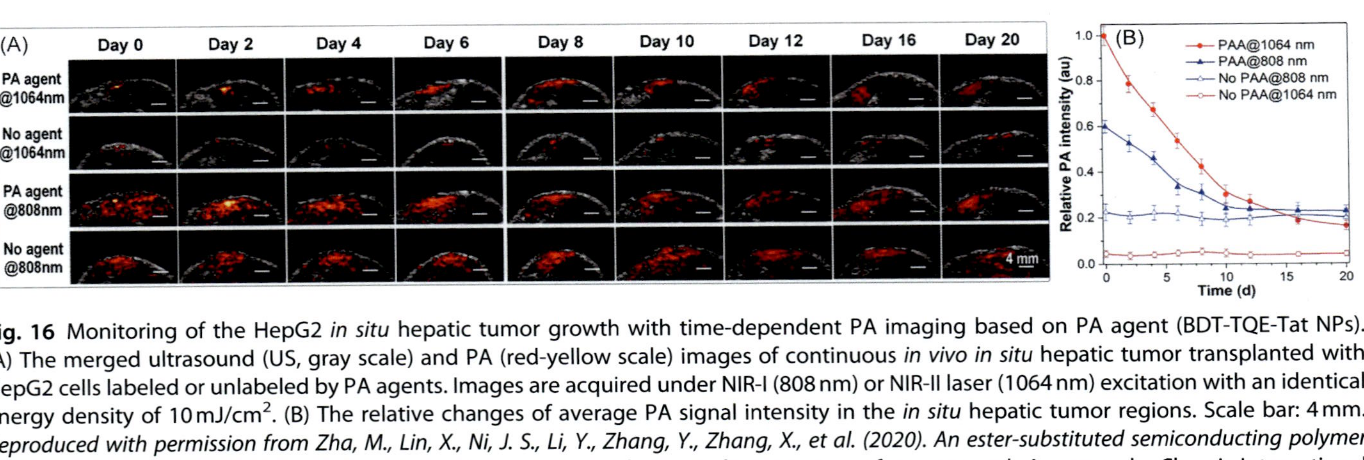

Fig. 16 Monitoring of the HepG2 *in situ* hepatic tumor growth with time-dependent PA imaging based on PA agent (BDT-TQE-Tat NPs). (A) The merged ultrasound (US, gray scale) and PA (red-yellow scale) images of continuous *in vivo in situ* hepatic tumor transplanted with HepG2 cells labeled or unlabeled by PA agents. Images are acquired under NIR-I (808 nm) or NIR-II laser (1064 nm) excitation with an identical energy density of 10 mJ/cm^2. (B) The relative changes of average PA signal intensity in the *in situ* hepatic tumor regions. Scale bar: 4 mm. *Reproduced with permission from Zha, M., Lin, X., Ni, J. S., Li, Y., Zhang, Y., Zhang, X., et al. (2020). An ester-substituted semiconducting polymer with efficient nonradiative decay enhances NIR-II photoacoustic performance for monitoring of tumor growth.* Angewandte Chemie International Edition, *59(51), 23268–23276. https://doi.org/10.1002/anie.202010228.*

7. Conclusion

In this study, we design a new SP system, simply adjusting the TICT effect in polymer chains for augmented PNRD property, to achieve the great enhancement of the photothermal conversion and photoacoustic performance. The obtained BDT-TQE-Tat NPs facilitate fast labeling of cancer cells and monitoring the growth of subcutaneous and *in situ* tumors in mice models. This study thus indicates that introducing the electron-deficient TQ-acceptor with ester-substituent (TQE) into SPs is beneficial for TICT and PNRD properties upon photoexcitation in this SP system.

Practically, the NIR-II PA signal for SPNs-labeled *in situ* subcutaneous and hepatic tumors can be enhanced up to 26.44 and 22.35 times, respectively, compared to that for unlabeled ones. The results suggest that NIR-II absorbing BDT-TQE-Tat NPs can serve as a promising long-term *in vivo* tumor photoacoustic tracking agent, showing great potentials in precisely monitoring the tumor growth with good contrast, especially in blood-rich tissues.

References

Attia, A. B. E., Balasundaram, G., Moothanchery, M., Dinish, U. S., Bi, R., Ntziachristos, V., et al. (2019). A review of clinical photoacoustic imaging: Current and future trends. *Photoacoustics*, *16*, 100144.

Bray, F., Ferlay, J., Soerjomataram, I., Siegel, R. L., Torre, L. A., & Jemal, A. (2018). Global cancer statistics 2018: GLOBOCAN estimates of incidence and mortality worldwide for 36 cancers in 185 countries. *CA: A Cancer Journal for Clinicians*, *68*(6), 394–424.

Dallos, T., Beckmann, D., Brunklaus, G., & Baumgarten, M. (2011). Thiadiazoloquinoxaline-acetylene containing polymers as semiconductors in ambipolar field effect transistors. *Journal of the American Chemical Society*, *133*(35), 13898–13901.

Graham, K. R., Cabanetos, C., Jahnke, J. P., Idso, M. N., El Labban, A., Ngongang Ndjawa, G. O., et al. (2014). Importance of the donor: Fullerene intermolecular arrangement for high-efficiency organic photovoltaics. *Journal of the American Chemical Society*, *136*(27), 9608–9618.

Hasegawa, T., Ashizawa, M., Aoyagi, K., Masunaga, H., Hikima, T., & Matsumoto, H. (2017). Thiadiazole-fused quinoxalineimide as an electron-deficient building block for n-type organic semiconductors. *Organic Letters*, *19*(12), 3275–3278.

Kim, T., Lemaster, J. E., Chen, F., Li, J., & Jokerst, J. V. (2017). Photoacoustic imaging of human mesenchymal stem cells labeled with Prussian blue–poly (l-lysine) nanocomplexes. *ACS Nano*, *11*(9), 9022–9032.

Li, J., & Pu, K. (2019). Development of organic semiconducting materials for deep-tissue optical imaging, phototherapy and photoactivation. *Chemical Society Reviews*, *48*(1), 38–71.

Liu, J., Li, K., & Liu, B. (2015). Far-red/near-infrared conjugated polymer nanoparticles for long-term in situ monitoring of liver tumor growth. *Advanced Science*, *2*(5), 1500008.

Lyu, Y., Li, J., & Pu, K. (2019). Second near-infrared absorbing agents for photoacoustic imaging and photothermal therapy. *Small Methods*, *3*(11), 1900553.

Mallidi, S., Luke, G. P., & Emelianov, S. (2011). Photoacoustic imaging in cancer detection, diagnosis, and treatment guidance. *Trends in Biotechnology*, *29*(5), 213–221.

Mantri, Y., & Jokerst, J. V. (2020). Engineering plasmonic nanoparticles for enhanced photoacoustic imaging. *ACS Nano*, *14*(8), 9408–9422.

Ni, J. S., Li, Y., Yue, W., Liu, B., & Li, K. (2020). Nanoparticle-based cell trackers for biomedical applications. *Theranostics*, *10*(4), 1923.

Ni, J. S., Liu, H., Liu, J., Jiang, M., Zhao, Z., Chen, Y., et al. (2018). The unusual aggregation-induced emission of coplanar organoboron isomers and their lipid droplet-specific applications. *Materials Chemistry Frontiers*, *2*(8), 1498–1507.

Nogueira, L. M., Yabroff, K. R., & Bernstein, A. (2020). Climate change and cancer. *CA: A Cancer Journal for Clinicians*, *70*(4), 239–244.

Rosenholm, J. M., Gulin-Sarfraz, T., Mamaeva, V., Niemi, R., Özliseli, E., Desai, D., et al. (2016). Prolonged dye release from mesoporous silica-based imaging probes facilitates long-term optical tracking of cell populations in vivo. *Small*, *12*(12), 1578–1592.

Wu, C., Zhang, R., Du, W., Cheng, L., & Liang, G. (2018). Alkaline phosphatase-triggered self-assembly of near-infrared nanoparticles for the enhanced photoacoustic imaging of tumors. *Nano Letters*, *18*(12), 7749–7754.

Xu, M., & Wang, L. V. (2006). Photoacoustic imaging in biomedicine. *Review of Scientific Instruments*, *77*(4), 305–598.

Zha, M., Lin, X., Ni, J. S., Li, Y., Zhang, Y., Zhang, X., et al. (2020). An ester-substituted semiconducting polymer with efficient nonradiative decay enhances NIR-II photoacoustic performance for monitoring of tumor growth. *Angewandte Chemie International Edition*, *59*(51), 23268–23276. https://doi.org/10.1002/anie.202010228.

Zhang, G., Fu, Y., Zhang, Q., & Xie, Z. (2010). Low bandgap EDOT-quinoxaline and EDOT-thiadiazol-quinoxaline conjugated polymers: Synthesis, redox, and photovoltaic device. *Polymer*, *51*(11), 2313–2319.

Zhang, S., Guo, W., Wei, J., Li, C., Liang, X. J., & Yin, M. (2017). Terrylenediimide-based intrinsic theranostic nanomedicines with high photothermal conversion efficiency for photoacoustic imaging-guided cancer therapy. *ACS Nano*, *11*(4), 3797–3805.

Zhou, E. Y., Knox, H. J., Reinhardt, C. J., Partipilo, G., Nilges, M. J., & Chan, J. (2018). Near-infrared photoactivatable nitric oxide donors with integrated photoacoustic monitoring. *Journal of the American Chemical Society*, *140*(37), 11686–11697.

CHAPTER NINE

A high-contrast photoacoustic agent with near-infrared emission

Weijie Chen, Fengying Ye, Jun Yin*, and Guang-Fu Yang*
Key Laboratory of Pesticide and Chemical Biology, Ministry of Education, International Joint Research Center for Intelligent Biosensing Technology and Health, College of Chemistry, Central China Normal University, Wuhan, PR China
*Corresponding authors: e-mail address: yinj@mail.ccnu.edu.cn; gfyang@mail.ccnu.edu.cn

Contents

Abstract

Benzobisthiadiazole as a typical electron acceptor, has been widely used to design fluorescent dyes and photoacoustic (PA) agents. With the strategy of constructing donor-acceptor-donor (D-A-D) type of electron characteristics, benzobisthiadiazole

Methods in Enzymology, Volume 657
ISSN 0076-6879
https://doi.org/10.1016/bs.mie.2021.06.037

derivatives tend to behave stable in near-infrared absorption and emission, which is beneficial to PA imaging. In this chapter, two molecular design strategies are combined to improve the photoacoustic imaging effects of new PA contrast agent **IR-1302 NPs**, by installing strengthened conjugated bridges and electron donors. The nanoparticles exhibit high-contrast noninvasive photoacoustic imaging in tumor models with longer wavelength absorption and emission and show potential as a clinic contrast agent.

1. Introduction

Photoacoustic imaging (PAI) is a new imaging technology that has been widely applied in the fields of chemical biology, biotechnology, and biomedical engineering (Chen, Wang, et al., 2021; Liba & de la Zerda, 2017; Ong, Zhang, Xiao, & Yao, 2021). By shining a laser on a target, the stimulated location could absorb the energy of light waves, causing its temperature to increase and volume to change. Through periodic laser irradiation, it will undergo regular thermoelastic volumetric changes, thus emitting ultrasonic waves, and the signals could propagate through the surrounding medium (Gröhl, Schellenberg, Dreher, & Maier-Hein, 2021). When applying this imaging method in living organisms, which own different levels of light absorption in diverse biological tissues, the inspected positions could reflect corresponding acoustic signals. Therefore, the light absorption distribution of different tissues can be distinguished by the feedback photoacoustic (PA) signals, achieving the purpose of observation and diagnosis in physiological researches (Moon et al., 2015). Compared with other traditional optical imaging methods, PAI combines the high spatial and temporal resolution of ultrasound with the high contrast of optical imaging. Moreover, it is less affected by light scattering, resulting in deep tissue penetration and low tissue damage, and provides volumetric information with a high signal-to-noise ratio without scanning procedures *in vivo* (Chaigne, Arnal, Vilov, Bossy, & Katz, 2017; Friedlein et al., 2020; Jiang, Du, Tang, Hsieh, & Zheng, 2019; Reza, Bell, Shi, Shapiro, & Zemp, 2018; Roberts et al., 2018).

Contrast agents are capable of improving the photoacoustic imaging effects of targeted tissues or lesion regions, thus it is significant to develop efficient contrast agents to increase the photoacoustic signals at the sites to be examined (Fathi et al., 2019; Fu, Zhu, Song, Yang, & Chen, 2018; Duan et al., 2018; Weber, Beard, & Bohndiek, 2016). Commonly-used contrast agents include endogenous photoacoustic factors such as hemoglobin and melanin (Danielli, Maslov, Favazza, Xia, & Wang, 2015; Longo et al., 2016), small molecular dyes such as indocyanine green

(Wilson et al., 2017), and various nanoparticles (NPs) (Lee et al., 2018; Mantri & Jokerst, 2020; Tian et al., 2016), etc. The endogenous photoacoustic factors are widely spread without control, which limits the practical use. Small molecular agents are to be modified with specific functions, but the disadvantages are evident in that they show low signals and short retention time. Taking these into account, nanoparticles own larger light absorption cross section and can bond with targeted ligands to generate stable photoacoustic signals *in vivo* (Kang et al., 2018).

One crucial process in PAI is the conversion from photon energy to heat energy, so the construction of novel contrast agents with obvious thermal change under laser irradiation is a key breakthrough point (Herrmann, Pech-May, & Retsch, 2021; Liu, Bhattarai, Dai, & Chen, 2019; Zhou et al., 2019). Numerous studies have found that some organic fluorescent nanoparticles that can emit fluorescence exhibit strong photothermal conversion function and give out photoacoustic signals in the meantime to realize more accurate detection by offering vast heat energy (Chen, Xia, et al., 2019; Hu, Prasad, & Huang, 2021; Ou et al., 2019; Ye et al., 2020; Yin et al., 2020). It provides a method to build new contrast agents with organic fluorescent dyes with high photothermal conversion efficiency. In clinical trials, however, this kind of PAI agent still faces problems of limited tissue penetration depth. To solve this problem, constructing fluorescent contrast agents with near-infrared (NIR) absorption and emission seems to be an effective strategy (Fu et al., 2020; He et al., 2019; Li, Jiang, et al., 2020; Liu, Shi, et al., 2019). Fluorophores located in the NIR region have obvious advantages over the visible region in reducing tissue scattering, decreasing spontaneous fluorescence and improving tissue penetration, and have been widely used in detection, labeling, diagnosis, and treatment (Chen, Li, Chen, Chen, & Yin, 2019; Chen, Zhang, et al., 2021; He, Liu, Yin, & Yoon, 2020; Ma, Zhang, et al., 2020; Xu et al., 2018, 2019). Meanwhile, the near-infrared absorption is suitable for the excitation conditions of the PAI agents to ensure the penetration ability of the excitation laser (Chen, Zhao, Yoon, Gambhir, & Emelianov, 2019; Guo et al., 2019; Jiang & Pu, 2017). Therefore, the development of near-infrared fluorescent contrast agents with strong absorption and efficient photothermal conversion is the key to improved performance of PA imaging.

Benzobisthiadiazole has been widely used in the construction of functional NIR fluorescent agents as a typical electron acceptor, which can even reach the second near-infrared window with appropriate modifications on both sides (Guo et al., 2018; Li, Cai, et al., 2020; Liu et al., 2020; Ye, Chen,

Pan, Liu, & Yin, 2019). In this chapter, we develop and describe the preparation of a novel fluorescent contrast agent **IR-1302** and its nanoparticles, which reach the NIR region and achieve remarkable PA imaging with high resolution in deep tissues and tumors.

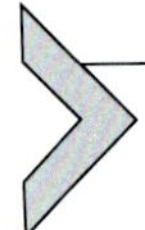

2. Preparation and properties of fluorescent agent IR-1302 NPs

2.1 Design principle of IR-1302 NPs

To access a new contrast agent with NIR absorption and emission, we selected benzobisthiadiazole as the electron-withdrawing core to construct a new donor-acceptor-donor (D-A-D) type fluorescent structure. As reported in the literature, the D-A-D electron distribution has been beneficial for fluorescent agents to reach far absorption and emission, revealing a deep relationship between molecular properties and structural donor/acceptor abilities (Ye et al., 2019). Hence, we have tried to make a summary of the effects on different electron donors and conjugated bridges installed to the modifiable side sites in benzobisthiadiazole.

Through investigating their optical properties with commonly-used conjugated bridges, we have found that the combination of conjugated groups efficiently broadens the wavelength distribution (Fig. 1). With the expansion of the electron-rich conjugated system at both ends of benzobisthiadiazole (such as phenyl and phenyl-thiophene), the molecular absorption wavelength significantly increased with maximum peak shifting from 640 to 860 nm, and the emission light was located at the NIR region (Li, Liu, et al., 2020; Lin et al., 2019). By introducing the electron-rich group fluorene and replacing the thiophene group with furan and selenophene, the

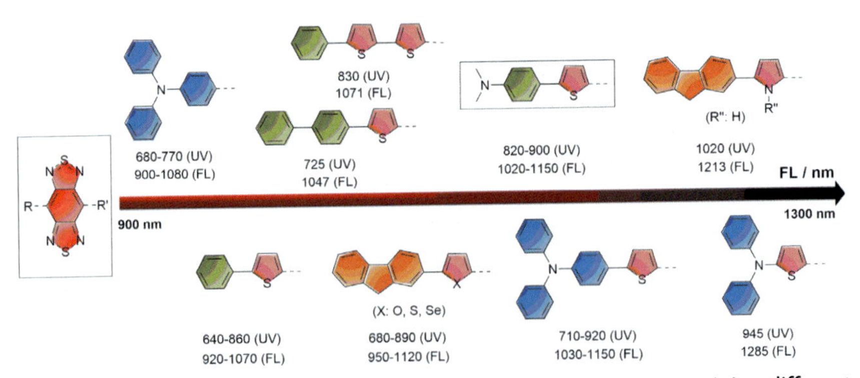

Fig. 1 The optical properties of benzobisthiadiazole-based agents containing different D-A-D systems.

emission wavelength can be further increased to 1100 nm (Ma, Liu, et al., 2020; Wang et al., 2020; Yang et al., 2018). In addition, the introduction of a typical electron-donating group, triphenylamine, showed subtle spectral changes. Since its conjugation degree was weaker than the above groups, its electron-donating ability made up the difference, rendering its optical property similar to that of the structure modified by a phenyl-thiophenyl bridge (Zhang et al., 2020). Therefore, some studies have attempted to introduce electron-donating groups and conjugated groups together to construct triarylamine-thiophen side chains, whose spectrum was further than that with either the electron-donating group or the conjugated group alone, indicating that the combination of two functional groups was conducive to the red shift of the wavelength (Liu, Zhou, et al., 2019). Moreover, it was found that the diarylamine-thiophene structure, as a side chain, enabled the molecule to have further emission, with the maximum absorption peak located at 945 nm and the fluorescence peak at 1285 nm (Qian et al., 2009). Other asymmetric derivatives, which consist of the above-mentioned phenyl-thiophene, fluorene, and triarylamine-thiophene groups on one side, and dimethylamine-phenylthiophene group with stronger donating capacity on the other side, all exhibit stable NIR absorption and emission (Qu et al., 2019). It is worth noting that when adding a pyrrole to a fluorenyl group at both ends of benzobisthiadiazole, the molecule showed a surprising far absorption that reached 1020 nm (Qian & Wang, 2010). It was derived from the formation of an intramolecular hydrogen bond between N-H in pyrrole and nitrogen atom in central ring. However, the formation of hydrogen bonds does not depend on the environment and is limited in practical applications. The above examples reveal that the strategy adopting large conjugated groups and strong electron-donating groups is expected to promote the optical characteristics of D-A-D structures. Based on this principle, the dimethylamino group with excellent electron-donating ability was selected as the end group, with bithiophene (containing a 3,4-ethylene-dioxythiophene which behaved strong conjugated property) and styrene as the conjugated bridge, we have successfully obtained **IR-1302**. To strengthen the biocompatibility and signal intensity, it was capsuled by a long-chain surfactant to form water-dispersible nanoparticles **IR-1302 NPs** with NIR absorption and emission.

2.2 Synthesis procedures

IR-1302 was synthesized in six steps from 3,4-ethyl-enedioxythiophene and 4,7-dibromobenzo[1,2-c:4,5-c′]bis([1,2,5]thiadiazole) (Fig. 2A). One

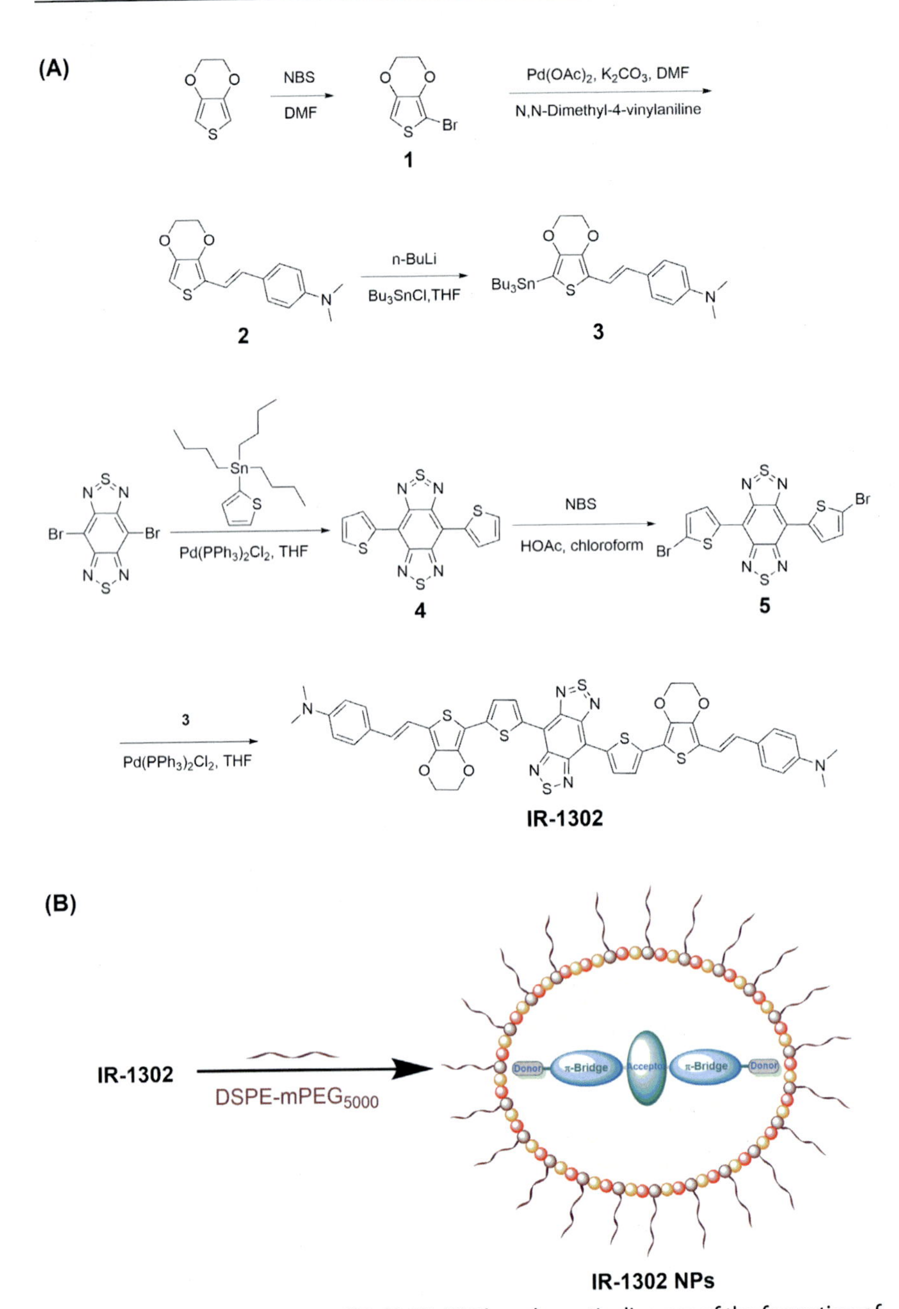

Fig. 2 (A) The synthetic route of **IR-1302**; (B) The schematic diagram of the formation of **IR-1302 NPs**.

additional step is required to prepare **IR-1302 NPs** (Fig. 2B). The synthesis and purification processes were carried out in a standard set of laboratory glassware, which contained three-necked flasks, two-necked flasks, condenser pipes, tee joints, injection syringes, round bottom flasks, chromatography equipment, Schlenk line glass tubes, and rotatory evaporators. The ^{1}H and ^{13}C NMR spectra were operated through an American Varian Mercury Plus 400 spectrometer (400MHz). The MALDI-TOF spectral data were recorded on a Bruker ultrafleXtreme MALDI-TOF-TOF mass spectrometer, DHB as matrix.

2.2.1 Reactants

1. 3,4-ethyl-enedioxythiophene (CAS# 126213-50-1)
2. N-Bromosuccinimide (NBS, CAS# 128-08-5)
3. Anhydrous sodium sulfate (Na_2SO_4, CAS# 7757-82-6)
4. Anhydrous potassium carbonate (K_2CO_3, CAS# 584–08-7)
5. Palladium (II) Acetate ($Pd(OAc)_2$, CAS# 3375-31-3)
6. *N*,*N*-dimethyl-4-vinylaniline (CAS# 2039-80-7)
7. n-butyllithium (n-BuLi, 1.6M in hexane, (CAS# 109-72-8)
8. Chlorotributyltin (CAS# 1461-22-9).
9. 4,7-dibromobenzo[1,2-c:4,5-c′]bis([1,2,5]thiadiazole) (CAS# 165617-59-4)
10. 2-(tributylstannyl)thiophene (CAS# 54663-78-4)
11. Bis(triphenylphosphine)palladium(II) chloride ($Pd(PPh_3)_2Cl_2$, CAS# 13965-03-2)
12. 1,2-distearoyl-sn-glycero-3-phosphoethanolamine-N-[methoxy (polyethylene glycol)-5000] (DSPE-$mPEG_{5000}$, CAS# 147867-65-0)

2.2.2 Solvents and solutions

1. Anhydrous *N*,*N*-dimethylformamide
2. Dichloromethane
3. Petroleum ether
4. Anhydrous tetrahydrofuran
5. Ethanol
6. Purified water
7. Acetic acid
8. Saturated potassium fluoride solution
9. Methanol
10. Chloroform
11. Hydrochloric acid

2.2.3 Compound 1

1. Prepare an ice bath and a 100 mL two-necked round bottom flask equipped with a magnetic stir bar. The tee joint was sealed by an elastic balloon in one pathway.
2. Add 3,4-ethyl-enedioxythiophene (4.00 g, 28.16 mmol) to the reaction vessel and seal with a rubber plug.
3. Use the Schlenk line glass tubes to create an environment flushed with nitrogen in the reaction flask. NOTE: A pump is needed to extract the air existing in the flask, then the inert gas nitrogen is released to the flask to make an oxygen-free reaction environment. The degassing behavior is operated for 15 min, and the nitrogen releasing behavior is operated every 5 min while degassing. The following procedures are operated after releasing nitrogen three times, and the nitrogen is continuously supplied until the reaction is complete. Unless otherwise stated, all reactions are performed under nitrogen.
4. Add anhydrous *N,N*-dimethylformamide (DMF, 20 mL) to the flask through an injector after degassing, wait for minutes in an ice bath until the reaction system cools to 0 °C and the 3,4-ethyl-enedioxythiophene is dissolved. NOTE: The anhydrous solvents are disposed with dehydration and stored with molecular sieves.
5. Cover the flask with tin foil paper to ensure a dark environment.
6. N-Bromosuccinimide (NBS, 5.00 g, 28.16 mmol) is dissolved in anhydrous DMF (5 mL) and then added to the reaction system by an injector dropwise under an ice bath in darkness. NOTE: NBS is dissolved with the help of ultrasonic concussion. The magnetic stir bar is stirred vigorously while injecting.
7. Remove the ice bath and then the reaction system is stirred at room temperature in darkness for 1 h.
8. The mixture is poured into water and extracted with dichloromethane. The organic layer is washed with water three times and dried over anhydrous Na_2SO_4. The collected organic layer is concentrated by rotary evaporation to obtain a crude product as an oil. NOTE: The obtained mixture oil must be kept under 0 °C and can be utilized for next steps without further purification.

2.2.4 Compound 2

1. Prepare an oil bath and a 100 mL two-necked round bottom flask equipped with a magnetic stir bar, and the tee joint is sealed by an elastic

balloon in one pathway and a condenser pipe. The flask is then covered by tin foil paper.

2. Add K_2CO_3 (4.99 g, 36.20 mmol), $Pd(OAc)_2$ (162.54 mg, 0.72 mmol) to the flask, and then it is sealed by a rubber plug.
3. Flush the flask with nitrogen for 15 min.
4. Add anhydrous DMF (30 mL) to the mixture through an injector after degassing.
5. Compound **1** (4.00 g) and *N,N*-dimethyl-4-vinylaniline (2.19 g, 21.72 mmol) are injected into the flask after covering at room temperature.
6. The reaction system is heated to 100 °C and stirred for 12 h in darkness.
7. After finishing the reaction, the reaction is cooled down room temperature. The mixture is quenched with water and extracted with dichloromethane. The organic layer is collected after washing with water three times and dried over anhydrous Na_2SO_4. The collected organic layer is purified by column chromatography on silica gel with petroleum ether/dichloromethane (v:v = 2:1) to obtain **2** in 15% yield as light yellow powder.

2.2.5 Compound 3

1. Prepare a 100 mL two-necked round bottom flask equipped with a magnetic stir bar with a sealed tee joint with an elastic balloon in one pathway. NOTE: The glassware is pre-treated with a drying oven for 30 min.
2. Compound **2** (500.00 mg, 1.74 mmol) is added to the flask and purified under nitrogen for 30 min.
3. Anhydrous tetrahydrofuran (25 mL) is added and cool to −78 °C in a cryogenic reactor.
4. n-BuLi (1.52 mL, 2.44 mmol) is slowly added dropwise to the solution and reacted for 1.5 h in −78 °C. NOTE: n-BuLi is a dangerous reagent that can easily release a lot of heat when touching water. Act quickly to draw and transfer n-BuLi to the reaction system in an injection syringe, avoiding long exposure to water in air. Meanwhile, the dehydration process of glassware and solvent tetrahydrofuran will directly decide the reaction efficiency. The used injection syringe should be exposed to air until the residuary n-BuLi reacts slowly with moisture in the air, and then quench with plenty of water.
5. Add chlorotributyltin (794.22 mg, 2.44 mmol) into the reaction system in −78 °C. Then the reaction slowly recovered to room temperature and stirred overnight. NOTE: The organic reagent containing tin has

neurovirulence. The experimenter should wear a suitable gas mask while operating and avoid skin contact with the use of gloves. All equipment that has touched the reagent including gloves, injected syringes, and others should be quenched with saturated potassium fluoride. Turn off the refrigeration function of cryogenic reactor and the temperature will gradually recover, but do not take the reaction flask out of the reactor.

6. The mixture is quenched with saturated potassium fluoride and extracted with dichloromethane, dried over anhydrous Na_2SO_4. The crude product is concentrated as brown oil without further purification. NOTE: The reaction equipment must be washed with saturated potassium fluoride.

2.2.6 Compound 4

1. Prepare an ice bath and a 50 mL two-necked round bottom flask equipped with a magnetic stir bar with a tee joint sealed by an elastic balloon in one pathway and a condenser pipe. The flask is then covered by tin foil paper.
2. Mix 4,7-dibromobenzo[1,2-c:4,5-c']bis([1,2,5]thiadiazole) (500.00 mg, 1.40 mmol), $Pd(PPh_3)_2Cl_2$ (490.70 mg, 0.70 mmol) in the flask and seal it with a rubber plug.
3. Flush the flask with nitrogen in 15 min.
4. Inject anhydrous tetrahydrofuran (25 mL) into the flask, followed by the addition of 2-(tributylstannyl)thiophene (1.31 g, 3.50 mmol). NOTE: The treatment to 2-(tributylstannyl)thiophene is the same as chlorotributyltin.
5. The reaction system is refluxed in 80 °C for 24 h.
6. Cool down to room temperature and pour it into methanol (200 mL). After forming a precipitate, filter to obtain a cake, wash it with plenty of methanol to obtain the crude product as a blue solid without further purification.

2.2.7 Compound 5

1. Prepare a 100 mL two-necked round bottom flask equipped with a magnetic stir bar with a tee joint that sealed by an elastic balloon in one pathway and a condenser pipe. The flask is covered by tin foil paper.
2. Add compound **4** (358.50 mg, 1.0 mmol) to the reaction vessel and seal it with a rubber plug.
3. Flush the flask with nitrogen for 15 min.

4. Inject chloroform (30 mL) into the flask and heat to reflux for 2 h.
5. Add NBS (392.00 mg, 2.2 mmol) in one portion in darkness, and then inject acetic acid (5 drops). The mixture is stirred for reflux for 3 h. NOTE: With nitrogen supply, NBS is quickly poured in through pulling the plug.
6. Another portion of NBS (196.00 mg, 1.1 mmol) is added to the system and stirred for 2 h.
7. After finishing the reaction, the mixture is cooled and poured to hydrochloric acid (180 mL, 2 M), stirred for another 2 h.
8. Filter to obtain a cake, wash it with water and methanol to get the crude product without further purification.

2.2.8 Compound *IR-1302*

1. Prepare a 100 mL two-necked round bottom flask equipped with a magnetic stir bar and a tee joint that is sealed by an elastic balloon in one pathway and a condenser pipe. The flask is covered by tin foil paper.
2. Add compound **5** (100 mg, 0.19 mmol) in the flask and seal it with a rubber plug.
3. Flush the flask with nitrogen for 15 min.
4. Inject anhydrous tetrahydrofuran (50 mL) and compound **3** (334.95 mg, 0.58 mmol) into the flask, followed by an addition of $Pd(PPh_3)_2Cl_2$ (6.67 mg, 0.0095 mmol).
5. The mixture is stirred at 80 °C for 48 h.
6. After that, the solvent is evaporated under vacuum and the residue **IR-1302** is washed with dichloromethane, water, and ethanol to give a brown solid (52.96 mg, 30%).

2.2.9 Compound *IR-1302* NPs

1. Dissolve DSPE-mPEG$_{5000}$ (9 mg) in purified water (9 mL) in an Eppendorf tube (50 mL).
2. Dissolve **IR-1302** (1 mg) in tetrahydrofuran (1 mL), and then the solution is added to aqueous solution containing DSPE-mPEG$_{5000}$ through a pipette.
3. The mixture is sonicated by a cell disruptor (WHSJXC-950E) for 2 min. NOTE: The sonicating process will produce a lot of heat and the Eppendorf tube must be settled with an ice bath.
4. The mixture is then bubbled with nitrogen, and the THF is removed by nitrogen flow until the whole volume is around 9 mL.

5. The solution is centrifuged at 3000 rpm/min for 30 min with 50 kDa centrifugal filter to separate extra DSPEG-mPEG$_{5000}$ and other non-capsulated organic molecules, and the residual concentrated solution is washed with water for 24 h.
6. The concentrated solution is finally titrated to 2 mL.
7. Pass through a 0.22 μm water syringe filter and obtain **IR-1302 NPs**. The nanoparticles should be stored at 4 °C for further use.

3. Optical characterization of IR-1302 NPs *in vitro*

Upon completing the synthetic procedures, we first investigated the fundamental optical properties of **IR-1302 NPs** *in vitro*. As displayed in Fig. 3A, an intense maximum absorption peak of the nanoparticles located at 942 nm appeared, revealing a NIR light-harvesting ability which was adaptive for PA imaging conditions. In addition, the maximum fluorescence peak of **IR-1302 NPs** was located at around 1302 nm which covered the range of 900–1500 nm, resulting in a large Stokes shift of 360 nm. It ensured that the autoabsorption was hard to generate through laser irradiation, meeting the requirements of a PA contrast agent. Besides, we have tested the stability of particles which were supposed to be unstable in the NIR region (Fig. 3B). After placing **IR-1302 NPs** at 37 °C for 6 days, we found no obvious changes to particle diameters, implying that they could stably exist in a simulative cellular environment. The excellent optical properties indicated that the NPs showed potential for imaging tests *in vivo*.

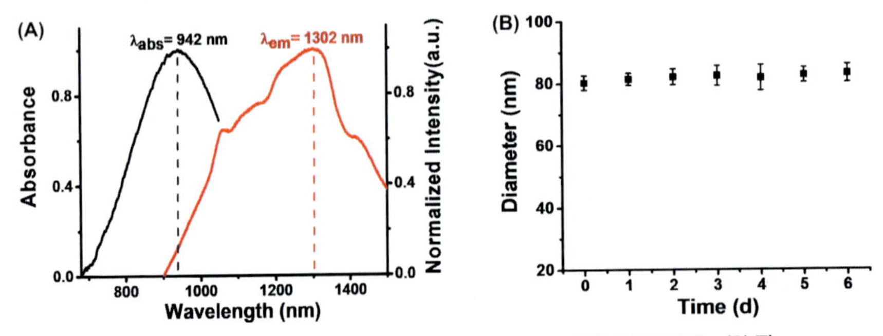

Fig. 3 (A) The NIR absorption and fluorescence spectra of **IR-1302 NPs**; (B) The average nanoparticles diameter of **IR-1302 NPs** in PBS for 6 days. *Reproduced with permission from Wiley: Ye, F., Huang, W., Li, C., Li, G., Yang W.-C., Liu, S. H., Yin, J., Sun, Y., Yang G.-F. (2020). Near-Infrared Fluorescence/Photoacoustic Agent with an Intensifying Optical Performance for Imaging-Guided Effective Photothermal Therapy. Advanced Therapeutics, 3, 2000170. https://doi.org/10.1002/adtp.202000170.*

3.1 Materials and equipment

- Hitachi U-3310 visible recording spectrophotometer
- Applied Nano Fluorescence spectrometer
- Excitation laser source of 980nm (Suzhou NIR-Optics Technologies Co., Ltd. NIR-II fluorescence spectrum)
- Malvern Zetasizer Nano ZS
- Quartz UV cuvette (4mL)
- Quartz fluorescent cuvette (4mL)
- Sterile Eppendorf tubes.
- Pipette and tips.
- Phosphate Buffer Saline (PBS, pH7.4, 10mM).

3.2 Measurement of absorption spectrum of IR-1302 NPs

1. Prepare the sample of **IR-1302 NPs** (500 $\mu g \cdot mL^{-1}$, 1.0mL) and PBS stored at 4 °C, wait until the samples recover to room temperature. NOTE: The solvated PBS is made with standard PBS powders and redistilled water, and the pH is adjusted to 7.4 with KOH and HCl before.
2. Add 30.0 μL of the NPs to 2.97mL of PBS in a 4mL Eppendorf tube to obtain a sample of 5 $\mu g \cdot mL^{-1}$. NOTE: The sample must be kept in the dark.
3. Prepare two samples of blank control solution with 3.00mL PBS in quartz cuvettes. NOTE: There are two kinds of surfaces of UV cuvette, and the experimenter can only touch the rough surface, since the smooth surface will be the path of light.
4. Measure the background absorption spectrum with two blank control samples to obtain a contrast curve which coincides with the base line. Scan wavelength range: 300–1400nm. NOTE: If the curve deviates from the base line too much, it must be measured again until generating a standard curve.
5. Replace one of the control solutions with the NPs sample to be tested in the Eppendorf tube. NOTE: The used cuvette must be washed with ethanol and dichloromethane in turn. Ethanol will clean the residual water with good water solubility, and dichloromethane can take away the ethanol. The residual dichloromethane has a low-boiling point and is easily evaporated to ensure a clean cuvette.
6. Measure the absorption spectrum of **IR-1302 NPs**. Scan wavelength range: 300–1400nm.
7. Plot the measured absorbance over the wavelength.

3.3 Measurement of the fluorescence spectra of IR-1302 NPs

1. Prepare the sample of **IR-1302 NPs** (500 μg · mL^{-1}, 1.0 mL) and wait until the sample reaches room temperature.
2. Add 30.0 μL of the NPs to 2.97 mL of PBS in a cuvette to obtain a sample with a concentration of 5 μg · mL^{-1}. NOTE: The cuvette has four smooth surfaces, and it should be held through holding edges without touching surfaces.
3. Measure the fluorescence spectrum of **IR-1302 NPs**. Excitation wavelength = 980 nm laser. NOTE: A laser light source is necessary for enough light energy.
4. Plot the measured fluorescent intensity over the wavelength. The data is normalized based on the maximum fluorescence peak intensity.

3.4 Stability assessment

1. Prepare the sample of **IR-1302 NPs** (500 μg · mL^{-1}, 1.0 mL) and wait until the sample recovers to room temperature.
2. Add around 1.00 mL of the NPs to a fluorescent cuvette. NOTE: Pay attention to adjust the amount of the NPs solution in the cuvette to meet the requirements of the instrument.
3. Measure the particle diameter three times to obtain three similar good quality reports. NOTE: When measuring, the related parameters must be set well, including the type of cuvette, type of solvent, test temperature, viscosity and refractive index of the solvent, etc.
4. Recycle the measured sample in a capped bottle and place it in the water bath at 37 °C in dark.
5. Measure the diameter on the 2nd, 3rd, 4th, 5th, 6th day consecutively, according to procedures 2–4. NOTE: The measurement settings are exactly the same as previous parameters.
6. Plot a histogram according to the distribution of diameters, figure out an average particle diameter every day. Gather the average diameter in 6 days and plot the diameter changes over days.

4. Photothermal functionality of IR-1302 NPs *in vitro*

As a precondition to PA imaging, **IR-1302 NPs** show excellent optical properties when applied to photothermal conversion, hence we explored related properties. The photothermal experiments showed that the NPs tended to produce heat upon laser irradiation (1 W · cm^{-2}) and caused an

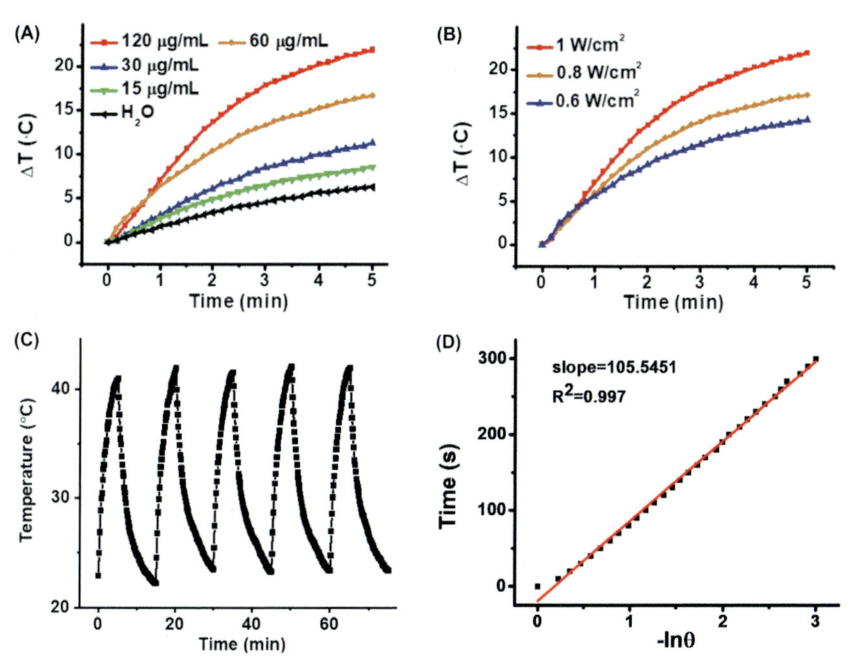

Fig. 4 (A) The temperature changing in photothermal conversion of **IR-1302 NPs** wih different concentrations (0–120 µg · mL^{-1}) under 980 nm laser irradiation (1 W · cm^{-2}); (B) The temperature changing in photothermal conversion of **IR-1302 NPs** (120 µg · mL^{-1}) under 980 nm laser irradiation with different laser powers (0.6–1 W · cm^{-2}); (C) The resistance to photo degradation **IR-1302 NPs** in photothermal cycles; (D) The time constant for **IR-1302 NPs** heat transfer from the system. *Reproduced with permission from Wiley: Ye, F., Huang, W., Li, C., Li, G., Yang W.-C., Liu, S. H., Yin, J., Sun, Y., Yang G.-F. (2020). Near-Infrared Fluorescence/Photoacoustic Agent with an Intensifying Optical Performance for Imaging-Guided Effective Photothermal Therapy. Advanced Therapeutics, 3, 2000170. https://doi.org/10.1002/adtp.202000170.*

increased temperature by approximately 23 °C in the concentration of 120 µg · mL^{-1} over 5 min (Fig. 4A). Meanwhile, in the control group containing water, the rise was about 5 °C. The temperature trend was found to be the same with other lesser concentrations and lower laser powers (Fig. 4B), revealing a feasible heat change that was favorable for PA signal generating. Additional heating circulation experiments have been carried out. In Fig. 4C, the temperature change was subtle and negligible in the circle of heating and cooling, and it could last at least five cycles. It showed a superior resistance to photo degradation. We also tested the photothermal conversion efficiency, and it was determined to be 28.6% (Fig. 4D).

4.1 Materials and equipment

- Excitation laser source of 980 nm equipped with optical cables
- Thermal imager (Flir-E4)
- Pipette and tips
- Sterile Eppendorf tubes
- Purified water

4.2 The photothermal conversion of IR-1302 NPs with concentrations

1. Prepare a sample of **IR-1302 NPs** (120 μg · mL^{-1}, 1.0 mL) with purified water in a 1.5 mL Eppendorf tube at room temperature.
2. Place the sample tube on an empty experimental bench and keep it fixed, turn on the 980 nm laser with the light focusing on the NPs solution. NOTE: The laser power is always 1 W · cm^{-2} in this test.
3. Record the temperature change of the sample in 5 min through the thermal imager. NOTE: The location of thermal imager and tubes containing NPs, the distance between the tested tube and laser need to be fixed until all thermal experiments are complete.
4. Add 1.0 mL purified water in a 1.5 mL Eppendorf tube as a control group and repeat procedures 2–3.
5. Prepare another sample in procedure 1, add 500.0 μL of this sample to 500.0 μL of purified water in a 1.5 mL Eppendorf tube to obtain a 60 μg · mL^{-1} solution. Repeat procedures 2–3.
6. Follow the guidance above to obtain other samples in 30 and 15 μg · mL^{-1} and measure the photothermal temperature changes.
7. Collect the highest temperature every 5 s and plot the temperature change conditions over time.

4.3 The photothermal conversion of IR-1302 NPs with laser powers

1. Prepare three samples of **IR-1302 NPs** (120 μg · mL^{-1}, 1.0 mL) with purified water in a 1.5 mL Eppendorf tube at room temperature.
2. Measure the temperature change of the samples in 5 min after fixing all equipment through the thermal imager with different laser powers, 1, 0.8, and 0.6 W · cm^{-2}, respectively.
3. Collect the highest temperature every 5 s and plot the temperature change conditions over time.

4.4 The cyclic photothermal conversion of IR-1302 NPs

1. Prepare a sample of **IR-1302 NPs** ($120\,\mu g \cdot mL^{-1}$, 1.0 mL) with purified water in a 1.5 mL Eppendorf tube at room temperature.
2. Measure the temperature heating changes of the sample in 5 min through the thermal imager with the laser ($1\,W \cdot cm^{-2}$) after fixing all equipment.
3. Turn off the laser and measure the following temperature cooling changes in 15 min without laser irradiation.
4. Repeat procedures 2–3 four times.
5. Prepare another two samples in procedure 1 and repeat procedures 2–4. NOTE: No obvious differences of the three tested samples are observed. If not, more repeated tests are necessary.
6. Plot the temperature change conditions over time.

4.5 Photothermal conversion efficiency of IR-1302 NPs

1. Prepare a sample of **IR-1302 NPs** ($120\,\mu g \cdot mL^{-1}$, 1.0 mL) with purified water in a 1.5 mL Eppendorf tube at room temperature.
2. Measure the temperature heating changes of the sample in 5 min through the thermal imager with the laser ($1\,W \cdot cm^{-2}$) after fixing all equipment.
3. Plot the time (t) over temperature driving force parameter (θ). The correlation is showed as the following:

$$t = -\tau_s ln\theta$$

$$\theta = \frac{T - T_{Surr}}{T_{Max} - T_{Surr}}$$

NOTE: τ_s—associated time constant.

T_{Max}—the maximum steady state temperature.

T_{Surr}—the environment temperature.

4. The photothermal conversion efficiency (η) is then calculated according to the reported method

$$\eta = \frac{hs(T_{Max} - T_{Surr}) - Q_{Dis}}{I(1 - 10^{-A})}$$

$$hs = \frac{mC_{water}}{\tau_s}$$

NOTE: Q_{Dis}—Heat dissipated from the laser mediated by the solvent and container.

I—Laser power.

A—Absorbance of NPs at 980 nm.

h—Heat transfer coefficient.

s—Surface area of the container.

m—Mass of the solution containing the photoactive material.

C_{water}—Specific heat capacity of the solution. $C_{water} = 4.2\text{J}/(\text{g}\cdot°\text{C})$.

5. Photoacoustic imaging of IR-1302 NPs *in vitro*

Inspired by all optical and photothermal characteristics, a PA spectrum of **IR-1302 NPs** has been conducted. Based on strong absorption in the NIR region, a strong PA signal covering 700–1200 nm was observed (Fig. 5A), revealing a designed NIR ultrasonic wave. In the meantime, we found that the PA signal was concentration-dependent of **IR-1302 NPs** in the range of 10–100 μg · mL^{-1}(Fig. 5B). The experiment results implied further imaging applications *in vivo*.

5.1 Materials and equipment

- Photoacoustic multispectral optical tomographer (MSOT)
- Pipette and tips
- Sterile Eppendorf tubes
- PBS (pH 7.4, 10 mM)

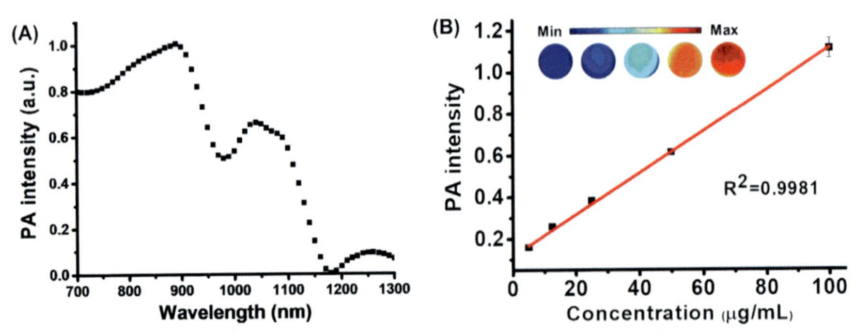

Fig. 5 The PA signals of **IR-1302 NPs** upon 980 nm laser (1 W · cm^{-2}) with different concentrations: (A) 50 μg · mL^{-1}; (B) 5–100 μg · mL^{-1}. *Reproduced with permission from Wiley: Ye, F., Huang, W., Li, C., Li, G., Yang W.-C., Liu, S. H., Yin, J., Sun, Y., Yang G.-F. (2020). Near-Infrared Fluorescence/Photoacoustic Agent with an Intensifying Optical Performance for Imaging-Guided Effective Photothermal Therapy. Advanced Therapeutics, 3, 2000170. https://doi.org/10.1002/adtp.202000170.*

5.2 Photoacoustic imaging of IR-1302 NPs with concentrations *in vitro*

1. Prepare a sample of **IR-1302 NPs** (500 μg · mL^{-1}, 1.0 mL) with PBS in a 1.5 mL Eppendorf tube at room temperature.
2. Add 500.0 μL of the NPs to 500.0 μL of PBS in a 1.5 mL Eppendorf tube to obtain a sample of 50 μg · mL^{-1}.
3. Transfer 200.0 μL of the NPs to a 200 μL Eppendorf tube. NOTE: Cap the tube tightly and seal the tube edge with sealing film.
4. Place the tube into the PA tomographer and capture PA imaging pictures of the NPs with laser (980 nm, 1 W · cm^{-2}), collect the imaging picture and the PA intensity data.
5. Plot the PA intensity as a function of wavelength.
6. Add 40.0, 20.0, 10.0, 5.0, and 2.0 μL of the NPs (500 μg · mL^{-1}) to 160.0, 180.0, 190.0, 195.0, and 198.0 μL of PBS in 200 μL Eppendorf tubes to obtain samples of 100, 50, 25, 12.5, and 5 μg · mL^{-1}.
7. Repeat procedure 4.
8. Plot the normalized maximum PA intensity of every PA imaging picture over different concentrations.

6. Photoacoustic imaging of IR-1302 NPs *in vivo*

PA imaging in 4T1 mice models was operated to evaluate the PAI effect of **IR-1302 NPs** *in vivo* (Fig. 6). After the intravenous injection of **IR-1302 NPs** (100 μg), the PAI signals in mice were monitored continuously. It can be observed that the PA signal quickly appeared in tumor sites and reached its strongest signal at 12 h after injection. These imaging signals provided high-spatial mapping of internal organs and tumors.

Animal experiments have been allowed by the Association for Assessment and Accreditation of Laboratory Animal Care International (AAALAC International), and the Care and Use of Laboratory Animals of the Chinese Animal Welfare Committee. The nude mice were purchased from Suzhou Belda BioPharmaceutical Co.

6.1 Materials and equipment

- Photoacoustic multispectral optical tomographer (MSOT)
- Cell incubator with 5% CO_2
- Vaporizer
- Insulin syringes

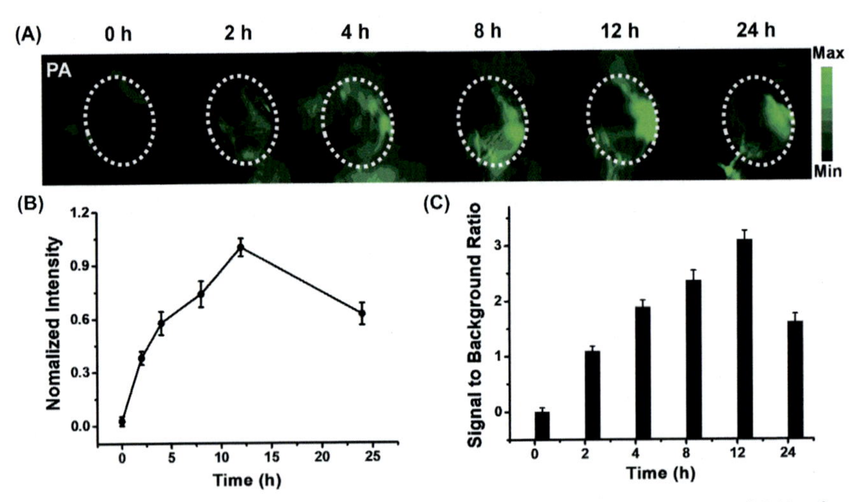

Fig. 6 (A) *In vivo* PA imaging captures with **IR-1302 NPs** at 0, 2, 4, 8, 12, and 24 h after injection, the circles highlight tumor regions; (B) The PA signal intensity in tumor over time. (C) The signal-to-background ratio over time for PA captures. *Reproduced with permission from Wiley: Ye, F., Huang, W., Li, C., Li, G., Yang W.-C., Liu, S. H., Yin, J., Sun, Y., Yang G.-F. (2020). Near-Infrared Fluorescence/Photoacoustic Agent with an Intensifying Optical Performance for Imaging-Guided Effective Photothermal Therapy. Advanced Therapeutics, 3, 2000170. https://doi.org/10.1002/adtp.202000170.*

6.2 Animals and cells

- 4T1 cells
- Nude female mice, 6–8 weeks old

6.3 Reagents

- Dulbecco's modified Eagle's medium (DMEM)
- Fetal bovine serum (FBS)
- PBS
- Isoflurane
- Diethyl ether

6.4 Procedures

1. Culture 4T1 cells in DMEM supplemented with 10% FBS at 37 °C in a cell incubator.
2. Wash 4T1 cells with PBS and inject cells (1×10^6 cells in 100 μL PBS) to the right hind limb of the mice. The 4T1 tumor-bearing mice are subjected to the following experiments when the tumor volume reached $80\,mm^3$.

3. Prepare a sample of **IR-1302 NPs** ($500\,\mu g \cdot mL^{-1}$, 1.0 mL) with PBS in a 200 μL Eppendorf tube at room temperature.
4. Warm up the photoacoustic multispectral optical tomographer and set a flow of gaseous isoflurane for anesthesia. NOTE: Set an appropriate and safe flow rate of isoflurane.
5. Fix the mice and put them to sleep with a moderate amount of diethyl ether gas.
6. Inject 200.0 μL of the NPs along the caudal vein through an insulin syringe.
7. Put mice in a tomographer and take PAI pictures at 0, 2, 4, 8, 12, and 24 h post-injection with laser (980 nm, $1\,W \cdot cm^{-2}$). NOTE: The mice are taken back at experiment intervals. Before taking an imaging picture, the mice are anesthetized with isoflurane flow.
8. Repeat the procedures in three experiment groups.
9. Plot the normalized maximum PA intensity over time.

7. Concluding remarks

Herein, we have presented the synthesis, optical properties, and PA imaging effects of **IR-1302 NPs**, a NIR contrast agent based on D-A-D skeleton. Through the introduction of two thiophene rings and styrene moieties as conjugated bridges, and the *N,N*-dimethylamino group as an electron donor to the benzobisthiadiazole core, the NIR absorption and emission show the strengthened conjugated degree and electron-donating ability. The special optical properties of the NPs guarantee an advantageous basic condition for high microscopic spatial resolution of PA imaging in deeper penetration depths and less light scattering *in vivo*. We believe this work may provide a universal route for designing novel near-infrared PA contrast agents in tumor imaging and therapy.

References

Chaigne, T., Arnal, B., Vilov, S., Bossy, E., & Katz, O. (2017). Super-resolution photoacoustic imaging via flow-induced absorption fluctuations. *Optica*, *4*(11), 1397–1404. https://doi.org/10.1364/OPTICA.4.001397.

Chen, Y., Li, L., Chen, W., Chen, H., & Yin, J. (2019). Near-infrared small molecular fluorescent dyes for photothermal therapy. *Chinese Chemical Letters*, *30*(7), 1353–1360. https://doi.org/10.1016/j.cclet.2019.02.003.

Chen, Y., Wang, M., Zheng, K., Ren, Y., Xu, H., Yu, Z., et al. (2021). Antimony nanopolyhedrons with tunable localized surface plasmon resonances for highly effective photoacoustic-imaging-guided synergistic photothermal/immunotherapy. *Advanced Materials*, *33*(18), 2100039. https://doi.org/10.1002/adma.202100039.

Chen, Z., Xia, Q., Zhou, Y., Li, X., Qi, L., Feng, Q., et al. (2019). 2-Dicyanomethylenethiazole based NIR absorbing organic nanoparticles for photothermal therapy and photoacoustic imaging. *Journal of Materials Chemistry B*, 7(25), 3950–3957. https://doi.org/10.1039/C9TB00808J.

Chen, W., Zhang, C., Chen, H., Zang, K., Liu, S. H., Xie, Y., et al. (2021). Near-infrared thienoisoindigos with aggregation-induced emission: molecular design, optical performance, and bioimaging application. *Analytical Chemistry*, *93*(7), 3378–3385. https://doi.org/10.1021/acs.analchem.0c04260.

Chen, Y.-S., Zhao, Y., Yoon, S. J., Gambhir, S. S., & Emelianov, S. (2019). Miniature gold nanorods for photoacoustic molecular imaging in the second near-infrared optical window. *Nature Nanotechnology*, *14*(5), 465–472. https://doi.org/10.1038/s41565-019-0392-3.

Danielli, A., Maslov, K., Favazza, C. P., Xia, J., & Wang, L. V. (2015). Nonlinear photoacoustic spectroscopy of hemoglobin. *Applied Physics Letters*, *106*(20), 203701. https://doi.org/10.1063/1.4921474.

Duan, Y., Xu, Y., Mao, D., Liew, W. H., Guo, B., Wang, S., et al. (2018). Photoacoustic and magnetic resonance imaging bimodal contrast agent displaying amplified photoacoustic signal. *Small*, *14*(42), 1800652. https://doi.org/10.1002/smll.201800652.

Fathi, P., Knox, H. J., Sar, D., Tripathi, I., Ostadhossein, F., Misra, S. K., et al. (2019). Biodegradable biliverdin nanoparticles for efficient photoacoustic imaging. *ACS Nano*, *13*(7), 7690–7704. https://doi.org/10.1021/acsnano.9b01201.

Friedlein, J. T., Baumann, E., Briggman, K. A., Colacion, G. M., Giorgetta, F. R., Goldfain, A. M., et al. (2020). Dual-comb photoacoustic spectroscopy. *Nature Communications*, *11*, 3152. https://doi.org/10.1038/s41467-020-16917-y.

Fu, Q., Li, Z., Ye, J., Li, Z., Fu, F., Lin, S.-L., et al. (2020). Magnetic targeted near-infrared II PA/MR imaging guided photothermal therapy to trigger cancer immunotherapy. *Theranostics*, *10*(11), 4997–5010. https://doi.org/10.7150/thno.43604.

Fu, Q., Zhu, R., Song, J., Yang, H., & Chen, X. (2018). Photoacoustic imaging: Contrast agents and their biomedical applications. *Advanced Materials*, *31*(6), 1805875. https://doi.org/10.1002/adma.201805875.

Gröhl, J., Schellenberg, M., Dreher, K., & Maier-Hein, L. (2021). Deep learning for biomedical photoacoustic imaging: A review. *Photoacoustics*, *22*, 100241. https://doi.org/10.1016/j.pacs.2021.100241.

Guo, B., Chen, J., Chen, N., Middha, E., Xu, S., Pan, Y., et al. (2019). High-resolution 3D NIR-II photoacoustic imaging of cerebral and tumor vasculatures using conjugated polymer nanoparticles as contrast agent. *Advanced Materials*, *31*(25), 1808355. https://doi.org/10.1002/adma.201808355.

Guo, B., Sheng, Z., Hu, D., Liu, C., Zheng, H., & Liu, B. (2018). Through scalp and skull NIR-II photothermal therapy of deep orthotopic brain tumors with precise photoacoustic imaging guidance. *Advanced Materials*, *30*(35), 1802591. https://doi.org/10.1002/adma.201802591.

He, Y., Liu, S. H., Yin, J., & Yoon, J. (2020). Sonodynamic and chemodynamic therapy based on organic/organometallic sensitizers. *Coordination Chemistry Reviews*, *429*, 213610. https://doi.org/10.1016/j.ccr.2020.213610.

He, S., Song, J., Liu, J., Liu, L., Qu, J., & Cheng, Z. (2019). Enhancing photoacoustic intensity of upconversion nanoparticles by photoswitchable Azobenzene-containing polymers for dual NIR-II and photoacoustic imaging in vivo. *Advanced Optical Materials*, 7(12), 1900045. https://doi.org/10.1002/adom.201900045.

Herrmann, K., Pech-May, N. W., & Retsch, M. (2021). Photoacoustic thermal characterization of low thermal diffusivity thin films. *Photoacoustics*, *22*, 100246. https://doi.org/10.1016/j.pacs.2021.100246.

Hu, W., Prasad, P. N., & Huang, W. (2021). Manipulating the dynamics of dark excited states in organic materials for phototheranostics. *Accounts of Chemical Research*, *54*(3), 697–706. https://doi.org/10.1021/acs.accounts.0c00688.

Jiang, X., Du, B., Tang, S., Hsieh, J.-T., & Zheng, J. (2019). Photoacoustic imaging of nanoparticle transport in the kidneys at high temporal resolution. *Angewandte Chemie International Edition*, *58*(18), 5994–6000. https://doi.org/10.1002/anie.201901525.

Jiang, Y., & Pu, K. (2017). Advanced photoacoustic imaging applications of near-infrared absorbing organic nanoparticles. *Small*, *13*(30), 1700710. https://doi.org/10.1002/smll.201700710.

Kang, J., Kim, D., Wang, J., Han, Y., Zuidema, J. M., Hariri, A., et al. (2018). Enhanced performance of a molecular photoacoustic imaging agent by encapsulation in mesoporous silicon nanoparticles. *Advanced Materials*, *30*(27), 1800512. https://doi.org/10.1002/adma.201800512.

Lee, D., Beack, S., Yoo, J., Kim, S.-K., Lee, C., Kwon, W., et al. (2018). In vivo photoacoustic imaging of livers using biodegradable hyaluronic acid-conjugated silica nanoparticles. *Advanced Functional Materials*, *28*(22), 1800941. https://doi.org/10.1002/adfm.201800941.

Li, Y., Cai, Z., Liu, S., Zhang, H., Wong, S. T. H., Lam, J. W. Y., et al. (2020). Design of AIEgens for near-infrared IIb imaging through structural modulation at molecular and morphological levels. *Nature Communications*, *11*, 1255. https://doi.org/10.1038/s41467-020-15095-1.

Li, S., Jiang, W., Yuan, Y., Sui, M., Yang, Y., Huang, L., et al. (2020). Delicately designed cancer cell membrane-camouflaged nanoparticles for targeted ^{19}F MR/PA/FL imaging-guided photothermal therapy. *ACS Applied Materials & Interfaces*, *12*(51), 57290–57301. https://doi.org/10.1021/acsami.0c13865.

Li, Y., Liu, Y., Li, Q., Zeng, X., Tian, T., Zhou, W., et al. (2020). Novel NIR-II organic fluorophores for bioimaging beyond 1550 nm. *Chemical Science*, *11*(10), 2621–2626. https://doi.org/10.1039/C9SC06567A.

Liba, O., & de la Zerda, A. (2017). Photoacoustic tomography: Breathtaking whole-body imaging. *Nature Biomedical Engineering*, *1*(5), 0075. https://doi.org/10.1038/s41551-017-0075.

Lin, J., Zeng, X., Xiao, Y., Tang, L., Nong, J., Liu, Y., et al. (2019). Novel near-infrared II aggregation-induced emission dots for in vivo bioimaging. *Chemical Science*, *10*(4), 1219–1226. https://doi.org/10.1039/C8SC04363A.

Liu, Y., Bhattarai, P., Dai, Z., & Chen, X. (2019). Photothermal therapy and photoacoustic imaging via nanotheranostics in fighting cancer. *Chemical Social Reviews*, *48*(7), 2053–2108. https://doi.org/10.1039/C8CS00618K.

Liu, S., Chen, C., Li, Y., Zhang, H., Liu, J., Wang, R., et al. (2020). Constitutional isomerization enables bright NIR-II AIEgen for brain-inflammation imaging. *Advanced Functional Materials*, *30*(7), 1908125. https://doi.org/10.1002/adfm.201908125.

Liu, F., Shi, X., Liu, X., Wang, F., Yi, H.-B., & Jiang, J.-H. (2019). Engineering an NIR rhodol derivative with spirocyclic ring-opening activation for high-contrast photoacoustic imaging. *Chemical Science*, *10*(40), 9257–9264. https://doi.org/10.1039/C9SC02764E.

Liu, S., Zhou, X., Zhang, H., Ou, H., Lam, J. W. Y., Liu, Y., et al. (2019). Molecular motion in aggregates: Manipulating TICT for boosting photothermal theranostics. *Journal of the American Chemical Society*, *141*(13), 5359–5368. https://doi.org/10.1021/jacs.8b13889.

Longo, D. L., Stefania, R., Callari, C., Rose, F. D., Rolle, R., Conti, L., et al. (2016). Water soluble melanin derivatives for dynamic contrast enhanced photoacoustic imaging of tumor vasculature and response to antiangiogenic therapy. *Advanced Healthcare Materials*, *6*(1), 1600550. https://doi.org/10.1002/adhm.201600550.

Ma, H., Liu, C., Hu, Z., Yu, P., Zhu, X., Ma, R., et al. (2020). Propylenedioxy thiophene donor to achieve NIR-II molecular fluorophores with enhanced brightness. *Chemistry of Materials*, *32*(5), 2061–2069. https://doi.org/10.1021/acs.chemmater.9b05159.

Ma, X., Zhang, C., Feng, L., Liu, S. H., Tan, Y., & Yin, J. (2020). Construction and bioimaging application of novel indole heptamethine cyanines containing functionalized tetrahydropyridine rings. *Journal of Materials Chemistry B*, *8*(43), 9906–9912. https://doi.org/10.1039/D0TB01890B.

Mantri, Y., & Jokerst, J. V. (2020). Engineering plasmonic nanoparticles for enhanced photoacoustic imaging. *ACS Nano*, *14*(8), 9408–9422. https://doi.org/10.1021/acsnano.0c05215.

Moon, H., Kang, J., Sim, C., Kim, J., Lee, H., Chang, J. H., et al. (2015). Multifunctional theranostic contrast agent for photoacoustics- and ultrasound-based tumor diagnosis and ultrasound-stimulated local tumor therapy. *Journal of Controlled Release*, *218*(28), 63–71. https://doi.org/10.1016/j.jconrel.2015.09.060.

Ong, S. Y., Zhang, C., Xiao, D., & Yao, S. Q. (2021). Recent advances in polymeric nanoparticles for enhanced fluorescence and photoacoustic imaging. *Angewandte Chemie International Edition*. https://doi.org/10.1002/anie.202101964.

Ou, H., Li, J., Chen, C., Gao, H., Xue, X., & Ding, D. (2019). Organic/polymer photothermal nanoagents for photoacoustic imaging and photothermal therapy in vivo. *Science China Materials*, *62*(11), 1740–1758. https://doi.org/10.1007/s40843-019-9470-3.

Qian, G., & Wang, Z. Y. (2010). Design, synthesis, and properties of benzobisthiadiazole-based donor-π-acceptor-π-donor type of low-band-gap chromophores and polymers. *Canadian Journal of Chemistry*, *88*(3), 192. https://doi.org/10.1139/v09-157.

Qian, G., Zhong, Z., Luo, M., Yu, D., Zhang, Z., Wang, Z. Y., et al. (2009). Simple and efficient near-infrared organic chromophores for light-emitting diodes with single electroluminescent emission above 1000 nm. *Advanced Materials*, *21*(1), 111–116. https://doi.org/10.1002/adma.200801918.

Qu, C., Xiao, Y., Zhou, H., Ding, B., Li, A., Lin, J., et al. (2019). Quaternary ammonium salt based NIR-II probes for in vivo imaging. *Advanced Optical Materials*, 7(15), 1900229. https://doi.org/10.1002/adom.201900229.

Reza, P. H., Bell, K., Shi, W., Shapiro, J., & Zemp, R. J. (2018). Deep non-contact photoacoustic initial pressure imaging. *Optica*, *5*(7), 814–820. https://doi.org/10.1364/OPTICA.5.000814.

Roberts, S., Seeger, M., Jiang, Y., Mishra, A., Sigmund, F., Stelzl, A., et al. (2018). Calcium sensor for photoacoustic imaging. *Journal of the American Chemical Society*, *140*(8), 2718–2721. https://doi.org/10.1021/jacs.7b03064.

Tian, C., Qian, W., Shao, X., Xie, Z., Cheng, X., Liu, S., et al. (2016). Plasmonic nanoparticles with quantitatively controlled bioconjugation for photoacoustic imaging of live cancer cells. *Advanced Science*, *3*(12), 1600237. https://doi.org/10.1002/advs.201600237.

Wang, W., Yang, Q., Du, Y., Zhou, X., Du, X., Wu, Q., et al. (2020). Metabolic labeling of peptidoglycan with NIR-II dye enables in vivo imaging of gut microbiota. *Angewandte Chemie International Edition*, *59*(7), 2628–2633. https://doi.org/10.1002/anie.201910555.

Weber, J., Beard, P., & Bohndiek, S. (2016). Contrast agents for molecular photoacoustic imaging. *Nature Methods*, *13*(8), 639–650. https://doi.org/10.1038/nmeth.3929.

Wilson, K. E., Bachawal, S. V., Abou-Elkacem, L., Jensen, K., Machtaler, S., Tian, L., et al. (2017). Spectroscopic photoacoustic molecular imaging of breast cancer using a B7-H3-targeted ICG contrast agent. *Theranostics*, 7(6), 1463–1476. https://doi.org/10.7150/thno.18217.

Xu, Z., Huang, X., Han, X., Wu, D., Zhang, B., Tan, Y., et al. (2018). A visible and near-infrared, dual-channel fluorescence-on probe for selectively tracking mitochondrial glutathione. *Chem*, *4*(7), 1609–1628. https://doi.org/10.1016/j.chempr.2018.04.003.

Xu, Z., Huang, X., Zhang, M.-X., Chen, W., Liu, S. H., Tan, Y., et al. (2019). Tissue imaging of glutathione-specific naphthalimide-cyanine dye with two-photon and near-infrared manners. *Analytical Chemistry*, *91*(17), 11343–11348. https://doi.org/10.1021/acs.analchem.9b02458.

Yang, Q., Hu, Z., Zhu, S., Ma, R., Ma, H., Ma, Z., et al. (2018). Donor engineering for NIR-II molecular fluorophores with enhanced fluorescent performance. *Journal of the American Chemical Society*, *140*(5), 1715–1724. https://doi.org/10.1021/jacs.7b10334.

Ye, F., Chen, W., Pan, Y., Liu, S. H., & Yin, J. (2019). Benzobisthiadiazoles: From structure to function. *Dyes and Pigments*, *171*, 107746. https://doi.org/10.1016/j.dyepig.2019.107746.

Ye, F., Huang, W., Li, C., Li, G., Yang, W.-C., Liu, S. H., et al. (2020). Near-infrared fluorescence/photoacoustic agent with an intensifying optical performance for imaging-guided effective photothermal therapy. *Advanced Therapeutics*, *3*(12), 2000170. https://doi.org/10.1002/adtp.202000170.

Ye, F., Liu, Y., Chen, J., Liu, S. H., Zhao, W., & Yin, J. (2019). Tetraphenylene-coated near-infrared benzoselenodiazole dye: AIE behavior, mechanochromism, and bioimaging. *Organic Letters*, *21*(18), 7213–7217. https://doi.org/10.1021/acs.orglett.9b02292.

Yin, C., Li, X., Wen, G., Yang, B., Zhang, Y., Chen, X., et al. (2020). Organic semiconducting polymer amphiphile for near-infrared-II light-triggered phototheranostics. *Biomaterials*, *232*, 119684. https://doi.org/10.1016/j.biomaterials.2019.119684.

Zhang, L., Liu, C., Zhou, S., Wang, R., Fan, Q., Liu, D., et al. (2020). Improving quantum yield of a NIR-II dye by phenylazo group. *Advanced Healthcare Materials*, *9*(4), 1901470. https://doi.org/10.1002/adhm.201901470.

Zhou, Y., Li, M., Liu, W., Sankin, G., Luo, J., Zhong, P., et al. (2019). Thermal memory based photoacoustic imaging of temperature. *Optica*, *6*(2), 198–205. https://doi.org/10.1364/OPTICA.6.000198.

CHAPTER TEN

Mitochondria-targeted photoacoustic probe for imaging of hydrogen peroxide in inflamed mouse model

Lele Zhang and Zijuan Hai*
Key Laboratory of Structure and Functional Regulation of Hybrid Materials, Ministry of Education, Institutes of Physical Science and Information Technology, Anhui University, Hefei, AH, China
*Corresponding author: e-mail address: zijuan@ahu.edu.cn

Contents

Abstract

In this chapter, we gave a brief introduction of hydrogen peroxide (H_2O_2) and its existing analytical methods and described the need of mitochondria-targeted photoacoustic (PA) probe for H_2O_2 detection *in vivo*. Then we provided the detailed protocols for the design and characterization of a mitochondria-targeted PA probe (**TPP-HCy-BOH**) to visualize H_2O_2 *in vivo*, which was developed in our previous work. Compared to control probe without mitochondria-targeted ability (**HCy-BOH**), **TPP-HCy-BOH** could efficiently accumulate in mitochondria and activate its PA signals toward overproduced H_2O_2 in inflamed mouse model with higher PA sensitivity.

Methods in Enzymology, Volume 657
ISSN 0076-6879
https://doi.org/10.1016/bs.mie.2021.06.036

1. Introduction

Hydrogen peroxide (H_2O_2) is produced by activation of NADPH oxidase complexes during cellular stimulation with cytokines, neurotransmitters and peptide growth (Giorgio, Trinei, Migliaccio, & Pelicci, 2007). It is involved in regulating a wide variety of physiological processes including host defense, immune responses and cellular signal transduction (Rhee, 2007). As a relatively stable reactive oxygen species (ROS), H_2O_2 can not only act as a precursor for the generation of other ROS, but also act as a byproduct from other ROS related reactions (Chen, Tian, Shin, & Yoon, 2011). The excessive production or accumulation of H_2O_2 can cause significant oxidative damage and results in severe diseases such as cancer, diabetes, cardiovascular and neurodegenerative diseases, which makes it a potential diagnostic marker for pathological states (Du et al., 2014; Ohshima, Tatemichi, & Sawa, 2003; Xiong et al., 2019). Mitochondria are the sites of most ATP generation in cells, which play a key role in cellular energy metabolism (Green & Kroemer, 2004). It is the primary cellular compartment of oxygen consumption and ROS generation (Dickinson & Chang, 2011). The ROS generated by mitochondria accounts for 90% of the total amount of ROS in the cell. (Yang & Pan, 2015) Therefore, precisely monitoring the level of H_2O_2 in the mitochondria is crucial to understand the biological roles of H_2O_2 in living systems.

Several analytical techniques including colorimetric assay, mass spectrometry, liquid chromatography and electrochemistry have been applied to detect H_2O_2 *in vitro* (Lu, Cagan, Munoz, Tangkuaram, & Wang, 2006; Lu et al., 2020). However, these traditional methods are incapable of dealing with living samples in real-time with the requirement of complicated sample preparation. In more recent years, abundant probes based on different imaging modes are developed to monitor H_2O_2 in living systems with either great sensitivity or great spatial resolution (Chen et al., 2017; Guo et al., 2015). Among them, fluorescence probes achieved remarkable progress based on diverse H_2O_2-responsive components such as sulfonyl ester, boric acid/borate ester, α-diketone and oxonium group (Dong et al., 2016). For example, based on H_2O_2-mediated transformation of borate ester to phenol, Lin and colleagues developed a fast responsive two-photon fluorescent probe for imaging H_2O_2 in lysosomes with a large turn-on fluorescence signal

(Ren et al., 2016). However, the majority of fluorescent probes emits in the visible range with low tissue penetration which are limited to cell study. Therefore, the development of new probes to visualize H_2O_2 *in vivo* is still of urgent necessity.

Photoacoustic (PA) imaging, which relies on ultrasound signals generated by photothermal expansion of light-absorbing tissues or contrast probes under pulsed laser irradiation, has emerged as a promising type of biomedical imaging technique combining advantages of both optical and ultrasound imaging (Chen et al., 2015; Kim, Favazza, & Wang, 2010; Xu & Wang, 2006). PA imaging has already moved into clinical trials owing to its significantly improved penetration depth (as deep as ~12 cm) and spatial resolution (100 μm) (Li et al., 2020; Zhou et al., 2016). A large variety of organic and inorganic materials with absorption in the near-infrared (NIR) "tissue transparent" optical window have been extensively explored as PA probes for bioimaging analysis (Cheng, Li, Tang, & Yoon, 2020; Wu et al., 2018). To date, only a few PA probes have been reported to visualize H_2O_2 *in vivo*, but they could not monitor the level of H_2O_2 at specific organelles (e.g., mitochondria) (Chen et al., 2017; Weber et al., 2019). Therefore, we rationally designed a mitochondria-targeted PA probe for imaging of H_2O_2 in inflammation *in vivo*.

2. Design of TPP-HCy-BOH and HCy-BOH

Mitochondria-targeted PA probe **TPP-HCy-BOH** contains three moieties: (1) the boric acid (BOH) moiety as a specific response site for H_2O_2; (2) the NIR hemicyanine (HCy) dye with outstanding absorption property provides activatable PA signal; (3) the triphenylphosphine (TPP) moiety is generally employed to target mitochondria. The PA signal of **TPP-HCy-BOH** are "OFF" due to the caged hydroxyl group of HCy with inhibited intramolecular charge transfer (ICT). After entering cells, **TPP-HCy-BOH** can accumulate at mitochondria effectively. Overproduced H_2O_2 in inflamed cells will quickly release the free HCy (**TPP-HCy**) by removing the BOH moiety. **TPP-HCy** with recovered ICT turns on its PA signal at mitochondria with superb sensitivity. **HCy-BOH** without mitochondria-targeted moiety (TPP group) is designed as a control probe (Fig. 1).

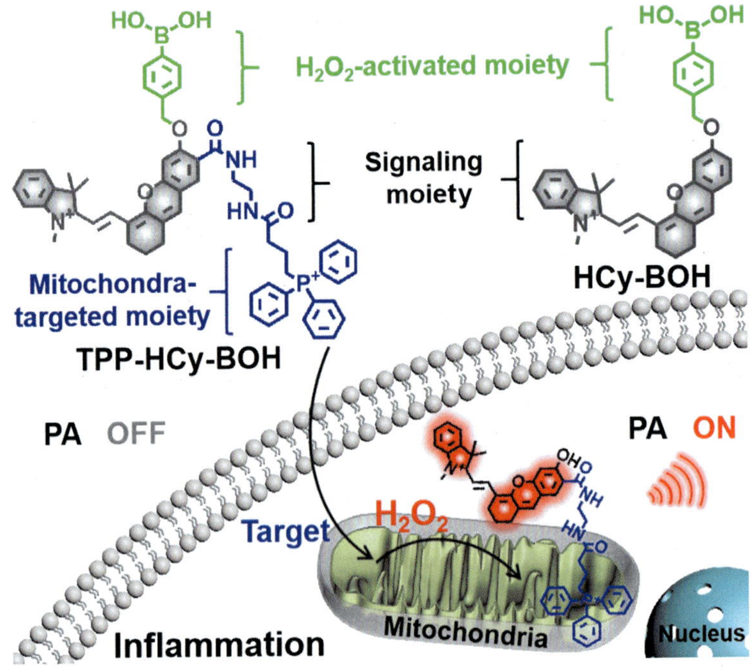

Fig. 1 Chemical structures of **TPP-HCy-BOH** and **HCy-BOH** and schematic illustration of mitochondria-targeted **TPP-HCy-BOH** for PA imaging of H_2O_2 in inflammation. *Reproduced with permission [Anal. Chem. 2020, 92, 14244.] Copyright 2020, American Chemical Society.*

3. Syntheses of TPP-HCy-BOH and HCy-BOH

We began the study with the syntheses of **TPP-HCy-BOH** and **HCy-BOH** (Fig. 2).

3.1 Procedure

1. Synthesis of **A**: (3-Carboxypropyl)triphenylphosphonium bromide (429 mg, 1.0 mmol), HBTU (455 mg, 1.2 mmol), and HOBT (160 mg, 1.2 mmol) in DMF (2 mL) were stirred for 0.5 h in presence of DIPEA (250 μL, 1.5 mmol). Then N-Boc-ethylenediamine (240 mg, 1.5 mmol) was added and stirred for 12 h at room temperature. The Boc protecting groups was removed with 50% TFA in CH_2Cl_2 for 3 h, then compound **A** (375.5 mg, yield 96%) was obtained after HPLC purification

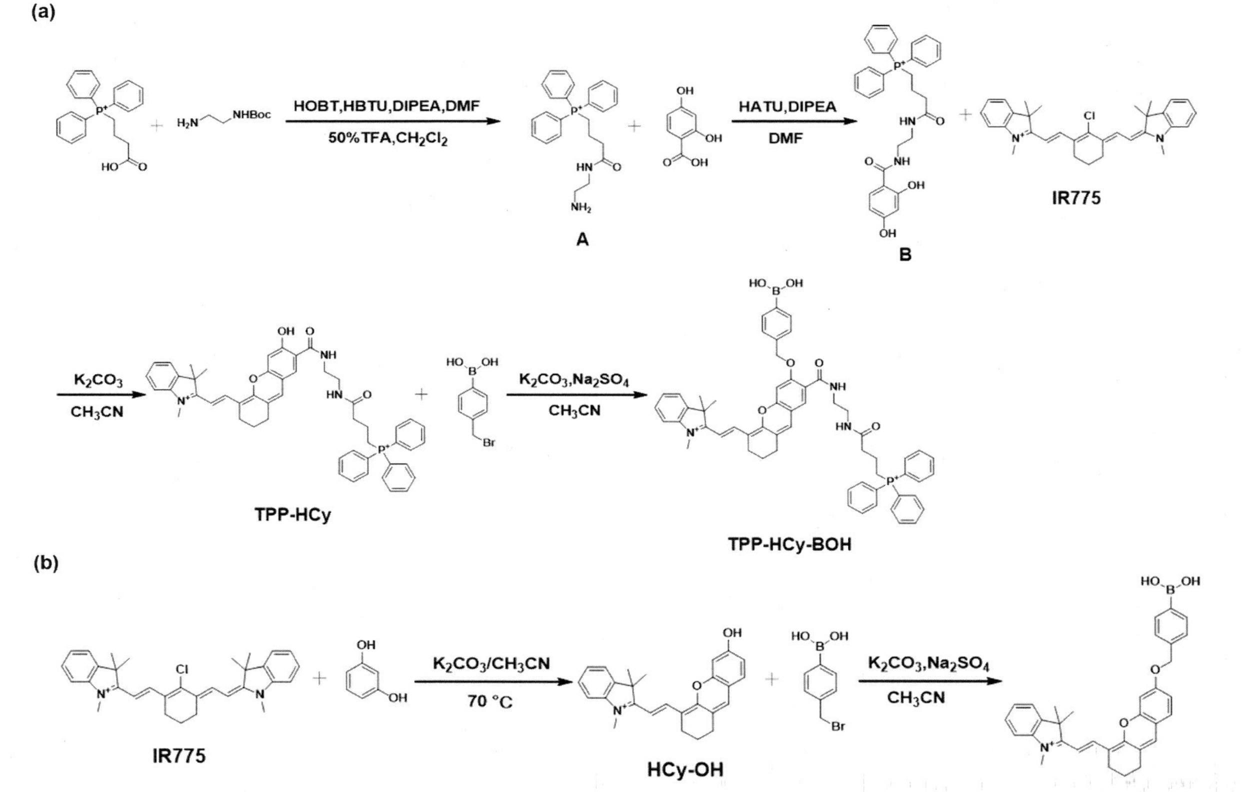

Fig. 2 The synthetic routes for **TPP-HCy-BOH** (A) and **HCy-BOH** (B). *Reproduced with permission [Anal. Chem. 2020, 92, 14244.] Copyright 2020, American Chemical Society.*

2. Synthesis of **B**: 2,4-Dihydroxybenzaic azide (296 mg, 1.9 mmol),HATU (722 mg, 1.9 mmol) in DMF (5 mL) were stirred for 0.5 h in presence of DIPEA (426 μL, 2.0 mmol). Then compound **A** (500 mg, 1.28 mmol) in DMF (4 mL) was added and stirred overnight at room temperature. Compound **B** (472.20 mg, yield 70%) was obtained after HPLC purification
3. Synthesis of **TPP-HCy**: K_2CO_3 (52 mg, 0.38 mmol) was added to a solution of Compound **B** (150 mg, 0.28 mmol) in CH_3CN (3 mL) at room temperature and stirred for 20 min under N_2 atmosphere. Then a solution of IR-775 chloride (99 mg, 0.19 mmol) in CH_3CN (5 mL) was added to the above solution and stirred for 24 h at 70 °C. The product **TPP-HCy** (54.5 mg, yield 24%) was purified by HPLC
4. Synthesis of **TPP-HCy-BOH**: The solution of 4-(Bromomethyl) phenylboronic acid (17 mg, 0.078 mmol) in CH_3CN was added to K_2CO_3 (11.1 mg, 0.078 mmol) and Na_2SO_4 (11.4 mg, 0.078) in a three-neck bottle under N_2 atmosphere. Then, **TPP-HCy** (21 mg, 0.026 mmol) was added into the mixture. The reaction mixture was stirred at room temperature for 24 h. Compound **TPP-HCy-BOH** (16.35 mg, yield 67%) was obtained after HPLC purification
5. Synthesis of **HCy-OH**: resorcinol (44.04 mg, 0.4 mmol) was dissolved in CH_3CN (500 μL) and was added rapidly to a two-neck flask containing K_2CO_3 (55.3 mg, 0.4 mmol) under nitrogen atmosphere. The reaction mixture was stirred at room temperature for 20 min. IR775 (104.0 mg, 0.2 mmol) in 5 mL CH_3CN was then introduced to the mixture. The reaction mixture was heated at 70 °C for 12 h. Compound **HCy-OH** (62.11 mg, yield 81%) was obtained after HPLC purification
6. Synthesis of **HCy-BOH**: 4-(Bromomethyl)phenylboronic acid (28.60 mg, 0.13 mmol) was dissolved in CH_3CN (400 μL) and was added rapidly to a two-neck flask containing K_2CO_3 (18.40 mg, 0.13 mmol) and Na_2SO_4 (7.20 mg, 0.07 mmol) under nitrogen atmosphere. Then, **HCy-OH** (17 mg, 0.04 mmol) was added and stirred overnight at room temperature. Compound **HCy-BOH** (15 mg, yield 75%) was obtained after HPLC purification

4. Characterization of TPP-HCy-BOH and HCy-BOH *in vitro*

4.1 UV–vis and PA properties of TPP-HCy-BOH and HCy-BOH

After that, we studied the UV–vis and PA spectra of **TPP-HCy-BOH** and **HCy-BOH**.

4.1.1 Equipment

PerkinElmer Lambda 25 UV–vis spectrometer
Multispectral Optoacoustic Tomographic Imaging System (MSOT in Vision 256, iThera Medical, Germany)

4.1.2 Reagents

TPP-HCy-BOH and **HCy-BOH** stock solution (2 mM in DMSO)
50 mM phosphate-buffered saline (PBS, pH 7.4)
H_2O_2 stock solution (1, 2, 3, 4, 6, and 8 mM in water)

4.1.3 Procedure

1. Prepare 20 μM **TPP-HCy-BOH** and **HCy-BOH** solution in 50 mM PBS (1% DMSO)
2. Incubate **TPP-HCy-BOH** and **HCy-BOH** solution (20 μM) with a series of H_2O_2 (final concentration: 0, 10, 20, 30, 40, 60 and 80 μM in 50 mM PBS) at 37 °C for 30 min
3. Measure UV–vis spectra (250–900 nm) of 20 μM **TPP-HCy-BOH** and **HCy-BOH** solution with 0 and 80 μM H_2O_2 in quartz cuvette
4. Measure PA spectra (660–900 nm) of 20 μM **TPP-HCy-BOH** and **HCy-BOH** solution with 0, 10, 20, 30, 40, 60 and 80 μM H_2O_2. Seal above solution into a 3.3 mm pipette and scan with a step size of 1 mm of longitudinal slices for PA measurement. The PA images of cell suspensions were obtained at a single wavelength at 685 nm

4.1.4 Results

In the presence of 80 μM H_2O_2, the visible absorption maximum of 20 μM **TPP-HCy-BOH** shifted from 592 to 690 nm (Fig. 3A) while the absorption of 20 μM **HCy-BOH** at 570 and 602 nm decreased and at 679 nm increased (Fig. 3D). In detail, the absorbance at 685 nm of **TPP-HCy-BOH** with H_2O_2 was 13.9-fold higher than that of **TPP-HCy-BOH** while the absorbance at 685 nm of **HCy-BOH** with H_2O_2 was 1.97-fold higher than that of **HCy-BOH**. As shown in Fig. 3B and E, the PA signal of 20 μM **TPP-HCy-BOH** with 80 μM H_2O_2 at 685 nm was 10.7-fold higher than **TPP-HCy-BOH** while the PA signal of 20 μM **HCy-BOH** with H_2O_2 at 685 nm was 1.7-fold higher than **HCy-BOH**, which were consistent with their absorption changes. The PA images at 685 nm also confirmed the much more increase of PA signal of **TPP-HCy-BOH** than **HCy-BOH** after H_2O_2 treatment. In addition, by studying the linear correlation between PA intensities at 685 nm and the concentrations of H_2O_2, the limits of detection (LOD) were calculated to be 0.872 μM for **TPP-HCy-BOH** and 3.369 μM for **HCy-BOH** (Fig. 3C and F), which are

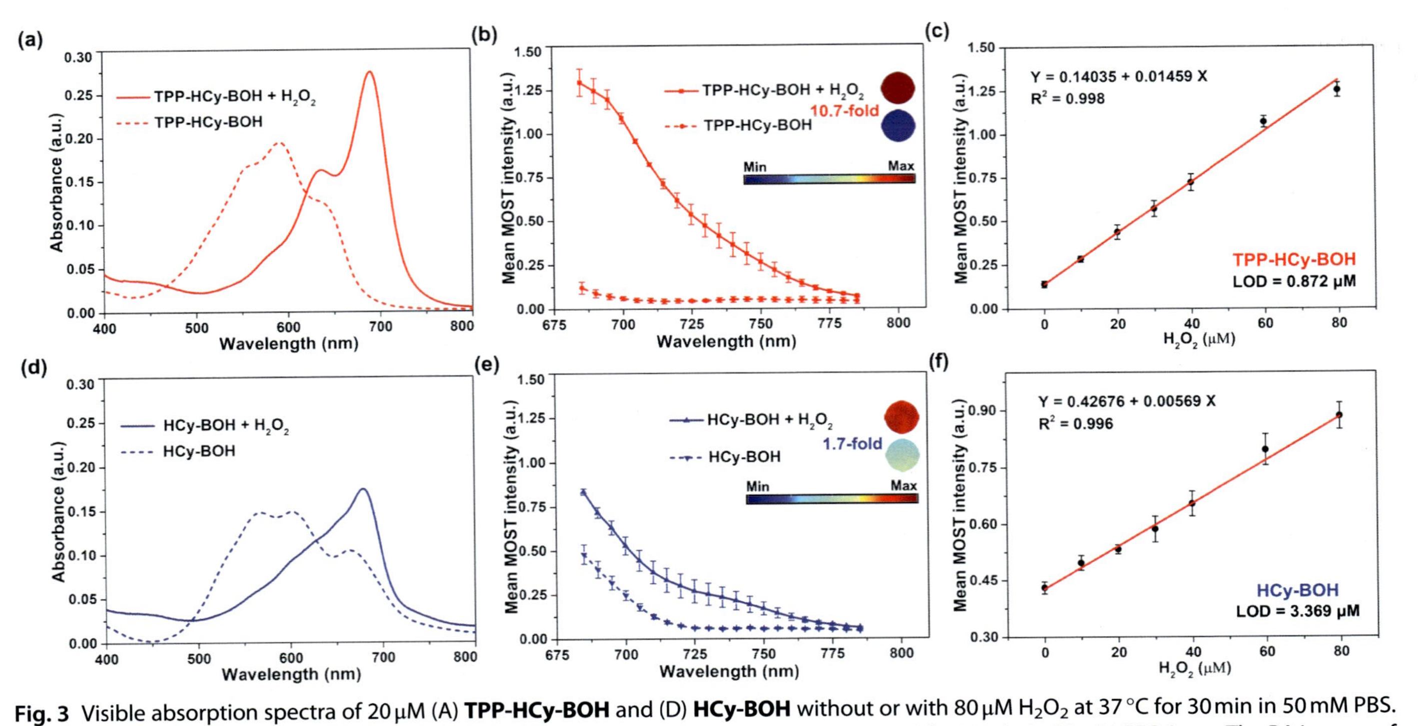

Fig. 3 Visible absorption spectra of 20 μM (A) **TPP-HCy-BOH** and (D) **HCy-BOH** without or with 80 μM H_2O_2 at 37 °C for 30 min in 50 mM PBS. PA spectra of 20 μM (B) **TPP-HCy-BOH** and (E) **HCy-BOH** without or with 80 μM H_2O_2 at 37 °C for 30 min in 50 mM PBS. Inset: The PA images of 20 μM **TPP-HCy-BOH** and **HCy-BOH** at 685 nm without or with 80 μM H_2O_2. The fitted calibration lines of 20 μM (C) **TPP-HCy-BOH** and (F) **HCy-BOH** in the linear region of 0–80 μM H_2O_2. The error bars represent the standard deviation from three separate measurements. *Reproduced with permission [Anal. Chem. 2020, 92, 14244.] Copyright 2020, American Chemical Society.*

Table 1 Summary of limits detection of **HCy-TPP-BOH**, **HCy-BOH** and recently reported H_2O_2 probes.

Probes	Limit of detection	Buffer solution
TPP-HCy-BOH	0.872 μM	50 mM PBS, pH 7.4,1% DMSO
HCy-BOH	3.369 μM	50 mM PBS, pH 7.4,1% DMSO
Mito-CD-PF1	0.75 μM	50 mM HEPES buffer, pH 7.0
Lyso-HP	1.21 μM	PBS buffer, pH 7.4, containing 50% DMF
MI-H_2O_2	80 nM	10 mM PBS, pH 8.0
ER-H_2O_2	120 nM	10 mM PBS, pH 7.4
LyNC	0.22 μM	50 mM PBS, pH 7.4, 0.5% DMSO
Mito-FBN	0.025 μM	10 mM phosphate buffers, pH 7.4
Lyso-B-L1	0.06 μM	10 mM acetate buffer, pH 4.5, 1% DMSO
pep3-NP1	2.90 μM	1 μM CT DNA,Tris-HCl, pH 7.2, 1% DMSO
SHP-Mito	4.6 μM	30 mM MOPS, 100 mM KCl, pH 7.4
NP1	0.17 μM	Mixed solvent of Me_2SO and PBS (5:1000, *v*/v), pH 7.2

comparable to those of recently reported probes for H_2O_2 detection (Table 1). These results proved that **TPP-HCy-BOH** is more sensitive to detect H_2O_2 than **HCy-BOH** *in vitro*.

4.2 Kinetic study of TPP-HCy-BOH and HCy-BOH toward H_2O_2

To further explore the higher sensitivity of **TPP-HCy-BOH** than **HCy-BOH**, we studied the kinetic response of probes toward H_2O_2.

4.2.1 Equipment

F4600-fluorescence spectrophotometer (Hitachi High Technologies Corporation, Japan)

Agilent 1200 HPLC system equipped with a G1322A pump and an in-line diode array UV detector using an Agilent Zorbax 300SB-C18 RP column, with CH_3CN (0.1% of TFA) and ultrapure water (0.1% of TFA) as the eluent

4.2.2 Reagents

TPP-HCy-BOH and **HCy-BOH** stock solution (1 mM in DMSO)
TPP-HCy and **HCy-OH** stock solution (1 mM in DMSO)
50 mM phosphate-buffered saline (PBS, pH 7.4)
H_2O_2 stock solution (4, 10 mM in water)

4.2.3 Procedure

1. Prepare 10 μM **TPP-HCy-BOH** and **HCy-BOH** solution in 50 mM PBS (1% DMSO)
2. Incubate **TPP-HCy-BOH** and **HCy-BOH** solution (10 μM) with H_2O_2 (final concentration: 1 mM in 50 mM PBS) at 37 °C in quartz cuvette, monitor the FL spectra (620–900 nm, λ_{ex} = 600 nm) of above solution every 30 s within 10 min
3. The pseudo-first-order rate constants (k_{obs}) were obtained from slope of the plot of ln [(F_{max}–F_t)/F_{max}] *vs* time. F_{max} is maximum fluorescence intensity during the measurement time, F_t is the fluorescence intensity of **TPP-HCy-BOH** at 716 nm and **HCy-BOH** at 706 nm at the corresponding time points
4. Prepare 10 μM **TPP-HCy** and **HCy-OH** solution in 50 mM PBS (1% DMSO)
5. Incubate **TPP-HCy-BOH** and **HCy-BOH** solution (10 μM) with H_2O_2 (final concentration: 40 μM in 50 mM PBS) at 37 °C for 30 min
6. Inject 1 mL above six solution (10 μM **TPP-HCy-BOH**, 10 μM **TPP-HCy-BOH** + 40 μM $\mathbf{H_2O_2}$, 10 μM **TPP-HCy**, 10 μM **HCy-BOH**, 10 μM **HCy-BOH** + 40 μM $\mathbf{H_2O_2}$, 10 μM **HCy-OH**) into HPLC system respectively to acquire HPLC traces

4.2.4 Results

According to the linear equation (In[(F_{max}-F_t)/F_{max}] = − k_{obs} t) under pseudo-first-order condition (10 μM **TPP-HCy-BOH** or **HCy-BOH**, 1 mM H_2O_2), the pseudo-first-order rate constants (k_{obs}) were calculated to be $4.72 \times 10^{-3}\ s^{-1}$ for **TPP-HCy-BOH** and $3.99 \times 10^{-3}\ s^{-1}$ for **HCy-BOH** (Fig. 4A and C), which were superior to those of recently reported H_2O_2 probes (Table 2). Reversed-phase high-performance liquid chromatography (RP-HPLC) was also employed to measure the response efficiency of 10 μM **TPP-HCy-BOH** and **HCy-BOH** toward 40 μM H_2O_2 at 37 °C for 30 min. At this condition, **TPP-HCy-BOH** could totally convert to **TPP-HCy** while only 82.5% of **HCy-BOH** converted to **HCy-OH** (Fig. 4B and D). Therefore, the much quicker response rate of **TPP-HCy-BOH** toward H_2O_2 endowed its superb sensitivity.

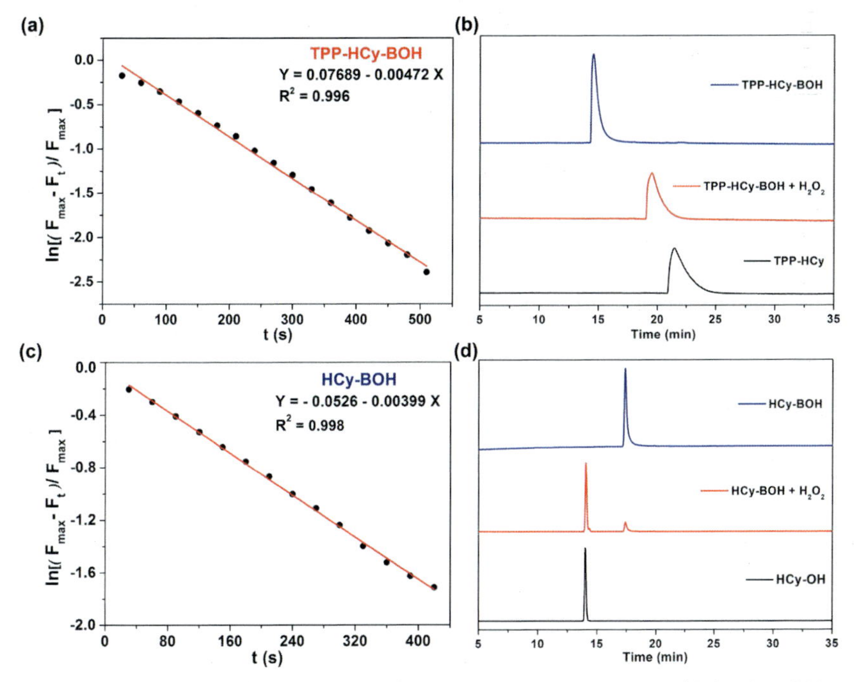

Fig. 4 Kinetic fluorescence responses of 10 μM (A) **TPP-HCy-BOH** and (C) **HCy-BOH** to 1 mM H_2O_2 in 50 mM PBS (1% DMSO). RP-HPLC traces of (B) **TPP-HCy** (10 μM), **TPP-HCy-BOH** (10 μM) in the absence or presence of H_2O_2 (40 μM) and (D) **HCy-OH** (10 μM), **HCy-BOH** (10 μM) in the absence or presence of H_2O_2 (40 μM) at 37 °C for 30 min in 50 mM PBS (1% DMSO). Wavelength: 600 nm. *Reproduced with permission [Anal. Chem. 2020, 92, 14244.] Copyright 2020, American Chemical Society.*

4.3 PA imaging of exogenous H_2O_2 in living cells

After *in vitro* investigation, we applied **TPP-HCy-BOH** and **HCy-BOH** for imaging of exogenous H_2O_2 in living cells.

4.3.1 Equipment

Zeiss LSM 880 laser scanning confocal microscope

Multispectral Optoacoustic Tomographic Imaging System (MSOT in Vision 256, iThera Medical, Germany)

4.3.2 Reagents

TPP-HCy-BOH and **HCy-BOH** stock solution (1, 2 mM in DMSO)

H_2O_2 stock solution (2.5, 5 mM in water)

MitoTracker Green (1 mM in DMSO)

Hoechst 33342 (1 mg/mL)

Table 2 Summary of pseudo-first-order rate constant (k_{obs}) of **HCy-TPP-BOH**, **HCy-BOH** and recently reported H_2O_2 probes.

Probes	k_{obs}	Concentration
TPP-HCy-BOH	$4.72 \times 10^{-3}\ s^{-1}$	10 μM **TPP-HCy-BOH**, 1 mM H_2O_2
HCy-BOH	$3.99 \times 10^{-3}\ s^{-1}$	10 μM **HCy-BOH**, 1 mM H_2O_2
pep3-NP1	$0.487 \times 10^{-3}\ s^{-1}$	1 μM pep3-NP1, 1 mM H_2O_2
SHP-Mito	$1.18 \times 10^{-3}\ s^{-1}$	1 μM SHP-Mito, 1 mM H_2O_2
Mito-CD-PL1	$3.0 \times 10^{-4}\ s^{-1}$	1 μM Mito-CD-PL1, 1 mM H_2O_2
MitoPY1	$2.0 \times 10^{-3}\ s^{-1}$	5 μM MitoPY1, 10 mM H_2O_2
NucPE1	$8.2 \times 10^{-3}\ s^{-1}$	5 mM NucPE1, 10 mM H_2O_2
NIR-H_2O_2	$2.95 \times 10^{-3}\ s^{-1}$	1 μM NIR-H_2O_2, 100 μM H_2O_2
MI-H_2O_2	$4.35 \times 10^{-3}\ s^{-1}$	10 μM MI-H_2O_2, 1 mM H_2O_2
ER-H_2O_2	$2.35 \times 10^{-3}\ s^{-1}$	10 μM ER-H_2O_2, 1 mM H_2O_2
DQHP	$1.4 \times 10^{-3}\ s^{-1}$	1.25 mM DQHP, 1 mM H_2O_2

4.3.3 Procedure

1. HeLa cells were cultured in Dulbecco's modified Eagle's medium with 10% fetal bovine serum. The culture dishes were maintained at 37 °C under a humid atmosphere with 5% CO_2
2. For FL imaging, HeLa cells were seeded into glass bottom cell culture dishes (3.5 cm) at 1×10^5/well and cultured overnight. In Group **TPP-HCy-BOH** and Group **HCy-BOH**, 10 μM **TPP-HCy-BOH** or **HCy-BOH** in 1 mL serum-free DMEM was incubated with HeLa cells for 30 min. HeLa cells in Group **H_2O_2 +TPP-HCy-BOH** and Group **H_2O_2 +HCy-BOH** were preincubated with 25 μM H_2O_2 for 30 min, and then incubated with 10 μM **TPP-HCy-BOH** or **HCy-BOH** for 30 min. These cells were stained with commercial MitoTracker Green (1 μM) and Hoechst 33342 (100 μg/mL) for 15 min before FL imaging
3. For PA imaging, in Group **TPP-HCy-BOH** and Group **HCy-BOH**, 20 μM **TPP-HCy-BOH** or **HCy-BOH** in 1 mL serum-free DMEM was incubated with HeLa cells for 30 min. HeLa cells in Group **H_2O_2 +TPP-HCy-BOH** and Group **H_2O_2 +HCy-BOH** were pre-incubated with 50 μM H_2O_2 for 30 min, and then incubated with 20 μM **TPP-HCy-BOH** or **HCy-BOH** for 30 min. HeLa cells in four

groups were detached from culture dishes with trypsin digestion solution after respective incubation. Cell suspensions in serum-free DMEM were sealed into a 3.3 mm pipette and scanned with a step size of 1 mm of longitudinal slices for PA imaging. PA signals were detected under different excitation wavelengths (660–900 nm). The PA images of cell suspensions were obtained at a single wavelength at 685 nm

4.3.4 Results

HeLa cells incubated with 10 μM **TPP-HCy-BOH** or **HCy-BOH** for 30 min only showed slight red fluorescence. When preincubated with 25 μM H_2O_2 for 30 min and then incubated with 10 μM **TPP-HCy-BOH** or **HCy-BOH**, the red fluorescence of HeLa cells obviously increased, suggesting the quick response of probes towards exogenous H_2O_2 in cells. The average FL intensity of Group **H_2O_2 +TPP-HCy-BOH** increased 6.6-fold while that of Group **H_2O_2 +HCy-BOH** increased 2.8-fold after H_2O_2 treatment (Fig. 5A). To investigate the mitochondrial targeting property of probes, costaining experiment was performed with the commercial MitoTracker Green. Pearson's colocalization coefficient (P) of **TPP-HCy-BOH** with MitoTracker Green was 0.93 while that of **HCy-BOH** was 0.71, validating that **TPP-HCy-BOH** was able to accumulate in mitochondria efficiently due to the TPP moiety. In addition, PA spectra and images of HeLa cells were studied. As shown in Fig. 5B, PA signals at 685 nm of HeLa cell suspensions in Group **H_2O_2 +TPP-HCy-BOH** (20 μM) remarkably enhanced 7.0-fold while that in Group **H_2O_2 +HCy-BOH** (20 μM) slightly enhanced 1.5-fold after preincubation with 50 μM H_2O_2. And the PA images at 685 nm of Group **H_2O_2 +TPP-HCy-BOH** also showed obvious change than Group **H_2O_2 +HCy-BOH** with H_2O_2 treatment. These results confirmed that **TPP-HCy-BOH** could enhance the FL signal by 2.4-fold and the PA signal by 4.7-fold additionally for more sensitive imaging of exogenous H_2O_2 in HeLa cells with good mitochondrial targeting ability than **HCy-BOH**.

4.4 PA imaging of endogenous H_2O_2 in living cells

Next, we further explored the feasibility of probes to detect endogenous H_2O_2 in murine macrophage (RAW 264.7) cells. Phorbol-12-myristate-13-acetate (PMA) was used to induce H_2O_2 generation through a cellular inflammation response.

4.4.1 Equipment

Multispectral Optoacoustic Tomographic Imaging System (MSOT in Vision 256, iThera Medical, Germany).

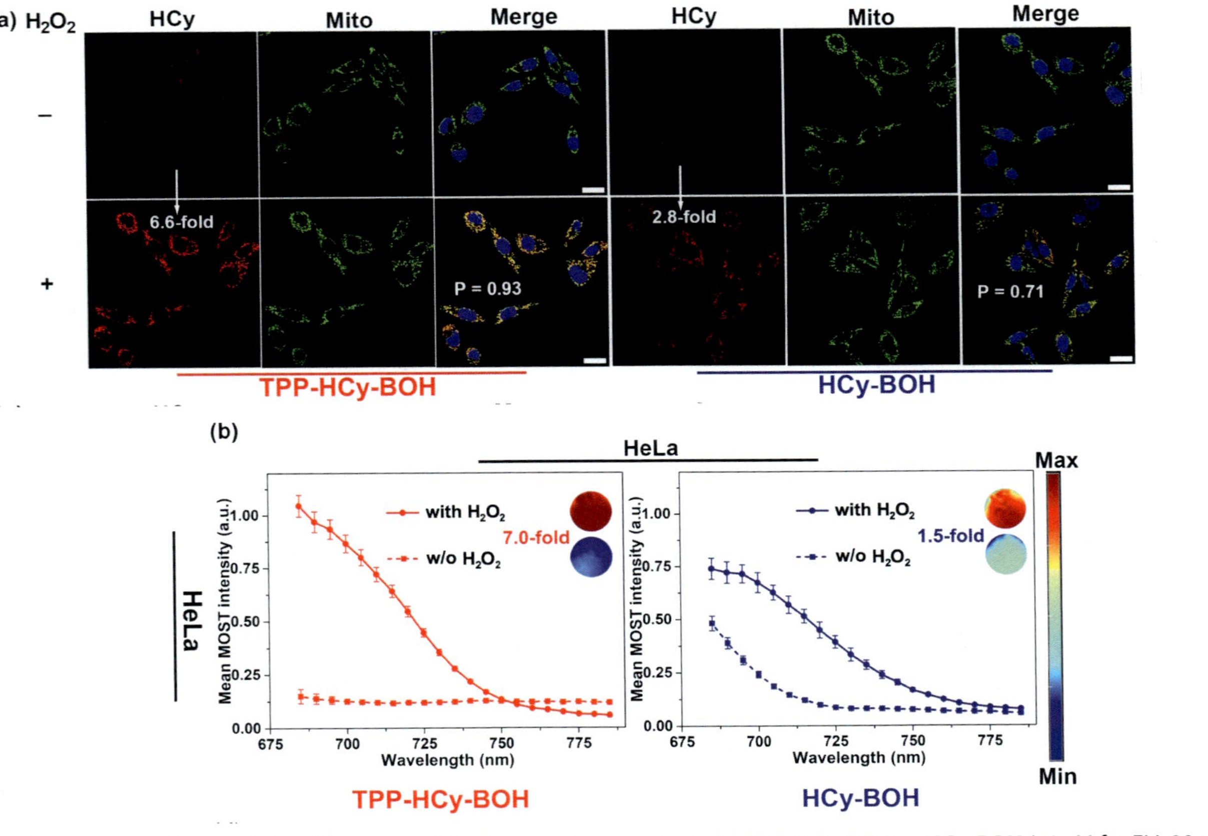

Fig. 5 Confocal fluorescence images (A) and PA spectra (B) of HeLa cells incubated with **TPP-HCy-BOH** or **HCy-BOH** (10 μM for FLI, 20 μM for PAI) for 30 min, preincubated with H_2O_2 (25 μM for FLI, 50 μM for PAI) for 30 min then **TPP-HCy-BOH** or **HCy-BOH** for 30 min in serum-free DMEM at 37 °C. Scale bar: 20 μm. Inset: The PA images of HeLa cell suspensions. Red is from HCy of **TPP-HCy-BOH** or **HCy-BOH**, green is staining of mitochondria, blue is staining of the nucleus. The error bars represent the standard deviation from three separate measurements. *Reproduced with permission [Anal. Chem. 2020, 92, 14244.] Copyright 2020, American Chemical Society.*

4.4.2 Reagents

TPP-HCy-BOH and **HCy-BOH** stock solution (2 mM in DMSO)
PMA stock solution (1 mg/mL in DMSO)

4.4.3 Procedure

1. RAW 264.7 cells were cultured in Dulbecco's modified Eagle's medium with 10% fetal bovine serum. The culture dishes were maintained at 37 °C under a humid atmosphere with 5% CO_2
2. For PA imaging, in Group **TPP-HCy-BOH** and Group **HCy-BOH**, 20 μM **TPP-HCy-BOH** or **HCy-BOH** in 1 mL serum-free DMEM was incubated with RAW 264.7 cells for 30 min. RAW 264.7 cells in Group **PMA + TPP-HCy-BOH** and Group **PMA + HCy-BOH** were preincubated with 1 μg/mL PMA for 30 min, and then incubated with 20 μM **TPP-HCy-BOH** or **HCy-BOH** for 30 min. RAW 264.7 cells in four groups were detached from culture dishes with trypsin digestion solution after respective incubation. Cell suspensions in serum-free DMEM were sealed into a 3.3 mm pipette and scanned with a step size of 1 mm of longitudinal slices for PA imaging. PA signals were detected under different excitation wavelengths (660–900 nm). The PA images of cell suspensions were obtained at a single wavelength at 685 nm

4.4.4 Results

As shown in Fig. 6, the MSOT intensity at 685 nm of RAW 264.7 cell suspensions in Group **PMA + TPP-HCy-BOH** or Group **PMA + HCy-**

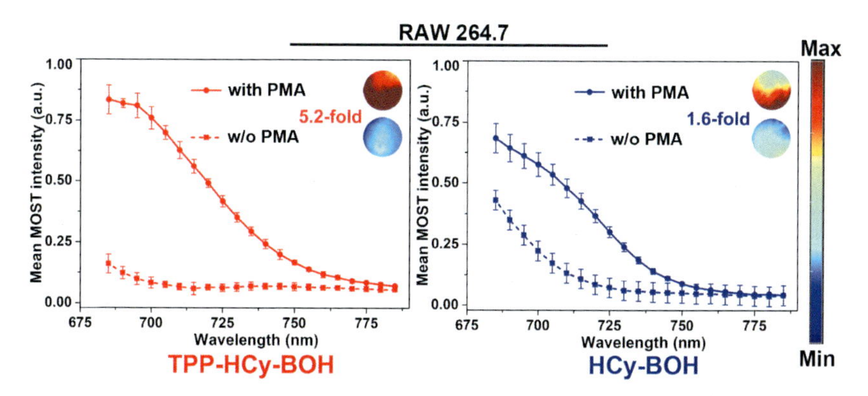

Fig. 6 PA spectra of RAW 264.7 cells incubated with **TPP-HCy-BOH** or **HCy-BOH** (20 μM) for 30 min, pretreated with 1 μg/mL PMA for 30 min and then **TPP-HCy-BOH** or **HCy-BOH** for 30 min in serum-free DMEM at 37 °C. Inset: The PA images of RAW 264.7 cell suspensions. The error bars represent the standard deviation from three separate measurements. *Reproduced with permission [Anal. Chem. 2020, 92, 14244.] Copyright 2020, American Chemical Society.*

BOH had a 5.2-fold or 1.6-fold enhancement after PMA stimulation, respectively. Therefore, probe **TPP-HCy-BOH** is more competent to image endogenous H_2O_2 in mitochondria with additional 3.3-fold PA increase than **HCy-BOH**.

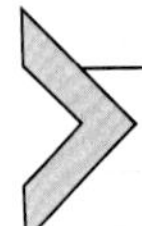

5. Characterization of TPP-HCy-BOH and HCy-BOH *in vivo*

5.1 Characterization of inflamed mouse model

After living cell imaging, we aimed to image endogenous H_2O_2 *in vivo* which generated by activated macrophages and neutrophils in a lipopolysaccharide (LPS) model of acute inflammation.

5.1.1 Equipment

Bio-Rad CFX384 real-time PCR system.

5.1.2 Reagents

TPP-HCy-BOH and **HCy-BOH** stock solution (25 mM in DMSO)
LPS stock solution (2 mg/mL in saline)
TRIzol reagent (Invitrogen)
HiScript III RT SuperMix (Vazyme, China)
SYBR Green qPCR Master Mix (Vazyme, China)

5.1.3 Procedure

1. 7-week-old (weighting 20 g) female BALB/c mice were divided into two groups ($n=3$ for each group): mice in Group **TPP-HCy-BOH** were intraperitoneally (i.p.) injected with **TPP-HCy-BOH** (250 μM, 200 μL) for 2 h and mice in Group **LPS + TPP-HCy-BOH** were first i.p. injected with LPS (2 mg/mL, 400 μL) for 6 h and then followed with **TPP-HCy-BOH**(250 μM, 200 μL) for 2 h. After that, we sacrificed the mice and took out peritoneal macrophage, intestine (ileum) and liver
2. The mRNA expression levels of TNFα, IL-1β, IL-6, CXCL10, ISG15, and iNOS were measured by Quantitative Real-Time PCR. Total RNA was isolated from peritoneal macrophage, intestine (ileum) and liver tissue using TRIzol reagent (Invitrogen) and reverse transcription was performed using HiScript III RT SuperMix (Vazyme, China) as manufacturer's instructions. The amplification was performed on the Bio-Rad CFX384 real-time PCR system in a 10 μL reaction system using SYBR Green qPCR Master Mix (Vazyme, China). GAPDH

was used as an endogenous control. The mRNA expression levels of the tested genes relative to GAPDH were determined using the $2^{-\Delta\Delta Ct}$ method and shown as fold induction. Primers were as follows (5′-3′): GAPDH (Forward: TGCACCACCA ACTGCTTAGC, Reverse: GGAAGGCCAT GCCAGTGA); TNFα (Forward: GACGTGGAAC TGGCAGAAGAG, Reverse: TTGGTGGTTT GTGAGTGTGAG); IL-6 (Forward: GATGGATGCT ACCAAACTGGAT, Reverse: CCAGGTAGCT ATGGTACTCCAGA); IL-1β (Forward: CTACAG GCTCC GAGATGAACAAC, Reverse: TCCATTGAGG TGGAGAG CTTTC); CXCL10 (Forward: GCCGTCATTT TCTGCCTCAT, Reverse: GCTTCCCTAT GGCCCTCATT); ISG15 (Forward: GGTG TCCGTG ACTAACTCCAT, Reverse: TGGAAAGGGT AAGAC CGTCCT); iNOS (Forward: GAGACAGGGA AGTCTGAAGCAC, Reverse: CCAGCAGTAG TTGCTCCTCTTC).

5.1.4 Results

We examined the inflammatory cytokines including tumor necrosis factor-α (TNF-α), interleukin 6 (IL-6), interleukin 1β (IL-1β), C-X-C motif chemokine 10 (CXCL10), interferon-stimulated gene 15 (ISG15) and nitric oxide synthase (iNOS) in peritoneal macrophage, intestine (ileum) and liver of Group **TPP-HCy-BOH** and Group **LPS + TPP-HCy-BOH**. Obviously, the level of inflammatory cytokines in Group **LPS + TPP-HCy-BOH** were much higher than those in Group **TPP-HCy-BOH** (Fig. 7), which confirmed the abdominal inflammation induced by LPS.

5.2 MSOT imaging of H_2O_2 in inflamed mouse model

Then, we employed MSOT to image overproduced H_2O_2 in the liver during LPS-induced inflammation.

5.2.1 Equipment

Multispectral Optoacoustic Tomographic Imaging System (MSOT in Vision 256, iThera Medical, Germany).

5.2.2 Reagents

TPP-HCy-BOH and **HCy-BOH** stock solution (25 mM in DMSO)
LPS stock solution (2 mg/mL in saline)

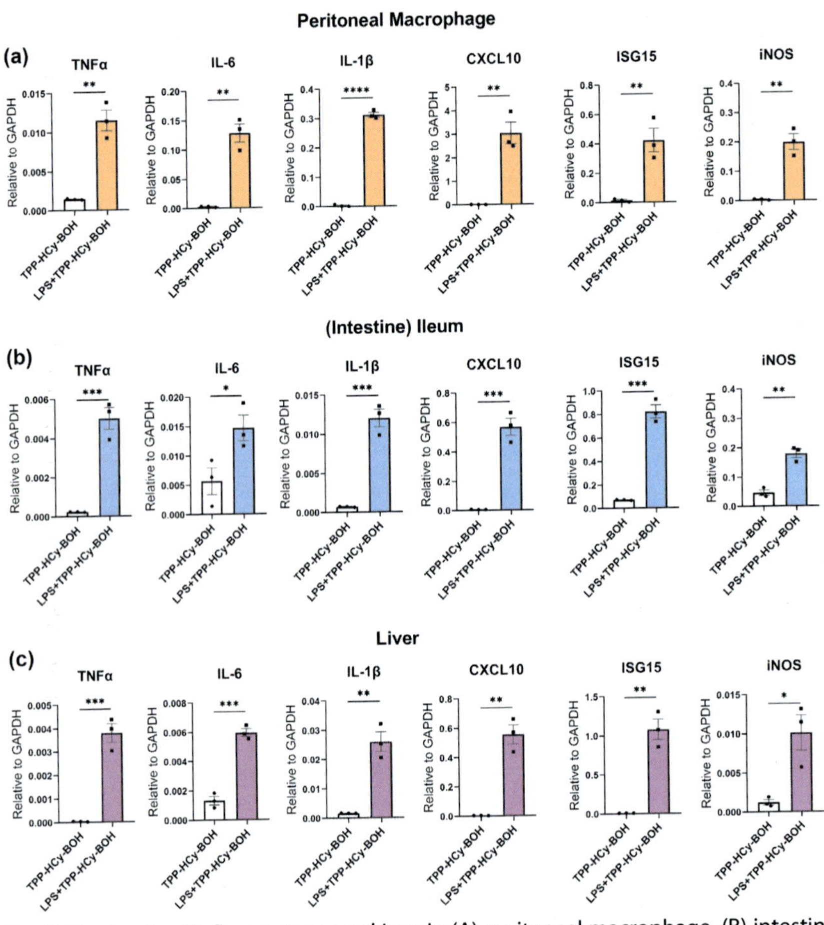

Fig. 7 The levels of inflammatory cytokines in (A) peritoneal macrophage, (B) intestine (Ileum) and (C) liver of mice in Group **TPP-HCy-BOH** and Group **LPS+TPP-HCy-BOH** were determined by qPCR. Data are shown as mean ± SEM. A value of $P<0.05$ was considered statistically significant. $^{*}P<0.05$, $^{**}P<0.01$, $^{***}P<0.001$, $^{****}P<0.0001$. *Reproduced with permission [Anal. Chem. 2020, 92, 14244.] Copyright 2020, American Chemical Society.*

5.2.3 Procedure

1. As shown in Fig. 8A, 7-week-old (weighting 20 g) female BALB/c mice were divided into four groups ($n=3$ for each group): mice in Group **TPP-HCy-BOH** and Group **HCy-BOH** were i.p. injected with **TPP-HCy-BOH** and **HCy-BOH** (250 μM, 200 μL) and mice in Group **LPS+TPP-HCy-BOH** and Group **LPS+HCy-BOH** were firstly i.p. injected with LPS (2 mg/mL, 400 μL) for 6 h and then followed with **TPP-HCy-BOH** and **HCy-BOH** (250 μM, 200 μL).

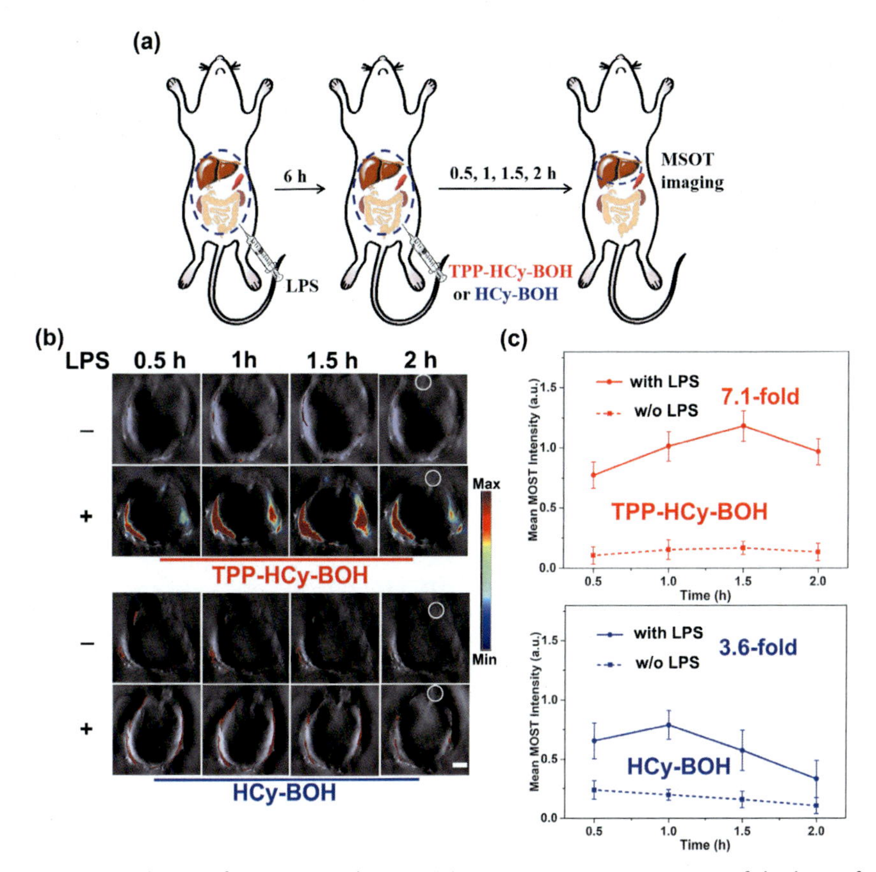

Fig. 8 (A) Scheme of experimental setup. (B) Time-course MSOT images of the liver of mice in Group **TPP-HCy-BOH** or **HCy-BOH** which was i.p. injected with **TPP-HCy-BOH** or **HCy-BOH** (250 μM, 200 μL) and Group **LPS + TPP-HCy-BOH** or **LPS + HCy-BOH** which was i.p. injected with LPS (2 mg/mL, 400 μL) for 6 h and followed by **TPP-HCy-BOH** or **HCy-BOH** (250 μM, 200 μL). White circles indicate the spinal cord. Scale bar: 3 mm. (C) Mean MSOT intensity in Fig. 8B. The error bars represent the standard deviation from three separate measurements. *Reproduced with permission [Anal. Chem. 2020, 92, 14244.] Copyright 2020, American Chemical Society.*

2. For MOST Imaging, mice in four groups were covered with ultrasound coupling agent and placed in the prone position. The livers of mice were scanned with a step size of 0.5 mm of longitudinal slices every 0.5 h within 2 h

5.2.4 Results

Fig. 8B showed the cross-sectional images reconstructed from the MSOT signals of the livers. MSOT signals of livers in mice of Group **LPS**

+ **TPP-HCy-BOH** obviously increased and reached maximum signal at 1.5 h while that of Group **LPS + HCy-BOH** slightly increased at 1 h and then quickly decreased. At 1.5 h, the mean MSOT intensity of Group **LPS + TPP-HCy-BOH** or **LPS + HCy-BOH** after LPS treatment was 7.1-fold or 3.6-fold higher than Group **TPP-HCy-BOH** or **HCy-BOH** (Fig. 8C). These results validated that mitochondria-targeted probe **TPP-HCy-BOH** could efficiently image the overproduced H_2O_2 in inflammation of mice with additional 2.0-fold higher sensitivity of PA in liver and longer retention time of 0.5 h than **HCy-BOH** *in vivo*.

6. Conclusions

In conclusion, we developed a mitochondria-targeted PA probe **TPP-HCy-BOH**, which could efficiently activate its PA signals toward overproduced H_2O_2 in inflamed mouse model. We systematically described the design and characterization of **TPP-HCy-BOH** and confirmed that it can specifically accumulate in mitochondria to image H_2O_2 with higher PA sensitivity than control probe **HCy-BOH** without mitochondria-targeted ability. We hope this work is useful for developing more activatable PA probes in the near future.

References

Chen, Q., Liang, C., Sun, X. Q., Chen, J. W., Yang, Z. J., Zhao, H., et al. (2017). H_2O_2-responsive liposomal nanoprobe for photoacoustic inflammation imaging and tumor theranostics via in vivo chromogenic assay. *Proceedings of the National Academy of Sciences of the United States of America*, *114*, 5343–5348.

Chen, Q., Liu, X. D., Chen, J. W., Zeng, J. F., Cheng, Z. P., Liu, Z., et al. (2015). A self-assembled albumin-based nanoprobe for in vivo ratiometric photoacoustic pH imaging. *Advanced Materials*, *27*, 6820–6827.

Chen, X., Tian, X., Shin, I., & Yoon, J. (2011). Fluorescent and luminescent probes for detection of reactive oxygen and nitrogen species. *Chemical Society Reviews*, *40*, 4783–4804.

Cheng, H. B., Li, Y. Y., Tang, B. Z., & Yoon, J. (2020). Assembly strategies of organic-based imaging agents for fluorescence and photoacoustic bioimaging applications. *Chemical Society Reviews*, *49*, 21–31.

Dickinson, B. C., & Chang, C. J. (2011). Chemistry and biology of reactive oxygen species in signaling or stress responses. *Nature Chemical Biology*, *7*, 504–511.

Dong, B., Song, X., Kong, X., Wang, C., Tang, Y., Liu, Y., et al. (2016). Simultaneous near-infrared and two-photon in vivo imaging of H_2O_2 using a ratiometric fluorescent probe based on the unique oxidative rearrangement of oxonium. *Advanced Materials*, *28*, 8755–8759.

Du, F. K., Min, Y. H., Zeng, F., Yu, C. M., Wu, S. Z., & Wu, S. Z. (2014). A targeted and fret-based ratiometric fluorescent nanoprobe for imaging mitochondrial hydrogen peroxide in living cells. *Small*, *10*, 964–972.

Giorgio, M., Trinei, M., Migliaccio, E., & Pelicci, P. G. (2007). Hydrogen peroxide: A metabolic by-product or a common mediator of ageing signals? *Nature Reviews Molecular Cell Biology*, *8*, 722–728.

Green, D. R., & Kroemer, G. (2004). The pathophysiology of mitochondrial cell death. *Science*, *305*, 626–629.

Guo, J. L., Wang, X. W., Henstridge, D. C., Richardson, J. J., Cui, J. W., Sharma, A., et al. (2015). Nanoporous metal-phenolic particles as ultrasound imaging probes for hydrogen peroxide. *Advanced Healthcare Materials*, *4*, 2170–2175.

Kim, C., Favazza, C., & Wang, L. V. (2010). In vivo photoacoustic tomography of chemicals: High-resolution functional and molecular optical imaging at new depths. *Chemical Reviews*, *110*, 2756–2782.

Li, Q., Li, S., He, S., Chen, W., Cheng, P., Zhang, Y., et al. (2020). An activatable polymeric reporter for near-infrared fluorescent and photoacoustic imaging of invasive cancer. *Angewandte Chemie International Edition*, *59*, 7018–7023.

Lu, D., Cagan, A., Munoz, R., Tangkuaram, T., & Wang, J. (2006). Highly sensitive electrochemical detection of trace liquid peroxide explosives at a prussian-blue 'artificial-peroxidase' modified electrode. *Analyst*, *131*, 1279–1281.

Lu, J. T., Zhang, H. W., Li, S., Guo, S. S., Shen, L., Zhou, T. T., et al. (2020). Oxygen-vacancy-enhanced peroxidase-like activity of reduced Co_3O_4 nanocomposites for the colorimetric detection of H_2O_2 and glucose. *Inorganic Chemistry*, *59*, 3152–3159.

Ohshima, H., Tatemichi, M., & Sawa, T. (2003). Chemical basis of inflammation-induced carcinogenesis. *Archives of Biochemistry and Biophysics*, *417*, 3–11.

Ren, M. G., Deng, B. B., Wang, J. Y., Kong, X. Q., Liu, Z. R., Zhou, K., et al. (2016). A fast responsive two-photon fluorescent probe for imaging H_2O_2 in lysosomes with a large turn-on fluorescence signal. *Biosensors & Bioelectronics*, *79*, 237–243.

Rhee, S. G. (2007). H_2O_2, a necessary evil for cell signaling. *Science*, *312*, 1882–1883.

Weber, J., Bollepalli, L., Belenguer, A. M., Di Antonio, M., De Mitri, N., & Joseph, J. (2019). An activatable cancer-targeted hydrogen peroxide probe for photoacoustic and fluorescence imaging. *Cancer Research*, *79*, 5407–5417.

Wu, Y. L., Huang, S. L., Wang, J., Sun, L. H., Zeng, F., Wu, S. Z., et al. (2018). Activatable probes for diagnosing and positioning liver injury and metastatic tumors by multispectral optoacoustic tomography. *Nature Communications*, *9*, 3983–3997.

Xiong, J. C., Xia, L. L., Li, L. S., Cui, M. Y., Gu, Y. Q., & Wang, P. (2019). An acetate-based NIR fluorescent probe for selectively imaging of hydrogen peroxide in living cells and in vivo. *Sensors and Actuators B: Chemical*, *288*, 127–132.

Xu, M. H., & Wang, L. V. (2006). Photoacoustic imaging in biomedicine. *Review of Scientific Instruments*, *77*, 305–598.

Yang, Y., & Pan, S. S. (2015). Latest advances in the research on the influence of reactive oxygen species-mediated oxidative stress on myocardial mitochondria and autophagy in cardiovascular stress and exercise. *China Sport Science*, *5*, 71–77.

Zhou, Y., Wang, D. P., Zhang, Y. M., Chitgupi, U., Geng, J. M., Wang, Y. H., et al. (2016). A phosphorus phthalocyanine formulation with intense absorbance at 1000 nm for deep optical imaging. *Theranostics*, *6*, 688–697.

Fluoro-photoacoustic polymeric renal reporter for real-time dual imaging of acute kidney injury

Penghui Cheng and Kanyi Pu*
School of Chemical and Biomedical Engineering, Nanyang Technological University, Singapore, Singapore
*Corresponding author: e-mail address: kypu@ntu.edu.sg

Contents

Abstract

In this chapter, we introduced a diagnostic approach for acute kidney injury (AKI) via photoacoustic imaging. We provided detailed synthetic procedures of a biomarker-activatable photoacoustic agent (FPRR) with high renal clearance efficiency. We also provided protocols for in vitro characterization, live-cell imaging, and in vivo imaging in a drug-induced AKI mice model. Compared to near-infrared fluorescence imaging, photoacoustic imaging is more sensitive with higher signal-to-background ratio.

Methods in Enzymology, Volume 657
ISSN 0076-6879
https://doi.org/10.1016/bs.mie.2021.06.020

As such, our approach serves as a general guideline in the development of photoacoustic agents for diagnosis of urological diseases. With this tool in hand, researchers in the field of optical imaging may be inspired to develop other photoacoustic agents for early stage disease diagnosis.

1. Introduction

Photoacoustic (PA) imaging that combines the advantages of both optical and ultrasound imaging has become an increasingly important tool in preclinical studies and even clinical applications (Deán-Ben, Gottschalk, Mc Larney, Shoham, & Razansky, 2017; Wang & Hu, 2012; Zackrisson, van de Ven, & Gambhir, 2014; Zhang, Maslov, Stoica, & Wang, 2006). By detecting acoustic signals from thermoelastic expansion followed by photon absorption, PA imaging overcomes tissue scattering, and thus possesses better spatial resolution and increased penetration depth as compared to optical imaging (Weber, Beard, & Bohndiek, 2016). Despite the fact that many materials have been exploited as contrast agents for PA imaging such as small-molecule dyes (Jeevarathinam, Lemaster, Chen, Zhao, & Jokerst, 2019; Zhang et al., 2014), gold nanoparticles (Jokerst, Thangaraj, Kempen, Sinclair, & Gambhir, 2012), carbon nanotubes (de la Zerda et al., 2010), semiconducting polymer nanoparticles (Cui et al., 2019; Jiang et al., 2019; Lyu et al., 2018; Miao & Pu, 2018; Yin et al., 2018), most of them rely on passive accumulation at the disease site to cause signal enhancement relative to signals from healthy tissue. These PA agents have been used to detect primary and metastatic tumors (Jiang et al., 2019; Lyu et al., 2018), vascular (Karlas et al., 2019), ocular and lymphatic diseases (de la Zerda et al., 2010; Kim, Galanzha, Shashkov, Moon, & Zharov, 2009). In contrast, activatable PA probes that only emit signal responding to disease biomarkers provide specific and accurate molecular information for disease diagnosis (Miao & Pu, 2016). Until now, a variety of biomarkers including pH (Miao, Lyu, Ding, & Pu, 2016), antioxidants (Zheng, Zeng, Zhang, Xing, & Zhang, 2019), reactive oxygen species (Pu et al., 2014; Zhang et al., 2017), enzymes (Yin et al., 2019; Zhen et al., 2018) and metal ions (Li, Zhang, Smaga, Hoffman, & Chan, 2015) have been detected by activatable PA probes. However, most of the PA agents show high uptake by the mononuclear phagocytic system (MPS) and undergo slow excretion via the hepatobiliary pathway (Gustafson, Holt-Casper, Grainger, & Ghandehari, 2015). This prolonged exposure can potentially result in acute or chronic toxicity to major organs, hindering the applications of these agents in clinics (Blanco, Shen, & Ferrari, 2015). Thus, developing activatable PA agents with high renal clearance is a promising approach toward future clinical practices (Liu, Yu, Zhou, & Zheng, 2013; Yu & Zheng, 2015).

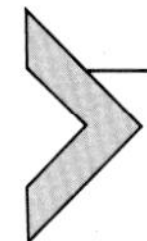

2. Selection of biomarkers for early diagnosis of acute kidney injury

2.1 Kidney injury and its clinical significance

Kidneys, as vital organs in the body that play important roles in regulating body fluid homeostasis and blood filtration, maintaining electrolyte balance, are vulnerable when exposed to drugs, imaging contrast agents and environmental hazards, resulting in kidney injury in an acute or chronic manner (Choudhury & Ahmed, 2006). In particular, drug-induced kidney injury (DIKI) is a frequent complication in intensive care unit (ICU) with both high risk of morbidity and mortality (VA/NIH Acute Renal Failure Trial Network, 2008). Moreover, DIKI is a significant contributor to acute kidney injury (AKI) and constitutes 19–25% of all cases in critically ill patients (Kellum & Prowle, 2018). Therefore, developing sensitive detection methods for DIKI is of great importance for pharmaceutical companies to screen nephrotoxic drugs in drug development process, and for clinicians to closely monitor patient safety in clinical care.

Current clinical diagnosis of DIKI relies on measurement of serum creatinine (sCr) and blood urea nitrogen (BUN), which are factors reflecting changes in glomerular filtration rate (GFR) (Darmon et al., 2017). However, sCr/BUN methods fail to sensitively detect kidney injury because substantial histological injuries could occur before measurable GFR changes (Waikar & Bonventre, 2009). Furthermore, factors such as patients' age and muscle mass can also cause fluctuations in the biomarker levels, hence affecting the accuracy of these tests (Baxmann et al., 2008). In contrast, urinalysis using biomarkers such as clusterin, Cystatin-C and β2-microglobulin has recently shown to be more sensitive and reliable than clinical diagnostic methods for detection of DIKI in animal studies (Dieterle et al., 2010). Till date, the detection methods for these urinary biomarkers are restricted to commercial immunoassays and matrix-assisted laser desorption ionization time of flight mass spectroscopy (MALDI-TOF-MS) that suffer from the following disadvantages. Commercial immunoassays generally suffer from low sensitivity and tedious measurement procedures (Noto et al., 1983);

while MALDI-TOF-MS is a highly specific method in biomarker detection, but its application has been limited by the number of reference MS spectra in the database (Duncan, Nedelkov, Walsh, & Hattan, 2016).

2.2 Our proposed approach

Herein, we report the synthesis of a fluoro-photoacoustic polymeric renal reporter (FPRR) for real-time non-invasive near-infrared fluorescence (NIRF) and PA dual imaging of drug-induced acute kidney injury (AKI) in murine mouse model. AKI is characterized by a rapid decline in kidney function and is associated with high morbidity and mortality (Myers & Moran, 1986). Among major AKI etiologies such as ischemia/reperfusion, sepsis, and nephrotoxin exposure, nephrotoxin-induced AKI accounts for 25% of all AKI cases (Cox et al., 2013). The sequential pathophysiological events induced by nephrotoxic drug (cisplatin) include renal epithelial cell damage and death, vigorous inflammatory reactions, GFR decline, and ultimately kidney dysfunction (Manohar & Leung, 2018). After proximal tubular epithelial cell damage, γ-glutamyl transferase (GGT), a brush border enzyme highly expressed in proximal tubular epithelial cells, is released by exocytosis or leakage and can be readily detected in the urine (Peres & da Cunha, 2013; Waikar & Bonventre, 2008). Thus, GGT is chosen as the targeted biomarker for the design of FPRR.

3. Design of FPRR

FPRR consists of three moieties, an optical imaging moiety, a biomarker-responsive moiety, and a renal clearance moiety (Scheme 1). The optical imaging moiety is a hemicyanine CyOH which emits both near-infrared fluorescence and photoacoustic signals. The biomarker-responsive moiety, γ-glutamate, is conjugated to the hydroxyl group of CyOH via a self-immolative linker. Initially, the electron-donating ability of the oxygen atom is diminished. In the presence of GGT, the amide bond next to γ-glutamate of FPRR can be specifically cleaved, followed by 1,6-self immolation and spontaneous release of the linker *p*-aminobenzyl alcohol (PABA). In the meantime, the electron-donating ability of the oxygen atom is recovered, with enhanced NIRF and PA signals. Therefore, both signals could be used to monitor the level of enzyme biomarker (GGT).

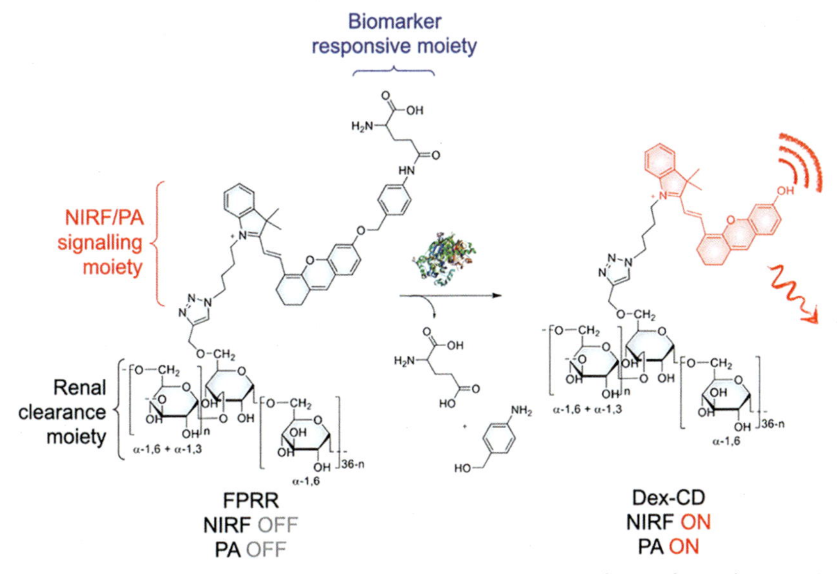

Scheme 1 Schematic illustration and molecular mechanism of FPRR for real-time NIRF and PA detection of γ-glutamyl transpeptidase.

4. Synthesis of FPRR

FPRR can be synthesized in 11 steps from commercially available starting materials, Fmoc-Glu-OtBu and 1-bromo-4-chlorobutane (Scheme 2). The following synthetic procedures require understanding of standard organic chemistry techniques and the availability of basic synthetic laboratory equipment. All chemicals and solvents are purchased from Sigma Aldrich unless otherwise indicated.

1-azido-4-chlorobutane (1)

1. Preheat an oil bath to 50 °C.
2. Add 1-bromo-4-chlorobutane (5.1 g, 30 mmol) and sodium azide (2.05 g, 31.5 mmol, 1.05 equiv.) into a 100 mL round-bottom flask equipped with a magnetic stir bar.
3. Dissolve the mixture with dimethylformamide.
4. Flush the reaction vessel with nitrogen.
5. Allow the reaction mixture to stir at 50 °C for 16 h.
6. After reaction, the residue is diluted with ethyl acetate and washed with deionized water for three times. The organic layer is dried over sodium sulfate and concentrated under reduced pressure.

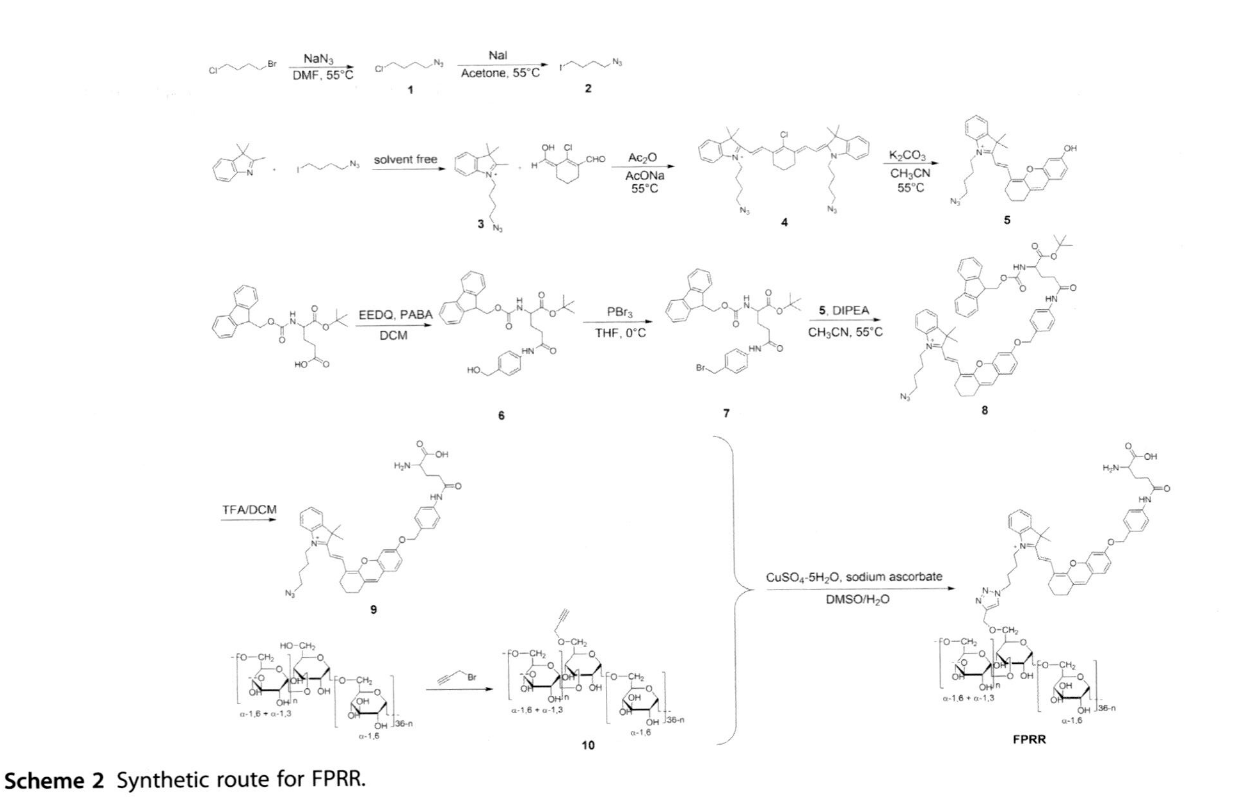

Scheme 2 Synthetic route for FPRR.

7. After silica gel chromatography (hexane), pure compound 1 (colorless oil) could be obtained (1.9 g, 48%).

1-azido-4-iodobutane (2)

1. Preheat an oil bath to 50 °C.
2. Add compound 1 (1 g, 7.5 mmol) and sodium iodide (1.2 g, 7.85 mmol, 1.05 equiv.) into a 100 mL round-bottom flask equipped with a magnetic stir bar.
3. Dissolve the mixture with acetone.
4. Flush the reaction vessel with nitrogen.
5. Allow the reaction mixture to stir at 50 °C for 24 h.
6. Concentrate the resulting residue with rotary evaporator, and pure compound 2 (light yellow oil, 1 g, 60%) could be obtained after purification with silica gel chromatography (petroleum ether). ^{1}H NMR (300 MHz, $CDCl_3$) δ: 3.32 (t, 2H), 3.21 (t, 2H), 1.87–1.96 (m, 2H), 1.69–1.76 (m, 2H). MS (ESI): m/z = 224.98 $[M+H]^+$.

1-(4-azidobutyl)-2,3,3-trimethyl-3*H*-indol-1-ium (3)

1. Preheat an oil bath to 90 °C.
2. Add 2,3,3-trimethylindolenine (318 mg, 2 mmol) and compound 2 (450 mg, 2 mmol) into a round-bottom flask with a magnetic stir bar.
3. Flush the reaction vessel with nitrogen.
4. Allow the reaction mixture to be vigorously stirred without solvent for 2 h, or until the reaction is complete as indicated by thin layer chromatography (TLC).
5. Upon completion, dilute the reaction mixture with 3 mL dichloromethane. Next, precipitate the mixture into excess diethyl ether, and washed three or more times with excess diethyl ether until crude product 3 is obtained with purity >95%. The product was directly used in the next step of synthesis without further purification.

CyN_3Cl (4)

1. Preheat an oil bath to 90 °C.
2. Add compound 3 (160 mg, 0.63 mmol), 2-chloro-3-(hydroxymethylidene) cyclodex-1-ene-1-carbaldehyde (53 mg, 0.3 mmol) and sodium acetate (25 mg, 0.3 mmol) into a round-bottom flask with a magnetic stir bar.
3. Dissolve the reaction mixture in 5 mL acetic anhydride.
4. Flush the reaction vessel with nitrogen.
5. Allow the reaction mixture to be vigorously stirred at 90 °C for 4 h, or until the reaction is complete as indicated by thin layer chromatography (TLC).
6. Concentrate the resulting residue with rotary evaporator.

7. Dilute the reaction residue with dichloromethane, and wash with saturated sodium bicarbonate for three times to remove residue acid. Wash the organic phase with brine (30 mL) and dry over anhydrous sodium sulfate before concentrated under reduced pressure.
8. Pure product 4 (120 mg, 70%) is obtained after silica gel chromatography ($CH_2Cl_2/CH_3OH=50:1$). ^1H NMR (300 MHz, $CDCl_3$) δ: 8.34 (d, $J=15$ Hz, 2H), 7.39 (m, 4H), 7.28 (s, 1H), 7.24 (d, 2H), 7.20 (d, $J=6$ Hz, 2H), 6.32 (d, $J=12$ Hz, 2H), 4.31 (t, 4H), 3.46 (t, 4H), 2.80 (t, 4H), 1.98 (m, 6H), 1.84 (m, 4H), 1.73 (s, 12H). MS (ESI): m/z = 649.35 $[M+H]^+$.

CyN_3OH (5)

1. Preheat an oil bath to 55 °C.
2. Add grinded potassium carbonate (130 mg, 0.9 mmol) and resorcinol (100 mg, 0.9 mmol) into a round-bottom flask with a magnetic stir bar.
3. Dissolve the reaction mixture with 5 mL anhydrous acetonitrile.
4. Flush the reaction vessel with nitrogen.
5. Allow the reaction mixture to stir at 55 °C for 20 min, with a color change from white to dark green.
6. Add compound 4 (200 mg, 0.3 mmol) dissolved in 5 mL acetonitrile dropwise into the mixture via a syringe.
7. Allow the reaction mixture to continuously stir for 4 h. or until the reaction is complete as indicated by thin layer chromatography (TLC).
8. Concentrate the reaction mixture with rotary evaporator.
9. Dissolve the reaction residue with dichloromethane and wash with brine for three times. The organic layer is then dried over sodium sulfate and concentrated under reduced pressure.
10. Pure compound 5 (100 mg, 60%) is obtained after purification with silica gel chromatography ($CH_2Cl_2/CH_3OH=25:1$).

^1H NMR (300 MHz, CD_3OD) δ: 8.63 (d, J = 15 Hz, 1H), 7.61 (s, 1H), 7.45 (m, 3H), 7.40 (m, 2H), 7.10 (t, 2H), 6.00 (d, J = 13 Hz, 1H), 4.09 (t, 1H), 3.46 (m, 2H), 2.77 (t, 2H), 2.64 (t, 2H), 1.95 (m, 3H), 1.78 (s, 5H), 1.57 (s, 6H). MS (ESI): m/z = 467.24 $[M+H]^+$.

Ph-Glu (6)

1. Add Fmoc-Glu-OtBu (CAS 84793-07-7, 300 mg, 0.7 mmol) to a 100 mL round-bottom flask equipped with a magnetic stir bar.
2. Flush the reaction vessel with nitrogen.
3. Add 10 mL anhydrous dichloromethane into the reaction vessel via a syringe with continuous magnetic stirring.

4. Dissolve *N*-Ethoxycarbonyl-2-ethoxy-1,2-dihydroquinoline (EEDQ, 691 mg, 2.8 mmol, 4equiv.) and *p*-aminobenzyl alcohol (PABA, 344 mg, 2.8 mmol, 4equiv.) in 20 mL anhydrous dichloromethane. The mixture was then added into the reaction vessel via a syringe.
5. The mixture was then stirred for 3 h at room temperature.
6. Upon completion, concentrate the reaction mixture by rotary evaporation to isolate the crude product as a solid.
7. Purify the reaction mixture by preparative HPLC to obtain pure product 6 (>95% purity). Pure product 1 (326 mg) could be obtained with a yield of 85%. ^{1}H NMR (300 MHz, $CDCl_3$) δ: 7.75 (d, J=6 Hz, 2H), 7.56 (d, J=9 Hz, 4H), 7.38 (m, 2H), 7.28 (m, 4H), 4.60 (s, 2H), 4.42 (m, 2H), 4.14 (m, 2H), 2.33 (m, 2H), 1.90 (m, 2H), 1.45 (s, 9H).

Br-Ph-Glu (7)

1. Prepare an ice bath.
2. Add pure product 6 (150 mg, 0.28 mmol) to a 50 mL round bottom flask equipped with a magnetic stir bar.
3. Flush the reaction vessel with nitrogen.
4. Add in 5 mL anhydrous tetrahydrofuran (THF) into the reaction vessel with continuous magnetic stirring via a syringe.
5. Add phosphorus tribromide (PBr_3, 53 µL, 0.56 mmol, 2 equiv.) dissolved in anhydrous tetrahydrofuran into the reaction vessel via a syringe.
6. Allow the reaction mixture to stir in ice bath for 2 h, or until the reaction is complete as indicated by thin layer chromatography (TLC).
7. Concentrate the reaction mixture by rotary evaporation to isolate the oily crude product.
8. Resuspend the crude product with ethyl acetate (30 mL). Wash the organic phase sequentially with saturated sodium bicarbonate (30 mL × 3) and brine (30 mL), followed by drying over anhydrous sodium sulfate before concentrated under reduced pressure.
9. The crude compound 7 (>95% purity) can be used directly in the next step synthesis without further purification.

CyGlu (8)

1. Preheat an oil bath to 55 °C.
2. Add compound 5 (32 mg, 0.07 mmol) into a round-bottom flask with a magnetic stir bar.
3. Dissolve *N*,*N*-Diisopropylethylamine (DIEA, 47 µL, 0.28 mmol, 4 equiv.) in anhydrous acetonitrile. The mixture is then added into the reaction vessel.

4. Flush the reaction vessel with nitrogen.
5. Allow the reaction mixture to stir at 55 °C under nitrogen atmosphere.
6. Add compound 7 (165 mg, 0.28 mmol, 4 equiv.) dissolved in anhydrous acetonitrile into the reaction mixture via a syringe.
7. Allow the reaction mixture to continuously stir at 55 °C for 8 h under nitrogen atmosphere, or until the reaction is complete as indicated by thin layer chromatography (TLC).
8. Upon completion, and resulting mixture was concentrated under reduced pressure and purified by preparative HPLC to obtain pure product 8 (48 mg, 70%). ^{1}H NMR (300 MHz, CD_3OD) δ: 8.64 (d, $J = 15$ Hz, 1H), 7.70 (m, 4H), 7.40 (m, 14H), 6.98 (s, 2H), 6.42 (d, $J = 15$ Hz, 1H), 5.23 (s, 2H), 4.33 (m, 6H), 3.44 (t, 2H), 2.48–2.71 (m, 5H), 2.18 (m, 2H), 1.80 (m, 11H), 1.42 (m, 11H).

CyG (9)

1. Prepare an ice bath.
2. Dissolve pure product 8 in 1 mL dichloromethane in a round-bottom flask, and place in the ice bath.
3. Add 1 mL trifluoroacetic acid (TFA/dichloromethane volumetric ratio = 1:1) dropwise into the reaction mixture.
4. Allow the reaction mixture to stir at 0 °C for 1.5 h, or until the reaction is complete as indicated by thin layer chromatography (TLC).
5. Upon completion, dilute the reaction mixture with dichloromethane (20 mL), and wash three times with saturated sodium bicarbonate.
6. Concentrate the organic phase under reduced pressure to yield crude product after dried over sodium sulfate.
7. Add piperidine (200 μL, 2.7 mmol, 5% v/v in dimethylformamide) to the above crude product dissolved in dimethylformamide (4 mL).
8. Allow the reaction mixture to stir at room temperature for 20 min.
9. Pure compound 9 is obtained (9 mg, 42%) after purification with preparative HPLC. ^{1}H NMR (300 MHz, CD_3OD) δ: 8.76 (d, $J = 15$ Hz, 1H), 7.66 (m, 3H), 7.48 (m, 7H), 7.07 (m, 2H), 6.53 (d, $J = 15$ Hz, 1H), 5.23 (s, 2H), 4.38 (t, 2H), 4.06 (t, 1H), 3.45 (t, 2H), 2.69 (m, 5H), 2.25 (m, 2H), 1.97 (m, 4H), 1.80 (m, 7H), 1.28 (m, 2H).

Dextran-alkyne (10)

1. Prepare an ice bath.
2. Add dextran (CAS 9004-54-0, 2 g, 0.34 mmol) and tetra-*tert*-butylammonium iodide (27 mg, 0.08 mmol, 0.2 equiv.) into a round bottom flask with magnetic stir bar.
3. Dissolve the reaction mixture in anhydrous dimethylformamide (30 mL), and flush the reaction vessel with nitrogen.

4. Allow the reaction mixture to stir in the ice bath.
5. A suspension of sodium hydride (120 mg, 5 mmol, 15 equiv.) in anhydrous dimethylformamide is added dropwise into the reaction mixture.
6. Allow the reaction mixture to continuously stir in ice bath under nitrogen atmosphere for 0.5 h.
7. Add propargyl bromide (52 μL, 0.68 mmol, 2 equiv.) into the reaction mixture via a syringe.
8. Remove the ice bath.
9. Allow the reaction mixture to continuously stir at room temperature for additional 24 h.
10. Upon completion, add diluted glacial acetic acid (10-times dilution with deionized water) dropwise to adjust pH to 7.4.
11. Add water into the mixture and dialyze the resulting solution against deionized water to remove any impurities.
12. After dialysis, wash the resulting inorganic solution three times with dichloromethane to further remove any impurities.
13. Pure Dextran-alkyne is obtained as white solid (1.8 g, 85%) after lyophilization: ^{1}H NMR (300 MHz, D_2O) δ: 4.95 (d, $J = 3$ Hz, 37H), 3.53–3.88 (m, 295H), 2.69 (s, 1H).

FPRR (11)

1. Add CyG (15 mg, 0.02 mmol) dissolved in dimethyl sulfoxide into a round bottom flask with a magnetic stir bar.
2. Add compound 10 (150 mg, 0.025 mmol), sodium ascorbate (1 mg, 0.004 mmol) and copper sulfate pentahydrate (1.25 mg, 0.005 mmol) dissolved in deionized water into the reaction mixture.
3. Flush the reaction vessel with nitrogen.
4. Allow the reaction mixture to stir at room temperature in the dark for 10 h.
5. Upon completion, add water into the mixture and dialyze the resulting solution against deionized water to remove any impurities.
6. Pure FPRR could be obtained as blue solid by further purification with preparative HPLC (113 mg, 85%): ^{1}H NMR (300 MHz, DMSO-d_6) δ: 6.0–8.0 (m, 14H), 4.67 (s, 37H), 3.44 (t, 298H). 1.0–2.5 (m, 20H).

5. In vitro characterization of FPRR

After successful synthesis of FPRR, it is necessary to (i) determine the photophysical properties of FPRR, (ii) demonstrate the detection capability, both sensitivity and specificity of FPRR toward the biomarker-of-interest, γ-glutamyl transpeptidase, (iii) confirm the biocompatibility of FPRR, and (iv) validate that FPRR is capable of detecting γ-glutamyl transpeptidase in living cells. To expedite our experiments and minimize the error involved in batch preparation, sample weighing, we recommend calculation of extinction coefficient in advance, which is described below.

Equipment

1—UV–visible range spectrophotometer
1—Quartz cuvettes with corresponding cuvette caps Microbalance
10—Microcentrifuge tube
3—Micropipettes and tips (P1000, P200, and P10)

Reagents

Compound of interest (>10mg)
Deionized water

Procedure (perform at least three times)

1. Label the microcentrifuge tubes (1, 2, 3, 4, and 5).
2. Zero the microbalance and acquire the tare weight of one microcentrifuge tube.
3. Add a small amount of solid compound into the tube (1-2mg).
4. Based on the actual weight of the compound, add corresponding volumes of deionized water to obtain a final concentration of 1 mg/mL.
5. Cap the tube and shake vigorously or sonicate until the solids are fully dissolved.
6. Perform serial dilution with deionized water to obtain 1 mL solutions with final concentrations of 0.5, 0.2, 0.1, and 0.05 mg/mL.
7. Convert mass concentration to molar concentration based on the molecular mass of FPRR.
8. Scan the sample in the spectrophotometer to identify the absorbance value at 604 nm (the λ_{max} of FPRR).
9. Using the Beer-Lambert law, determine the extinction coefficient of FPRR from molar concentration and absorption.
10. The extinction coefficient of FPRR is 40,000 M^{-1} cm^{-1} at 604 nm.

5.1 Photophysical characterization and detection capability of FPRR

This experiment identifies the basic optical properties of FPRR, including absorption, fluorescence, and photoacoustic spectra. The spectral change is recorded and compared in the absence or presence of the biomarker-of-interest, γ-glutamyl transpeptidase. Besides, the detection capability toward γ-glutamyl transpeptidase will be characterized, including enzyme kinetics and response rate.

Equipment

1—UV–visible range spectrophotometer
1—Fluorolog 3-TCSPC spectrofluorometer (Horiba Jobin Yvon)
1—Multispectral optoacoustic tomographic (MSOT) imaging system (iThera Medical GmbH)
1—High performance liquid chromatography
1—Quartz cuvettes with corresponding cuvette caps Microbalance
10—Microcentrifuge tube
3—Micropipettes and tips (P1000, P200, and P10)
1—Water bath with temperature control

Reagents

Compound of interest (>10 mg)
Phosphate buffered solution (PBS, 10 mM, pH = 7.4)
γ-glutamyl transpeptidase (Sigma G9270)
Biological grade water
Trifluoroacetic acid

Procedure (perform at least three times)

1. Preheat the water bath to 37 °C.
2. Freshly prepare enzyme solution by dissolving lyophilized enzyme powder in biological grade water to a final concentration of 5 μg/μL.
3. Prepare the FPRR stock solution by dissolving 3 mg FPRR in 1 mL PBS and measuring absorption spectra.
4. Calculate molar concentration of the stock solution according to the extinction coefficient.
5. Dilute the FPRR stock solution with PBS to final concentration of 10 μM.
6. Label two microcentrifuge tubes, with "−" and "+" respectively.
7. Pipette 1 mL 10 μM FPRR solution into the two tubes; add 10 μg enzyme into the "+" tube.
8. Incubate two tubes in the 37 °C water bath for 2 h. Terminate the enzyme incubation by adding 5% trifluoroacetic acid.
9. Measure absorption, fluorescence and photoacoustic spectra, perform HPLC analysis of FPRR in the absence or presence of enzyme (as shown in Fig. 1).

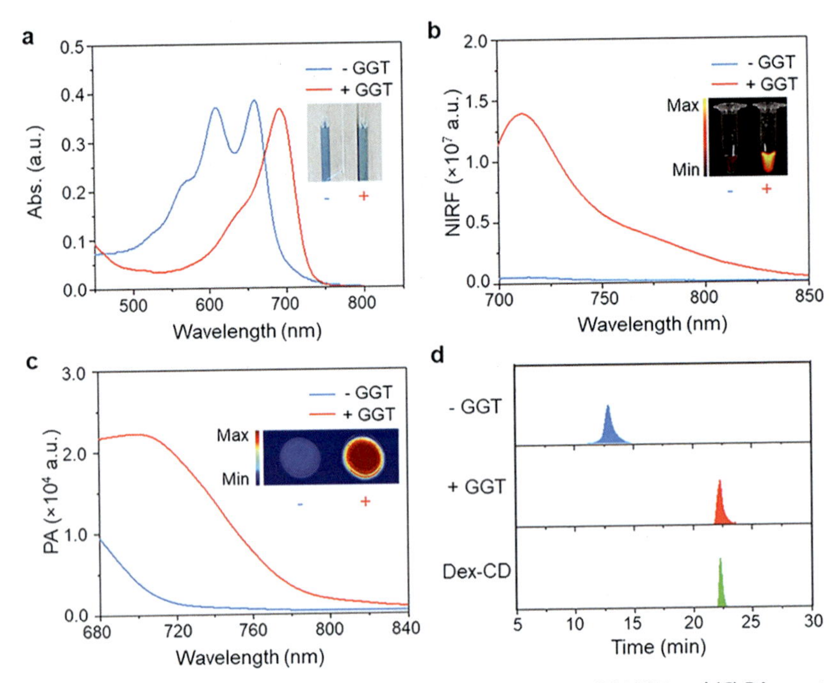

Fig. 1 In vitro characterization of FPRR. (A) UV/vis absorption, (B) NIRF and (C) PA spectra of FPRR (25 μM) in the absence or presence of GGT (20 μg) in PBS (10 mM, pH = 7.4) at 37 °C for 2 h. (D) HPLC traces of FPRR in the absence or presence of GGT, and the traces of pure Dex-CD.

10. To determine the enzymatic cleavage rate changes with time, carry out fluorescence measurement at incubation time, $t = 5$, 10, 20, 30, 60, 120 min.

5.2 Enzyme kinetic and selectivity studies

Equipment

1—UV–visible range spectrophotometer
1—Fluorolog 3-TCSPC spectrofluorometer (Horiba Jobin Yvon)
1—Quartz cuvettes with corresponding cuvette caps Microbalance
10—Microcentrifuge tube
3—Micropipettes and tips (P1000, P200, and P10)
1—Water bath with temperature control
1—Timer

Reagents

Compound of interest (>10 mg)
Phosphate buffered solution (PBS, 10 mM, pH = 7.4)

γ-glutamyl transpeptidase (Sigma G9270)
Alanine aminopeptidase
Caspase-3
Furin
N-acetyl-β-D-glucosaminidase
Urokinase
Biological grade water
Trifluoroacetic acid

Procedure (perform at least three times)

1. Preheat the water bath to 37 °C.
2. Freshly prepare enzyme solution by dissolving enzyme lyophilized powder in biological grade water to a final concentration of 5 μg/μL.
3. Dissolve 3 mg FPRR in 1 mL PBS and measure absorption spectra, this is the stock solution.
4. Calculate molar concentration according to the extinction coefficient.
5. Prepare various concentrations of FPRR (6, 15, 40, 65, 80, 160 μM) in PBS (10 mM, pH = 7.4) in microcentrifuge tubes.
6. Add GGT (12.5 μg) into each tube.
7. Incubate the tubes at 37 °C for 5 min. Terminate the enzyme incubation by adding 5% trifluoroacetic acid.
8. Measure fluorescence spectra of the enzyme incubation mixtures in the microcentrifuge tubes.
9. Take a dozen of tubes, prepare CyOH solution with different concentrations (3, 5, 10, 15, 20, 40, 60 μM).
10. Measure fluorescence spectra of CyOH with excitation at 640 nm. Plot a calibration curve of fluorescence signals at 720 nm against different CyOH concentrations.
11. Calculate the enzyme conversion, which is percentage of CyOH production according to the calibration curve.
12. Calculate the initial reaction velocity (μmol/s), which is plotted against FPRR concentration, and fitted to a Michaelis-Menten curve. The kinetic parameters were calculated using Michaelis-Menten equation: $V = Vmax^*[S]\ (K_M + [S])$, where V is initial velocity, and [S] is substrate concentration. The calculated parameters are as follows: $V_{max} = 0.03\ \mu M/s$, $K_M = 10.21\ \mu M$, $k_{cat} = 158.74\ s^{-1}$, $k_{cat}/K_M = 15.6\ \mu M^{-1}\ s^{-1}$ (as shown in Fig. 2A).
13. For detection selectivity study, incubate FPRR with other proteases like alanine aminopeptidase, caspase-3, furin, *N*-acetyl-β-D-glucosaminidase (NAG), urokinase (uPA), and etc. (as shown in Fig. 2B). Measure absorption and fluorescence spectra after incubation for 2 h.

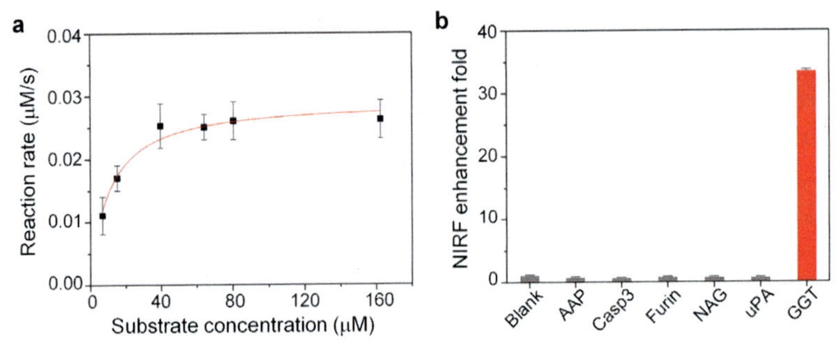

Fig. 2 (A) Enzyme kinetics of GGT (20 μg) toward FPRR (6, 15, 40, 65, 80, 160 μM). $K_M = 10.2\,\mu M$, $k_{cat} = 158.7\,s^{-1}$, $k_{cat}/K_M = 15.6\,\mu M^{-1}\,s^{-1}$. (B) Relative NIRF (710 nm) enhancement after incubation with indicated enzymes in PBS (10 mM, pH = 7.4) at 37 °C for 2 h. Alanine aminopeptidase (AAP), caspase-3 (casp3), *N*-acetyl-β-D-glucosaminidase (NAG), urokinase (uPA).

5.3 Cell culture

Equipment

3—Micropipettes and tips (P1000, P200, and P10)
1—Cell incubator
10—Cell culture flask
1—Water bath with temperature control
1—Biosafety Cabinet
1—Centrifuge machine
1—Cell incubator
1—White light microscope

Reagents

Human Kidney-2 (HK-2) cell line
Normal Dermal Fibroblast (NDF) cell line
Compound of interest (>10 mg)
Dulbecco's Modified Eagle's Medium
Fetal Bovine Serum
Penicillin-Streptomycin

Procedure

1. Preheat the water bath to 37 °C.
2. Prepare cell culture medium, DMEM supplemented with heat-inactivated fetal bovine serum (10%) and antibiotics penicillin/streptavidin (1%).
3. Thaw HK-2 and NDF cells in cryogenic vials from liquid nitrogen storage at 37 °C.
4. Centrifuge the cryogenic vials at 1000 rpm, 3 min.

5. Remove the supernatant, resuspend the cell pellet in cell culture medium.
6. Pipette the cell suspension into cell culture flask and add adequate volume of cell culture medium.
7. Let the cells grow in a humidified environment containing 5% CO_2 and 95% air at 37 °C.
8. Replace the cell culture medium every 2 to 3 days.
9. When the cell confluency reaches 80–90%, subculture the cells following the principle of one to three.
10. When the cells are in good condition, we can proceed to cell cytotoxicity test and fluorescence imaging.

5.4 FPRR cell cytotoxicity test

Equipment

3—Micropipettes and tips (P1000, P200, and P10)
1—Cell incubator
1—Biosafety Cabinet
1—Microplate reader
1—96-well plate
1—White light microscope

Reagents

Human Kidney-2 cell line
Normal Dermal Fibroblast cell line
Compound of interest (>10 mg)
Cell culture medium
MTS assay (Promega Cat. no. G3581, 100 mL, 0.1 mg/mL)

Procedure

Day 1

1. Seed NDF and HK-2 cells in 96-well plates with density of 6×10^3 cells per well overnight.

Day 2

1. Check the cells under microscope, ensure the cells are adherent and well-grown.
2. Prepare different concentrations of FPRR (5, 10, 20, 40, 80 μM) in cell culture medium.
3. Remove the cell culture medium in the 96-well, and pipette FPRR solutions with different concentrations into the 96-well plates, the cells receive only cell culture media is the control group. Carry out the assay in triplicates.
4. Incubate the cells for 24 h.

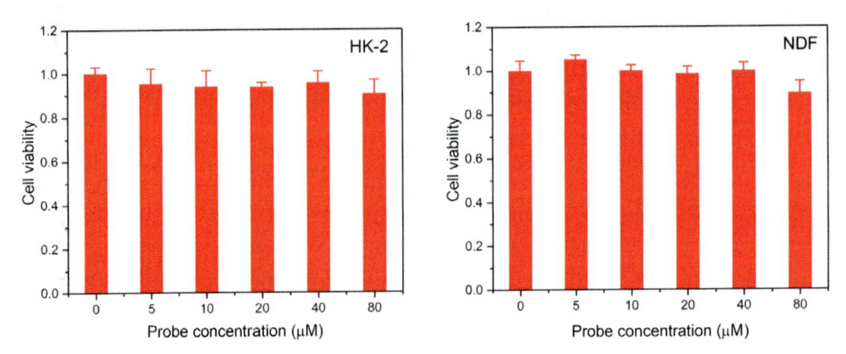

Fig. 3 FPRR cytotoxicity test by MTS assay in HK-2 and NDF living cells. Data are presented as mean ± SD, n = 5 independent measurement.

Day 3

1. Remove the cell culture medium from 96-well plate.
2. Add MTS assay diluted with cell culture medium and incubate for 4 h.
3. Read the absorbance at 490 nm by using a microplate reader.
4. As shown in Fig. 3, the cytotoxic effects (VR) of FPRR is then calculated using the following equation: $V_R = A/A_0 \times 100\%$, where A and A_0 are the absorbance of the experimental groups and control group, respectively.

5.5 FPRR live-cell imaging

Equipment

3—Micropipettes and tips (P1000, P200, and P10)
1—Cell incubator
1—Biosafety Cabinet
1—Laser Scanning Microscope
10—Cell culture dish with glass bottom (dia. 35 mm)
1—White light microscope

Reagents

Human Kidney-2 cell line
Normal Dermal Fibroblast cell line
Compound of interest (>10 mg)
Cell culture medium
2-amino-4{[3-(carboxymethyl)phenyl](methyl)phosphono}butanoic acid (GGsTop, GGT inhibitor)
Hoechst 33342 (NucBlue Live ReadyProbes Reagent, Thermo Fisher)
4% polyformaldehyde solution.

Procedure

Day 1

1. Seed the NDF and HK-2 cells in culture dishes with density of 6×10^4 cells per well overnight.

Day 2

1. Check the cell conditions under microscope.
2. Randomly divide the cells into three groups: (1) untreated cells, (2) cells treated with FPRR, and (3) cells treated with GGsTop before incubation with FPRR.
3. For group 3, remove cell culture medium, and incubate the cells with GGsTop (100 μM) for 30 min.
4. Remove the medium, wash the cells with PBS buffer three times.
5. Incubate cells from group 2 and 3 with FPRR (25 μM) in DMEM medium for 1 h.
6. Remove the medium, wash the cells with PBS buffer three times.
7. Stain the cell nucleus with Hoechst 33342 for 20 min.
8. Fix the cells with 4% polyformaldehyde solution (optional).
9. Acquire cell fluorescence images. Set the excitation and emission wavelengths to 640/655–710 nm for FPRR and 405/410–470 nm for Hoechst.
10. Use ImageJ software to remove background signals and quantify cellular fluorescence intensity (see Fig. 4).

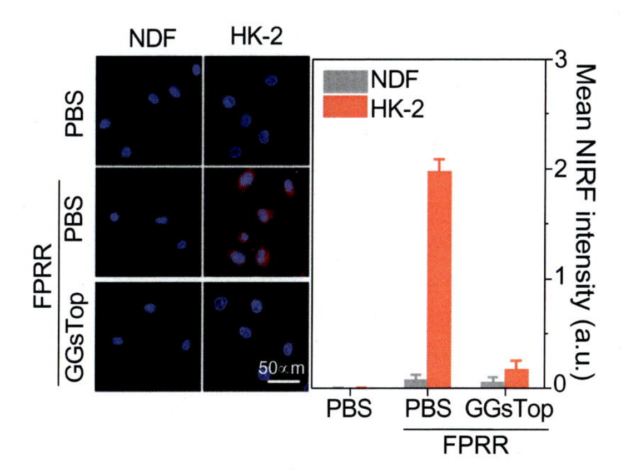

Fig. 4 NIRF cell images of normal dermal fibroblasts (NDF) and proximal tubule epithelial cells (HK-2), and mean NIRF intensity of single cells after different treatments. From top to bottom: untreated cells, cells treated with FPRR, and cells treated with GGsTop before incubation with FPRR. GGsTop (100 μM, 30 min), FPRR (25 μM, 1 h). The cell nucleus stained with 4′6-diamidino-2-phenylindole (DAPI) showed a blue fluorescence signal and FPRR showed a pseudo-red fluorescence signal.

6. In vivo characterization of FPRR

After confirming the capability of FPRR to detect endogenous biomarker, γ-glutamyl transpeptidase in living cells, we proceeded with in vivo applications. It is necessary to demonstrate that (i) FPRR is biocompatible, (ii) FPRR could achieve sensitive detection of γ-glutamyl transpeptidase in living mice, (iii) FPRR could achieve early diagnosis of drug-induced acute kidney injury in mice models. Furthermore, taking advantage of the dual channel imaging capability of FPRR, we will carry out both real-time fluorescence and photoacoustic imaging. Such multi-modal imaging capability demonstrates the clinical translation potential of FPRR.

Please note that all animal experiments were performed in accordance with the Guidelines for Care and Use of Laboratory Animals of the Nanyang Technological University-Institutional Animal Care and Use Committee (NTU-IACUC) and approved by the Institutional Animal Care and Use Committee (IACUC) for Animal Experiment, Singapore. Female nude mice (Tac: Cr:(NCr)-Fox1nu with age of 4–6 weeks) are used in the following experiment.

6.1 FPRR biocompatibility study

Equipment

1—Laser Scanning Microscope
1—Cryostat
1—Paraffin section machine
10—Syringe and needles for intravenous injection
5—Biological-grade scissors and tweezers

Reagents

Phosphate buffered solution
4% polyformaldehyde solution
Compound of interest (>10 mg)
FITC-conjugated anti-mouse caspase-3 antibody
FPRR (20 mM solution in PBS buffer)
Hematoxylin and eosin

Procedure

Day 1

1. Female nude mice (Tac:Cr:(NCr)-Fox1nu, 16–18 g) are housed in a temperature-controlled (22 °C) room with 12 h dark light cycles (0700 h on and 1900 h off). The animals are provided ad libitum with food and water.

2. Intravenously inject 0.2 mL FPRR solution (6.5 μmol/kg body weight (bw)) into living mice.

Day 2

1. At 24 h post-injection, dissect the mice and harvest all major organs (heart, liver, spleen, lung, kidney, skin, muscle, intestine) for histological analysis.
2. Perform both cryo-sectioning and paraffin-embedded sectioning.
3. For cryo-sections, stain tissue sections with Hoechst 33342 and anti-mouse caspase-3 antibody. Perform fluorescence imaging under Laser Scanning Microscope and analyze the images with ImageJ.
4. For paraffin-embedded sections, stain tissue sections with both hematoxylin and eosin. Image the slides under white light microscope to identify any tissue injury (see Fig. 5).

6.2 FPRR biodistribution study

Equipment

3—Mice metabolic cages
10—Microcentrifuge tubes
1—Homogenizer
1—Analytical high performance liquid chromatography
10—Syringe and needles for intravenous injection

Reagents

Phosphate buffered solution
FPRR (20 mM solution in PBS buffer)

Procedure

1. Intravenously inject 0.2 mL FPRR solution (6.5 μmol/kg bw) into living mice.
2. Press the injection point for 10 s to stop bleeding.
3. Immediately place the mice into metabolic cages.
4. Collect mice urine at 3, 6 and 24 h post-injection.

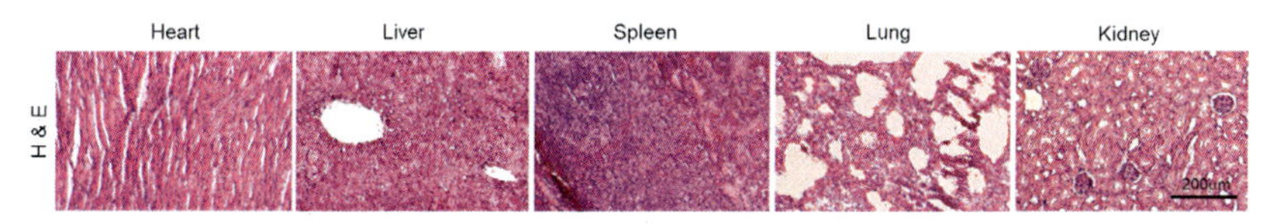

Fig. 5 Representative hematoxylin and eosin staining of major organ tissues from healthy mice after i.v. injection of FPRR. Mice were dissected 24 h after injection and organs were taken out for sectioning.

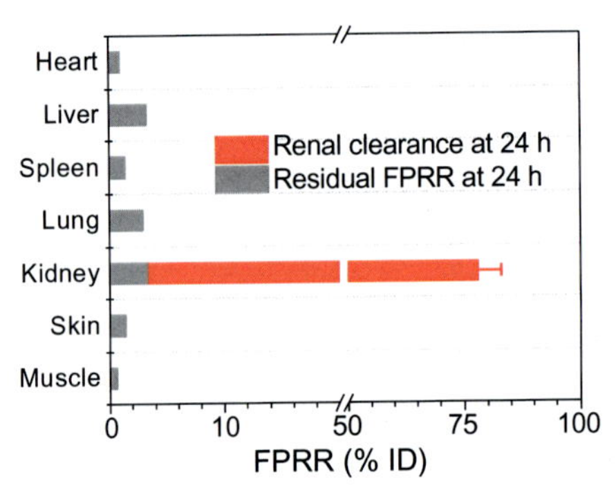

Fig. 6 HPLC quantitative analysis of residual FPRR in major organs (gray bar) and FPRR excreted from kidney into urine (red bar) 24 h after i.v. injection. Data is presented as the mean ± SD, n = 3 independent experiments.

5. Centrifuge the urine samples at 4500 r.p.m. for 10 min and filter through a 0.22 μm syringe filter.
6. After 24 h urine collection, dissect the mice and harvest all major organs (heart, liver, spleen, lung, kidney, skin, muscle, intestine). Homogenize the organs in PBS buffer and centrifuged at 4500 r.p.m. for 15 min to remove insoluble components.
7. As shown in Fig. 6, the amount of FPRR in the urine excretion and major organs are then quantitatively studied using analytical HPLC.

6.3 FPRR pharmacokinetic study

Equipment

10—Heparinized capillary tubes
1—Timer
1—Ice box
1—Centrifuge machine
1—Analytical high performance liquid chromatography
10—Syringe and needles for intravenous injection

Reagents

Phosphate buffered solution
FPRR (20 mM solution in PBS buffer)

Procedure

1. Intravenously inject 0.2 mL FPRR solution (6.5 μmol/kg bw) into living mice.
2. Start the timer immediately. Meanwhile, press the injection point for 10 s to stop bleeding.

3. Use scissor to make a small cut at the tail vein.
4. At post-injection time points $t = 3, 5, 8, 15, 30, 60, 90, 120$ min, use heparinized capillary tubes to collect 20 μL blood sample from mouse tail vein. Note that the total blood volume of a mouse is only 1.5–2 mL, and mouse will die if there is too much blood loss.
5. Store the blood samples in the ice box to prevent clotting.
6. Centrifuge the blood samples at 3500 r.p.m. for 10 min.
7. Separate the serum (the supernatant) from red blood cells.
8. Quantify the amount of FPRR in the serum using HPLC, and plot the % ID g^{-1} against post-injection time. As shown in Fig. 7, the points were then fitted by a bi-exponential decay curve to estimate elimination half-life ($t_{1/2}$).

6.4 Real-time imaging in drug-induced acute kidney injury mice model

Equipment

1—IVIS® Spectrum In Vivo Imaging System
1—MSOT photoacoustic imaging system
1—Timer
10—Syringe and needles for intravenous injection

Reagents

Saline
FPRR (20 mM solution in PBS buffer)
Cisplatin (1 mg/mL in 0.9% saline solution)
N-acetyl-L-cystein (NAC, 20 mg/mL in 0.9% saline solution)

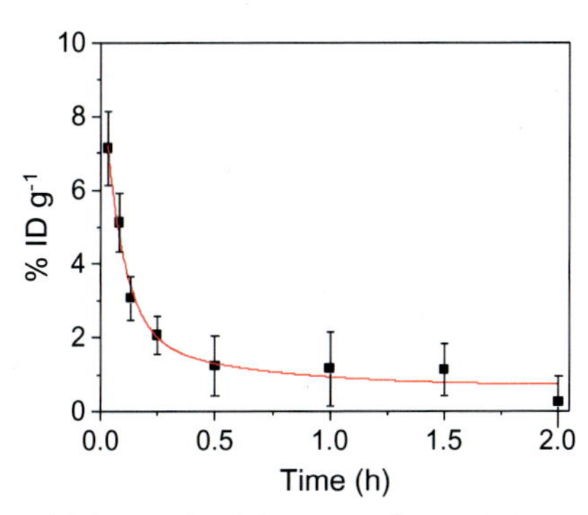

Fig. 7 Pharmacokinetics of FPRR in healthy mice after i.v. injection of FPRR (6.5 μmol/kg body weight). The blood elimination half-life ($t_{1/2}$) is determined to be 23 min.

Procedure

Part 1. Establishment of cisplatin-induced AKI.

1. Randomly divide mice into three groups: (i) control group, (ii) drug-treated group, (iii) antioxidant-pretreated group.
2. Treat the control groups with saline only (0.2 mL).
3. For the drug-treated group, treat the mice with cisplatin at reported nephrotoxic dosage (intraperitoneal injection, 20 mg/kg bw) without anesthesia.
4. For antioxidant-pretreated group, intravenously inject the mice with NAC (400 mg/kg bw) 30 min before cisplatin (intraperitoneal injection, 20 mg/kg bw) administration.

Part 2. Real-time NIRF and PA imaging.

5. Real-time imaging will be conducted at different post-drug treatment time points, $t = 16$, 24, 48 h (see Fig. 8A and B).
6. At these time points, FPRR will be intravenously injected, and images will be acquired every 30 min for a duration of 2 h after i.v. injection of FPRR (6.5 μmol/kg bw). Mice are anesthetized with inhaled isoflurane throughout real-time imaging (2 h).
7. As shown in Fig. 8C, conduct fluorescence imaging using IVIS system fluorescence mode (Ex/Em = 675/720 nm, exposure time: 0.1 s).
8. As shown in Fig. 8D, conduct in vivo photoacoustic imaging with MSOT photoacoustic imaging system. Position the mice perpendicular to the scanning plane and anesthetize them by 2% isoflurane during the PA imaging experiment. Perform real-time MSOT imaging from 680 to 850 nm. The in vivo PA spectrum is shown in Fig. 8E.
9. At the end of imaging, dissect the mice and harvest all major organs for biodistribution studies.
10. Store organs samples at −80 °C for histological studies (see Section 6.6).
11. Process the ROI intensities at the kidney sides using IVIS or iThera software for fluorescence and photoacoustic images, respectively (see Fig. 8F and G).

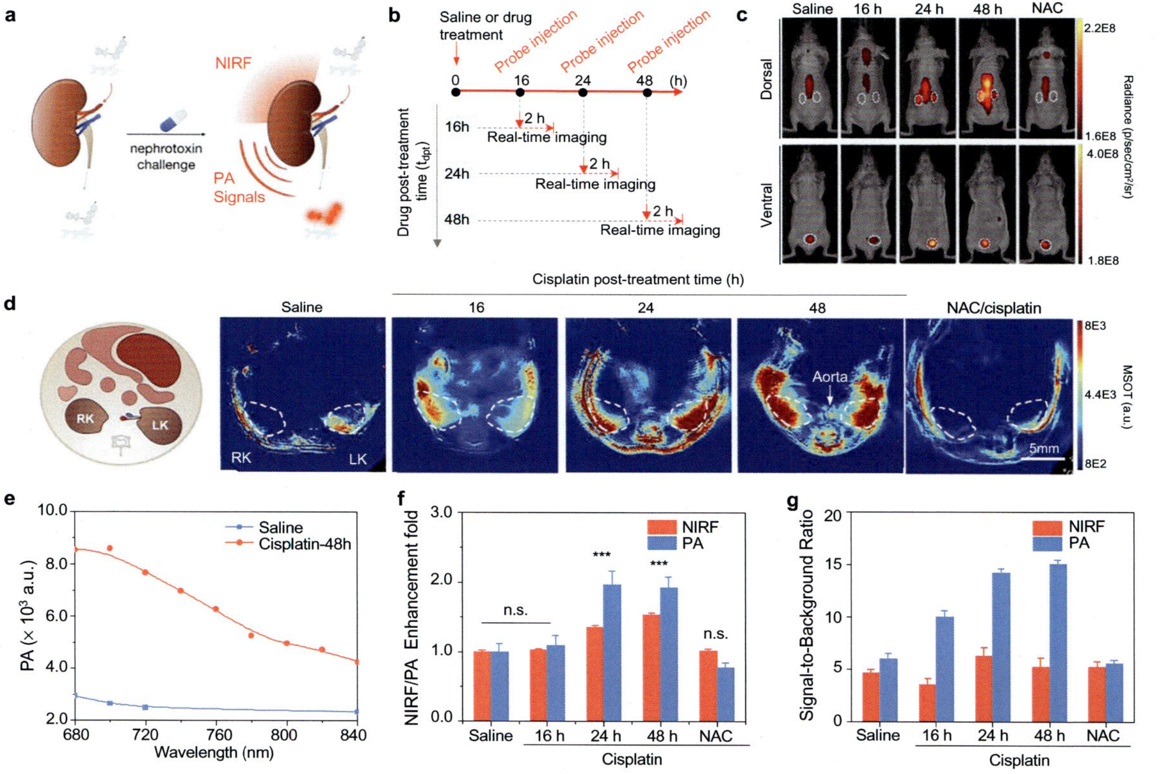

Fig. 8 Real-time NIRF and PA imaging of cisplatin-induced AKI. (A) Schematic illustration of using FPRR for bimodal imaging of AKI. (B) Timeline for development of cisplatin-induced AKI and bimodal imaging. (C) Representative NIRF images of living mice at 60 min after i.v. injection of FPRR in different treatment groups (Ex/Em = 675/720 nm). The white circles indicate the site of kidneys and bladder respectively in dorsal and ventral side. (D) Representative PA images of mice transverse section at 120 min after i.v. injection of FPRR in different treatment groups (700 nm). The white circles indicate the site of kidneys. Right kidney (RK), left kidney (LK). (E) In vivo PA spectra of kidney areas after intravenous administration of FPRR at 120 min. (F) Relative NIRF and PA signals enhancement at kidney site for different treatment groups as compared to control. (G) Comparison of signal-to-background ratio between NIRF and PA imaging. Data in (F and G) are presented as the mean ± SD, n = 3 independent mice. *: $p < 0.05$, **: $p < 0.01$, ***: $p < 0.001$, n.s.: not significant.

6.5 Online urinalysis studies

Equipment

3—Mice metabolic cages
1—Fluorolog 3-TCSPC spectrofluorometer (Horiba Jobin Yvon)
1—UV–visible range spectrophotometer
1—Quartz cuvettes with corresponding cuvette caps
10—Microcentrifuge tubes
10—Syringe and needles for intravenous injection

Reagents

Saline
FPRR (20 mM solution in PBS buffer)
Cisplatin (1 mg/mL in 0.9% saline solution)

Procedure

1. Establish cisplatin-induced acute kidney injury model as previously introduced.
2. For control group and cisplatin-treated mice at drug post-treatment time points $t = 16$, 24, and 48 h, mice will receive intravenous injection of FPRR (6.5 μmol/kg bw).
3. Press the injection point for 10 s to stop bleeding and put the mice into metabolic cages immediately.
4. Collect urine samples at 6 h post-injection.
5. Centrifuge the urine samples at 4500 r.p.m. for 10 min, and then filter through a 0.22 μm syringe filter.
6. Take the clear urine samples for UV/vis absorption and NIRF measurement (see Fig. 9).

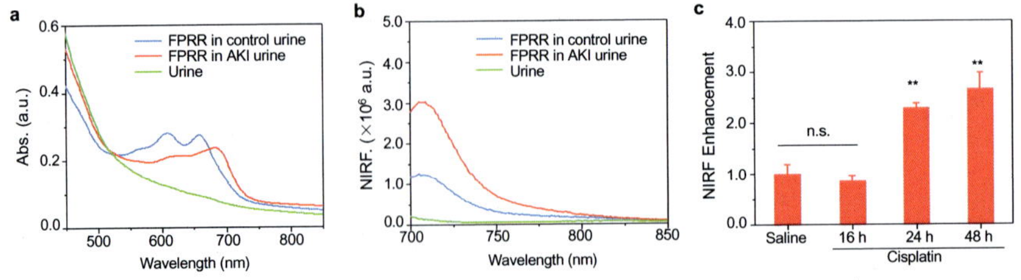

Fig. 9 (A) UV/vis absorption and (B) NIRF spectra of FPRR in urine samples excreted from both control and AKI mice after FPRR i.v. injection, and urine samples from untreated mice. (C) Relative NIRF signals (710 nm) enhancement of urine samples excreted from drug treated mice as compared to control mice after FPRR i.v. injection. Data are presented as mean ± SD, n = 3 independent experiment.

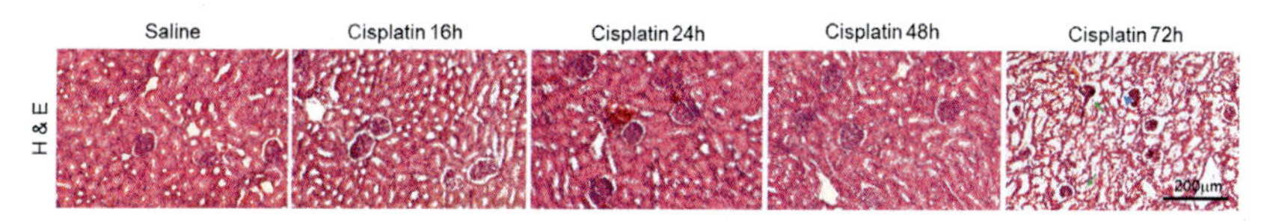

Fig. 10 Representative hematoxylin and eosin staining of kidney tissues from different drug post-treatment groups. Green arrows and blue triangle indicate renal cell debris and loss of brush border, respectively.

6.6 Kidney histological studies

Equipment

1—Laser Scanning Microscope
1—Cryostat
1—Paraffin section machine
5—Biological-grade scissors and tweezers

Reagents

Collected kidney samples (from Section 6.4)
Phosphate buffered solution
4% polyformaldehyde solution
FITC-conjugated anti-mouse caspase-3 antibody
Hematoxylin and eosin

Procedure

1. Perform cryo-sectioning or paraffin sectioning of kidney tissue samples.
2. Stain the cryo-sections with Hoechst 33342 for cell nuclei and FITC-conjugated anti-mouse caspase-3 antibody.
3. Stain the paraffin sections with Hematoxylin and eosin.
4. Identify any histological injuries (see Fig. 10).

7. Conclusions

In this chapter, we described the synthesis and development of a highly renal-clearable polymeric reporter (FPRR) with simultaneously activatable NIRF and PA signals toward an AKI biomarker (GGT), and applied it for real-time imaging of cisplatin-induced AKI in mouse model. We validated the capability of FPRR to detect GGT both in vitro and in living cells. Furthermore, we applied FPRR to detect the onset of AKI in living mice models. At 24h post-cisplatin treatment, both NIRF and PA signals at the kidney sites were detected. Such detection time point was 48h earlier than increase in serum creatinine and GFR decline.

Additionally, PA imaging shows 2.3-fold higher signal-to-background ratio as compared to NIRF imaging, and thus could potentially provide more sensitive and accurate disease information. As such, the molecular design serves as a general strategy for the development of activatable photoacoustic probes to detect urological disorders. We hope that this approach can provide useful guidelines for researchers in the field of optical imaging and molecular biology to probe more possibilities.

References

Baxmann, A. C., Ahmed, M. S., Marques, N. C., Menon, V. B., Pereira, A. B., Kirsztajn, G. M., et al. (2008). Influence of muscle mass and physical activity on serum and urinary creatinine and serum cystatin C. *Clinical Journal of the American Society of Nephrology*, *3*(2), 348–354. https://doi.org/10.2215/Cjn.02870707.

Blanco, E., Shen, H., & Ferrari, M. (2015). Principles of nanoparticle design for overcoming biological barriers to drug delivery. *Nature Biotechnology*, *33*(9), 941–951. https://doi.org/10.1038/nbt.3330.

Choudhury, D., & Ahmed, Z. (2006). Drug-associated renal dysfunction and injury. *Nature Clinical Practice. Nephrology*, *2*(2), 80–91. https://doi.org/10.1038/ncpneph0076.

Cox, Z. L., McCoy, A. B., Matheny, M. E., Bhave, G., Peterson, N. B., Siew, E. D., et al. (2013). Adverse drug events during AKI and its recovery. *Clinical Journal of the American Society of Nephrology*, *8*(7), 1070–1078. https://doi.org/10.2215/CJN.11921112.

Cui, D., Li, P., Zhen, X., Li, J., Jiang, Y., Yu, A., et al. (2019). Thermoresponsive semiconducting polymer nanoparticles for contrast-enhanced photoacoustic imaging. *Advanced Functional Materials*, *29*.

Darmon, M., Ostermann, M., Cerda, J., Dimopoulos, M. A., Forni, L., Hoste, E., et al. (2017). Diagnostic work-up and specific causes of acute kidney injury. *Intensive Care Medicine*, *43*(6), 829–840. https://doi.org/10.1007/s00134-017-4799-8.

de la Zerda, A., Liu, Z., Bodapati, S., Teed, R., Vaithilingam, S., Khuri-Yakub, B. T., et al. (2010). Ultrahigh sensitivity carbon nanotube agents for photoacoustic molecular imaging in living mice. *Nano Letters*, *10*(6), 2168–2172. https://doi.org/10.1021/nl100890d.

de la Zerda, A., Paulus, Y. M., Teed, R., Bodapati, S., Dollberg, Y., Khuri-Yakub, B. T., et al. (2010). Photoacoustic ocular imaging. *Optics Letters*, *35*(3), 270–272. https://doi.org/10.1364/OL.35.000270.

Deán-Ben, X., Gottschalk, S., Mc Larney, B., Shoham, S., & Razansky, D. (2017). Advanced optoacoustic methods for multiscale imaging of in vivo dynamics. *Chemical Society Reviews*, *46*(8), 2158–2198.

Dieterle, F., Perentes, E., Cordier, A., Roth, D. R., Verdes, P., Grenet, O., et al. (2010). Urinary clusterin, cystatin C, β2-microglobulin and total protein as markers to detect drug-induced kidney injury. *Nature Biotechnology*, *28*(5), 463–469.

Duncan, M. W., Nedelkov, D., Walsh, R., & Hattan, S. J. (2016). Applications of MALDI mass spectrometry in clinical chemistry. *Clinical Chemistry*, *62*(1), 134–143. https://doi.org/10.1373/clinchem.2015.239491.

Gustafson, H. H., Holt-Casper, D., Grainger, D. W., & Ghandehari, H. (2015). Nanoparticle uptake: The phagocyte problem. *Nano Today*, *10*(4), 487–510. https://doi.org/10.1016/j.nantod.2015.06.006.

Jeevarathinam, A. S., Lemaster, J. E., Chen, F., Zhao, E., & Jokerst, J. (2019). Photoacoustic imaging quantifies drug release from nanocarriers via redox chemistry of dye-labeled cargo. *Angewandte Chemie, International Edition in English*, *59*, 4678–4683. https://doi.org/10.1002/anie.201914120.

Jiang, Y., Upputuri, P. K., Xie, C., Zeng, Z., Sharma, A., Zhen, X., et al. (2019). Metabolizable semiconducting polymer nanoparticles for second near-infrared photoacoustic imaging. *Advanced Materials*, *31*(11), e1808166. https://doi.org/10.1002/adma.201808166.

Jokerst, J. V., Thangaraj, M., Kempen, P. J., Sinclair, R., & Gambhir, S. S. (2012). Photoacoustic imaging of mesenchymal stem cells in living mice via silica-coated gold nanorods. *ACS Nano*, *6*(7), 5920–5930.

Karlas, A., Fasoula, N. A., Paul-Yuan, K., Reber, J., Kallmayer, M., Bozhko, D., et al. (2019). Cardiovascular optoacoustics: From mice to men—A review. *Photoacoustics*, *14*, 19–30. https://doi.org/10.1016/j.pacs.2019.03.001.

Kellum, J. A., & Prowle, J. R. (2018). Paradigms of acute kidney injury in the intensive care setting. *Nature Reviews Nephrology*, *14*(4), 217–230. https://doi.org/10.1038/nrneph.2017.184.

Kim, J.-W., Galanzha, E. I., Shashkov, E. V., Moon, H.-M., & Zharov, V. P. (2009). Golden carbon nanotubes as multimodal photoacoustic and photothermal high-contrast molecular agents. *Nature Nanotechnology*, *4*(10), 688.

Li, H., Zhang, P., Smaga, L. P., Hoffman, R. A., & Chan, J. (2015). Photoacoustic probes for ratiometric imaging of copper(II). *Journal of the American Chemical Society*, *137*(50), 15628–15631. https://doi.org/10.1021/jacs.5b10504.

Liu, J., Yu, M., Zhou, C., & Zheng, J. (2013). Renal clearable inorganic nanoparticles: A new frontier of bionanotechnology. *Materials Today*, *16*(12), 477–486.

Lyu, Y., Zeng, J., Jiang, Y., Zhen, X., Wang, T., Qiu, S., et al. (2018). Enhancing both biodegradability and efficacy of semiconducting polymer nanoparticles for photoacoustic imaging and photothermal therapy. *ACS Nano*, *12*(2), 1801–1810.

Manohar, S., & Leung, N. (2018). Cisplatin nephrotoxicity: A review of the literature. *Journal of Nephrology*, *31*(1), 15–25. https://doi.org/10.1007/s40620-017-0392-z.

Miao, Q., Lyu, Y., Ding, D., & Pu, K. (2016). Semiconducting oligomer nanoparticles as an activatable photoacoustic probe with amplified brightness for in vivo imaging of pH. *Advanced Materials*, *28*(19), 3662–3668.

Miao, Q., & Pu, K. (2016). Emerging designs of activatable photoacoustic probes for molecular imaging. *Bioconjugate Chemistry*, *27*(12), 2808–2823. https://doi.org/10.1021/acs.bioconjchem.6b00641.

Miao, Q., & Pu, K. (2018). Organic semiconducting agents for deep-tissue molecular imaging: Second near-infrared fluorescence, self-luminescence, and photoacoustics. *Advanced Materials*, *30*(49), e1801778. https://doi.org/10.1002/adma.201801778.

Myers, B. D., & Moran, S. M. (1986). Hemodynamically mediated acute renal failure. *The New England Journal of Medicine*, *314*(2), 97–105. https://doi.org/10.1056/NEJM198601093140207.

Noto, A., Ogawa, Y., Mori, S., Yoshioka, M., Kitakaze, T., Hori, T., et al. (1983). Simple, rapid spectrophotometry of urinary N-acetyl-Beta-D-glucosaminidase, with use of a new chromogenic substrate. *Clinical Chemistry*, *29*(10), 1713–1716. Retrieved from <Go to ISI>://WOS:A1983RM71800001.

Peres, L. A., & da Cunha, A. D., Jr. (2013). Acute nephrotoxicity of cisplatin: Molecular mechanisms. *Jornal Brasileiro de Nefrologia*, *35*(4), 332–340. https://doi.org/10.5935/0101-2800.20130052.

Pu, K., Shuhendler, A. J., Jokerst, J. V., Mei, J., Gambhir, S. S., Bao, Z., et al. (2014). Semiconducting polymer nanoparticles as photoacoustic molecular imaging probes in living mice. *Nature Nanotechnology*, *9*(3), 233.

VA/NIH Acute Renal Failure Trial Network. (2008). Intensity of renal support in critically ill patients with acute kidney injury. *New England Journal of Medicine*, *359*(1), 7–20.

Waikar, S. S., & Bonventre, J. V. (2008). Biomarkers for the diagnosis of acute kidney injury. *Nephron. Clinical Practice*, *109*(4), c192–c197. https://doi.org/10.1159/000142928.

Waikar, S. S., & Bonventre, J. V. (2009). Creatinine kinetics and the definition of acute kidney injury. *Journal of the American Society of Nephrology*, *20*(3), 672–679. https://doi.org/10.1681/Asn.2008070669.

Wang, L. V., & Hu, S. (2012). Photoacoustic tomography: In vivo imaging from organelles to organs. *Science*, *335*(6075), 1458–1462. https://doi.org/10.1126/science.1216210.

Weber, J., Beard, P. C., & Bohndiek, S. E. (2016). Contrast agents for molecular photoacoustic imaging. *Nature Methods*, *13*(8), 639–650. https://doi.org/10.1038/nmeth.3929.

Yin, L., Sun, H., Zhang, H., He, L., Qiu, L., Lin, J., et al. (2019). Quantitatively visualizing tumor-related protease activity in vivo using a ratiometric photoacoustic probe. *Journal of the American Chemical Society*, *141*(7), 3265–3273.

Yin, C., Wen, G., Liu, C., Yang, B., Lin, S., Huang, J., et al. (2018). Organic semiconducting polymer nanoparticles for photoacoustic labeling and tracking of stem cells in the second near-infrared window. *ACS Nano*, *12*(12), 12201–12211. https://doi.org/10.1021/acsnano.8b05906.

Yu, M., & Zheng, J. (2015). Clearance pathways and tumor targeting of imaging nanoparticles. *ACS Nano*, *9*(7), 6655–6674. https://doi.org/10.1021/acsnano.5b01320.

Zackrisson, S., van de Ven, S., & Gambhir, S. S. (2014). Light in and sound out: Emerging translational strategies for photoacoustic imaging. *Cancer Research*, *74*(4), 979–1004. https://doi.org/10.1158/0008-5472.CAN-13-2387.

Zhang, Y., Jeon, M., Rich, L. J., Hong, H., Geng, J., Zhang, Y., et al. (2014). Non-invasive multimodal functional imaging of the intestine with frozen micellar naphthalocyanines. *Nature Nanotechnology*, *9*(8), 631–638. https://doi.org/10.1038/nnano.2014.130.

Zhang, H. F., Maslov, K., Stoica, G., & Wang, L. V. (2006). Functional photoacoustic microscopy for high-resolution and noninvasive in vivo imaging. *Nature Biotechnology*, *24*(7), 848–851. https://doi.org/10.1038/nbt1220.

Zhang, J., Zhen, X., Upputuri, P. K., Pramanik, M., Chen, P., & Pu, K. (2017). Activatable photoacoustic nanoprobes for in vivo ratiometric imaging of peroxynitrite. *Advanced Materials*, *29*(6), 1604764. https://doi.org/10.1002/adma.201604764.

Zhen, X., Zhang, J., Huang, J., Xie, C., Miao, Q., & Pu, K. (2018). Macrotheranostic probe with disease-activated near-infrared fluorescence, photoacoustic, and photothermal signals for imaging-guided therapy. *Angewandte Chemie, International Edition in English*, *57*(26), 7804–7808. https://doi.org/10.1002/anie.201803321.

Zheng, J., Zeng, Q., Zhang, R., Xing, D., & Zhang, T. (2019). Dynamic-reversible photoacoustic probe for continuous ratiometric sensing and imaging of redox status in vivo. *Journal of the American Chemical Society*, *141*(49), 19226–19230. https://doi.org/10.1021/jacs.9b10353.

CHAPTER TWELVE

ALP-activated probe for diagnosis of liver injury by multispectral optoacoustic tomography

Yinglong Wu, Fang Zeng*, Lihe Sun, Junjie Chen, and Shuizhu Wu*
State Key Laboratory of Luminescent Materials and Devices, Guangdong Provincial Key Laboratory of Luminescence from Molecular Aggregates, College of Materials Science and Engineering, South China University of Technology, Guangzhou, China
*Corresponding authors: e-mail address: mcfzeng@scut.edu.cn; shzhwu@scut.edu.cn

Contents

Methods in Enzymology, Volume 657
ISSN 0076-6879
https://doi.org/10.1016/bs.mie.2021.06.019

Abstract

In this chapter, we highlight the advantages of multispectral optoacoustic tomography (MSOT) technique and the activatable photoacoustic probes in the biomedical field, and give a brief introduction to enzyme-activated probes for disease diagnosis and therapeutic outcome evaluation. We also present a detailed description of the procedures for the synthesis of an activatable small molecule probe $C_1X\text{-}OR_1$ and confirmation of its specific response to alkaline phosphatase in solution and cells. With MSOT, the liposomal $C_1X\text{-}OR_1$ can be utilized for detection of hepatic ALP as well as for in vivo diagnosis of drug-induced liver injury in a three-dimensional manner.

1. Introduction

Drug-related toxicity leads to millions of emergency department visits and even hospitalizations every year, thus causing an increasing concern in the field of modern medicine (Wang, Avorn, & Kesselheim, 2013). Liver is the primary organ for concentrating and metabolizing drugs (Corsini & Bortolini, 2013), and drug-induced liver injury (DILI) with clinical manifestations ranging from asymptomatic biochemical changes to liver failure (Sakatis et al., 2012), becomes one of the most common complications and the main reason of post-marketing drug withdrawal (Kola & Landis, 2004). Therefore, a preclinical or clinical screening method for drug-induced liver injury is urgently needed to improve patient safety and therapeutic outcomes.

It has been reported that several biomarkers especially the circulating ones have been utilized for diagnosis of DILI in clinical trials and clinical practices (Yang, Salminen, & Schnackenberg, 2012). Among these, alkaline phosphatase (ALP) which can hydrolyze phosphate monoesters into corresponding hydroxyl products and inorganic phosphates shows a vital role in maintaining the balance of dephosphorylation and phosphorylation processes in cells (Fedarko, Bianco, Vetter, & Robey, 1990). Typically, elevation of ALP levels in serum is considered to be associated with liver disease (Kaplan & Righetti, 1970). However, in addition to being present in the liver, ALP is also expressed in other tissues and organs including intestine, bone, placenta, etc. (Coleman, 1992). Only the elevated level of ALP in the liver can sufficiently serve as the indicator for liver dysfunction. Hence, spatiotemporal detection of ALP level in liver is critical to diagnosis of drug-induced liver injury.

For in vivo identification and quantification of disease-specific biomarkers at molecular levels, the nonionizing and noninvasive photoacoustic (PA, also known as optoacoustic) imaging which combines the superb contrast and high sensitivity of optical imaging with the deep-tissue penetration and high-resolution attainable in ultrasound imaging, is an advantageous approach (Wang & Yao, 2016). Particularly, multispectral optoacoustic tomography (MSOT), which is an emerging functional photoacoustic imaging technique, can map the distribution of specific contrast agents with unique spectra by acquiring multispectral data and subsequently performing spectral unmixing using the preset algorithms (Taruttis & Ntziachristos, 2015). Furthermore, MSOT can realize three-dimensional (3D) visualization using volumetric imaging technique or through reconstruction of a maximum intensity projection (MIP) image from a stack of two-dimensional cross-sectional images (Gottschalk et al., 2019), which can achieve accurate location of the biomarkers in tissues of interest and hence displays great potential for preclinical research and clinical practice.

To achieve functional PA imaging, various contrast agents including inorganic nanomaterials, semiconducting polymer nanoparticles and organic small-molecule dyes have been developed and used in PA imaging for disease diagnosis, therapy guidance, evaluation of treatment outcomes, etc. (Jeevarathinam, Lemaster, Chen, Zhao, & Jokerst, 2020; Wu, Sun, Zeng, & Wu, 2019). Among them, some activatable contrast agents (probes) with the ability of specifically respond to biological stimuli and subsequently eliciting changes in photoacoustic signal are quite advantageous, as they can sensitively and specifically locate disease foci and track its dynamic processes with low background noise (Knox et al., 2017).

In this chapter, we highlight the strategy for designing activatable PA probes for specific biomarker, and briefly introduce enzyme-activated probes through concentrating on their photophysical properties, specific response to enzymes and biomedical applications in MSOT imaging. Also, we give a detailed overview of a small-molecule PA probe $C_1X\text{-}OR_1$, which is developed by our group and able to specifically respond to the hepatic ALP level and thus achieving in vivo diagnosis of drug-induced liver injury using MSOT (Wu et al., 2018).

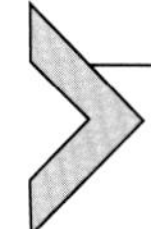

2. Construction of activatable photoacoustic probes for detection of specific biomarkers

When constructing an activatable PA probe for detection of a specific biomarker with MSOT imaging, several issues should be carefully considered. First, the signaling chromophore should have a characteristic absorption peak with a quite high molar extinction coefficient at near-infrared

(NIR) wavelengths, which affords sufficient contrast to endogenous absorbers (such as hemoglobin, melanin, etc.) upon spectral unmixing even at diluted concentrations. Second, the recognition moiety of a PA probe must exhibit high sensitivity and specificity to its corresponding biomarker in tissues of interest, thereby reducing undesirable background noise and false-positive signals. Third, once the signaling chromophore and recognition moiety are selected or constructed, the activation mechanism should be reasonably designed. Typically, the activation mechanisms for eliciting changes in photoacoustic signal are usually achieved by the variations in photophysical properties of the probes, which are caused by a number of changes in molecular structure, such as conversion of intramolecular electronic state, alteration of intramolecular movement/rotation degree, transformation between aggregate and nonaggregate states, and so on.

In recent years, a great number of activatable photoacoustic probes have been developed for specific identification of various biologically-relevant species such as nucleic acids, ions, ROS/RNS, enzymes, pH, etc. (Zlitni, Gowrishankar, Steinberg, Haywood, & Gambhir, 2020). In particular, enzyme-activated probes are ideal ones because enzyme-mediated reactions exhibit some attractive advantages such as high efficiency and substrate specificity, thus contributing to rapid and precise disease diagnosis (Cheng et al., 2020). For instance, hyaluronidase (HAase), which can selectively degrade its substrate hyaluronan (HA) into small fragments, has been used as a diagnostic and prognostic indicator for multiple cancers. For example, by employing a hyaluronidase (HAase)-activated PA probe fabricated by self-assembly of a positively charged tricyanofuran-containing polyene (TCHM) as the signaling chromophore and the negatively charged HA as the response moiety, the lymphatic metastasis and orthotopic bladder tumors can be spatiotemporally located and diagnosed using MSOT (Wu, Chen, Sun, Zeng, & Wu, 2019). In addition, a nitroreductase-reponsive PA probe has also been developed through conjugation of an electron donor dihydroxanthene, an electron acceptor quinolinium and the recognition element nitrobenzyloxydiphenylamino. The overexpressed nitroreductase in tumor cells can specifically turn on the probe's PA signal, thus achieving visualization of metastases from the orthotopic breast tumors to lymph nodes and then lung, and monitor of the chemotherapy efficacy (Ouyang et al., 2020). Moreover, many other probes are designed to be activated by enzymes in the tumor microenvironment for tumor-related applications. Among these activatable PA probes, the hepatic enzyme-activated probe for diagnosis of liver disease is still limited, and this prompted our group to develop an ALP-activated photoacoustic probe for imaging of drug-induced liver injury.

3. Design of the ALP-activated small molecule photoacoustic probe (C_1X-OR_1)

The ALP-activated photoacoustic probe C_1X-OR_1 consists of an identification moiety phosphate (R_1) and a signaling chromophore C_1X-OH which is also the activated product of the ALP-mediated reaction. Upon responding to ALP, the enzyme-mediated hydrolysis transforms the electron-withdrawing phosphate (R_1) into the electron-donating hydroxyl, thus converting C_1X-OR_1 into its activated product (C_1X-OH). Different from C_1X-OR_1, the activated product C_1X-OH displays a relatively high extinction coefficient ($7.7 \times 10^4 M^{-1} cm^{-1}$) at 684 nm, which is suitable for MSOT imaging. Such an ALP-mediated structural transformation leads to variation in absorption followed by a significant change in PA signal, thereby realizing diagnosis of drug-induced liver injury with MSOT imaging (Scheme 1). To achieve higher water dispersibility and hepatocyte targetability, a phospholipid DSPE-PEG2000-ChA was synthesized and formed into liposomes with the probe.

ALP

C_1X-OR_1 → C_1X-OH

Scheme 1 Illustration for response of C_1X-OR_1 to ALP and its activation mechanism.

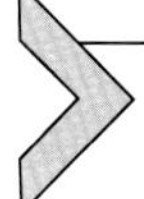

4. Synthesis of the probe C_1X-OR_1 and the hepatocyte-targeting phospholipid DSPE-PEG_{2000}-ChA

The probe C_1X-OR_1 and hepatocyte-targeting phospholipid DSPE-PEG_{2000}-ChA were synthesized according to the synthetic routes depicted in Scheme 2, and well-characterized by ^1H NMR and mass spectrometry.

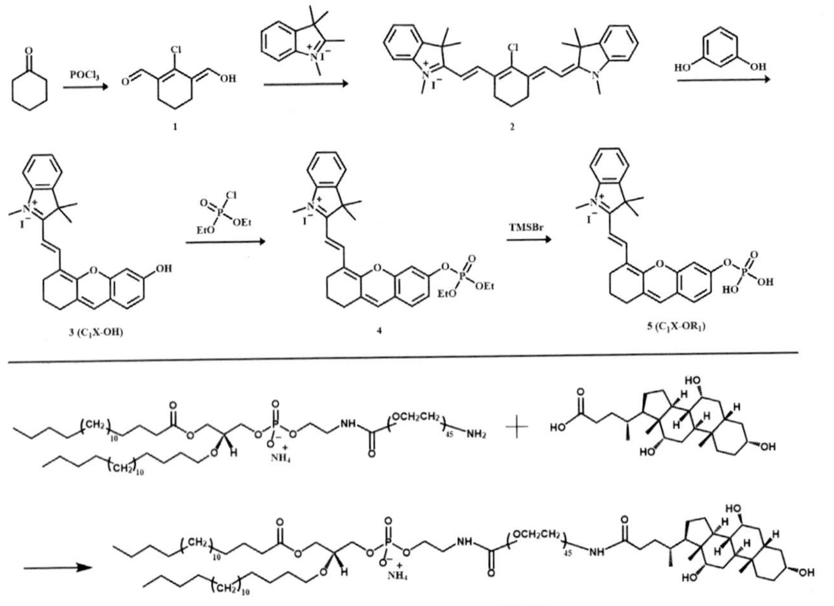

Scheme 2 Synthetic routes of the probe C_1X-OR_1 and the hepatocyte-targeting phospholipid DSPE-PEG_{2000}-ChA.

The familiarity with basic synthetic laboratory equipment and standard organic chemistry techniques are needed to understand the following synthetic procedures. All solvents for silica gel chromatography are defined as the ratio of volume to volume.

4.1 Equipment

1. Magnetic stirrer
2. Vacuum pump
3. Rotary evaporator
4. Vacuum oven
5. Refrigerator
6. Freeze-dryer
7. An oil bath
8. Two-neck round flasks
9. Pressure equalizing dropping funnels
10. Condensers
11. Buchner funnels
12. Columns for chromatography

13. Thin layer chromatography (TLC) silica gel
14. Dialysis membranes (MwCO = 1000)
15. Rubber stoppers
16. Syringes
17. Magnetic PTFE stir bars

4.2 Reagents

1. Phosphorus oxychloride
2. Cyclohexanone
3. 1,2,3,3-tetramethyl-3H-indolium iodide
4. Resorcinol
5. Diethyl chlorophosphate
6. Bromotrimethylsilane
7. Cholic acid
8. *N*-(3-Dimethylaminopropyl)-*N*′-ethylcarbodiimide hydrochloride crystalline (EDC·HCl)
9. *N*-Hydroxysulfosuccinimide sodium salt (Sulfo-NHS)
10. 1,2-distearoyl-*sn*-glycero-3-phosphoethanolamine-N-[amino (polyethylene glycol)-2000] (DSPE-PEG_{2000}-NH_2)
11. Acetic anhydride
12. Sodium acetate
13. Potassium carbonate
14. Organic solvents: dichloromethane, acetonitrile, dimethylformamide, methanol
15. Deionized water

4.3 Procedures

4.3.1 2-Chloro-3-(hydroxymethylene)cyclohex-1-ene-1-carbaldehyde (compound 1)

1. Add anhydrous dimethylformamide (20 mL, 273 mmol, 2.1 equiv.) into a 250 mL two-neck round flask equipped with a magnetic PTFE stir bar
2. Dissolve phosphorus oxychloride (20 mL, 131 mmol, 1.0 equiv.) with the organic solvent dichloromethane (20 mL) in a pressure equalizing dropping funnel equipped with a rubber stopper
3. Place the dropping funnel onto the round flask
4. Cool the reaction mixture in an ice bath
5. Vacuum and purge the reaction vessel with nitrogen five times
6. Open the stopcock and add phosphorus oxychloride solution dropwise within 20 min at 0 °C under nitrogen atmosphere

7. Stir the reaction mixture at room temperature for another 30 min
8. Add cyclohexanone (5.0 g, 51 mmol, 0.4 equiv.) drop by drop into the mixture via a syringe
9. Reflux the reaction mixture with a condenser under vigorously stirring in an oil bath for 1.5 h at 80 °C
10. Cool down the reaction mixture, pour it into ice-cold water, and then allow it to stand in a refrigerator overnight
11. Filter the resulting suspension and collect the precipitate
12. Dry the product Compound 1 in a vacuum oven as a yellow solid (5.2 g, yield: 59%).

4.3.2 2-(-2-(-2-Chloro-3-(2-(-1,3,3-trimethylindolin-2-ylidene) ethylidene)cyclohex-1-en-1-yl)vinyl)-1,3,3-trimethyl-3H-indol-1-ium (compound 2)

1. Add Compound 1 (0.86 g, 5 mmol, 1.0 equiv.) and 1,2,3,3-tetramethyl-3H-indolium iodide (3.01 g, 10 mmol, 2.0 equiv.) into a 100 mL two-neck round flask equipped with a magnetic PTFE stir bar
2. Dissolve the mixed solids with anhydrous acetic anhydride (40 mL)
3. Add sodium acetate (0.82 g, 10 mmol, 2.0 equiv.) into the mixture
4. Vacuum and purge the reaction vessel with nitrogen five times
5. Reflux the mixture in an oil bath set at 130 °C for 1 h
6. Remove the solvent by vacuum-rotary evaporation
7. Purify the resulting green solid mixture by silica gel chromatography with dichloromethane/methanol (*v*/v = 20:1) as the eluent
8. Dry the target product compound 2 in a vacuum oven as a dark green solid (2.8 g, yield: 91%).

4.3.3 2-(2-(6-Hydroxy-2,3-dihydro-1H-xanthen-4-yl)vinyl)-1,3,3-trimethyl-3H-indol-1-ium (compound 3, C_1X-OH)

1. Dissolve resorcinol (0.88 g, 8 mmol, 2.0 equiv.) with anhydrous acetonitrile (30 mL) in a 100 mL two-neck round-bottom flask equipped with a magnetic PTFE stir bar
2. Add potassium carbonate (1.1 g, 8 mmol, 2.0 equiv.) into the flask
3. Stir the mixture at room temperature for 30 min
4. Dissolve Compound 2 (2.44 g, 4 mmol, 1.0 equiv.) with the organic solvent acetonitrile (20 mL) in a pressure equalizing dropping funnel equipped with a rubber stopper
5. Place the dropping funnel onto the flask
6. Vacuum and purge the reaction vessel with nitrogen five times

7. Open the stopcock and add Compound 2 solution dropwise at room temperature
8. Heat the mixture in an oil bath at 55 °C for 4 h
9. Remove the solvent by vacuum-rotary evaporation
10. Purify the crude product by silica gel column chromatography with dichloromethane/methanol (*v*/v = 20:1) as eluent
11. Dry the target product compound 3 (C_1X-OH) in a vacuum oven as a greenish blue solid (1.44 g, yield: 70%).

4.3.4 2-(2-(6-((Diethoxyphosphoryl)oxy)-2,3-dihydro-1H-xanthen-4-yl)vinyl)-1,3,3-trimethyl-3H-indol-1-ium (compound 4)

1. Dissolve Compound 3 (1.0 g, 1 mmol, 1.0 equiv.) and potassium carbonate (0.68 g, 5 mmol, 5.0 equiv.) with anhydrous acetonitrile (20 mL) in a 50 mL round-bottom flask equipped with a magnetic PTFE stir bar
2. Vacuum and purge the reaction vessel with nitrogen five times
3. Stir the mixture at room temperature for 30 min
4. Add a solution of diethyl chlorophosphate (3.44 g, 20 mmol, 20 equiv.) in acetonitrile (10 mL) into the above mixture drop by drop via a syringe
5. Heat the reaction mixture at 40 °C overnight
6. Remove the solvent by vacuum-rotary evaporation
7. Purify the crude product by silica gel column chromatography with dichloromethane/methanol (*v*/v = 30:1) as eluent
8. Dry the target product compound 4 in a vacuum oven as dark blue solid (1.08 g, yield: 83%).

4.3.5 1,3,3-Trimethyl-2-(2-(6-(phosphonooxy)-2,3-dihydro-1H-xanthen-4-yl)vinyl)-3H-indol-1-ium (compound 5, $C_1X\text{-}OR_1$)

1. Dissolve compound 4 (0.64 g, 1 mmol, 1.0 equiv.) with anhydrous acetonitrile (15 mL) in a 50 mL round-bottom flask equipped with a magnetic PTFE stir bar
2. Vacuum and purge the reaction vessel with nitrogen five times
3. Inject bromotrimethylsilane (3.06 g, 20 mmol, 20 equiv.) into the above solution via a syringe
4. Stir the mixture at room temperature for 2 days
5. Add methanol (6 mL) into the mixture to quench the reaction
6. Stir the resulting mixture for another 1 day
7. Remove the solvent by vacuum-rotary evaporation
8. Purify the resulting residue by silica gel column chromatography with dichloromethane/methanol (*v*/v = 15:1) as eluent

9. Dry the target product compound 5 (C_1X-OR_1) in a vacuum oven as dark blue solid (0.2g, yield: 45%).

4.3.6 DSPE-PEG$_{2000}$-ChA (compound 6)

1. Dissolve cholic acid (41mg, 0.1mmol, 10 equiv.), EDC·HCl (38mg, 0.2mmol, 20 equiv.) and Sulfo-NHS (43mg, 0.2mmol, 20 equiv.) with anhydrous dimethylformamide (2mL) in a 10mL round-bottom flask equipped with a magnetic PTFE stir bar
2. Stir the mixture for 2h at room temperature
3. Add DSPE-PEG$_{2000}$-NH_2 (27.5mg, 0.01mmol, 1.0 equiv.) into the flask
4. Vacuum and purge the reaction vessel with nitrogen five times
5. Stir the reaction mixture at room temperature for 2 days
6. Monitor the conjugation efficiency by thin-layer chromatography (TLC).
7. Dilute the reaction mixture with deionized water (8mL).
8. Dialyze the mixture against deionized water in the dark at room temperature for 2 days
9. Lyophilize the above dialysate to obtain the target product Compound 6 (DSPE-PEG$_{2000}$-ChA) as white powder (18.2mg, yield: 58%).

5. Response of C_1X-OR_1 toward ALP in solution

After completing the synthesis, it is necessary to (i) investigate the changes in optical spectra and photoacoustic signal of C_1X-OR_1 in response to ALP, (ii) evaluate the probe C_1X-OR_1's selectivity toward ALP and anti-interference to other possible interfering species, and (iii) verify the proposed mechanism that C_1X-OR_1 produces the product C_1X-OH upon ALP treatment. Note that we prepared the stock solutions of the compounds in advance and diluted them into solutions with appropriate concentrations for tests to minimize experimental error caused by weighing trace amounts of compound each time.

5.1 Equipment

1. Hitachi U-3010 UV–vis spectrophotometer
2. InVision128 multispectral optoacoustic tomographic (MSOT) imaging system (iThera Medical GmbH) equipped with a phantom
3. Hitachi F-4600 fluorescence spectrophotometer
4. Agilent 1260 Infinity liquid chromatograph (with diode-array detector, DAD)

5. Magnetic stirrer
6. Freeze-dryer
7. Refrigerator
8. Microbalance
9. Ultrasonic cleaner
10. Centrifuge
11. A water bath
12. Quartz cuvettes
13. NMR tubes
14. PTFE syringe filters (0.22 μm)
15. Vials
16. Micropipettes and tips
17. Syringes
18. Spatulas

5.2 Reagents

1. Synthetic C_1X-OR_1 and C_1X-OH
2. ALP
3. Acetylcholinesterase (AchE)
4. Glutamyl transferase (GGT)
5. Carboxylesterase (CaE)
6. Nitroreductase (NTR)
7. Cysteine (Cys)
8. Homocysteine (Hcy)
9. Glutathione (GSH)
10. Ascorbic acid (AA)
11. Hydrogen peroxide solution (H_2O_2)
12. Organic solvents: DMSO, methanol, acetonitrile (HPLC grade)
13. Aqueous buffer: 10 mM Tris–HCl (pH = 8.0)
14. Deionized water

5.3 Procedure

5.3.1 Preparation of stock solutions of C_1X-OR_1 and C_1X-OH

1. Label the vials with the name of compounds (C_1X-OR_1 or C_1X-OH).
2. Place the vials onto the weighing platform and zero the microbalance
3. Weigh a certain amount of solid compound into its corresponding vial with a spatula

4. Add the calculated volume of DMSO to each vial with a micropipette for preparation of a 1 mM stock solution
5. Cap the vials and sonicate until the solid compounds are completely dissolved
6. Store the prepared stock solutions in a refrigerator at −20 °C. Note: The stock solutions need to be fully thawed and shaken before use

5.3.2 *Measurement of optical properties of $C_1X\text{-}OR_1$ and its proposed activated product $C_1X\text{-}OH$*

The experiments give absorption spectra of $C_1X\text{-}OR_1$ as well as its proposed activated product $C_1X\text{-}OH$, and the extinction coefficient for $C_1X\text{-}OH$ at the maximum absorption wavelength. This information is important, as the photoacoustic performance of a signaling chromophore is closely associated with its absorption properties. The absorption spectra of $C_1X\text{-}OR_1$ and its proposed activated product $C_1X\text{-}OH$ are shown in Fig. 1A and the extinction coefficient at the maximum absorbance wavelength (684 nm) for $C_1X\text{-}OH$ is determined as $7.7 \times 10^4\, M^{-1}\, cm^{-1}$. In addition, PA signal of $C_1X\text{-}OR_1$ and $C_1X\text{-}OH$ at different wavelengths (Fig. 1B) are determined using MSOT imaging system. The results are conducive to verification of the possibility that $C_1X\text{-}OR_1$ generates PA signal after responding to ALP and turning into $C_1X\text{-}OH$.

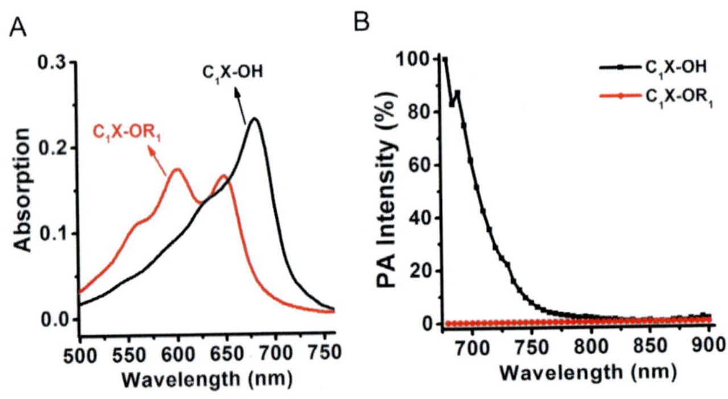

Fig. 1 Optical properties of $C_1X\text{-}OR_1$ and its activated product $C_1X\text{-}OH$. (A) Absorption spectra and (B) Normalized PA signal intensities as a function of wavelength for $C_1X\text{-}OR_1$ (5 μM) and its activated product $C_1X\text{-}OH$ (3 μM). *Reproduced with permission from Springer Nature: Wu, Y., Huang, S., Wang, J., Sun, L., Zeng, F., & Wu, S. (2018). Activatable probes for diagnosing and positioning liver injury and metastatic tumors by multispectral optoacoustic tomography. Nature Communications, 9, 3983. https://doi.org/10.1038/s41467-018-06499-1.*

(A). Measurement of absorption spectra of $C_1X\text{-}OR_1$ and its proposed activated product $C_1X\text{-}OH$

1. Warm up the Hitachi U-3010 UV–vis spectrophotometer in advance
2. Set up the parameters to collect absorbance spectra from 500 to 760 nm, at 1 nm increments
3. Correct the systemic baseline for the UV–vis spectrophotometer without any sample
4. Correct the solvent baseline for the UV–vis spectrophotometer by using two quartz cuvettes with Tris-HCl buffer (1 mL) containing 5% DMSO
5. Remove the cuvette from the sample holder while maintain the cuvette in the reference holder still
6. Prepare the test solution of $C_1X\text{-}OR_1$ and $C_1X\text{-}OH$ by diluting their stock solution with DMSO and Tris–HCl buffer under stirring. The concentration of test solution (containing 5% DMSO) of $C_1X\text{-}OR_1$ and $C_1X\text{-}OH$ is 5 μM and 3 μM, respectively
7. Pipette 1 mL test solution of $C_1X\text{-}OR_1$ or $C_1X\text{-}OH$ into the quartz cuvette, respectively, and then put it in the sample holder
8. Start measurement to collect the absorbance spectrum and record the wavelength of maximum absorbance for $C_1X\text{-}OR_1$ and $C_1X\text{-}OH$
9. Measure four other test solutions of $C_1X\text{-}OH$ at different final concentrations for determination of its extinction coefficient. Noted that the peak absorbance value of all the test samples should fall between 0.1 and 1.0 a.u
10. Create a linear regression line by use of the peak absorbance value of each sample and its corresponding concentration, and then calculate the extinction coefficient of $C_1X\text{-}OH$ by the Beer-Lambert Law

(B). Measurement of photoacoustic signal of $C_1X\text{-}OR_1$ and its proposed activated product $C_1X\text{-}OH$

1. Switch on the MSOT imaging system and fill the internal water tank to the marking position with deionized water
2. Warm up the system to 34 °C in advance of this experiment
3. Set up the parameters to collect photoacoustic signal from 680 to 900 nm at 5 nm increments, 3 frames per wavelengths
4. As described in Procedure 5.3.2-6, prepare test solutions of $C_1X\text{-}OR_1$ and $C_1X\text{-}OH$

5. Fully fill commercial NMR tubes with test solutions of C_1X-OR_1 and C_1X-OH, respectively, and then place the NMR tubes into the phantom
6. Fix the phantom in the holder of the instrument and then put the holder with samples into the instrument
7. Start measurement to acquire original data
8. Reconstruct MSOT images at each wavelength using a standard back-projection algorithm within the ViewMSOT software
9. Draw an equal circle line on the sample position in the image and select the circular area as region of interest (ROI).
10. Record the mean PA intensity at each wavelength of ROI

5.3.3 Optical and acoustic response of C_1X-OR_1 toward ALP

In order to investigate the response of C_1X-OR_1 toward ALP, the absorbance, photoacoustic and fluorescence change after incubating with varied amount of ALP for 30 min at 37 °C are recorded. The results are shown in Fig. 2.

(A). Measurement of changes in absorbance spectrum of C_1X-OR_1 upon incubation with varied amounts of ALP at 37 °C for 30 min

1. Repeat Procedure 5.3.2 (A) steps 1–5

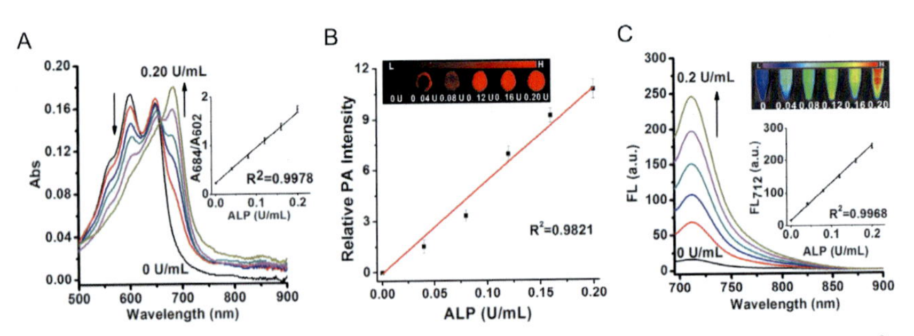

Fig. 2 Optical and acoustic response of C_1X-OR_1 toward ALP. (A) Absorption spectra for C_1X-OR_1 upon incubation with varied concentrations of ALP. The insets display the plots of absorbance ratio A_{684}/A_{602} vs ALP concentration. (B) PA response of C_1X-OR_1 to ALP at different concentrations. The insets display the representative PA images of C_1X-OR_1 at 684 nm in phantom after reacting with different concentrations of ALP. (C) Fluorescent spectra for C_1X-OR_1 upon incubation with varied concentration of ALP. The insets display the plots of fluorescence intensity at 712 nm vs ALP concentration. *Reproduced with permission from Springer Nature: Wu, Y., Huang, S., Wang, J., Sun, L., Zeng, F., & Wu, S. (2018). Activatable probes for diagnosing and positioning liver injury and metastatic tumors by multispectral optoacoustic tomography. Nature Communications, 9, 3983. https://doi.org/10.1038/s41467-018-06499-1.*

2. Prepare the test solution by incubating $C_1X\text{-}OR_1$ (5 μM) with varied amounts of ALP (0, 0.04, 0.08 0.12, 0.16, and 0.20 U/mL) at 37 °C in Tris–HCl buffer solution containing 5% DMSO for 30 min
3. Pipette 1 mL test solution into the quartz cuvette and then put it in the sample holder
4. Start measurement to collect the absorbance spectrum and record the absorbance intensity at 602 and 684 nm (the maximum absorbance wavelength of $C_1X\text{-}OR_1$ and its activated product $C_1X\text{-}OH$, respectively)
5. Analyze the data and plot absorbance ratio A_{684}/A_{602} as a function of ALP concentration

(B). Measurement of changes in photoacoustic signal of $C_1X\text{-}OR_1$ upon reaction with varied amounts of ALP at 37 °C for 30 min
1. Repeat Procedure 5.3.2 (B) steps 1–2
2. Set up the parameters to collect optoacoustic signal at 684 nm (the maximum optoacoustic signal wavelength of the activated product $C_1X\text{-}OH$), 3 frames for each test
3. Prepare the test solution by incubating $C_1X\text{-}OR_1$ (5 μM) with varied concentrations of ALP (0, 0.04, 0.08 0.12, 0.16, and 0.20 U/mL) at 37 °C in Tris–Hcl buffer solution containing 5% DMSO for 30 min
4. Fully fill a commercial Wilmad NMR tube with the test solution, place the tube into the phantom, and then fix the phantom on the instrument holder
5. Repeat Procedure 5.3.2 (B) steps 5–9
6. Analyze the data and plot relative PA intensity as a function of ALP concentration

(C). Measurement of fluorescence spectra of $C_1X\text{-}OR_1$ upon reaction with varied concentrations of ALP at 37 °C for 30 min
1. Warm up the Hitachi F-4600 fluorescence spectrophotometer in advance
2. Set up the parameters (PMT voltage = 600 V, Ex = 680 nm, EM = 695–900 nm at 1 nm increasement, Ex slit = 5 nm, Em slit = 5 nm, Response time = 4 s, Scan speed = 240 nm/min) to collect fluorescence spectrum
3. Prepare the test solution by incubating $C_1X\text{-}OR_1$ (5 μM) with varied amounts of ALP (0, 0.04, 0.08 0.12, 0.16, and 0.20 U/mL) at 37 °C in Tris–Hcl buffer solution containing 5% DMSO for 30 min

4. Pipette 1 mL test solution into the quartz cuvette and then put it in the sample holder
5. Start measurement to collect the fluorescence spectrum and record the fluorescence intensity at 712 nm (the maximum fluorescence emission wavelength of the activated product C_1X-OH)
6. Analyze the data and plot fluorescence intensity at 712 nm as a function of ALP concentration

5.3.4 Evaluation of C_1X-OR_1's selectivity toward ALP and anti-interference to other possible interfering species

To study the response selectivity, we parallelly examine various potential interfering substances by MSOT imaging under the same test conditions. As for anti-interference experiments, C_1X-OR_1 is incubated with ALP in the presence of one potential interfering species. The results are displayed in Fig. 3.

1. Repeat Procedure 5.3.3 (B) steps 1–2
2. Prepare the test solutions by mixing C_1X-OR_1 (5 μM) with ALP (0.20 U/mL) and a potential interference substance at 37 °C in Tris–Hcl buffer solution containing 5% DMSO for 30 min. The concentration of AchE, GGT, CaE or NTR is 0.20 U/mL, and that of Cys, Hcy or GSH is 1 mM, and that of H_2O_2 or AA is 100 μM
3. Repeat Procedure 5.3.3 (B) steps 4–5
4. Analyze the data

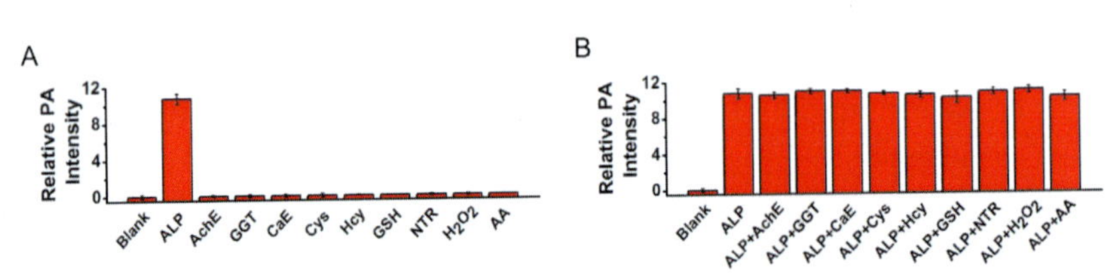

Fig. 3 Selectivity and anti-interference ability of C_1X-OR_1. (A) PA response of C_1X-OR_1 in the presence of 0.2 U/mL ALP or a potential interference substance (upper panel), and (B) PA response of the probes in the presence of 0.2 U/mL ALP and a potential interference substance (lower panel). *Reproduced with permission from Springer Nature: Wu, Y., Huang, S., Wang, J., Sun, L., Zeng, F., & Wu, S. (2018). Activatable probes for diagnosing and positioning liver injury and metastatic tumors by multispectral optoacoustic tomography. Nature Communications, 9, 3983. https://doi.org/10.1038/s41467-018-06499-1.*

5.3.5 Verification of the proposed activation mechanism

The proposed response mechanism that C_1X-OR_1 produces the product C_1X-OH upon treatment with ALP is verified by HPLC, as shown in Fig. 4.

1. Prepare the sample of enzyme-mediated reaction by mixing C_1X-OR_1 and ALP for 20 min. The procedure is given as follows: 5 μM C_1X-OR_1 and 0.2 U/mL ALP is stirred at 37 °C in 2 mL Tris–Hcl buffer solution containing 5% DMSO for 20 min, and the reaction mixture is lyophilized and re-dissolved in 2 mL methanol. After that, the solution is filtered by a syringe with a PTFE filter. The resultant solution is used as the sample for HPLC test
2. Purge each solvent (acetonitrile, methanol and deionized water) channel for 10 min at a flow rate of 1 mL/min
3. Set up the parameters (Mobile phase: methanol/acetonitrile/water = 75/15/10, Flow rate: 1 mL/min, Injection volume: 20 μL, Stop time: 30 min, DAD wavelength channel: 684 nm, Temperature: 25 °C).
4. Balance the baseline

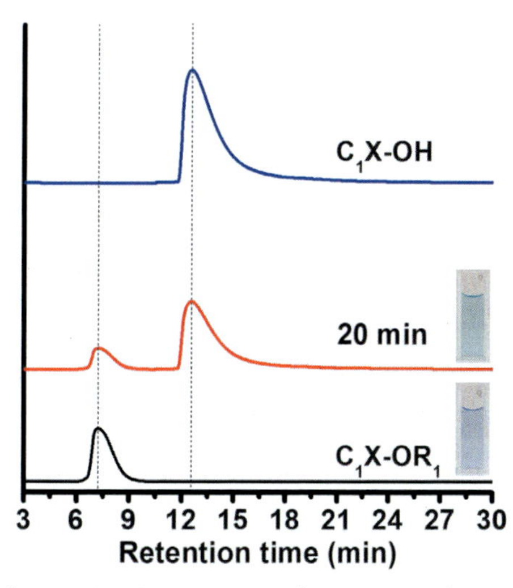

Fig. 4 HPLC data of pure C_1X-OR_1, mixture of C_1X-OR_1 and ALP upon incubation for 20 min and pure C_1X-OH. Mobile phase: methanol/acetonitrile/water = 75/15/10. *Reproduced with permission from Springer Nature: Wu, Y., Huang, S., Wang, J., Sun, L., Zeng, F., & Wu, S. (2018). Activatable probes for diagnosing and positioning liver injury and metastatic tumors by multispectral optoacoustic tomography. Nature communications, 9, 3983. https://doi.org/10.1038/s41467-018-06499-1.*

5. Add test solutions into 1 mL sample vials, respectively. Sample 1: C_1X-OH (5 μM) in methanol; Sample 2: mixture of C_1X-OR_1 (5 μM) and ALP (0.2 U/mL) upon reaction for 20 min in methanol; Sample 3: C_1X-OR_1 (5 μM) in methanol
6. Run samples and collect their data

6. Response of C_1X-OR_1 toward ALP in cells

Since C_1X-OR_1 is designed to be used as an activatable PA probe for detection of ALP in a biological environment, it is vital to demonstrate that C_1X-OR_1 is noncytotoxic to cell lines (Fig. 5). The murine aneuploid fibrosarcoma cells (L929) and human hepatocarcinoma cells (HepG2) are cultured in complete medium at 37 °C under 5% of CO_2. When the cell density reaches 70–80% of confluence, subculturing is considered complete. The medium is changed approximately every 1–2 days.

6.1 Equipment

1. Cell incubator
2. Biological safety cabinet
3. Thermo MK3 ELISA reader
4. BD CS6 flow cytometer

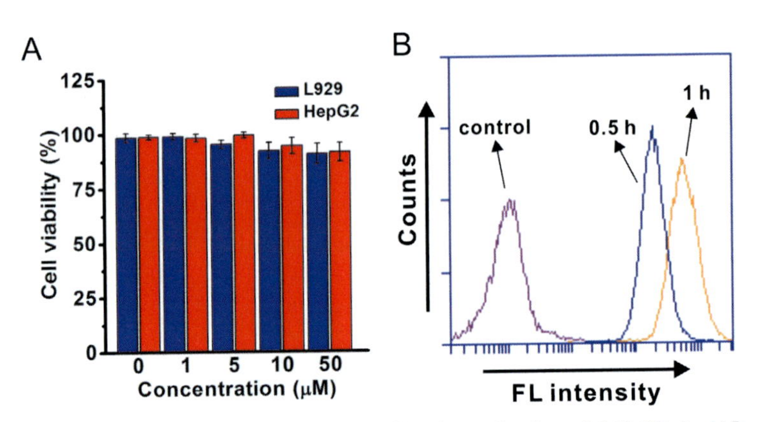

Fig. 5 Cell viabilities and flow cytometry profiles for activation of C_1X-OR_1 in ALP over-expressed cells. (A) Viabilities of L929 and HepG2 cells upon 24 h of incubation with C_1X-OR_1 of different concentrations (0, 1, 5, 10 and 50 μM). (B) Flow cytometry profiles for HepG2 cells in the absence (the control) and presence of C_1X-OR_1 for 0.5 or 1 h. *Reproduced with permission from Springer Nature: Wu, Y., Huang, S., Wang, J., Sun, L., Zeng, F., & Wu, S. (2018). Activatable probes for diagnosing and positioning liver injury and metastatic tumors by multispectral optoacoustic tomography. Nature communications, 9, 3983. https://doi.org/10.1038/s41467-018-06499-1.*

5. Olympus IX 71 fluorescence microscope with a DP72 color CCD
6. Centrifuge
7. 96-well plates
8. 6-well plates
9. 30-mm glass culture dishes
10. Polylysine-coated cell culture glass slides
11. Micropipettes and tips

6.2 Reagents

1. Stock solution (1 mM) of C_1X-OR_1
2. L929 cells and HepG2 cells
3. RPMI 1640 Medium supplemented with 10% FBS and 1% penicillin and streptomycin
4. Dulbecco's modified eagle medium (DMEM) supplemented with 10% FBS and 1% penicillin and streptomycin
5. Methylthiazolyldiphenyl-tetrazolium bromide (MTT)
6. Organic solvents: DMSO
7. Aqueous buffer: 1 × PBS (pH 7.4)

6.3 Procedures

6.3.1 Validation that C_1X-OR_1 is noncytotoxic to cells

1. Culture L929 cells and HepG2 cells in RPMI1640 medium and DMEM, respectively
2. Seed 25 wells of a 96-well plate with 100 μL (5×10^4 cells/mL of media) L929 or HepG2 cells
3. Incubate the plate at 37 °C for 24 h
4. Wash the cells with 1 × PBS solution, and replace the PBS with fresh culture medium (100 μL) containing C_1X-OR_1 at different concentration (0, 1, 5, 10, and 50 μM) and 0.2% *v*/v DMSO
5. Incubate the plate at 37 °C for another 24 h
6. Wash the cells with 1 × PBS solution, and then replace the PBS with fresh culture medium (100 μL) containing 0.5 mg/mL MTT
7. Incubate the plate at 37 °C for 2 h
8. Discard the culture medium, and then add 150 μL of DMSO into each well to dissolve the precipitates
9. Place the plate into the Thermo MK3 ELISA Reader, and measure the absorbance value at 570 nm. Note that the statistical mean and standard deviation were employed to estimate cell viability

6.3.2 Activation of C_1X-OR_1 in ALP overexpressed cells

1. Culture HepG2 cells in DMEM
2. Seed the wells of a 6-well plate with 1 mL (2×10^5 cells/mL of media) HepG2 cells
3. Incubate the plate at 37 °C for 24 h
4. Add 5 μL stock solution of C_1X-OR_1 to each well
5. Incubate cells with C_1X-OR_1 at 37 °C for another 0.5 or 1 h
6. Wash the cells with 1 × PBS solution for three times, and then discard the PBS
7. Trypsinize the cells for 2 min at 37 °C and then add 1 mL prewarmed medium to each well for trypsin neutralization
8. Blow the cells gently with a pipette, and then centrifuge the cells at 1000 rpm for 5 min
9. Discard the supernatant and disperse the cell pellet into a single cell suspension with prewarmed PBS
10. Analyze 1×10^4 cells with a BD CS6 flow cytometer and collet the signals in FL4 channel

6.3.3 Imaging endogenous ALP in cells by use of C_1X-OR_1

1. Culture L929 cells or HepG2 cells in RPMI1640 medium or DMEM, respectively
2. Pass 2 mL L929 and HepG2 cells (2×10^5 cells/mL of media) on the cell culture glass slides modified with polylysine in the 30-mm petri dishes
3. Incubate the cells at 37 °C under 5% CO_2 for 24 h
4. Add 10 μL stock solution of C_1X-OR_1 to each dish. NOTE: In one dish, HepG2 cells were inhibited with Na_3VO_4 (10 mM) for 30 min before addition of C_1X-OR_1
5. Incubate cells with C_1X-OR_1 at 37 °C for another 1 h
6. Wash the cells with 1 × PBS solution for three time and then take the glass slides out of the dishes
7. Place the glass slides onto an Olympus IX71 inverted fluorescence microscope with a DP72 color CCD and acquire fluorescence images. (Excitation filter: 530–550 nm, Emission filter: >590 nm) (Fig. 6)

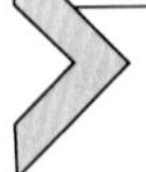

7. In vivo detection of ALP for imaging drug-induced liver injury by use of liposomal C_1X-OR_1

Since the activation of C_1X-OR_1 by cellular ALP has been confirmed in vitro, its ability for diagnosing drug-induced liver injury should also be fully investigated in vivo. For in vivo experiments, to enhance the

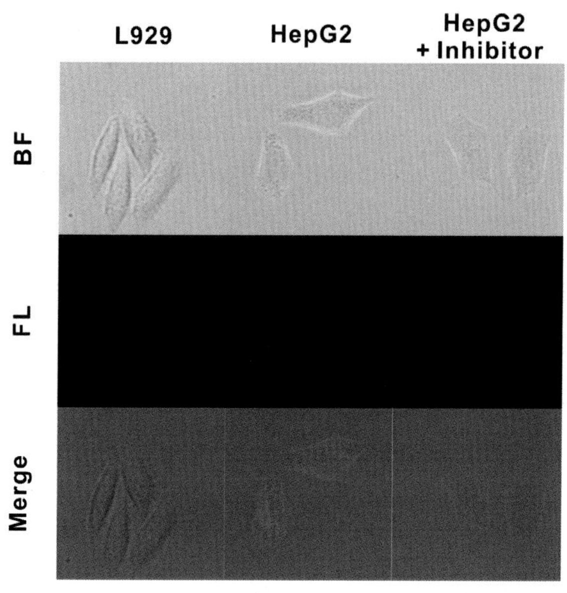

Fig. 6 Fluorescent microscopic images for C_1X-OR_1. Bright field (top row) and fluorescence images (middle row) for L929 and HepG2 cells incubated with C_1X-OR_1 for 1 h, and images for HepG2 cells pretreated with Na_3VO_4 (an ALP inhibitor) and then incubated with C_1X-OR_1 for 1 h. *Reproduced with permission from Springer Nature: Wu, Y., Huang, S., Wang, J., Sun, L., Zeng, F., & Wu, S. (2018). Activatable probes for diagnosing and positioning liver injury and metastatic tumors by multispectral optoacoustic tomography. Nature communications, 9, 3983. https://doi.org/10.1038/s41467-018-06499-1.*

dispersibility in blood stream and the accumulation and retention of C_1X-OR_1 in liver, we fabricate liposomal C_1X-OR_1 by encapsulating it into the synthetic hepatocyte-targeting phospholipid DSPE-PEG_{2000}-ChA together with several other commercial phospholipids. We first confirmed that liposomal C_1X-OR_1 is nontoxic in vivo, and then use it to image Drug induced liver injury (DILI) via photoacoustic and fluorescent imaging. We also investigate the relationship between dose of administrated drug and liver injury by determining serum levels of some hepatic biomarkers, which further indicates the increased hepatic ALP level as the result of liver injury can be spatiotemporally detected by liposomal C_1X-OR_1 with MSOT imaging.

Before conducting any in vivo experiment, institutional approval of the animal experiments must be obtained, and the training in the ethics and practice of animal handling must be completed. All our animal experiments were conducted under SPF environment in the Laboratory Animal Center of South China Agricultural University. All the experimental protocols have been approved by the Animal Ethics Committee of South China

Agricultural University in accordance with the Guidelines for Care and Use of Laboratory Animals of China, the Regulations on the Management of Laboratory Animals of China, and the Regulations on the Administration of Laboratory Animals of Guangdong Province. 7–8 weeks old male BALB/c nude mice (16–20 g) are purchased from Guangdong Medical Laboratory Animal Center (GDMLAC). Mice are housed in a specific pathogen-free room with laminar airflow hoods and 12 h light/12 h dark cycle. Every three mice are assigned in one sterile cage with bedding material of sterilized wood chips and fed autoclaved chow and water ad libitum. Prior to dissection operation, the mice are exposed to a rising concentration of carbon dioxide gas for euthanasia.

7.1 Equipment

1. JEM-1400 transmission electron microscopy
2. Malvern Nano-ZS90 particle size analyzer
3. Hitachi U-3010 UV–vis spectrophotometer
4. InVision128 multispectral optoacoustic tomographic (MSOT) imaging system (iThera Medical GmbH)
5. AMI small animal fluorescence imaging system
6. Thermo MK3 ELISA reader
7. Centrifuge
8. Refrigerator
9. Freeze-dryer
10. Microscope
11. Microbalance
12. Vacuum pump
13. Rotary evaporator
14. Single-neck round bottom flasks
15. Dialysis membranes (MwCO = 1000)
16. Rubber stoppers
17. Syringes
18. Magnetic stirrer
19. Magnetic PTFE stir bar

7.2 Regents

1. C_1X-OR_1 solid
2. Synthetic DSPE-PEG_{2000}-ChA powder
3. 1,2-Dipalmitoyl-*sn*-glycero-3-phosphocholine (DPPC)

4. 1,2-Distearoyl-*sn*-glycero-3-phosphocholine (DSPC)
5. ELISA Kits of alkaline phosphatase (ALP), aspartate transaminase (AST), alanine aminotransferase (ALT)
6. Hematoxylin and Eosin Stain Kit
7. Formalin solution (36.5–38%)
8. Isoflurane
9. Organic solvents: THF, DMSO
10. Aqueous buffer: 1 × PBS (pH 7.4)
11. Deionized water

7.3 Procedure

7.3.1 Preparation of liposomal $C_1X\text{-}OR_1$ for in vivo experiments

Liposomal $C_1X\text{-}OR_1$ is prepared according to a procedure stated below:

1. Dissolve DPPC (12.46 mg), DSPC (2.8 mg) and DSPE-PEG_{2000}-ChA (1.7 mg) at a molar ratio of 85/10/5 with the organic solvent tetrahydrofuran (1 mL) in a 5 mL single-neck round bottom flask equipped with a magnetic PTFE stir bar
2. Add a solution of $C_1X\text{-}OR_1$ (20 mg) in THF (0.5 mL) into the flask under continuous stirring
3. Inject the above mixture dropwise into the deionized water (10 mL) in 50 mL single-neck round bottom flask equipped with a magnetic PTFE stir bar under continuous stirring
4. Remove the organic solvent was removed by vacuum-rotary evaporation
5. Dialyze the mixture against the deionized water in the dark at room temperature for 24 h
6. Centrifuge the suspension at 4 °C to collect the nanoparticles
7. Wash the resulting solid with the deionized water for three times
8. Resuspend the nanoparticles in 1 × PBS (or water for the concentration determination) of designated volume (20 mL) to obtain liposomal $C_1X\text{-}OR_1$

After preparation of the liposomal $C_1X\text{-}OR_1$, its average diameter is determined to be ~105 nm by transmission electron microscopic (TEM) and dynamic light scattering (DLS). For determination of concentration of liposomal $C_1X\text{-}OR_1$ in water and $C_1X\text{-}OR_1$ loading, liposomal $C_1X\text{-}OR_1$ solution is freeze-dried and weighed. After that, lyophilized liposomal $C_1X\text{-}OR_1$ is dissolved in 1 mL of DMSO. The loading efficiency of the $C_1X\text{-}OR_1$ by phospholipids is calculated based on the absorbance intensity at 684 nm according to a predetermined standard calibration curve. The concentration of liposomal

$C_1X\text{-}OR_1$ in water is calculated to be 0.7 mg/mL and $C_1X\text{-}OR_1$ loading efficiency is determined as 42.7% based on the following two formulas:

(1) Concentration of liposomal probe = (mass of nanoparticles)/(volume of solution)

(2) Loading efficiency = (mass of loaded $C_1X\text{-}OR_1$)/(mass of nanoparticles) × 100%

7.3.2 Confirmation of liposomal $C_1X\text{-}OR_1$'s nontoxicity in vivo

Before being used for animal experiments, it is necessary to verify that liposomal $C_1X\text{-}OR_1$ is nontoxic in vivo by measuring changes in serum biochemistry and histopathology, and the procedure is given below:

1. Anesthetize the mice in an isoflurane/oxygen chamber, and maintain an anesthetized state by use of 1% isoflurane delivered via a breathing mask
2. Inject liposomal $C_1X\text{-}OR_1$ (8.75 mg/kg) or saline intravenously into mice using insulin syringe and feed them for another day as usual
3. Collect blood samples of those mice from their orbital sinus, and euthanize them via exposure to carbon dioxide gas in a rising concentrations and then excise their main organs including heart, liver, spleen, lung and kidney
4. Centrifuge the blood samples and collect serum for ALT, AST and ALP assay using ELISA kits
5. Flush the excised organs with sterile PBS solution, fix the organs with formalin solution, and embed them in paraffin. Then use hematoxylin and eosin stain the paraffin sections for fluorescence imaging under a microscope
6. Analyze the data

7.3.3 MSOT imaging of drug-induced liver injury in mouse model using liposomal $C_1X\text{-}OR_1$

This experiment is to demonstrate that the drug-induced liver injury in mouse model can be diagnosed by the systemically administered liposomal $C_1X\text{-}OR_1$ with MSOT. Herein, we utilize *N*-acetyl-*p*-aminophenol (APAP) to establish mouse model of drug-induced liver injury. APAP is a clinical medicine for fever and pain treatment, but overdose may cause hepatotoxicity accompanied by elevation in hepatic ALP (Simon & Sutherland, 1977). For APAP-induced liver injury, the mice are intraperitoneally administrated with varied dosage of APAP predissolved in sterilized saline solutions at a concentration of 15 mg/mL (Fig. 7).

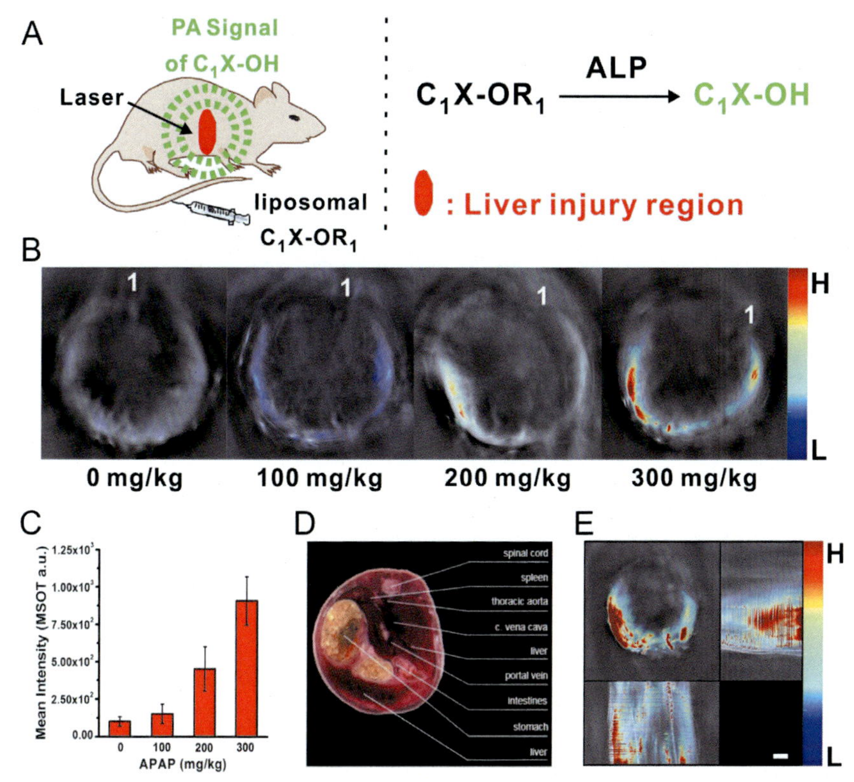

Fig. 7 In vivo PA imaging drug-induced liver injury by using C_1X-OR_1. (A) Schematic illustration for PA imaging of DILI with liposomal C_1X-OR_1. (B) Typical MSOT images of the mice administrated with different amount of APAP at 30 min upon injection of liposomal C_1X-OR_1. (C) Mean PA intensities at ROI in liver area for the mice administrated with different amount of APAP. (D) A cryosection image of a male mouse with the cross section's location comparable to those shown in panel (B). (E) Representative 3-D MSOT images for the mice pretreated 300 mg/kg APAP after intravenous injection of liposomal C_1X-OR_1 for 30 min. *Reproduced with permission from Springer Nature: Wu, Y., Huang, S., Wang, J., Sun, L., Zeng, F., & Wu, S. (2018). Activatable probes for diagnosing and positioning liver injury and metastatic tumors by multispectral optoacoustic tomography. Nature Communications, 9, 3983. https://doi.org/10.1038/s41467-018-06499-1.*

1. Repeat Procedure 5.3.2 (B) steps 1–2
2. Import the absorption spectra of C_1X-OH into the system
3. Set up the parameters to collect optoacoustic signals at 680, 684, 700, 715, 730, 760, 800 (background), and 850 nm (NOTE: These wavelengths are chosen for correspondence with main turning points in

the absorption spectra of endogenous hemoglobin and activated product C_1X-OH), 10 frames per wavelength

4. Anesthetize the mice preadministered with varied dose of APAP in an isoflurane/oxygen chamber, and maintain an anesthetized state by using 1% isoflurane delivered through a breathing mask during the experiments
5. Place the mice inside the sample holder and insert a catheter connected with a syringe into the tail vein
6. Seal the holder with the ultrathin plastic membrane to prevent water from leaking into the animal holder during experiments, and remove air between mice and membrane with medical ultrasonic couplant
7. Place the holder with mice into the system and inject liposomal C_1X-OR_1 (8.75 mg/kg) into the mice using the syringe
8. Start measurement to acquire data 30 min after probe injection
9. Reconstruct MSOT images using a standard back-projection algorithm and separate PA signals of the activated C_1X-OR_1 from those of endogenous absorbers such as hemoglobin using guided ICA spectral unmixing protocol
10. Select a region of interest (ROI) volume consisting of transverse slices with a step size of 0.2 mm, spanning through the liver region
11. Render the z-stack cross-sectional images into an orthogonal MIP image
12. Record the mean PA intensity of ROI
13. Analyze the data

7.3.4 Fluorescence imaging of drug-induced liver injury in mouse model using liposomal C_1X-OR_1

1. Switch on the AMI small animal fluorescence imaging system and start AMIView software
2. Initialize the instrument and cool the CCD camera to operating temperature at −90 °C
3. Set up the parameters to collect fluorescence images. (excitation filter: 675 nm; emission filter of 715 nm; Power: 30%; exposure: 0.5 s; Binning: none; FOV: 20 cm; Objective height: 1.5 cm)
4. Anesthetize the mice preadministered with varied dose of APAP in an isoflurane/oxygen chamber and inject liposomal C_1X-OR_1 (8.75 mg/kg) into the mice via tail vein
5. Place the mice into the chamber of the imaging system, maintain an anesthetized state by use of 1% isoflurane delivered via a breathing mask and close the instrument door

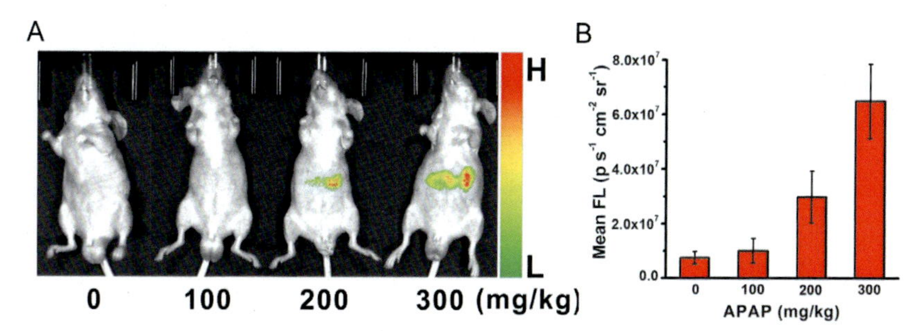

Fig. 8 In vivo fluorescence imaging drug-induced liver injury by using $C_1X\text{-}OR_1$. (A) Representative fluorescence images indicating biodistribution of the activated $C_1X\text{-}OR_1$ for the mice administrated with varied dose of APAP. (B) Mean fluorescent intensities at ROI of the mice administrated with different amount of APAP. *Reproduced with permission from Springer Nature: Wu, Y., Huang, S., Wang, J., Sun, L., Zeng, F., & Wu, S. (2018). Activatable probes for diagnosing and positioning liver injury and metastatic tumors by multispectral optoacoustic tomography. Nature Communications, 9, 3983. https:/doi.org/10.1038/s41467-018-06499-1.*

6. Start measurement to acquire images 30 min after injection
7. Select a region of interest (ROI) spanning through the liver region
8. Record the mean FL intensity of ROI
9. Analyze the data (Fig. 8)

7.3.5 Verification that treatment with overdosed APAP causes liver injury

To confirm the relationship between APAP overdose and liver injury, several clinically representative parameters including serum ALT, AST and ALP for liver dysfunctions are measured by using ELISA kits. APAP overdose leads to a significant increase in serum enzyme activity (Fig. 9), thus further indicating that the mice administrated with overdose APAP in the imaging experiments suffer liver damage. All these results suggest that, drug-induced liver injury accompanied by the elevated hepatic ALP level can be spatiotemporally detected by MSOT.

These mice are sacrificed, and their serum is collected 12 h after APAP administration. The procedures for measuring serum ALT/AST/ALP using ELISA kits are strictly performed according to the instructions.

1. Dilute standard into five different concentrations for working curve
2. Add the five diluted standards (50 μL) and fivefold diluted serum samples (50 μL) into the wells of ELISA plate
3. Cover the plate and incubate it for 30 min at 37 °C after gentle shaking
4. Dilute concentrated wash solution 30 times with deionized water

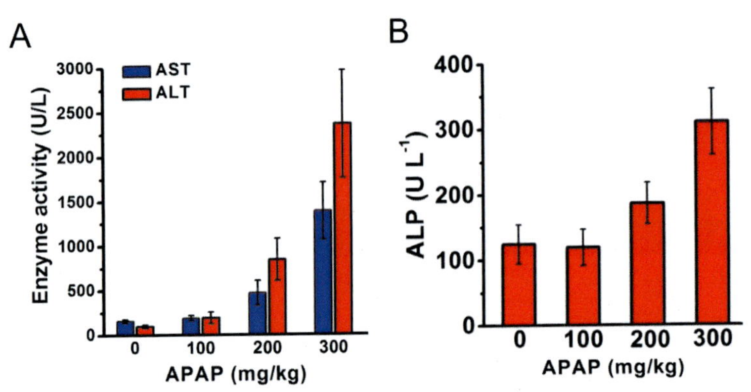

Fig. 9 Serum biochemistry evaluation. (A) Serum levels of AST as well as ALT and (B) Serum ALP levels for the mice at 12h upon administration with different amount of APAP. Columns represent means ± SD. Error bars represent the standard deviation (SD). *Reproduced with permission from Springer Nature: Wu, Y., Huang, S., Wang, J., Sun, L., Zeng, F., & Wu, S. (2018). Activatable probes for diagnosing and positioning liver injury and metastatic tumors by multispectral optoacoustic tomography. Nature communications, 9, 3983. https://doi.org/10.1038/s41467-018-06499-1.*

5. Discard the liquid in the wells and wash five times with 30-fold diluted wash solution
6. Add 50 μL of enzyme-labeled reagent to each well
7. Repeat the incubating procedure as in step 3
8. Repeat the wash procedure as in step 5
9. Add 50 μL of Color Development Reagent A and 50 μL of Color Development Reagent B to each well successively. Cover the plate and incubate it for 15 min in the dark at 37 °C after gentle shaking
10. Add 50 μL of Stop Solution to each well. NOTE: The solution will immediately change from blue to yellow once terminating the reaction
11. Zero the blank well. Read absorbance at 450 nm within 15 min by using a Thermo MK3 ELISA Reader

8. Summary

Herein, we detail the synthesis and validation of $C_1X\text{-}OR_1$, an ALP-activated PA probe that can be used for spatiotemporal detection and imaging of liver injury in vivo. We believe this probe $C_1X\text{-}OR_1$ may be applied in preclinical screening for hepatotoxicity to improve patient safety and be utilized as a convenient and cost-effective tool in pharmaceutical industry to avoid post-marketing drug withdrawal. More generally, we demonstrate a systematic and comprehensive approach for design, preparation as well as application of activatable PA probes for disease diagnosis with multispectral photoacoustic imaging in a 3D manner and hopefully this approach could be used to accurately diagnose and locate various diseases by construction of other activatable probes.

Acknowledgment

This work was supported by the National Natural Science Foundation of China (21788102) and the Fund of Guangdong Provincial Key Laboratory of Luminescence from Molecular Aggregates (2019B030301003).

References

Cheng, P., Chen, W., Li, S., He, S., Miao, Q., & Pu, K. (2020). Fluoro-photoacoustic polymeric renal reporter for real-time dual imaging of acute kidney injury. *Advanced Materials*, *32*, 1908530.

Coleman, J. E. (1992). Structure and mechanism of alkaline phosphatase. *Annual Review of Biophysics and Biomolecular Structure*, *21*, 441–483.

Corsini, A., & Bortolini, M. (2013). Drug-induced liver injury: The role of drug metabolism and transport. *The Journal of Clinical Pharmacology*, *53*, 463–474.

Fedarko, N. S., Bianco, P., Vetter, U., & Robey, P. G. (1990). Human bone cell enzyme expression and cellular heterogeneity: Correlation of alkaline phosphatase enzyme activity with cell cycle. *Journal of Cellular Physiology*, *144*, 115–121.

Gottschalk, S., Degtyaruk, O., Mc Larney, B., Rebling, J., Hutter, M. A., Deán-Ben, X. L., et al. (2019). Rapid volumetric optoacoustic imaging of neural dynamics across the mouse brain. *Nature Biomedical Engineering*, *3*, 392–401.

Jeevarathinam, A. S., Lemaster, J. E., Chen, F., Zhao, E., & Jokerst, J. V. (2020). Photoacoustic imaging quantifies drug release from nanocarriers via redox chemistry of dye-labeled cargo. *Angewandte Chemie International Edition*, *59*, 4678–4683.

Kaplan, M. M., & Righetti, A. (1970). Induction of rat liver alkaline phosphatase: The mechanism of the serum elevation in bile duct obstruction. *The Journal of Clinical Investigation*, *49*, 508–516.

Knox, H. J., Hedhli, J., Kim, T. W., Khalili, K., Dobrucki, L. W., & Chan, J. (2017). A bioreducible N-oxide-based probe for photoacoustic imaging of hypoxia. *Nature Communications*, *8*, 1794.

Kola, I., & Landis, J. (2004). Can the pharmaceutical industry reduce attrition rates? *Nature Reviews Drug Discovery*, *3*, 711–716.

Ouyang, J., Sun, L., Zeng, Z., Zeng, C., Zeng, F., & Wu, S. (2020). Nanoaggregate probe for breast cancer metastasis through multispectral optoacoustic tomography and aggregation-induced NIR-I/II fluorescence imaging. *Angewandte Chemie International Edition*, *59*, 10111–10121.

Sakatis, M. Z., Reese, M. J., Harrell, A. W., Taylor, M. A., Baines, I. A., Chen, L., et al. (2012). Preclinical strategy to reduce clinical hepatotoxicity using in vitro bioactivation data for > 200 compounds. *Chemical Research in Toxicology*, *25*, 2067–2082.

Simon, F. R., & Sutherland, E. (1977). Hepatic alkaline phosphatase isoenzymes: Isolation, characterization and differential alteration. *Enzyme*, *22*, 80–90.

Taruttis, A., & Ntziachristos, V. (2015). Advances in real-time multispectral optoacoustic imaging and its applications. *Nature Photonics*, *9*, 219–227.

Wang, B., Avorn, J., & Kesselheim, A. S. (2013). Clinical and regulatory features of drugs not initially approved by the FDA. *Clinical Pharmacology & Therapeutics*, *94*, 670–677.

Wang, L. V., & Yao, J. (2016). A practical guide to photoacoustic tomography in the life sciences. *Nature Methods*, *13*, 627.

Wu, Y., Chen, J., Sun, L., Zeng, F., & Wu, S. (2019). A nanoprobe for diagnosing and mapping lymphatic metastasis of tumor using 3D multispectral optoacoustic tomography owing to aggregation/deaggregation induced spectral change. *Advanced Functional Materials*, *29*, 1807960.

Wu, Y., Huang, S., Wang, J., Sun, L., Zeng, F., & Wu, S. (2018). Activatable probes for diagnosing and positioning liver injury and metastatic tumors by multispectral optoacoustic tomography. *Nature Communications*, *9*, 3983.

Wu, Y., Sun, L., Zeng, F., & Wu, S. (2019). A conjugated-polymer-based ratiometric nanoprobe for evaluating in-vivo hepatotoxicity induced by herbal medicine via MSOT imaging. *Photoacoustics*, *13*, 6–17.

Yang, X., Salminen, W. F., & Schnackenberg, L. K. (2012). Current and emerging biomarkers of hepatotoxicity. *Current Biomarker Findings*, *2*, 43–55.

Zlitni, A., Gowrishankar, G., Steinberg, I., Haywood, T., & Gambhir, S. S. (2020). Maltotriose-based probes for fluorescence and photoacoustic imaging of bacterial infections. *Nature Communications*, *11*, 1250.

CHAPTER THIRTEEN

Photoacoustic probe of targeting intracellular *Staphylococcus aureus* infection with signal-enhanced by self-assembly

Xin Jin, Yue Fei, Jingmei Ma, Li-Li Li, and Hao Wang*
CAS Center for Excellence in Nanoscience, CAS Key Laboratory for Biomedical Effects of Nanomaterials and Nanosafety, National Center for Nanoscience and Technology (NCNST), University of Chinese Academy of Sciences, Beijing, China
*Corresponding author: e-mail address: wanghao@nanoctr.cn

Contents

Methods in Enzymology, Volume 657
ISSN 0076-6879
https://doi.org/10.1016/bs.mie.2021.06.027

Abstract

In this chapter, we introduce the photoacoustic imaging probe in detecting the infection of bacterial in host cells. We give detailed protocols for the synthesis and confirmation of a PA probe which response to caspase-1 and then self-assembly to enhance the signal. With this PA contrast agent, we may provide a new approach for an intracellular bacterial infection with the selective and sensitive diagnosis.

1. Introduction

Staphylococcus aureus (*S. aureus*) not only damage the health of human beings severely but caused huge economic losses (Zhou et al., 2018). *Staphylococcus aureus* could invade and survive in host cells (e.g., macrophages and osteoblasts) and then evade immune surveillance. The survival way of *S. aureus* has been reported as one of the mechanisms that produce potential antibiotic resistance (Fraunholz & Sinha, 2012; Yang et al., 2018). The innate immune system acts quick response upon infecting by bacteria or virus (Foley et al., 2012; Qadri, Almadani, Jay, & Elsaid, 2018). Classically activated macrophages (M1 macrophages) produce numerous reactive oxygen species (ROS) and inflammatory cytokines to protect host from infection by bacteria or virus (Wang, Liu, & Zhao, 2020; Xuan, Qu, Zheng, Xiong, & Fan, 2015). When host cells infected by bacteria, macrophages move into tissue and kill bacteria, which is a systematic process response to a chemoattractant at host cells (Guerra, Borgogna, Patel, Sward, & Voyich, 2017). However, *Staphylococcus aureus* enables secrete many kinds of factors to evade immune surveillance and prevent themselves from being killed by antibiotics, alongside long-term chronic inflammation (Kalinka et al., 2014; Zhou et al., 2018). Under such conditions, it is important to detect bacterial infection that hided inside the macrophage in diagnosis and treatment, which will be a great help for reducing antibiotic abuse and potential resistance.

Fluorescent probes have made significant progress in monitoring microenvironment changes including ROS, pH, polarity (Sun, Li, Fan, & Peng, 2019). However, these fluorescent probes in optical imaging showed the poor resolution with maximum depth of ~1 mm owing to the poor tissue penetration (Ntziachristos, 2010). Photoacoustic (PA) imaging is an attracting imaging technology that overcome the disadvantages that fluorescence imaging showed. PA has high spatial resolution and deep tissue penetration, although the sensitivity and resolution of PA imaging are lower (Chen et al., 2019; Weber, Beard, & Bohndiek, 2016). Photoacoustic imaging has been developed in biomedical and clinical applications (Fu, Zhu, Song, Yang, & Chen, 2019), for example, detecting H_2S (Chen et al., 2019), NO (Reinhardt, Zhou, Jorgensen, Partipilo, & Chan, 2018), ROS (Zheng, Zeng, Zhang, Xing, & Zhang, 2019), important biomarkers (Wang, Hu, et al., 2019; Yin et al., 2019) and metal ions (Wang, Sheng, et al., 2019; Zhang et al., 2020). It provides us a possibility to enhance the contrast and sensitivity of PA probes which through the aggregation process of small molecules in situ.

In our living systems, many supramolecular organizations play important roles in regulating their biofunctions, like, the DNA, protein complexes. Natural chlorophyll molecules could also form supramolecular through assembly in the process of photosynthesis (Mirkovic et al., 2017; Song & Tezcan, 2014; Wang et al., 1979; Zheng et al., 2019). Our group have reported artificially modified chlorophyll derivative with functional peptides could self-assembly in specific conditions *in vivo* (Li, Ma, et al., 2016, Li, Zeng, et al., 2016). In our previous work, we demonstrated that the typical J-type self-assembly of chlorophyll derivatives simultaneously self-assembled into nanofibers in tumor site that exhibited AIR effect, thus exhibiting an increased photoacoustic signal and improved signal-to-noise ratio (Zhang et al., 2015). Furthermore, the derivatives of chlorophyll showed good biocompatibility and biodegradation properties in bioimaging (Li et al., 2017; Liang, Tang, & Liu, 2015). Recently, researchers have developed materials to delivery drugs at infection sites, which avoid toxicity to host, increase pesticide effect, and across blood-brain barriers (Batrakova, Gendelman, & Kabanov, 2011; Pang et al., 2017). It will be a good idea to apply them into developing photoacoustic probe for bacterial infection diagnosis and to improve the sensitivity and resolution of probe through self-assembly *in vivo*.

In this chapter, we introduced a novel photoacoustic probe which be activated by caspase-1 and induced photoacoustic enhancement by assembly when infected by *Staphylococcus aureus*. We then provide a detailed guide to our group's development of a probe (**MPC**) that could response to intracellular *S. aureus* infection. The agent target activated macrophage cells and moved into macrophage cells. Tailoring by caspase-1 and enhancing the photoacoustic (PA) signal through molecular self-assembly.

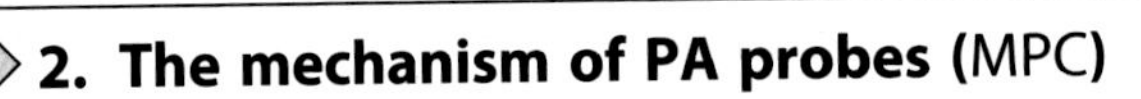

2. The mechanism of PA probes (MPC)

We synthesis a photoacoustic agent (**MPC**) based on chlorophyll-peptide that can response to bacterial infections and bioimage *in vivo*.

When macrophages infected by bacteria, it turns to the classically activated M1 phenotype by polarized. Mannose receptors are highly over-expressed on polarized macrophages (Hatami et al., 2019). Nanomaterials decorated with mannosides showed remarkably higher entry, minimal cytotoxicity, and almost negligible cytokine production in macrophages. In this aspect, we choose the mannose motif targeting to the macrophages. MPC was uptaken by macrophage and tailored by caspase-1 at the infectious site in a highly efficient manner. Inflammatory caspases (i.e., caspase-1, caspase-4, caspase-5, caspase-11) are used by the host to control bacterial, viral, fungal or protozoan pathogens (Man, Karki, & Kanneganti, 2017). In our previous work, we have set a noninvasive way to quantitatively measure the activity of caspase-1, which reflected the infection of bacterial in the early stage (Li, Zeng, et al., 2016). In our system, caspase-1 is immediately activated and efficiently cleaves our designed peptide substrate. The monomeric simultaneously self-assemble into J-type aggregates and accumulate inside macrophage. The assembly/aggregation-induced-retention (AIR) effect significantly enhances **MPC** accumulation at the infectious site. Consequently, the infection-induced AIR effect remarkably increases the contrast and the chlorophyll aggregate induced photoacoustic signal enhancement and contribute to specific and sensitive bacterial infection.

3. Before you begin

What buffers need to be made, what you need to consider when choosing a cell line, etc.

1. buffer
   - **(a)** PBS (phosphate buffer saline)
2. Cell line

   RAW 264.7

4. Key resources table

Note that not all areas will be used in every protocol.

Bacterial and Virus Strains		
Staphylococcus aureus	Institute of Microbiology, Chinese Academy of Sciences	ATCC 6538
Biological Samples		
Macrophage	Cell Resource Center (IBMS, CAMS/PUMC, China)	RAW 264.7
Chemicals, Peptides, and Recombinant Proteins		
Gentamycin	J&K Scientific	1403-66-3
Caspase 1 activity detection kit	Beyotime	C1101

5. Materials and equipment

[List materials, equipment and reagents needed for this procedure]

- Ultra-high resolution small animal photoacoustic imaging system (M-128)
- Time-of-flight mass spectrometer (Autoflex_MAX)
- Circular Di chromatograph (J-810)
- UV–visible spectrophotometer (UV-2600)

6. Step-by-step method details

6.1 The synthesis of MPC

1. Molecule 1a
   - **(a)** Weighed 200 mg Fmoc-Lys (Dde) Wang resin to a dry and clean peptide synthesis tube and then swelled it with *N*, *N*-Dimethylformamide (DMF) for 2 h. This process was carried out on a shaker.
   - **(b)** Removed the peptide synthesis tube from the shaker, Washed three times alternately with Dichloromethane (DCM) and DMF.
     Note: when rinsing, try to wash up the resin each time.
   - **(c)** The protective group Fmoc was removed using a piperidine/DMF (v/v = 2/8) mixture for 15 min. At the same time, the next amino acids and 10 times equivalent of O-Benzotriazole-*N*, *N*, *N*′, *N*′-tetramethyl-uronium-hexafluorophosphate (HBTU) were added into 4-Methylmorpholine/DMF (v/v = 5/95) solution to react. The vortex can be used to accelerate the dissolution.

(d) Preheated water bath.
(e) Washed three times alternately with DCM and DMF after deprotection, added detection reagents to a 1.5 mL centrifuge tube, and removed a small amount of resin with a capillary tube into the centrifuge tube. Configuration of detection reagents: took one drop of each of the three detection reagents A, B, and C. Reagent A is 0.5 g of ninhydrin dissolved in 10 mL of absolute ethanol; Reagent B is that 20 g of phenol is dissolved in 5 mL of absolute ethanol; Reagent C is that of 0.1 g Vc dissolved in 5 mL of absolute ethanol.
(f) Quickly removed the centrifuge tube and put it in a water bath to heat for 30 s to observe the color of the resin.

Note: over the course of the reaction, the resin turns purple-black to indicate that there are exposed amino groups, and the resin does not turn purple-black to indicate that there are no exposed amino groups.
(g) Added the mixture to peptide synthesis tube for 45 min.

Note: the product obtained in the above steps can be directly used in the next reaction.
(h) Repeated steps 3–7 until the last amino acid is coupled with 5-hexanoic acid in the same way.
(i) A mixture of hydrazine hydrate/DMF (v/v=2/98) was used to remove the protective group Dde of Lys.
(j) Linked Ala to the side chain of Lys.
(k) The purpurin 18 was attached to Ala in the same way for 3 h.
(l) Rinsed three times with methanol to shrank the polypeptide, drained it and poured it into an ampoule.

Note: in order to prevent agglomeration, a straw can be used to stir.
(m) Placed the ampoule containing the peptide in an ice bath, added 5 mL of lysate in a fume hood, and stirred for 2.5 h at 300 rpm.

Note: mix 95% trifluoroacetic acid (TFA)/2.5% H2O/2.5% Triisopropylsilane (TIPS) (v/v) and peptides in an ice bath for 2.5 h, which can protect most amino acid side chains and resin cleavage. The remaining protection groups were removed with dithiothreitol (DTT).
(n) Soaked ether with ice-water mixture in advance.
(o) The sample was suction filtered with a small funnel, and the obtained filtrate was blown with nitrogen at 37 °C.

- **(p)** Put the product in ice bath and let stand for 3 min.
- **(q)** Added ether to the product and transferred it to a 1.5 mL centrifuge tube, washed it three times (gradually integrated the products of the six centrifuge tubes into three centrifuge tubes), centrifuged for 3 min at 8000 rpm.
- **(r)** The synthesized peptide was dried under vacuum.

2. Molecule 2a
- **(a)** Product 1a was compatible with Cu2+ under N2 protection in dimethyl sulfoxide (DMSO) for 24 h.
- **(b)** he product 2a was poured into the prepared dialysis bags, changed the water every hour, and freeze-dried after about five water changes.

3. MPC
- **(a)** Put the acetylated mannose into methanol/1 M NaOH (v/v = 9:1) mixed solution for azide modification for 3 h. Then, add 0.5 M HCl to stop the reaction with excess NaOH.
- **(b)** Removal of solvents in a vacuum environment.
- **(c)** Put 2-azidoethyl β-D-mannopyranoside into the prepared dialysis bag for dialysis and freeze-drying.
- **(d)** Molecule 2a, 2-Azidoethyl β-D-mannopyranoside and CuSO4·5H2O were added in the mixture of THF/H2O (v/v = 1/1) and then stirred for 30 min under N2 protection.
- **(e)** Added 0.5 mL (0.5 equiv) sodium ascorbate aqueous solution to the mixed solution, and continued to react for 6 h in the dark under the protection of N2.
- **(f)** The MPC was dialyzed in the same way as above, and freeze-dried.

6.2 The synthesis of MPsC

4. Molecule 1b
- **(a)** Weighed 200 mg Fmoc-Lys (Dde) Wang resin to a dry and clean peptide synthesis tube and then swelled it with DMF for 2 h. This process was carried out on a shaker.
- **(b)** Removed the peptide synthesis tube from the shaker, Washed three times alternately with DCM and DMF.

  Note: When rinsing, try to wash up the resin each time.
- **(c)** The protective group Fmoc was removed using a piperidine/DMF (v/v = 2/8) mixture for 15 min. At the same time, the next amino acids and 10 times equivalent of HBTU were added into 4-Methylmorpholine/DMF (v/v = 5/95) solution to react. The vortex can be used to accelerate the dissolution.

(d) Preheated water bath.

(e) Washed three times alternately with DCM and DMF after deprotection, added detection reagents to a 1.5 mL centrifuge tube, and removed a small amount of resin with a capillary tube into the centrifuge tube. Configuration of detection reagents: took one drop of each of the three detection reagents A, B, and C. Reagent A is 0.5 g of ninhydrin dissolved in 10 mL of absolute ethanol; Reagent B is that 20 g of phenol is dissolved in 5 mL of absolute ethanol; Reagent C is that of 0.1 g Vc dissolved in 5 mL of absolute ethanol.

(f) Quickly removed the centrifuge tube and put it in a water bath to heat for 30 s to observe the color of the resin.

Note: over the course of the reaction, the resin turns purple-black to indicate that there are exposed amino groups, and the resin does not turn purple-black to indicate that there are no exposed amino groups.

(g) Added the mixture to peptide synthesis tube for 45 min.

Note: the product obtained in the above steps can be directly used in the next reaction.

(h) Repeated steps 3–7 until the last amino acid was coupled with 5-hexanoic acid in the same way.

(i) A mixture of hydrazine hydrate/DMF (v/v = 2/98) was used to remove the protective group Dde of Ly.

(j) Linked Ala to the side chain of Lys.

(k) The purpurin 18 was attached to Ala in the same way for 3 h.

(l) Rinsed three times with methanol to shrank the polypeptide, drained it and poured it into an ampoule.

Note: in order to prevent agglomeration, a straw can be used to stir.

(m) Placed the ampoule containing the peptide in an ice bath, added 5 mL of lysate in a fume hood, and stirred for 2.5 h at 300 rpm.

Note: mix 95% TFA/2.5% H2O/2.5% TIPS (v/v) and peptides in an ice bath for 2.5 h, which can protect most amino acid side chains and resin cleavage. The remaining protection groups were removed with DTT.

(n) Soaked ether with ice-water mixture in advance.

(o) The sample was suction filtered with a small funnel, and the obtained filtrate was blown with nitrogen at 37 °C.

(p) Put the product in ice bath and let stand for 3 min.

   (q) Added ether to the product and transferred it to a 1.5 mL centrifuge tube, washed it three times (gradually integrated the products of the six centrifuge tubes into three centrifuge tubes), centrifuged for 3 min at 8000 rpm.
   (r) The synthesized peptide was dried under vacuum.

5. Molecule 2b
   (a) Molecule 2b was compatible with $Cu2+$ under N2 protection in DMSO for 24 h at room temperature.
   (b) The Molecule 2b was poured into the prepared dialysis bags, changed the water every hour, and freeze-dried after about five water changes.

6. MP_SC
   (a) Put the acetylated mannose into methanol/1 M NaOH (v/v = 9:1) mixed solution for azide modification for 3 h. Then, add 0.5 M HCl to stop the reaction with excess NaOH.
   (b) Removal of solvents in a vacuum environment.
   (c) Put 2-azidoethyl β-D-mannopyranoside into the prepared dialysis bag for dialysis and freeze-drying.
   (d) Molecule 2b, 2-Azidoethyl β-D-mannopyranoside and CuSO4·5H2O were added in the mixture of THF/H2O (v/v = 1/1) and then stirred for 30 min under N2 protection.
   (e) Added 0.5 mL (0.5 equiv) sodium ascorbate aqueous solution to the mixed solution, and continued to react for 6 h in the dark under the protection of N2.
   (f) The MPC was dialyzed in the same way as above, and freeze-dried.

6.3 The synthesis of PRC

7. PRC
   (a) Weighed 200 mg Fmoc-Lys (Dde) Wang resin to a dry and clean peptide synthesis tube and then swelled it with DMF for 2 h. This process was carried out on a shaker.
   (b) Removed the peptide synthesis tube from the shaker, Washed three times alternately with DCM and DMF.
   Note: when rinsing, try to wash up the resin each time.
   (c) The protective group Fmoc was removed using a piperidine/DMF (v/v = 2/8) mixture for 15 min. At the same time, the next amino acids and 10 times equivalent of HBTU were added into 4-Methylmorpholine/DMF (v/v = 5/95) solution to react. The vortex can be used to accelerate the dissolution.

(d) Preheated water bath.

(e) Washed three times alternately with DCM and DMF after deprotection, added detection reagents to a 1.5 mL centrifuge tube, and removed a small amount of resin with a capillary tube into the centrifuge tube. Configuration of detection reagents: took one drop of each of the three detection reagents A, B, and C. Reagent A is 0.5 g of ninhydrin dissolved in 10 mL of absolute ethanol; Reagent B is that 20 g of phenol is dissolved in 5 mL of absolute ethanol; Reagent C is that of 0.1 g Vc dissolved in 5 mL of absolute ethanol.

(f) Quickly removed the centrifuge tube and put it in a water bath to heat for 30 s to observe the color of the resin.

Note: over the course of the reaction, the resin turns purple-black to indicate that there are exposed amino groups, and the resin does not turn purple-black to indicate that there are no exposed amino groups.

(g) Added the mixture to peptide synthesis tube for 45 min.

Note: the product obtained in the above steps can be directly used in the next reaction.

(h) Repeated steps 3–7 until the Lys was coupled.

(i) A mixture of hydrazine hydrate/DMF (v/v = 2/98) was used to remove the protective group Dde of Lys.

(j) Linked Ala to the side chain of Lys.

(k) The purpurin 18 was attached to Ala in the same way for 3 h.

(l) Molecule was compatible with Cu2+ under N2 protection in DMSO for 24 h at room temperature.

(m) The Molecule was poured into the prepared dialysis bags, changed the water every hour, and freeze-dried after about five water changes.

(n) Soaked ether with ice-water mixture in advance.

(o) The sample was suction filtered with a small funnel, and the obtained filtrate was blown with nitrogen at 37 °C.

(p) Put the product in ice bath and let stand for 3 min.

(q) Added ether to the product and transferred it to a 1.5 mL centrifuge tube, washed it three times (gradually integrated the products of the six centrifuge tubes into three centrifuge tubes), centrifuged for 3 min at 8000 rpm.

(r) The synthesized peptide was dried under vacuum.

6.4 The synthesis of PC

8. PC

(a) Weighed 200 mg Fmoc-Lys (Dde) Wang resin to a dry and clean peptide synthesis tube and then swelled it with DMF for 2 h. This process was carried out on a shaker.

(b) Removed the peptide synthesis tube from the shaker, Washed three times alternately with DCM and DMF.

Note: when rinsing, try to wash up the resin each time.

(c) The protective group Fmoc was removed using a piperidine/DMF (v/v = 2/8) mixture for 15 min. At the same time, the next amino acids and 10 times equivalent of HBTU were added into 4-Methylmorpholine/DMF (v/v = 5/95) solution to react. The vortex can be used to accelerate the dissolution.

(d) Preheated water bath.

(e) Washed three times alternately with DCM and DMF after deprotection, added detection reagents to a 1.5 mL centrifuge tube, and removed a small amount of resin with a capillary tube into the centrifuge tube. Configuration of detection reagents: took one drop of each of the three detection reagents A, B, and C. Reagent A is 0.5 g of ninhydrin dissolved in 10 mL of absolute ethanol; Reagent B is that 20 g of phenol is dissolved in 5 mL of absolute ethanol; Reagent C is that of 0.1 g Vc dissolved in 5 mL of absolute ethanol.

(f) Quickly removed the centrifuge tube and put it in a water bath to heat for 30 s to observe the color of the resin.

Note: over the course of the reaction, the resin turns purple-black to indicate that there are exposed amino groups, and the resin does not turn purple-black to indicate that there are no exposed amino groups.

(g) Added the mixture to peptide synthesis tube for 45 min.

Note: the product obtained in the above steps can be directly used in the next reaction.

(h) Repeated steps 3–7 until the last amino acid was coupled with 5-hexanoic acid in the same way.

(i) A mixture of hydrazine hydrate/DMF (v/v = 2/98) was used to remove the protective group Dde of Lys.

(j) Linked Ala to the side chain of Lys.

(k) The purpurin 18 was attached to Ala in the same way for 3 h.

(l) Rinsed three times with methanol to shrank the polypeptide, drained it and poured it into an ampoule.

Note: in order to prevent agglomeration, a straw can be used to stir.

(m) Placed the ampoule containing the peptide in an ice bath, added 5 mL of lysate in a fume hood, and stirred for 2.5 h at 300 rpm.

Note: mix 95% TFA/2.5% H2O/2.5% TIPS (v/v) and peptides in an ice bath for 2.5 h, which can protect most amino acid side chains and resin cleavage. The remaining protection groups were removed with DTT.

(n) Soaked ether with ice-water mixture in advance.

(o) The sample was suction filtered with a small funnel, and the obtained filtrate was blown with nitrogen at 37 °C.

(p) Put the product in ice bath and let stand for 3 min.

(q) Added ether to the product and transferred it to a 1.5 mL centrifuge tube, washed it three times (gradually integrated the products of the six centrifuge tubes into three centrifuge tubes), centrifuged for 3 min at 8000 rpm.

(r) The synthesized peptide was dried under vacuum.

6.5 Characterization of molecule *in vitro*

$\mathbf{P_RC}$ was dissolved in DMSO to prepare 1×10^{-3} M solution of $\mathbf{P_RC}$. The experiment groups were set as Table 1.

Table 1 The configuration of the experimental group.

Group	Solution of P_RC	DMSO	H_2O
1	50 μL	950 μL	0 μL
2	50 μL	850 μL	100 μL
3	50 μL	750 μL	200 μL
4	50 μL	650 μL	300 μL
5	50 μL	550 μL	400 μL
6	50 μL	450 μL	500 μL
7	50 μL	350 μL	600 μL
8	50 μL	250 μL	700 μL
9	50 μL	150 μL	800 μL
10	50 μL	50 μL	900 μL
11	50 μL	0 μL	950 μL

Tested the absorption of each group from 600–900 nm.
The CD spectra of each group was tested from 300–900 nm.

6.5.1 Transmission electron microscope (TEM) studies

The samples of TEM images were prepared as follow:

(a) Took 10 μL solution of group 11 and dripped onto the copper grids slowly.
(b) Waited for 30 s.
(c) Removed the excess droplets carefully.
(d) Tested samples on TEM (Tecnai G2-TWIN electron microscope), the accelerating voltage was 200 keV.
NOTE: Make sure the copper grids stay dry.

6.6 Characterization of molecule *in vivo*

6.6.1 Infected macrophage lysate

(a) Cultured cells of RAW 264.7(Using DMEM medium).
(b) Added RAW 264.7 to 96-well plates and kept 104 cells of each well.
(c) Added 2 × 105 *S. aureus* and cultured for 1 h to infect RAW 264.7.
(d) Added gentamycin (50 μg/mL) and cultured the mixture for 30 min.
(e) Removed extracellular medium and collected RAW 264.7 cells.
(f) Lysed with 1/4 PBS to prepare macrophage cell lysates.

6.6.2 Photoacoustic imaging of RAW 264.7

(a) Sample preparation:
1. PRC dimers: 200 μM DMSO solution of PRC.
2. PRC assemblies: 200 μM DMSO/H2O (1: 19) solution of PRC.
3. MPC in buffer: 200 μM PBS solution of MPC.
4. MPSC in buffer: 200 μM PBS solution of MPSC.
5. MPC treated with lysates: 200 μM infected macrophage lysate of MPC and incubated for 2 h.
6. MPSC treated with lysates: 200 μM infected macrophage lysate of MPSC and incubated for 2 h.

(b) Left for 10 min before analysis.
(c) Tested samples on MSOT 128 Multi-Spectral Optoacoustic Tomography (temperature 25 °C), and collected PA signal.
(d) Through the ratio image analysis of the PA signal at 680 and 710 nm, the ratio PA imaging of each group was obtained.

6.6.3 *Photoacoustic imaging* in vivo

(a) Seven-weeks old female mice (Balb/c) were used to build mice models (purchased from Vital River Laboratory Animal Technology Co., Ltd).

(b) Three different mice models were built as following methods:

1. Intramuscular infection of macrophage cells (1 × 107 cells and infected by *S. aureus*) for 12 h.
2. Intramuscular infection of *S. aureus* (1 × 108 cfu), after 12 h, i.v. injected with 100 μg gentamycin for 12 h to eliminated extracellular infection.
3. Intramuscular infection of *S. aureus* (1 × 108 cfu) for 12 h.

(c) Injected MPC or MPSC (35 mg/kg) though i.v. administration.

(d) Obtained and collected data of PA tomography images from 1 to 36 h.

7. Expected outcomes

We expected that we could detect bacterial infection in the early stage through the signal of PA in vivo. In this way, we could treat the infection as soon as possible to avoid the antibiotic resistance and long-term chronic inflammation.

8. Quantification and statistical analysis

[Details of analysis and related statistics. For example: Measurement, Interpretation (how to accurately interpret, how to avoid over-interpretation), Representation, New Approaches, Statistical Approaches].

The transmission electron microscope (TEM) image was carried out in aqueous solution (H2O/DMSO; 95/5; v/v) on a Tecnai G2-TWIN electron microscope operating at an accelerating voltage of 200 keV. The samples were prepared by placing P_RC fibers droplets on copper grids for 30 s and removing the excess droplets.

9. Advantages

In this method, we synthesize the material easily by solid-phase peptide synthesis technique which response to caspase-1 in host cells. Through the process of self-assembly in vivo, the signal of this probe could be enhanced that give us the direction of treatment.

10. Limitations

None.

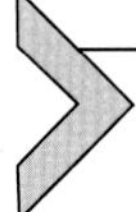

11. Optimization and troubleshooting

11.1 Problem

None.

11.2 Potential Solution to optimize the procedure

None.

12. Safety considerations and standards

[Details of relevant safety considerations and international or national safety standards/regulations to be considered]

1. Peptide synthesis was carried out in a fume hood.
2. Containers that have been in contact with TFA need to be immersed in NaOH aqueous solution before cleaning.
3. Bacteria related experiments need to be carried out in a vertical flow clean bench.

13. Alternative methods/procedures

[Details of any relevant alternative methods or procedures]

1. The protective group Fmoc can be removed by 1,8-Diazabicyclo [5.4.0] undec-7-ene/DMF (v/v=2/8).
2. TFA can be removed by vacuum distillation or nitrogen blowing.

14. Concluding remarks

Herein we describe the synthesis and validation of novel PA probe targeting macrophage for bioimaging detection of bacterial infection. This photoacoustic contrast agent can specifically self-assembly and accumulate inside the activated macrophage with a dynamic process in the presence of caspase-1. This method could provide a possibility for sensitive alert in the early stage of bacterial infection through PA detection *in vivo*.

References

Batrakova, E. V., Gendelman, H. E., & Kabanov, A. V. (2011). Cell-mediated drug delivery. *Expert Opinion on Drug Delivery*, *8*, 415–433.

Chen, Z., Mu, X., Han, Z., Yang, S., Zhang, C., Guo, Z., et al. (2019). An optical/photoacoustic dual-modality probe: Ratiometric in/ex vivo imaging for stimulated H2S upregulation in mice. *Journal of the American Chemical Society*, *141*, 17973–17977.

Foley, J. P., Lam, D., Jiang, H., Liao, J., Cheong, N., McDevitt, T. M., et al. (2012). Toll-like receptor 2 (TLR2), transforming growth factor-beta, hyaluronan (HA), and receptor for HA-mediated motility (RHAMM) are required for surfactant protein A-stimulated macrophage chemotaxis. *The Journal of Biological Chemistry*, *287*, 37406–37419.

Fraunholz, M., & Sinha, B. (2012). Intracellular Staphylococcus aureus: Live-in and let die. *Frontiers in Cellular and Infection Microbiology*, *2*, 43.

Fu, Q., Zhu, R., Song, J., Yang, H., & Chen, X. (2019). Photoacoustic imaging: Contrast agents and their biomedical applications. *Advanced Materials*, *31*, e1805875.

Guerra, F. E., Borgogna, T. R., Patel, D. M., Sward, E. W., & Voyich, J. M. (2017). Epic immune battles of history: Neutrophils vs. staphylococcus aureus. *Frontiers in Cellular and Infection Microbiology*, 7, 286.

Hatami, E., Mu, Y., Shields, D. N., Chauhan, S. C., Kumar, S., Cory, T. J., et al. (2019). Mannose-decorated hybrid nanoparticles for enhanced macrophage targeting. *Biochemistry and Biophysics Reports*, *17*, 197–207.

Kalinka, J., Hachmeister, M., Geraci, J., Sordelli, D., Hansen, U., Niemann, S., et al. (2014). Staphylococcus aureus isolates from chronic osteomyelitis are characterized by high host cell invasion and intracellular adaptation, but still induce inflammation. *International Journal of Medical Microbiology*, *304*, 1038–1049.

Li, L. L., Ma, H. L., Qi, G. B., Zhang, D., Yu, F., Hu, Z., et al. (2016). Pathological-condition-driven construction of supramolecular nanoassemblies for bacterial infection detection. *Advanced Materials*, *28*, 254–262.

Li, L. L., Qiao, S. L., Liu, W. J., Ma, Y., Wan, D., Pan, J., et al. (2017). Intracellular construction of topology-controlled polypeptide nanostructures with diverse biological functions. *Nature Communications*, *8*, 1276.

Li, L. L., Zeng, Q., Liu, W. J., Hu, X. F., Li, Y., Pan, J., et al. (2016). Quantitative analysis of caspase-1 activity in living cells through dynamic equilibrium of chlorophyll-based nano-assembly modulated photoacoustic signals. *ACS Applied Materials & Interfaces*, *8*, 17936–17943.

Liang, J., Tang, B. Z., & Liu, B. (2015). Specific light-up bioprobes based on AIEgen conjugates. *Chemical Society Reviews*, *44*, 2798–2811.

Man, S. M., Karki, R., & Kanneganti, T. D. (2017). Molecular mechanisms and functions of pyroptosis, inflammatory caspases and inflammasomes in infectious diseases. *Immunological Reviews*, *277*, 61–75.

Mirkovic, T., Ostroumov, E. E., Anna, J. M., van Grondelle, R., Govindjee, & Scholes, G. D. (2017). Light absorption and energy transfer in the antenna complexes of photosynthetic organisms. *Chemical Reviews*, *117*, 249–293.

Ntziachristos, V. (2010). Going deeper than microscopy: The optical imaging frontier in biology. *Nature Methods*, 7, 603–614.

Pang, L., Zhang, C., Qin, J., Han, L., Li, R., Hong, C., et al. (2017). A novel strategy to achieve effective drug delivery: Exploit cells as carrier combined with nanoparticles. *Drug Delivery*, *24*, 83–91.

Qadri, M., Almadani, S., Jay, G. D., & Elsaid, K. A. (2018). Role of CD44 in regulating TLR2 activation of human macrophages and downstream expression of proinflammatory cytokines. *Journal of Immunology*, *200*, 758–767.

Reinhardt, C. J., Zhou, E. Y., Jorgensen, M. D., Partipilo, G., & Chan, J. (2018). A ratiometric acoustogenic probe for in vivo imaging of endogenous nitric oxide. *Journal of the American Chemical Society*, *140*, 1011–1018.
Song, W. J., & Tezcan, F. A. (2014). A designed supramolecular protein assembly with in vivo enzymatic activity. *Science*, *346*, 1525–1528.
Sun, W., Li, M., Fan, J., & Peng, X. (2019). Activity-based sensing and theranostic probes based on photoinduced electron transfer. *Accounts of Chemical Research*, *52*, 2818–2831.
Wang, Y., Hu, X., Weng, J., Li, J., Fan, Q., Zhang, Y., et al. (2019). A photoacoustic probe for the imaging of tumor apoptosis by caspase-mediated macrocyclization and self-assembly. *Angewandte Chemie (International Ed. in English)*, *58*, 4886–4890.
Wang, Y., Liu, H., & Zhao, J. (2020). Macrophage polarization induced by probiotic bacteria: A concise review. *Probiotics and Antimicrobial Proteins*, *12*, 798–808.
Wang, A. H., Quigley, G. J., Kolpak, F. J., Crawford, J. L., van Boom, J. H., van der Marel, G., et al. (1979). Molecular structure of a left-handed double helical DNA fragment at atomic resolution. *Nature*, *282*, 680–686.
Wang, S., Sheng, Z., Yang, Z., Hu, D., Long, X., Feng, G., et al. (2019). Activatable small-molecule photoacoustic probes that cross the blood-brain barrier for visualization of copper(II) in mice with Alzheimer's disease. *Angewandte Chemie (International Ed. in English)*, *58*, 12415–12419.
Weber, J., Beard, P. C., & Bohndiek, S. E. (2016). Contrast agents for molecular photoacoustic imaging. *Nature Methods*, *13*, 639–650.
Xuan, W., Qu, Q., Zheng, B., Xiong, S., & Fan, G. H. (2015). The chemotaxis of M1 and M2 macrophages is regulated by different chemokines. *Journal of Leukocyte Biology*, *97*, 61–69.
Yang, S., Han, X., Yang, Y., Qiao, H., Yu, Z., Liu, Y., et al. (2018). Bacteria-targeting nanoparticles with microenvironment-responsive antibiotic release to eliminate intracellular Staphylococcus aureus and associated infection. *ACS Applied Materials & Interfaces*, *10*, 14299–14311.
Yin, L., Sun, H., Zhang, H., He, L., Qiu, L., Lin, J., et al. (2019). Quantitatively visualizing tumor-related protease activity in vivo using a ratiometric photoacoustic probe. *Journal of the American Chemical Society*, *141*, 3265–3273.
Zhang, C., Gao, R., Zhang, L., Liu, C., Yang, Z., & Zhao, S. (2020). Design and synthesis of a ratiometric photoacoustic probe for in situ imaging of zinc ions in deep tissue in vivo. *Analytical Chemistry*, *92*, 6382–6390.
Zhang, D., Qi, G. B., Zhao, Y. X., Qiao, S. L., Yang, C., & Wang, H. (2015). In situ formation of nanofibers from Purpurin18-peptide conjugates and the assembly induced retention effect in tumor sites. *Advanced Materials*, *27*, 6125–6130.
Zheng, J., Zeng, Q., Zhang, R., Xing, D., & Zhang, T. (2019). Dynamic-reversible photoacoustic probe for continuous ratiometric sensing and imaging of redox status in vivo. *Journal of the American Chemical Society*, *141*, 19226–19230.
Zhou, K., Li, C., Chen, D., Pan, Y., Tao, Y., Qu, W., et al. (2018). A review on nanosystems as an effective approach against infections of Staphylococcus aureus. *International Journal of Nanomedicine*, *13*, 7333–7347.

CHAPTER FOURTEEN

Bioengineered bacterial vesicles for optoacoustics-guided phototherapy

Vipul Gujrati[a,b] and Vasilis Ntziachristos[a,b,*]
[a]Center for Translational Cancer Research (TranslaTUM), School of Medicine, Technical University of Munich, Munich, Germany
[b]Institute of Biological and Medical Imaging (IBMI), Helmholtz Zentrum München (GmbH), Neuherberg, Germany
*Corresponding author: e-mail address: v.ntziachristos@tum.de

Contents

Abstract

Genetically engineered bacterial outer membrane vesicles (OMVs) offer promising applications for gene therapy, immunotherapy, and vaccine delivery. Importantly, OMVs are biocompatible, biodegradable, and easy to engineer and produce on a large scale. In this chapter, we discuss the development and application of bioengineered OMVs for optoacoustics-guided phototherapy applications (theranostics). We provide detailed protocols for OMVs preparation, characterization, and *in vitro* and *in vivo* validation. The engineered OMVs carry the biopolymer melanin, which generates a strong optoacoustic (OA) signal and intense heat upon absorption of near-infrared (NIR) light, enabling optoacoustics-guided cancer diagnosis and photothermal therapy *in vivo*.

Methods in Enzymology, Volume 657
ISSN 0076-6879
https://doi.org/10.1016/bs.mie.2021.06.030

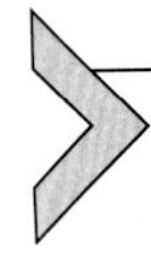

1. Introduction

1.1 Bacterial outer membrane vesicles (OMVs)

Gram negative-bacteria continuously release outer membrane vesicles (OMVs) during normal growth (Lee, Choi, Kim, & Gho, 2008). OMVs are released from the bacterial outer membrane and therefore consist of a spherical lipid bilayer carrying lipopolysaccharides (LPS) and proteins from the outer membrane as well as periplasmic and cytoplasmic spaces (Kuehn & Kesty, 2005; Lee et al., 2008). OMVs act as messengers by trafficking genetic information and virulence factors to the surrounding prokaryotic or eukaryotic environment (Kim et al., 2008; Kuehn & Kesty, 2005). Additionally, OMVs are small (20–250-nm diameter) and exhibit a rigid and stable structure, making them suitable for *in vivo* drug delivery through the enhanced permeability and retention effect (EPR) (Gujrati et al., 2014, 2019; Kim et al., 2008). Furthermore, OMVs can be engineered to reduce the endotoxic potential of LPS and to ensure cancer-specific delivery of a payload by expressing cancer-targeting ligands on the surface (Gujrati et al., 2014, 2019). To develop OMVs for the purpose of *in vivo* drug delivery and imaging, it is essential to address the safety and immunogenicity issues. Using OMVs derived from bacterial strains with no substantial pathogenicity is one possible approach. For example, *Escherichia coli* (*E. coli*) W3110, a K12-derived strain carrying mutations in the msbB gene, was previously used to develop OMVs (Gujrati et al., 2014, 2019). The msbB gene plays a crucial role in lipid A biosynthesis, an endotoxic component of LPS. Mutational inactivation of the msbB gene reduces the biosynthesis of LPS and, in turn, the ability of LPS to stimulate Toll-like receptor pathways in immune cells, causing a significant loss in virulence (Hajjar, Ernst, Tsai, Wilson, & Miller, 2002; Somerville, Cassiano, & Darveau, 1999). Current research on engineered OMVs has largely focused on their use of for vaccine development (against cancer and infectious diseases) and delivery (Gerritzen, Martens, Wijffels, van der Pol, & Stork, 2017; Grandi et al., 2017; Kim et al., 2009) by utilizing the adjuvant property of LPS, lipoproteins, and displaying antigens of interest on the surface (Gerritzen et al., 2017; Grandi et al., 2017; Kim et al., 2009). In addition to the vaccine platform, OMVs are being explored for tumor microenvironment (TME) reprogramming (Qing et al., 2020), optical imaging (by expressing luciferase or GFP for bioluminescence or fluorescence (Huang et al., 2019), gene delivery (Gujrati et al., 2014), and optoacoustics-guided phototherapy (Gujrati et al., 2019) applications

1.2 Melanin for optoacoustic imaging and photothermal therapy

We developed OMVs from engineered *E. coli* to carry the biopolymer melanin (Gujrati et al., 2019) (Fig. 1). Melanin is found naturally in many living organisms and it absorbs strongly in the wide (visible and NIR) range (Simon, 2000) to produce strong contrast for optoacoustic imaging (Fan et al., 2014; Gujrati et al., 2019; Stritzker et al., 2013). Additionally, melanin has a high photothermal conversion efficiency and is consequently highly suitable for localized photothermal therapy to kill cancer cells (Gujrati et al., 2019; Liu et al., 2013). The ability of melanin to serve as a contrast enhancement as well as a therapeutic agent makes it highly suitable for optoacoustics-guided theranostics (combination of diagnostics and therapy in a single agent). We engineered the K12-derived strain W3110 to overexpress tyrosinase, which produces melanin that is passively incorporated into the cytosol and membrane of OMVs (OMV^{Mel}) (Gujrati et al., 2019). We then tested the ability of OMV^{Mel} to produce OA contrast and local heating *in vitro* and *in vivo* after irradiation with a pulsed NIR light source. We showed that a single dose of OMV^{Mel} after phototherapy inhibits tumor growth while triggering only mild, short-term systemic inflammation. In the future, it may be possible to replace melanin with other naturally derived theranostic agents (peptides or proteins) to generate a platform for combined imaging and therapy applications against cancer and other diseases.

1.3 Principle of multispectral optoacoustic tomography

Advances in illumination technology, detectors, and reconstruction methods have given photo/optoacoustic imaging (OAI) the capability of going beyond

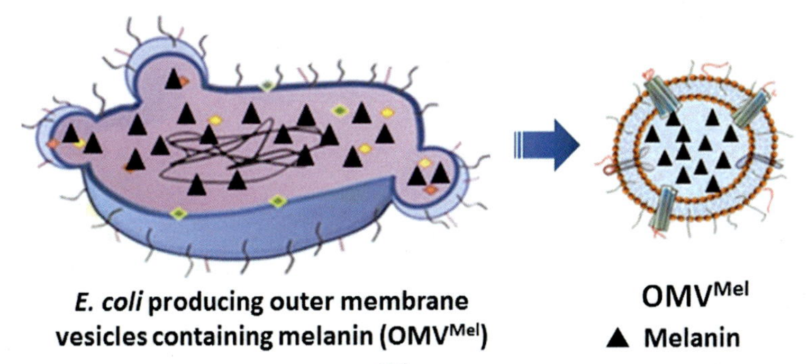

Fig. 1 Schematic representation of OMV^{Mel} generation. *Reproduced with the permission from Gujrati, V., Prakash, J., Malekzadeh-Najafabadi, J., Stiel, A., Klemm, U., Mettenleiter, G., et al. (2019). Bioengineered bacterial vesicles as biological nano-heaters for optoacoustic imaging.* Nature Communications, 10*(1), 1–10. http://creativecommons.org/licenses/by/4.0/.*

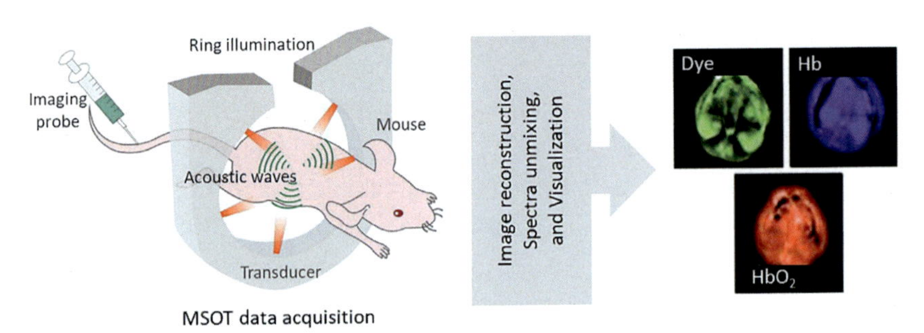

Fig. 2 Principle of optoacoustic tomography.

microscopic interrogation and enable high-fidelity mesoscopic and macroscopic imaging (Li et al., 2020; Ntziachristos & Razansky, 2010). Advanced OAI allows real-time application of multispectral optoacoustic tomography (MSOT), a technique that illuminates tissue at multiple wavelengths within tens of milliseconds (Ntziachristos & Razansky, 2010; Razansky, Buehler, & Ntziachristos, 2011). OA signal generation occurs in three steps: (1) a sample (tissue or object) absorbs light, (2) the absorbed optical energy is converted into heat which increases the local temperature, and (3) thermoelastic expansion takes place to generate acoustic waves (Fig. 2). Based on the principles of OA, MSOT has been shown to resolve several molecular entities such as hemoglobin, melanin, myoglobin, water, lipids, and a range of exogenous absorbers (dyes, nanoparticles, reporter biomolecules). Resolving the biodistribution of these molecules leads to images of vascularization, tissue oxygenation, and tumor hypoxia and can delineate general morphological features of tissue (Ntziachristos & Razansky, 2010; Razansky et al., 2011). Methods are employed to process the images and resolve the distribution of different photo-absorbers based on their characteristic absorption spectral profiles, with the resolved unmixed images yielding the localization of the absorbing molecules. The accuracy and the sensitivity of molecular detection increases with the number of wavelengths employed: the more wavelengths are used, the more molecules can be resolved and the higher the quantification accuracy achieved.

2. Protocol

2.1 OMVs generation, purification, and characterization

To ensure OMV^{Mel} and OMV^{WT} are free from parent bacterial contamination, a thorough stepwise purification is needed (Fig. 3). However, these

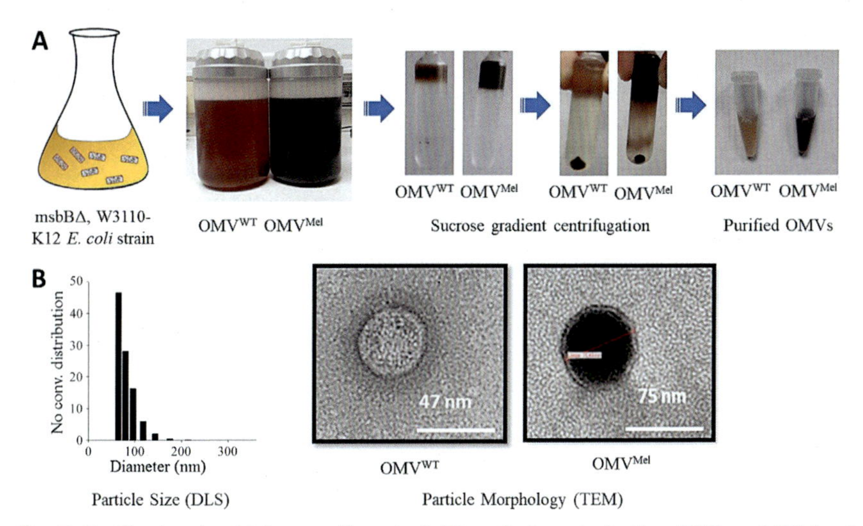

Fig. 3 Purification (multiple centrifugation) (A) and characterization (DLS and TEM) of OMVs (B). *Reproduced with the permission from Gujrati, V., Prakash, J., Malekzadeh-Najafabadi, J., Stiel, A., Klemm, U., Mettenleiter, G., et al. (2019). Bioengineered bacterial vesicles as biological nano-heaters for optoacoustic imaging.* Nature Communications, 10*(1), 1–10. http://creativecommons.org/licenses/by/4.0/.*

methods of production and purification can be modified depending on the available facility for the isolation and purification of OMVs.

Equipment:

1. Polymerase chain reaction (PCR) thermal cycler
2. Bacteria culture incubator (for plate and flask)
3. Centrifuge (table-top and floor)
4. Ultracentrifuge
5. Vacuum pump
6. Magnetic stirrer
7. Weighing balance
8. Microwave oven

Reagents:

1. Sterile phosphate buffer (PBS)
2. Lysogeny/Luria broth (LB) and agar plates (to grow bacteria, relevant antibiotics are added depending on the selection markers)
3. Isopropyl β-D-1-thiogalactopyranoside (IPTG)
4. $CuSO_4$
5. L-tyrosine
6. Sucrose solution of 2.5, 1.6, and 0.6 M in ultra-pure water (prepare a desired volume by adding desired amount of sucrose in water and by

heating in a microwave oven and stirring; final solution can be stored in a sealed flask at 4 °C)

7. Ammonium sulfate

Procedure:

1. PCR-amplify and clone the MelA gene of *Rhizobium etli* encoding tyrosinase (we received ours from Prof. Guillermo Gosset, Universidad Nacional Autónoma de México, Mexico) into the pGEX-4 T-1 vector or another vector suitable for the selected bacterial strain (we selected *Eco*NI/*Xho*I). (Note: we removed the GST tag from the pGEX-4 T-1 vector which was used for MelA cloning).
2. Transform the pGEX-4 T-1-melA construct into a W3110-K12 *E. coli* strain (or *E. coli* you are willing to use for OMVs production)
3. Select the positive clones by colony PCR and gene sequencing and prepare the plasmid and cell stock of positive clones
4. Culture the bacteria (positive clone) in 1.5 L flask (LB broth) at 30 °C and 180 rpm (between 3 and 5 days), and induce tyrosinase production at OD 0.6 using isopropyl β-D-1-thiogalactopyranoside (IPTG) at a final concentration of 0.5 mM. As the wild type OMVs (OMV^{WT}) control, culture the untransformed W3110 *E. coli*
5. To support melanin production, the media should be supplemented with 94.5 mg $CuSO_4$ as well as 1 g L-tyrosine
6. Collect the bacterial culture (1.5 L) from step 4. Number of cultures can be increased as per requirement and available facilities
7. Centrifuge the culture at 7500 × *g* for 30 min at 4 °C to pellet the bacteria
8. Collect the supernatant in separate flasks and discard the bacterial pellet following bio-waste management protocols of the institution
9. Filter the supernatant through 0.45-μm filters (*e.g.*, Nalgene, Thermo Scientific with 500 mL capacity) by applying vacuum and remove cell debris
10. Concentrate the supernatant to 100 mL using 100 K Ultrafiltration Membranes (*e.g.*, Millipore) and collect in a 1 L glass bottle
11. Next, precipitate the OMVs using ammonium sulfate (final concentration 400 g/L) by adding it to the 1 L flask from step 10. Add ammonium sulfate slowly with continuous stirring using a magnetic stirrer
12. Store the flask at 4 °C overnight
13. The following day, recover the crude OMVs (OMV^{Mel} and OMV^{WT}) by centrifugation and washing with PBS, as described below:

14. Collect the precipitated solution in 50 mL round bottom centrifugal tubes (*e.g.*, Nalgene centrifuge tubes) and centrifuge at 12,000 × g for 30 min
15. Remove the supernatant and resuspend the OMVs pellet in PBS using 1–2 mL per vial
16. Collect all pellets in one round-bottom centrifugal tube, resuspend by repeated pipetting, fill to 50 mL with PBS and centrifuge as in step 9
17. Resuspend the pellet of OMVs in 1 mL PBS and layer it over a 3 mL sucrose gradient (1 mL each of 2.5, 1.6, 0.6 M sucrose)

Note: Add an equal volume of each sucrose solution (usually 1 mL, however volume can be changed as per the size of the sample holder) from the side of the centrifugation tube used for gradient preparation. Here, first add the high density sucrose solution followed by lower density solutions, and finally place OMVs on the top.

18. Carry out the ultracentrifugation at 150,000 × *g* for 3–5 h at 4 °C to separate the free melanin and other impurities
19. Collect the OMVs fraction by carefully removing the impurity and sucrose layers with the help of a 1 mL pipet
20. Wash the OMVs fraction twice with PBS (1 mL) by ultracentrifugation at 150,000 × *g* for 1–2 h at 4 °C
21. Finally, resuspend the OMVs pellet in 1 mL 15% glycerol, and pass through Detoxi-Gel Endotoxin Removing Columns (Thermo Scientific) to remove free endotoxins
22. Filter the recovered OMVs suspension through 0.20 or 0.45-μm cellulose acetate filters (DISMIC-25cs), and store at −20 °C until use
23. The total protein concentration of OMV^{WT} should be determined by the bicinchoninic acid assay (BCA protein assay kit; Thermo Scientific) according to manufacturer's instructions

(Note: due to its black color, OMV^{Mel} cannot be detected by this method. Instead one should ultracentrifuge an equal volume of OMV^{WT} and OMV^{Mel} (as in step 20), which allows an indirect estimation of OMVs based on pellet size. Additionally, intensities of major protein bands can be compared by carrying out sodium dodecyl sulfate–polyacrylamide gel electrophoresis (SDS-PAGE).

24. Finally, characterize the OMVs for size and morphology by using an electrophoretic light-scattering apparatus (DLS) and transmission electron microscope (TEM). (Note: To visualize OMV^{WT}, the sample

should be placed on a copper grid and stained with 2% uranyl acetate for 10 min and washed by dipping into distilled water five times. No staining is needed for OMV^{Mel}).

2.2 Photophysical characterization of OMVs

The following experiments are carried out to identify the optoacoustic spectrum of the OMV^{Mel}. Here one has to use the multispectral optoacoustic tomography (MSOT) scanner (discussed in Section 2.4) to scan the phantoms with OMVs samples placed in the center. The cylindrical agar phantoms that are used for the measurements contain Indian ink and intralipid to mimic the light scattering that occurs in tissue.

Equipment:

1. Spectrophotometer
2. MSOT scanner
3. Microwave oven

Reagents:

1. Indian ink
2. Agar (molecular biology grade)
3. Intralipid
4. Deionized water

Procedure:

1. Cylindrical phantoms of 2 cm in diameter are generally prepared using a gel made from distilled water containing Indian ink (0.1 OD), Agar (1.3% w/w) and an intralipid 20% emulsion (6% v/v) (Fig. 4). In a glass beaker, mix the agar in deionized water containing Indian ink and intralipid (weight as per the size of the phantom to be made) and heat in the microwave

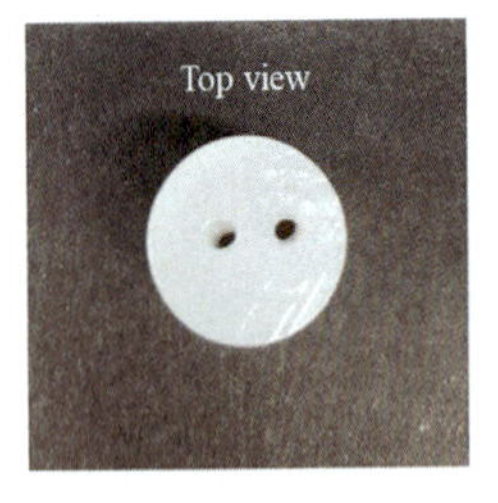

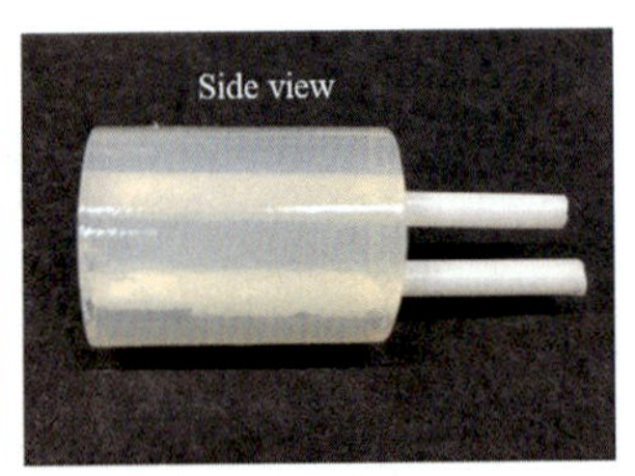

Phantom

Phantom with samples (in tubing)

Fig. 4 Tissue-mimicking phantoms.

2. Remove beaker and stir carefully (no bubbles should generate) to a homogeneous solution
3. Pour the hot solution into a cylindrical plastic holder (falcon tube can be used), and cool down for 1–2 h at 4 °C
4. Carefully remove the phantom from the holder and make 3 mm diameter holes in the center using the plastic tubing that is used to load the samples in the next step
5. Place two cylindrical pieces of tubing of approximately 3 mm in diameter, containing the OMV samples (OMV^{WT} and OMV^{Mel}) in water, in the middle of the phantom
6. MSOT data can be acquired as discussed in the section (Section 2.4), at a single position, located approximately in the middle of the phantom

2.3 Photothermal heating and cytotoxicity test

Equipment:
1. Fluorescence microscope
2. NIR light source (750 nm, 650 mW cm^{-2}; Tunable Optical Parametric Oscillator Laser)

Reagents:
1. Calcein AM and ethidium homodimer-1 (Invitrogen)
2. RPMI medium with 10% fetal bovine serum (FBS)
3. OptiMEM
4. Sterile phosphate buffer (PBS)

Cell line:
1. 4T1 mammary gland carcinoma (CRL-2539, ATCC, Manassas, VA, USA)

Procedure:

2.3.1 Photothermal heating

1. Place the samples (PBS or OMV^{WT} or OMV^{Mel}) in in a quartz cuvette
2. Expose solutions to an NIR light source at 750 nm (650 mW cm^{-2}; Tunable Optical Parametric Oscillator Laser) and record the temperature changes with a digital thermometer for up to 10 min. Concentration-response experiments should be done to validate the results (Fig. 5A).

2.3.2 Cytotoxicity

1. Seed the cancer cells in a 96 well plate (~10,000 cells per well)
2. Next day, replace the media with optiMEM (100 μL) containing OMV^{WT} or OMV^{Mel} (~75 μg) and incubate for 6 h

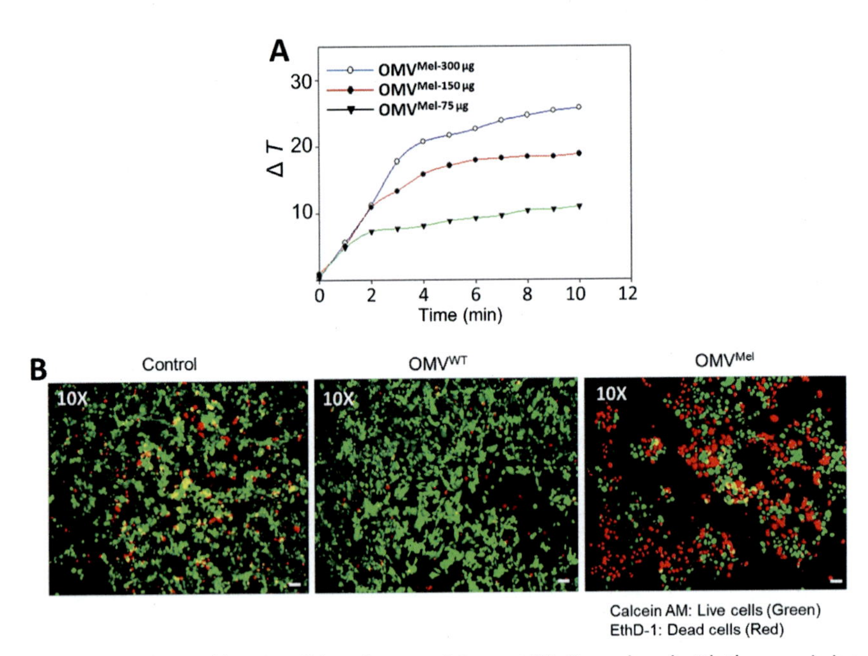

Fig. 5 Photothermal heating (A) and cytotoxicity test (B). *Reproduced with the permission from Gujrati, V., Prakash, J., Malekzadeh-Najafabadi, J., Stiel, A., Klemm, U., Mettenleiter, G., et al. (2019). Bioengineered bacterial vesicles as biological nano-heaters for optoacoustic imaging.* Nature Communications, 10*(1), 1–10. http://creativecommons.org/licenses/by/4.0/.*

3. Replace the media containing OMVs with fresh media, and expose the cells to a nanosecond-pulsed laser treatment (730–830 nm, $650\,\mathrm{mW\,cm^{-2}}$) for 6 min
4. Stain the cells with calcein AM and ethidium homodimer-1 following the manufacturer's instructions
5. Examine cells under a fluorescence microscope to determine numbers of live (stained green) and dead cells (stained red) and capture the images (Fig. 5B).

2.4 *In vivo* multispectral optoacoustic tomography (MSOT)

The phantom and mice OAI can be performed using a commercially available MSOT scanner (MSOT256-TF, iThera Medical GmbH, Munich, Germany), or another suitable optoacoustic tomography system. In MSOT, nanosecond-pulsed light is generated from a tunable optical parametric oscillator (OPO) laser and delivered to the sample with a fiber bundle having 10 outputs placed in a ring-shape. The wavelength used for imaging ranges from 680 to 900 nm, with a tunable step size (usually 10 nm). Light is absorbed by the sample and generates an acoustic signal that propagates through the sample and is detected outside the sample. Acoustic signals are detected using a cylindrically focused, 256-element transducer array, which has a central frequency of 5 MHz with a radius of curvature of 40 mm and an angular coverage of 270°. Optoacoustic images are then reconstructed using a commercial MSOT inVision software (by back-projection or model-based reconstruction methods). These spectral opto-acoustic images can be used to unmix the different biological chromophores through a linear unmixing method. Longitudinal melanin quantification can be performed using the unmixed melanin distribution. Specifically, melanin from the tumor region can be quantified by plotting the mean intensity from a defined region of interest. Signals from other biomolecules like oxy- and deoxyhemoglobin can also be unmixed.

Equipment:

1. Multispectral optoacoustic tomography scanner (MSOT256-TF, iThera Medical GmbH, Munich, Germany)
2. Isoflurane and oxygen supply (for anesthesia)
3. Sterile insulin syringe
4. Cell incubator (37 °C in 5% CO_2)

Cell line:

4 T1 mammary gland carcinoma (CRL-2539, ATCC, Manassas, VA, USA).

Animals:

1. Female athymic Fox—N—1 nude 6-week old mice

Note: Tumor volume calculation: $(\text{width})^2 \times (\text{length}) \times 1/2$, and relative tumor volume can be calculated as V/V_0 (where V_0 is tumor volume when the treatment is initiated), volume can be expressed as group mean ± standard deviation (SD). Tumor growth inhibition (TGI) can be determined on the final day as % TGI: $100\% \times ({T_{vol}}^{PBS} - {T_{vol}}^{OMV}) \times ({T_{vol}}^{PBS})^{-1}$, where T_{vol} is final tumor volume − initial tumor volume.

Reagents:

1. Sterile phosphate buffer (PBS)
2. OMVs samples (~75 μg; OMV^{WT} or OMV^{Mel}) resuspended in PBS (150 μL)
3. RPMI 1640 medium supplemented with 10% fetal bovine serum (FBS) and antibiotics

Procedure:

1. Implant 4 T1 murine breast cancer cells (0.8×10^6 cells/animal) subcutaneously in the nude mice following the institutional animal care

guidelines. It usually takes approximately 8–10 days for 4 T1 tumors to grow to ~150 mm³ (measurements should be done using a digital caliper).

2. Warm up the MSOT chamber for approximately 15 min until the water temperature reaches 34 °C. Pre-set the imaging protocol to: imaging wavelengths from 680 to 900 nm with a step size of 10 nm and 10 averages
3. Anesthetize the tumor-bearing mouse in a isoflurane chamber, and maintain the anesthesia and oxygen flow through the nasal route
4. Using an insulin syringe, inject the 150 μL (~75 μg) OMVs through the tail vein
5. Apply the ultrasound gel on the mouse body to enable coupling between the tissue and water medium
6. Place the mouse in an animal holder and transfer to the MSOT water chamber
7. Scan the mouse at different time intervals (pre-injection and post-injection) based on the study design
8. Repeat the experiment with more mice, with an appropriate sample size chosen for statistics
9. Process the data using MSOT inVison software, and select the regions of interest in the tumor and vital organs (liver, kidney *etc.*) to estimate OA signal due to melanin. Fig. 6 shows the biodistribution of OMV^{Mel} in the tumor, liver and kidney (monitored at different time intervals).

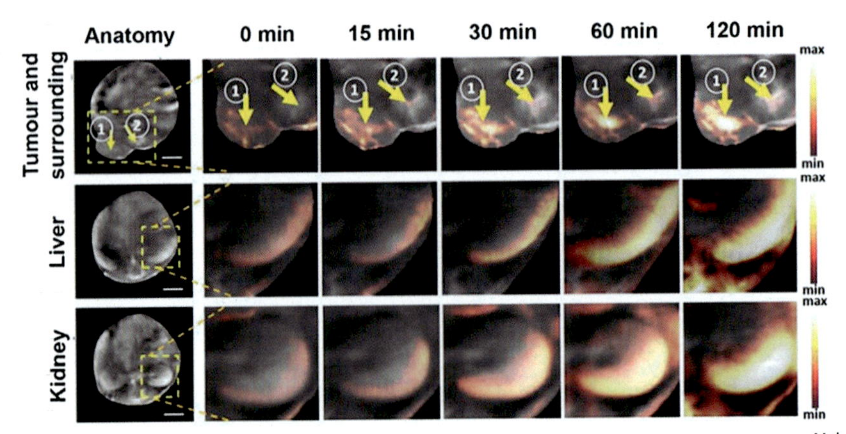

Fig. 6 *In vivo* MSOT imaging of tumor and vital organs after injecting OMV^{Mel}. *Reproduced with the permission from Gujrati, V., Prakash, J., Malekzadeh-Najafabadi, J., Stiel, A., Klemm, U., Mettenleiter, G., et al. (2019). Bioengineered bacterial vesicles as biological nano-heaters for optoacoustic imaging.* Nature Communications, 10*(1), 1–10. http://creativecommons.org/licenses/by/4.0/.*

10. Carry out melanin quantification on the unmixed melanin image from the tumor region by plotting the mean intensity from a defined region of interest (corresponding to a similar tissue area in all animals) over time

2.5 *In vivo* photothermal therapy

The goal of this experiment is to verify that near infrared (NIR) light irradiation can generate localized heat (in the tumor region) due to presence of OMV^{Mel}. Melanin generates strong heat to induce necrosis in the tumor region and retards tumor progression (Fig. 7). Greater heating is expected if OMV^{Mel} is injected directly into the tumor (i.t.) compared to the tail vein (i.v.).

Equipment:

1. Continuous wave laser (1.5 W cm^{-2}, 800 nm)
2. IR thermal camera (*e.g.*, FLIR i60)
3. Isoflurane and oxygen supply (for anesthesia)
4. Sterile insulin syringe
5. Cell incubator (37 °C in 5% CO_2)

Cell line:

1. 4 T1 mammary gland carcinoma (CRL-2539, ATCC, Manassas, VA, USA)

Animals:

1. Female athymic Fox—N—1 nude 6-week old mice

Note: Prepare the tumor animal model as discussed in the previous section of Section 2.4 imaging.

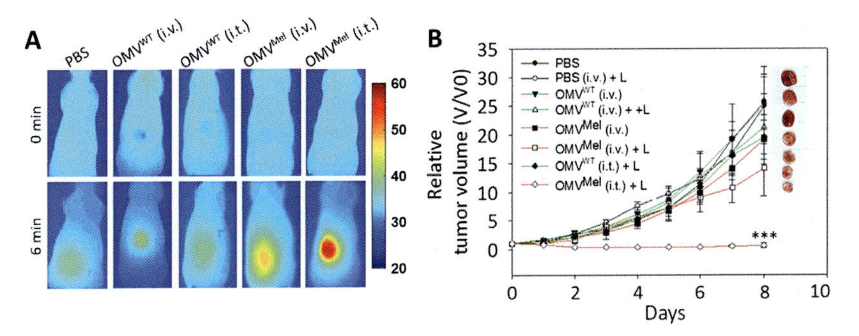

Fig. 7 *In vivo* photothermal therapy (A) and tumor growth curve (B). *Reproduced with the permission from Gujrati, V., Prakash, J., Malekzadeh-Najafabadi, J., Stiel, A., Klemm, U., Mettenleiter, G., et al. (2019). Bioengineered bacterial vesicles as biological nano-heaters for optoacoustic imaging.* Nature Communications, 10*(1), 1–10. http://creativecommons.org/licenses/by/4.0/.*

Procedure:

1. Anesthetize the mouse in an isoflurane chamber, and maintain the anesthesia and oxygen flow through the nasal route
2. Using an insulin syringe, inject the OMVs sample (~75 μg) intravenously through the tail vein or intratumorally as per the study design
3. At the predetermined time point (based on biodistribution MSOT information), expose the tumor region to a continuous wave laser ($1.5\,W\,cm^{-2}$, 800 nm) for 6 min
4. Record the tumor surface temperature during light irradiation using an IR thermal camera
5. Record the tumor growth rate and animal body weight at fixed intervals (usually animals are sacrificed when tumor size reaches ~1.5 cm).
6. Compare the tumor growth retardation in OMV^{Mel} treated animals with all controls

3. Summary

Herein we describe the production of bacterial outer membrane derived vesicles (OMVs) that are engineered to package the biopolymer melanin. Such melanin-carrying OMVs (OMV^{Mel}) can be used for opto-acoustic imaging and photothermal applications in early diagnosis and image-guided therapy of diverse solid tumors. Furthermore, we believe that OMVs will be engineered to package diverse payloads for a variety of theranostic applications in the future.

Acknowledgments

This project received funding from the European Research Council (ERC) under the European Union's Horizon 2020 research and innovation programme under grant agreement no. 694968 (PREMSOT). The research leading to these results was supported by the Deutsche Forschungsgemeinschaft (DFG), Germany (Gottfried Wilhelm Leibniz Prize 2013, NT 3/10-1) as well as by the DFG as part of the CRC 1123 (Z1). We thank all the authors of our manuscript on OMV^{Mel} *Nature Communications* publication (2019; 10(1), 1–10) for their contribution to the work.

We thank the publishers of *Nature Communications* (Nature Research, http://creativecommons.org/licenses/by/4.0/) and ACS Nano (American Chemical Society) who allowed us to use the content (figures) for the reprint.

References

Fan, Q., Cheng, K., Hu, X., Ma, X., Zhang, R., Yang, M., et al. (2014). Transferring biomarker into molecular probe: Melanin nanoparticle as a naturally active platform for multimodality imaging. *Journal of the American Chemical Society*, *136*(43), 15185–15194.

Gerritzen, M. J., Martens, D. E., Wijffels, R. H., van der Pol, L., & Stork, M. (2017). Bioengineering bacterial outer membrane vesicles as vaccine platform. *Biotechnology Advances*, *35*(5), 565–574.

Grandi, A., Tomasi, M., Zanella, I., Ganfini, L., Caproni, E., Fantappiè, L., et al. (2017). Synergistic protective activity of tumor-specific epitopes engineered in bacterial outer membrane vesicles. *Frontiers in Oncology*, 7, 253.

Gujrati, V., Kim, S., Kim, S.-H., Min, J. J., Choy, H. E., Kim, S. C., et al. (2014). Bioengineered bacterial outer membrane vesicles as cell-specific drug-delivery vehicles for cancer therapy. *ACS Nano*, *8*(2), 1525–1537.

Gujrati, V., Prakash, J., Malekzadeh-Najafabadi, J., Stiel, A., Klemm, U., Mettenleiter, G., et al. (2019). Bioengineered bacterial vesicles as biological nano-heaters for optoacoustic imaging. *Nature Communications*, *10*(1), 1–10.

Hajjar, A. M., Ernst, R. K., Tsai, J. H., Wilson, C. B., & Miller, S. I. (2002). Human toll-like receptor 4 recognizes host-specific LPS modifications. *Nature Immunology*, *3*(4), 354–359.

Huang, Y., Beringhs, A. O. R., Chen, Q., Song, D., Chen, W., Lu, X., et al. (2019). Genetically engineered bacterial outer membrane vesicles with expressed nanoluciferase reporter for *in vivo* bioluminescence kinetic modeling through noninvasive imaging. *ACS Applied Bio Materials*, *2*(12), 5608–5615.

Kim, J.-Y., Doody, A. M., Chen, D. J., Cremona, G. H., Shuler, M. L., Putnam, D., et al. (2008). Engineered bacterial outer membrane vesicles with enhanced functionality. *Journal of Molecular Biology*, *380*(1), 51–66.

Kim, S.-H., Kim, K.-S., Lee, S.-R., Kim, E., Kim, M.-S., Lee, E.-Y., et al. (2009). Structural modifications of outer membrane vesicles to refine them as vaccine delivery vehicles. *Biochimica et Biophysica Acta (BBA)-Biomembranes*, *1788*(10), 2150–2159.

Kuehn, M. J., & Kesty, N. C. (2005). Bacterial outer membrane vesicles and the host–pathogen interaction. *Genes & Development*, *19*(22), 2645–2655.

Lee, E. Y., Choi, D. S., Kim, K. P., & Gho, Y. S. (2008). Proteomics in gram-negative bacterial outer membrane vesicles. *Mass Spectrometry Reviews*, *27*(6), 535–555.

Li, J., Chekkoury, A., Prakash, J., Glasl, S., Vetschera, P., Koberstein-Schwarz, B., et al. (2020). Spatial heterogeneity of oxygenation and haemodynamics in breast cancer resolved *in vivo* by conical multispectral optoacoustic mesoscopy. *Light: Science & Applications*, *9*(1), 1–15.

Liu, Y., Ai, K., Liu, J., Deng, M., He, Y., & Lu, L. (2013). Dopamine-melanin colloidal nanospheres: An efficient near-infrared photothermal therapeutic agent for *in vivo* cancer therapy. *Advanced Materials*, *25*(9), 1353–1359.

Ntziachristos, V., & Razansky, D. (2010). Molecular imaging by means of multispectral optoacoustic tomography (MSOT). *Chemical Reviews*, *110*(5), 2783–2794.

Qing, S., Lyu, C., Zhu, L., Pan, C., Wang, S., Li, F., et al. (2020). Biomineralized bacterial outer membrane vesicles potentiate safe and efficient tumor microenvironment reprogramming for anticancer therapy. *Advanced Materials*, *32*(47), 2002085.

Razansky, D., Buehler, A., & Ntziachristos, V. (2011). Volumetric real-time multispectral optoacoustic tomography of biomarkers. *Nature Protocols*, *6*(8), 1121–1129.

Simon, J. D. (2000). Spectroscopic and dynamic studies of the epidermal chromophores trans-urocanic acid and eumelanin. *Accounts of Chemical Research*, *33*(5), 307–313.

Somerville, J. E., Cassiano, L., & Darveau, R. P. (1999). Escherichia coli msbB gene as a virulence factor and a therapeutic target. *Infection and Immunity*, *67*(12), 6583–6590.

Stritzker, J., Kirscher, L., Scadeng, M., Deliolanis, N. C., Morscher, S., Symvoulidis, P., et al. (2013). Vaccinia virus-mediated melanin production allows MR and optoacoustic deep tissue imaging and laser-induced thermotherapy of cancer. *Proceedings of the National Academy of Sciences*, *110*(9), 3316–3320.

CHAPTER FIFTEEN

A practical guide to photoswitching optoacoustics tomography

Mariia Stankevych[a], Kanuj Mishra[a], Vasilis Ntziachristos[a,b], and Andre C. Stiel[a,*]

[a]Institute of Biological and Medical Imaging (IBMI), Helmholtz Zentrum München (GmbH), Neuherberg, Germany

[b]Center for Translational Cancer Research (TranslaTUM), School of Medicine, Technical University of Munich, Munich, Germany

*Corresponding author: e-mail address: andre.stiel@helmholtz-muenchen.de

Contents

Abstract

Photochromic proteins and photoswitching optoacoustics (OA) are a promising combination, that allows OA imaging of even small numbers of cells in whole live animals and thus can facilitate a more wide-spread use of OA in life-science and preclinical research. The concept relies on exploiting the modulation achieved by the photoswitching to discriminate the agents' signal from the non-modulating background. Here we share our analysis approaches that can be readily used on data generated with commercial OA tomography imaging instrumentation allowing—depending on the used photoswitching agent and sample—routine visualizations of as little as several hundreds of transgene labeled cells per imaging volume in the live animal.

Methods in Enzymology, Volume 657
ISSN 0076-6879
https://doi.org/10.1016/bs.mie.2021.06.031

1. Introduction

Application of Photo- or Optoacoustics (OA) imaging is advancing with an increasing number of breakthrough clinical studies (Knieling et al., 2017; Regensburger et al., 2019) and use in biomedical research (Märk et al., 2018; Reber et al., 2018; Wang et al., 2018). Most OA studies presently rely on endogenous hemoglobin as contrast, while some make use of dyes and nanoparticles (Gujrati, Mishra, & Ntziachristos, 2017). In contrast the number of studies employing transgene contrast agents (Brunker, Yao, Laufer, & Bohndiek, 2017; Yao & Wang, 2018) is effectively limited by the number of suitable agents. This shortcoming largely prevents OA imaging from providing insight into life's functioning with imaging volumes and depth beyond intravital microscopy (IVM) and resolution beyond bioluminescence (BLI) or whole animal fluorescence (FLI). The major obstacle in the development of suitable agents stems from tissue absorbance by hemoglobin, lipids or melanin; while it provides rich intrinsic anatomical information to OA imaging it also poses a massive background that any targeted contrast agents needs to overcome.

While approaches exist which rely on transgene encoded pigment synthesis (melanin, Jathoul et al., 2015; violacein Jiang et al., 2015; etc.) or near infrared (NIR) absorbing chromoproteins (Filonov, Krumholz, Xia, Yao, & Verkhusha, 2012), we (Mishra et al., 2020; Stiel et al., 2015; Vetschera et al., 2018) and others (Chee, Li, Zhang, Campbell, & Zemp, 2018; Li et al., 2018; Märk et al., 2018) advocate the use of so called photochromic proteins to overcome the background in the temporal—rather than the spectral-domain. Such transgene contrast agents can be photoswitched by light between an ON- and OFF-state by illumination with light of two distinct wavelengths (Fig. 1A) allowing a modulation that can be used to separate the signal of the labeled cells from the non-modulating background (e.g., blood hemoglobin). While a number of protein classes show photoswitching behavior, proteins of the Bacteriophytochrome (BphPs) family are most promising due to their NIR absorbance derived from a biliverdin chromophore which is readily available in many mammalian cells (Fig. 1B). *Cis/trans* isomerization of the chromophore results in the photophysical behavior, foremost the photochromic spectral change which affords the photoswitching signal (Fig. 1C).

In an OA imaging experiment the sample is illuminated with a **light-schedule** composed of repetitions (**cycles**) of several **pulses** of the

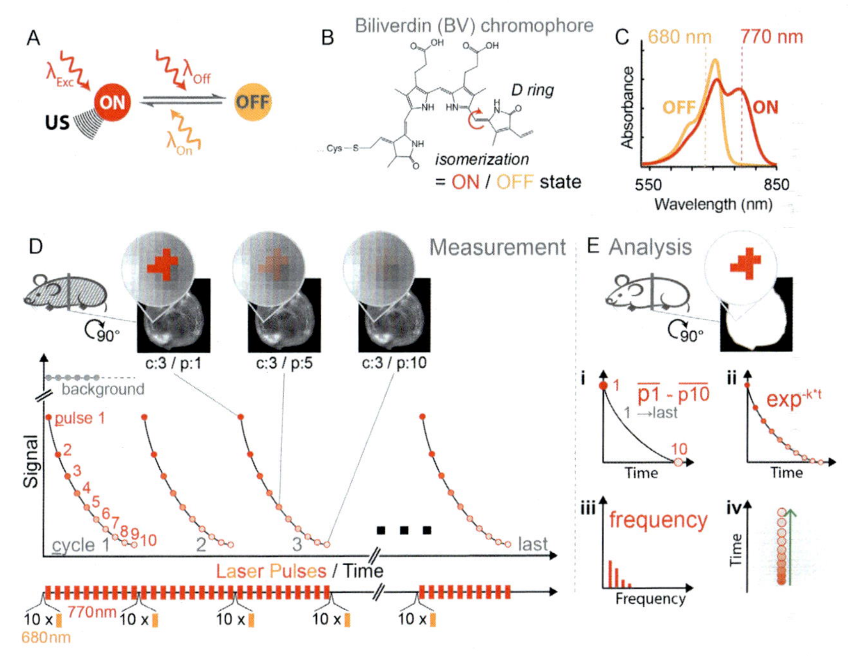

Fig. 1 Principle of photochromic agents and photoswitching OA. (A) Reversible switching between an ON- and OFF-state by light of distinct wavelengths. (B) Biliverdin chromophore of BphPs and its *cis/trans* isomerization (red arrow). (C) Exemplary absorbance spectra of BphP *Re*BphP-PCM (Mishra et al., 2020). (D) Schematics of a photoswitching OA measurement. The sample is illuminated by a pulsed light of a specific schedule (bottom) composed of a number of cycles (c) of OFF-switching (red) and ON-switching (orange) light pulses (p). This results in repetitive exponential decays and recoveries of the agents' signal. For simplification the signal for the OFF-switching is shown only (red circles filled in shades). The top panel shows the vanishing of a mock photoswitching OA signal. (E) The time series generated in d can be analyzed by a number of approaches or combinations thereof, from top-left to bottom-right: (i) differential image between the fully ON- and fully OFF-state, (ii) fitting of the exponential decay, (iii) extraction of the modulation frequency, (iv) VCA. Such analysis can result in a virtually background free map of the agents' distribution.

OFF-switching wavelength, resulting in an exponential decay of the agents' signal (**OFF-switching kinetic**), followed by pulses of the **ON-switching** wavelength to transfer the agent back to the ON-state (Fig. 1D). The wavelengths are best chosen by the absorbance peaks of the ON- and OFF-state of the agent (Table 1) and by minimizing the expected cross-talk between ON- and OFF-switching; additional combinations of wavelengths to achieve discrimination by a combination of the photoswitching and spectral

Table 1 Photoswitching proteins used in OA yet.

	Optical characteristics						Optoacoustics characteristics				
Protein	Photoswitching state	Absorption maxima (nm)	Molar absorption coefficient ($M^{-1} cm^{-1}$)	Emission maxima (nm)	Fluo. QY[a]	Absorption coefficient dynamic range (Pfr/Pr)	OA switching dynamic range	Photoacoustic generation efficiency (PGE)	Switch-off halftime $t_{1/2}^{off}$ (s)[b]	Switch-on halftime $t_{1/2}^{off}$ (s)	Availability/ coding sequence
Rhizobium etli (*Re*BphP-PCM)[c]	Pfr(ON)	708/760	92,287/84,520	None	None	48.0	4.82 ± 0.25[d,e]	1.12 ± 0.02697[d,e]	0.38 ± 0.01[d,e]	0.24[d,e]	Stiel lab, Addgene #156463
	Pr(OFF)	704	11,272	720	0.029						
Deinococcus radiodurans (*Dr*BphP-PCM)[c]	Pfr(ON)	750	87,030	None	None	9.5	2.87 ± 0.01[d,e]	1.074 ± 0.01[d,e]	2.12 ± 0.04[d,e]	0.24[d,e]	Verkhusha lab, Addgene #158986
	Pr(OFF)	700	98,246	720	0.029						
Rhodopseudomonas palustris (*Rp*BphP1-PCM)[c]	Pfr(ON)	758	74,812	None	None	6.7	2.22 ± 0.14[d,e]	0.83 ± 0.065[d,e]	0.82 ± 0.04[d,e]	0.29[d,e]	As described in Mishra et al.
	Pr(OFF)	688	41,197	720	0.018						
Rhodopseudomonas palustris (*Rp*BphP1)[f]	Pfr(ON)	756	78,300	None	None	7.0	n.d.	n.d.	0.44[f]	n.d.	Verkhusha lab, Addgene #79845
	Pr(OFF)	678	87,500								
Agp1[g]	Pfr(ON)	749	n.d.	None	None	n.d.	3.5[g]	n.d.	~0.5[g]	~0.06[g]	Pubmed, EGP56260
	Pr(OFF)	702									

sGPC2[h]	Po	630	n.d.	None	None	n.d.	n.d.	n.d.	n.d.	n.d.	
	Pfr	700	87,200	730	0.3						
sGPC3[h]	Po	630	n.d.	None	None	n.d.	n.d.	n.d.	n.d.	n.d.	
	Pfr	700	89,100	730	0.2						
Dronpa[i]	ON	503	95,000	518	0.85	n.a.	0.57[j]	0.03[j]	19.78[j]	n.d.	Pubmed, BAD72874
	OFF	~390	n.d.		0.67						
Dronpa-M159T[i], Dronpa-2[i]	ON	489	61,732, 56,000	515	0.23, 0.33, 0.28	n.a.	0.89[j]	0.68[j]	0.64[j]	n.d.	
	OFF	~390	n.d.								

[a]For bactriophytochromes $\lambda_{ex} = 680$ nm and Dronpa like protein $\lambda_{ex} = 488$ nm.
[b]After 300 cycles switching sufficiently to the OFF-state and back ON (Mishra et al., 2020), ± represents standard deviation.
[c]Values from (Mishra et al., 2020).
[d]Data shown is from MSOT where 680/770 nm was used for photoswitching.
[e]Pulsed illumination (energy at 770 nm, ~90 mJ/pulse, 7 ns pulse length, 10 Hz repetition rate).
[f]Values from (Li et al., 2018).
[g]Values from (Märk et al., 2018).
[h]Values from (Chee et al., 2018).
[i]Values from (Vetschera et al., 2018).
[j]Multimodal laser spectrometer (MLS) with 1.3 mJ/pulse, 7 ns pulse length, 50 Hz repetition rate was used for photoswitching.

characteristics have been reported (Chee et al., 2018). The result of recording data with a given light-schedule is a **time series of images** encoding the modulation of the agent with a known frequency governed by the light-schedule. In all agents used so far, the OFF-switching kinetic is more informative (slower, red shifted, low residual signal in the fully OFF-state and thus higher dynamic range), however, conceptually, the ON-switching can be used in a similar way for analysis. For an ideal case, concatenation of all OFF-switching cycles of the time series for a given pixel yields a **trajectory** with a characteristic "saw tooth pattern" if a photoswitching agent is localized at this position in the sample (Fig. 1D), while for other positions the trajectory might shows a constant signal or random noise indicating **background**. Importantly, the photophysics of different photoswitching proteins result in **intrinsic time-constants** which allow **multiplexing** of different agents based on the different exponential decays constants (Li et al., 2018; Mishra et al., 2020; Stiel et al., 2015).

To achieve the **classification** between agents' signal and background the time series of images can be analyzed in various ways (Fig. 1E):

(i). **Differential imaging** is performed by pixelwise subtraction of the ON- and OFF-images of every cycle or of the mean cycle (Yao et al., 2016) or variants of this approach (Chee et al., 2018; Dortay et al., 2016; Märk et al., 2018). Regularly the first and last pulse of a cycle are used. The difference we call the **photoswitching amplitude**.

(ii). **Frequency analysis** is performed by pixel-wise transformation of the trajectories into frequency space (Fourier transform) and extraction of the amplitude at the fundamental photoswitching frequency and its harmonics governed by the light-schedule (Li et al., 2018).

(iii). Pixel-wise **exponential fitting** of trajectories of mean cycles yields the quality of fit (r-square) and the decay constant which can be matched against the expected decay constant for the photoswitching agent

(iv). **Vertex component analysis** considers the OA images for each pulse as separate data planes, which are geometrically projected iteratively through the whole time series. This results in canceling stable or randomly changing background, while pixels with photoswitching patterns are extracted. The extracted patterns are effectively de-noised trajectories per pixel (Stiel et al., 2015).

Differential imaging is the simplest approach, it considers only the change of signal and uses the repetitiveness only to decrease noise. Frequency analysis considers the repetitiveness of the signal and also allows separation against other periodic changes (respiratory movement or heart beat) due to the known frequency of photoswitching governed by the light-schedule. However, frequency analysis only considers repetitiveness of the pattern and not its form; this can be established by exponential fitting which matches the signal changes against the expected exponential decay form governed by the photophysics of the agent and light-fluence. Since each method exploits a different aspect of the photoswitching for discrimination it is apparent that their **combination** yields the most reliable classification. For such combinations the classification of pixels as background or photoswitching agent can be performed either manually by applying a set of thresholds derived for each method, or automatically by pre-trained **machine learning (ML)-based models** (Mishra et al., 2020).

In the following we will focus on the analysis using combination-based approaches while also trying to highlight aspects relevant for the analysis using individual methods. We will structure our text in Sections 2–7.

Overall, all methods and approaches presented herein are under constant development and we try to share our experiences and give some tentative initial suggestions towards an effective analysis.

2. General experimental considerations

Due to their NIR absorbance BphPs are most suited for photoswitching OA imaging in vivo (Vetschera et al., 2018) (Table 1). While most proteins fundamentally should also work in a mammalian setting this has so far only been shown for BphP1 (Yao et al., 2016), *Re*BphP-PCM (Mishra et al., 2020), Agp1 (Märk et al., 2018) and *Dr*BphP-PCM (Li et al., 2018; Mishra et al., 2020). Of those proteins *Re*BphP-PCM show the best ratio of switching dynamic range, signal and resistance to photo-fatigue. In mammalian cells *Re*BphP-PCM and *Dr*BphP-PCM (Mishra et al., 2020) as well as *Dr*BphP-PCM and BphP1 have been already successfully used as pairs for multiplexed photoswitching OA (Li et al., 2018).

We have implemented the concept on data acquired with the commercial iThera Medical InVision MSOT 256-TF, however, the general approach is readily applicable to other OA tomography devices. Beyond that, also OA scanning concepts can be used—most easily for the differential imaging method (Dortay et al., 2016).

Key to each experiment is a light-schedule. The schedule is comprised of cycles of ON and OFF-switching pulses. The number of pulses needs to be chosen in a way to adequately capture the photoswitching characteristics of the agent, i.e. establish the existence of a trend in the signal. Thus, the number of pulses does not need to yield complete switching of the agent, we rather found that more repetitions of partial switching lead to better discrimination than less repetitions of full switching (i.e. **cycles over pulses**) (Mishra et al., 2020). Since the photoswitching is light-fluence dependent the exact numbers are system/laser dependent (Table 1) and are best established based on simple phantom experiments (e.g., the agent in a straw surrounded by intra-lipid and blood agar; Beziere et al., 2015). The necessary number of pulses of OFF-light depends on the signal-to-noise ratio and a good starting is $t_{1/2}^{off}$, but normally it should not be less than 6 pulses to establish sufficient data points for analysis. A sufficient number of pulses of ON-light must be chose to ensure complete back-switching to the ON-state. The number of cycles cannot be easily determined and is best adjusted based on a number of trial experiments. While more cycles are generally favorable, this extends the dwell time and can result in photo-fatigue of the agent. As a rule of thumb, depending on the strength of the expected signals, 20–100 cycles can be expected to yield a good discriminative effect. Up to now we saw no increase in classification accuracy above 100 cycles which is likely due to non-random noise. For a protein like *Re*BphP-PCM this results in dwell-times between ~40 and 200 s per slice (10 Hz laser repetition rate). Most commercial systems do not have dedicated settings for such a light-schedule, yet. For our instrument the use of "shots", normally used for averaging, and "repetitions", normally used for time-trace measurements can be used to achieve a light-schedule with pulses and cycles, respectively.

The recorded data (pressure waves recorded by ultrasound detectors (transducers)) is then reconstructed to form the time series of images. In our example the commercial ViewMSOT software is used (iThera Medical). While both, backprojection and model-based reconstructed data work for subsequent analysis, we found that it can happen that in model-based the photoswitching dynamics are distorted due to regularization. Hence, we suggest to compare both methods or reconstruct the data for analysis by simple backprojection and keep model-based reconstruction for anatomy pending further developments in dedicated regularization approaches which better preserve the signal changes introduced by photoswitching. Averaging before reconstruction is not used since this confounds the transient change of the photoswitching signal. The output of reconstruction is a time series of images used as input to our analysis. We performed all subsequent analysis using Matlab. Data conversion from the proprietary format of the instrument is done using a library provided by the manufacturer (iThera Medical) that allows to load the reconstructed data as arrays. We perform all analysis on the OFF-switching trajectory (770 nm) but depending on the agents also other photoswitching trajectories can be used (e.g., 680 nm on-switching). Many aspects of the analysis are implemented in our analysis scripts and we occasionally refer to the use of those scripts in `Courier font` (see code availability statement below); they include analysis as well as pre-trained basic ML models together with scripts to build and train new models. In general, the analysis can be performed on any time series of images capturing the switching processes but, from our side, is only tested on the instrumentation mentioned above.

3. Data preprocessing

We routinely use the following preprocessing steps that pertain to the overall quality of the dataset:

(i) To compensate for OA intensity fluctuations in-between pulses, we **normalize** all data per-pulse to the mean intensity of an area that can be considered a homogenous invariant background outside the specimen. However, the area must be chosen with care, e.g., not too close to the detectors since those regions are generally prone to artifacts. If no suitable area can be found we refrain from normalization.

(ii) For **fluence correction** we presently employ a simple exponential model calibrated on a phantom (2% (w/v) agar, 3% (v/v) intralipid, 5% (v/v) blood) and using the manually defined outer boundary of the specimen as starting point. This is far from being accurate but served us well in homogenizing the switching amplitudes across the specimen.

(iii) For **motion correction** we use phase correlation registration together with affine and rigid (Che, Mathai, & Galeotti, 2017) or non-rigid (Thirion, 1998) diffusion-based approaches as standard preprocessing (both in `get_movement_corrected_recon`). We suggest the use of rigid approaches since they are based on line-edges detections rather than intensities profiles and as such preserve the intensity changes induced by photoswitching while compensating for breathing-motion expansions. This is especially relevant for small areas of photoswitching where non-rigid approaches show distortion (less than 5×5 pixels regions for 300×300 pixels image).

(iv) Switching trajectories per pixel are reduced, by **averaging** over all cycles as a simple method of de-noising (Fig. 2A). Hereafter called **mean cycle**.

(v) Mean cycles are **re-scaled** per pixel to allow a unified treatment of signals of different photoswitching amplitudes (Fig. 2B) and improve exclusion of false positives in areas with high OA signals in general (Fig. 3, arrows iv and vi). This step also helps to some extent to overcome illumination dependent inhomogeneities and allows an easier adjustment of uniform thresholds for all aspects of the dataset.

(vi) For the analysis it can help if the data is **trimmed** to only the number of pulses in each cycle which carry photoswitching information. This is especially relevant if no prior optimization of the number of pulses on the level of the light-schedule has been made. The stretch to be trimmed (**noise tail**) is established based on a number of pixels, identified in a first (un-trimmed) analysis, that shows a comparably reliable signal. These trajectories are split in two segments of varying number of pulses and both parts are fit. The maximal length of the second segment that shows no exponential change is considered the noise tail (Fig. 2C). In the next step this noise tail is trimmed for all data and the data is re-analyzed. In case of several agents and also mixtures a second strategy is to trim individually a noise tail for every pixel and then resample all data to the same cycle length (Fig. 2D).

4. "Manual" data analysis

Several methods of data analysis exist and have been explored by different groups (see above), we believe that it is generally advisable to use as much information from the photoswitching trajectory for discriminative effect as possible. Thus, we pixel-wise analyze several **characteristics** of the data either based on all cycles of one wavelength concatenated or the mean cycle. The respective thresholds for those characteristics determine if a given position in the sample is classified as containing photoswitching agent. Some characteristics directly describe the photoswitching phenomenon:

(i) The **mean cycle amplitude** (`get_basic_analysis_values`), which is the change of signal introduced by the photoswitching. i.e. a differential image, de-noised by averaging over cycles.

(ii) The **amplitude in the frequency** spectrum (Fourier Transform) at the modulation frequency and its harmonics as given by the light-schedule or the adjusted switching cycle length after pre-processing steps

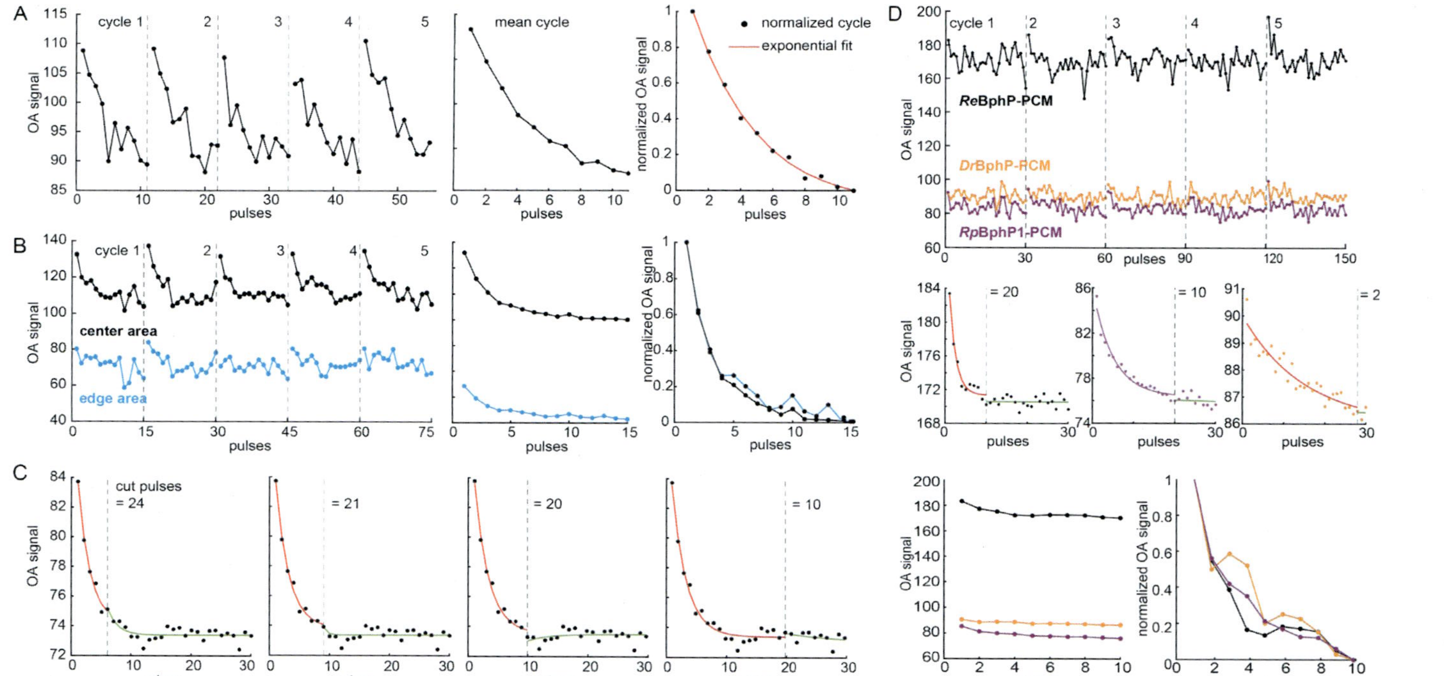

Fig. 2 Example data and processing steps. (A) Data from 500 *Re*BphP-PCM expressing Jurkat cells/μL in a blood-phantom (~4 cells per $200 \times 200 \times 200$ μm imaging volume) shown are 5 raw cycles, the corresponding mean cycle (of 50 raw cycles) and the exponential fit of the normalized mean cycle. (B) Data from *Re*BphP-PCM expressing 4T1 cells growing as *sub cutaneous* tumor; data shown at two different positions in the center and at the edge of the area that shows photoswitching signal together with the mean cycles (of 50 raw cycles) and the normalized mean cycles. (C) Data from 1.4×10^5 *Re*BphP-PCM expressing *E. coli*/μL *sub cutaneous* implanted, shown with different extend of noise trimming on the mean cycles. For 24 and 21 pulses the trimmed part of the cycle (green) shows still an exponential behavior indicating too extensive trimming, while 10 is not cropped extensive enough incorporating noise to the data for further processing (red), 20 trimmed pulses are optimal for this data. (D) Processing of data from three different photoswitching agents (Jurkat T-cells and *E. coli* expressing *Re*BphP-PCM, *Dr*BphP-PCM and *Rp*BphP1-PCM, respectively, implanted into a 4T1 tumor). The different kinetics are requiring different trimmings of the mean cycles (of 50 raw cycles, middle panel). Subsequently the data is resampled to allow processing with ML approaches trained only on a single photoswitching protein (in this case the fastest). Note that the assessment of the different kinetics for classification is performed on the data not resampled.

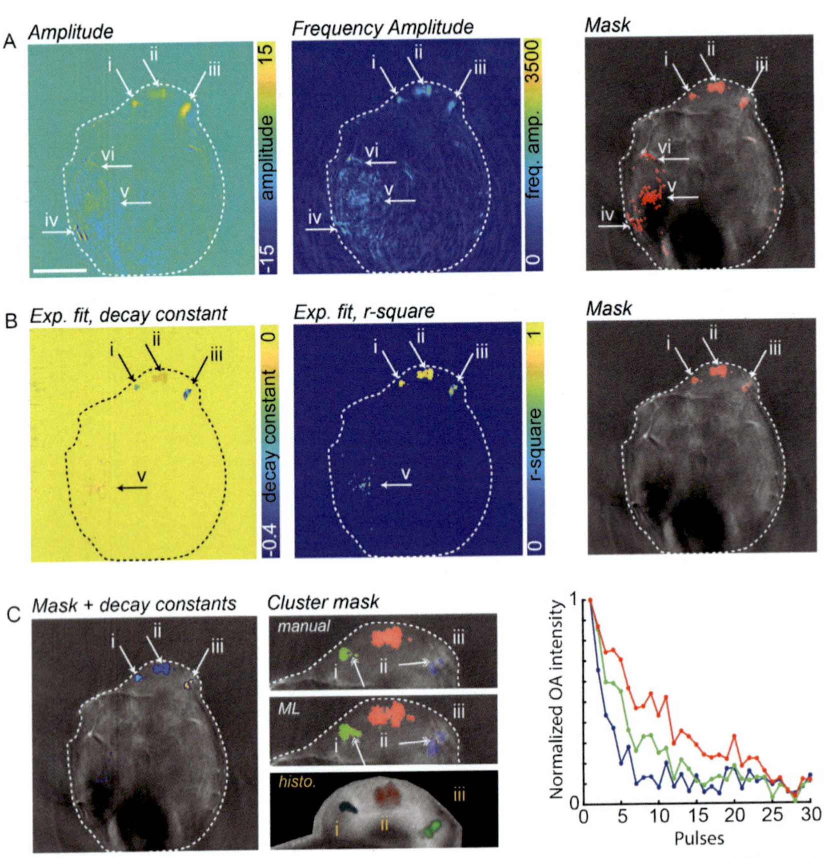

Fig. 3 Examples of classification masks. (A) Basic analysis values showing differential image of mean cycle (photoswitching amplitude, left) and the amplitude of frequency (middle) specified by the light-schedule, together with the binary mask obtained by applying thresholds for photoswitching amplitudes and frequency amplitudes overlaying an anatomy image (right). Arrows (i–iii) denote photoswitching agents with different decay constants (*Rp*BphP1-PCM, *Dr*BphP-PCM and *Re*BphP-PCM). Arrows (iv–vi) show different types of noise. (B) Exponential fitting, based on mask from (A) showing exponential decay constant (left) fitting quality, r-square (middle) together with the binary mask obtained by applying thresholds for both exponential metrics (right). (C) Multiplexing by multiplication of binary mask from (B) and exponential decay constants from (B) (left). Result of clustering of decay constants by different methods together with histology (bottom). Note that the areas identified are different depending on the clustering approach (double arrow). Colors in histology do not match cluster colors but represent Cy5 (cyan), mCherry (red) and GFP (green) channel. Mean trajectories of the individual clusters (right). Scale bar is 5 mm.

(get_basic_analysis_values, e.g., for 11 OFF-switching pulses per cycle and a 10 Hz system, 1.1 Hz). Note however, false-positives often show an unrealistically high amplitude in the frequency spectrum ($\sim 50 \times$ mean amplitude over all pixels) and are best excluded by a maximum threshold, considering expected amplitudes' range ($\sim 3 \times$ mean amplitude over all pixels).

(iii) The **exponential decay constant and the quality of fit (r-square)** of the mean cycle (get_exp_fit). The decay constant can be matched against prior phantom experiments to yield an expected range (for initial values with our InVision MSOT 256-TF system refer to Table 1). Since exponential fitting considers the characteristic form of the photoswitching trajectories it allows to further reduce false positives (Fig. 3 arrow v). For the shown cases we use 0.9 as a threshold for the r-square, however, it depends on the expected signal strength; but not less than 0.65 is advisable.

Other characteristics describe the data as a whole or in respect to the specimen:

(iv) The **mean intensity** and **overall mean amplitude** (get_basic_analysis_values) over all cycles per pixel (concatenated full time trajectories) is a good indicator for false positives, since very strong intensities often indicate confounding blood signals rather than true photoswitching signals ($> \sim 100 \times$ mean image intensity), while too low signals indicate varying noise (in the range of the mean background intensity). Those two characteristics allow to estimate the relation between OA intensities and noise for the specific pixel.

(v) **Statistic values, describing the behavior of trajectories** on a pixel-wise basis (get_cycles_analysis_values). I.e., mean and standard deviation of the minima and maxima over all individual cycles, the length of the useful cycle part (without noise tail, see Section 3 (vi)) and the number of cycles in the whole trajectories, showing the trend within the trimmed part. We found that these metrics describes the "quality" of the full switching trajectory very well and can be substituted for a consideration of individual cycles.

(vi) **Distance to the sample's surface** (manual ROI definition, get_sample_surface). This metric allows to put the photoswitching kinetics and observed amplitudes into perspective to the expected light-fluence since the kinetics are fluence dependent.

While some characteristics like the amplitude of modulation frequency and exponential decay constant can be used as clear identification of photoswitching-alike behavior, further characteristics, such as fitting quality, mean intensities etc., can be used for discrimination of false positives by adjustment of suitable thresholds. Further, usability of those characteristics for each concrete dataset as well as characteristics' dimension reduction can be analyzed and performed by principal component analysis (PCA), if ground-truth assignment is available, and can be subsequently used for the analysis of similar scans. The whole set of characteristics can be pre-calculated for each new scan and then their sub-set or combinations will be used according to the analysis model in use, allowing characteristics' dimension reduction for more efficient data usage in training of ML-models.

5. Data analysis using ML

While we found that the extensive set of parameters clearly improves classification accuracy they are challenging to adjust manually, thus, we use simple models to classify pixels as photoswitching agent or background. Presently, we used two supervised learning, feature-based ML classification models (**feature** $\sim$ **characteristic**), that both are suitable for operating in high-dimensional feature spaces:

1. Support vector machine (SVM) builds a hyperplane in feature space to divide values into classes (Burges, 1998). SVM provides a very strict and comprehensive separation of signal from background scheme, that is suitable in cases where the analyzed data is very similar in nature to the data used for the training dataset. We found it not applicable for more diverse cases since it has limitations in flexibility and generalization quality.
2. Random Forrest allows to build more generalized classification schemes, by combining decision trees' ability of forming complex hierarchies of classification rules with techniques to prevent overfitting (Biau & Scornet, 2016). This is also suitable for processing data that differs in signal strength and position of photoswitching agents from the ones used in the training datasets.

To generate the necessary **training data** representative datasets need to be created by establishing highly trustworthy ground truth based on histology. We use co-expression of conventional fluorescent proteins (e.g., via an IRES or P2A element). Presently we use manual expert selection on histology images that are superimposed with OA imaging data by diffusion-based non-rigid methods (Ashburner & Friston, 1997) (`get_registered_img`). Since it is challenging to precisely attribute the borders of the fluorescence signals in histology and to not taint the training dataset we exclude 3 pixels of border area. The OA data used for training should be clear data (r-square of exponential fits over ~0.7), since in our experience training with noisy data does not increase the accuracy of classification in datasets with similar levels of noise. Moreover, the ratio between training data showing photoswitching and data showing background needs to be balanced by random selection of background points. Along with accepted approaches we always divide training/validation datasets randomly as 70%/30%. Finally, we use PCA for dimension reduction. PCA forms a combination of features (principal components) to use instead of full number of features and excludes features not included in the principal component or have very low weights in them. However, it is essential to always confirm that the ability to fully describe the data by the PCA derived components is not lost. Thus, every time the training dataset is changed the PCA needs to be re-run as the number of components may change as well as the components themselves. For example, in case of a dataset extension, more variety in data may require more components. Regarding the number of necessary scans for a training dataset we can only comment based on our present experience, that addition of alike scans raising the number of individual photoswitching trajectories over 3000 does not lead to further model performance improvement. Along with our scripts we also provide a model which was pretrained on different scans of agent expressing 4T1 tumors and *subcutaneous* bacteria implants on a ViewMSOT 256-TF device with a 10 Hz laser. We found this model to explain a number of similar sample types but also agent expressing HCT116 tumors broadly distributed throughout the mouse body as well as cell implants with concentrations well below the training-data. It can be expected to at least provide initial results for similar instrumentation.

On the long run it will be interesting to implement ML methods operating on the image level like convolutional neural networks (CNN) instead of a pixel-wise treatment.

6. Multiplexing

Similar to different spectral characteristics in spectral-unmixing also for photoswitching OA distinct photophysical characteristics allow multiplexing of a number of agents. Different photoswitching agents have different intrinsic photoswitching likelihoods (i.e. the quantum yield of photoswitching) which results in effectively different exponential decay constants at similar light-fluency. This can be exploited for separation by fitting the exponential decays. We so far have explored two approaches:

(i) We first identify all regions, showing photoswitching behavior by applying a **binary classification** based either on manual thresholding or ML-based binary classification model (Section 5). To apply the model uniformly to different switching trajectories, pixel-wise trimming and resampling has to be performed as described in Section 3(vi) (Fig. 2D). After establishing the binary mask, we cluster the exponential decay constants fitted on original mean cycles (not resampled) of regions classified as photoswitching either using previously estimated cluster centers (expected decay constant values of every protein, e.g., from phantoms (`get_clusters_by_values`), or number of clusters (`get_clusters_by_kmeans`). The latter is especially relevant if mixtures of different agents are expected. Those mixtures will result in mean decay constants depending on the quantity of the respective agent. However, since in this approach we classify only by exponential decay constants, it can lead to classification errors for pixels on the edges of switching areas, which are more affected by noise. High levels of noise compared to the photoswitching amplitude somewhat flatten the fitted curve, so that those areas can be misclassified as agents with a different decay constant. Those regions can be post-processed by checking the r-square values of exponential fit and area homogeneity. We found binary classification using ML, together with unified treatment of trajectories of different length by trimming and resampling allows this method to yield better performance than manual unmixing, that is especially relevant for agents with short $t_{1/2}{}^{off}$ (Fig. 3C, compare middle panel top and middle).

(ii) We directly include the number of different agents as **multiple classes** in the **classification model** (and balanced training data based on histology ground truth, e.g., by co-expressing different fluorescent proteins for the different agents). Presently we only tried this on a limited number of datasets (always only two different agents) but found it to be more faithful especially in representation of the edges of areas with photoswitching agents

As described above two doublets of agents expressed in mammalian cells have been use in multiplexing, yet. From those we infer that a difference of ~5 in the $t_{1/2}^{off}$s appear to be ideal.

7. Pitfalls and troubleshooting

(i) Often false-positives pixels directly neighboring a region of true-positives can be eliminated by observing the mean intensities together with the amplitude of the signals. In such false-positive regions one can often observe a drastic decrease in intensities and amplitudes but no apparent change in other characteristics (exponential fit and frequency component). The same is true for false-positives in neighboring slices due to the often poor out-of-plane resolution.

(ii) If jumps in intensity are encountered (checking by assessing the mean signal over the complete imaging dataset) we suggest to exclude the whole respective cycle. If only a single pulse is affected the data can be omitted and replaced by an interpolation.

(iii) For very low signal conditions or high noise scenarios we found that often only a number of cycles are responsible for poor data quality. This can be improved by deleting randomly selected individual cycles and check for improvement in quality measures like r-square of the exponential fits of the mean without the omitted cycle (`select_good_cycles`).

(iv) Similar to other unmixing tasks in OA tomography we suggest to work with small step-size between slices and corroborate positively identified areas in one slice by their neighbors, since either the area of agent expands over several slices or, due to the generally limited out-of-plane resolution the agents signal is also visible in neighboring slices.

Acknowledgments

The authors thank Antonia Longo for discussions on the manuscript.

Funding

K.M. and A.C.S. received funding from DFG STI 656/1-1.
Code availability: https://gitlab.lrz.de/ga45huk/rsoap_analysis

References

Ashburner, J., & Friston, K. (1997). Multimodal image Coregistration and partitioning—A unified framework. *NeuroImage*, *6*(3), 209–217. https://doi.org/10.1006/nimg.1997.0290.

Beziere, N., Lozano, N., Nunes, A., Salichs, J., Queiros, D., Kostarelos, K., et al. (2015). Dynamic imaging of PEGylated indocyanine green (ICG) liposomes within the tumor microenvironment using multi-spectral optoacoustic tomography (MSOT). *Biomaterials*, *37*, 415–424. https://doi.org/10.1016/j.biomaterials.2014.10.014.

Biau, G., & Scornet, E. (2016). A random forest guided tour. *TEST*, *25*(2), 197–227. https://doi.org/10.1007/s11749-016-0481-7.

Brunker, J., Yao, J., Laufer, J., & Bohndiek, S. E. (2017). Photoacoustic imaging using genetically encoded reporters: A review. *Journal of Biomedical Optics*, *22*(7), 70901. https://doi.org/10.1117/1.JBO.22.7.070901.

Burges, C. J. C. (1998). A tutorial on support vector machines for pattern recognition. *Data Mining and Knowledge Discovery*, *2*(2), 121–167.

Che, C., Mathai, T. S., & Galeotti, J. (2017). Ultrasound registration: A review. *Methods, 115*, 128–143. https://doi.org/10.1016/j.ymeth.2016.12.006.

Chee, R. K. W., Li, Y., Zhang, W., Campbell, R. E., & Zemp, R. J. (2018). In vivo photoacoustic difference-spectra imaging of bacteria using photoswitchable chromoproteins. *Journal of Biomedical Optics, 23*(10), 1. https://doi.org/10.1117/1.JBO.23.10.106006.

Dortay, H., Märk, J., Wagener, A., Zhang, E., Grötzinger, C., Hildebrandt, P., et al. (2016). Dual-wavelength photoacoustic imaging of a photoswitchable reporter protein. In *Proc. SPIE 9708. Photons plus ultrasound: Imaging and sensing 2016, 970820*. https://doi.org/10.1117/12.2208259.

Filonov, G. S., Krumholz, A., Xia, J., Yao, J., Wang, L. V., & Verkhusha, V. V. (2012). Deep-tissue photoacoustic tomography of a genetically encoded near-infrared fluorescent probe. *Angewandte Chemie (International Edition in English), 51*(6), 1448–1451. https://doi.org/10.1002/anie.201107026.

Gujrati, V., Mishra, A., & Ntziachristos, V. (2017). Molecular imaging probes for multispectral optoacoustic tomography. *Chemical Communications, 53*(34), 4653–4672. https://doi.org/10.1039/C6CC09421J.

Jathoul, A. P., Laufer, J., Ogunlade, O., Treeby, B., Cox, B., Zhang, E., et al. (2015). Deep in vivo photoacoustic imaging of mammalian tissues using a tyrosinase-based genetic reporter. *Nature Photonics, 9*(4), 239–246. https://doi.org/10.1038/nphoton.2015.22.

Jiang, Y., Sigmund, F., Reber, J., Luís Deán-Ben, X., Glasl, S., Kneipp, M., et al. (2015). Violacein as a genetically-controlled, enzymatically amplified and photobleaching-resistant chromophore for optoacoustic bacterial imaging. *Scientific Reports, 5*, 11048. https://doi.org/10.1038/srep11048.

Knieling, F., Neufert, C., Hartmann, A., Claussen, J., Urich, A., Egger, C., et al. (2017). Multispectral optoacoustic tomography for assessment of Crohn's disease activity. *The New England Journal of Medicine, 376*(13), 1292–1294. https://doi.org/10.1056/NEJMc1612455.

Li, L., Shemetov, A. A., Baloban, M., Hu, P., Zhu, L., Shcherbakova, D. M., et al. (2018). Small near-infrared photochromic protein for photoacoustic multi-contrast imaging and detection of protein interactions in vivo. *Nature Communications, 9*(1), 2734. https://doi.org/10.1038/s41467-018-05231-3.

Märk, J., Dortay, H., Wagener, A., Zhang, E., Buchmann, J., Grötzinger, C., et al. (2018). Dual-wavelength 3D photoacoustic imaging of mammalian cells using a photoswitchable phytochrome reporter protein. *Communications on Physics, 1*(1), 3. https://doi.org/10.1038/s42005-017-0003-2.

Mishra, K., Stankevych, M., Fuenzalida Werner, J. P., Grassmann, S., Gujrati, V., Klemm, U., et al. (2020). Multiplexed whole animal imaging with reversibly switchable optoacoustic proteins. *Science Advances, 6*, eaaz6293. https://doi.org/10.1101/2020.02.01.930222.

Reber, J., Willershäuser, M., Karlas, A., Paul-Yuan, K., Diot, G., Franz, D., et al. (2018). Non-invasive measurement of brown fat metabolism based on optoacoustic imaging of hemoglobin gradients. *Cell Metabolism, 27*(3), 689–701 (e4) https://doi.org/10.1016/j.cmet.2018.02.002.

Regensburger, A. P., Fonteyne, L. M., Jüngert, J., Wagner, A. L., Gerhalter, T., Nagel, A. M., et al. (2019). Detection of collagens by multispectral optoacoustic tomography as an imaging biomarker for Duchenne muscular dystrophy. *Nature Medicine, 25*(12), 1905–1915. https://doi.org/10.1038/s41591-019-0669-y.

Stiel, A. C., Deán-Ben, X. L., Jiang, Y., Ntziachristos, V., Razansky, D., & Westmeyer, G. G. (2015). High-contrast imaging of reversibly switchable fluorescent proteins via temporally unmixed multispectral optoacoustic tomography. *Optics Letters, 40*(3), 367. https://doi.org/10.1364/ol.40.000367.

Thirion, J.-P. (1998). Image matching as a diffusion process: An analogy with Maxwell's demons. *Medical Image Analysis*, *2*(3), 243–260. https://doi.org/10.1016/S1361-8415(98)80022-4.
Vetschera, P., Mishra, K., Fuenzalida-Werner, J. P., Chmyrov, A., Ntziachristos, V., & Stiel, A. C. (2018). Characterization of reversibly switchable fluorescent proteins in optoacoustic imaging. *Analytical Chemistry*, *90*(17), 10527–10535. https://doi.org/10.1021/acs.analchem.8b02599.
Wang, J., Jeevarathinam, A. S., Humphries, K., Jhunjhunwala, A., Chen, F., Hariri, A., et al. (2018). A mechanistic investigation of methylene blue and heparin interactions and their photoacoustic enhancement. *Bioconjugate Chemistry*, *29*(11), 3768–3775. https://doi.org/10.1021/acs.bioconjchem.8b00639.
Yao, J., Kaberniuk, A. A., Li, L., Shcherbakova, D. M., Zhang, R., Wang, L., et al. (2016). Multiscale photoacoustic tomography using reversibly switchable bacterial phytochrome as a near-infrared photochromic probe. *Nature Methods*, *13*(1), 67–73. https://doi.org/10.1038/nmeth.3656.
Yao, J., & Wang, L. V. (2018). Recent progress in photoacoustic molecular imaging. In *Current opinion in chemical biology* Elsevier Ltd. https://doi.org/10.1016/j.cbpa.2018.03.016.

Oxygen-embedded quinoidal acene based semiconducting chromophore nanoprobe for amplified photoacoustic imaging

Baoli Yin, Yanpei Wang, Zhifei Ye, Shuangyan Huan*, and Guosheng Song*
State Key Laboratory of Chemo/Biosensing and Chemometrics, College of Chemistry and Chemical Engineering, Hunan University, Changsha, PR China
*Corresponding authors: e-mail address: syhuan@hnu.edu.cn; songgs@hnu.edu.cn

Contents

Methods in Enzymology, Volume 657
ISSN 0076-6879
https://doi.org/10.1016/bs.mie.2021.06.034

Abstract

In this chapter, we summarize the advantages of photoacoustic imaging and the current methods of enhancing photoacoustic. We then provide detailed procedures for the synthesis and characterization of a photoacoustic imaging molecule, Nano(O-Nonacene)-PEG, developed in our research group. At the same time, we proved that the incorporation of $Zn_{0.4}Fe_{2.6}O_4$ can enhance the photoacoustic imaging effect of Nano(O-Nonacene)-PEG. This provides a new material for photoacoustic imaging to guide tumor treatment.

1. Introduction

The precise treatment of tumors is inseparable from imaging the tumors (Liu, Bu, & Shi, 2017). But existing imaging methods hold weak points in some aspects. Fluorescence (FL) or bioluminescence has high sensitivity, but the depth of tissue penetration is poor (Huang & Pu, 2020; Li, Li, et al., 2020; Owens, Henary, El Fakhri, & Choi, 2016). Magnetic resonance imaging (MRI) or computed tomography (CT) has high resolution but poor sensitivity (Debroye & Parac-Vogt, 2014; Ni, Bu, Ehlerding, Cai, & Shi, 2017; Yan et al., 2019). Radionuclide imaging, although the sensitivity is very high, the spatial resolution is poor and will produce ionizing radiation (Boros & Packard, 2019; Pellico, Gawne, & de Rosales, 2021). However, for photoacoustic imaging, there are the following advantages: (1) Photoacoustic imaging is a new type of biological imaging for diagnosis that is noninvasive resulting in less damage to biological tissues (Knox & Chan, 2018; Zha et al., 2020); (2) Because the excitation light is usually a near-infrared light source, it can reduce the scattering of biological tissues, making it possible to reach a deeper layer of biological tissues, thus having a deeper tissue penetration depth (Boros & Packard, 2019; Chitgupi et al., 2019; Li, Li, et al., 2020; Weber, Beard, & Bohndiek, 2016); (3) Compared with pure optical imaging, it has higher contrast, and compared with pure acoustic imaging, it has higher spatial resolution (Fu, Zhu, Song, Yang, & Chen, 2019; Jiang, Du, Tang, Hsieh, & Zheng, 2019; Li, Li, et al., 2020; Li, Park, et al., 2020); (4) Compared with other imaging instruments, the photoacoustic imaging system has the advantages of miniaturization and integration, simple operation, low cost, and better promotion and popularity (Ong, Zhang, Dong, & Yao, 2021; Zha et al., 2020).

The photoacoustic phenomenon was first discovered by Alexander G. Bell in 1880 (Nie & Chen, 2014; Wang & Yao, 2016). The PA effect is based on the fact that acoustic wave generation can be induced by optical excitation (Zeng, Ma, Lin, & Huang, 2018). First, a beam of near-infrared pulsed light is emitted through a laser light source. After the contrast agent receives this pulsed laser, the excited electrons in the material will undergo a transition that converting the light energy into the form of heat. This photothermal effect will cause the expansion of the volume of biological tissues, which will further generate ultrasonic waves, and the image of biological tissues can be obtained through proper decoding of these ultrasonic waves (Men et al., 2020; Wang et al., 2021; Wilson, Homan, & Emelianov, 2012). Generally, it is necessary to select a suitable excitation light source according to the contrast agent, so as to obtain the maximum light-to-heat conversion efficiency. In organisms, melanin (Fan et al., 2014), hemoglobin, (Haedicke et al., 2020; Wang et al., 2003; Weber et al., 2016) etc., also produce photoacoustic signals. These are usually collectively referred to as endogenous contrast agents. However, these agents are distributed all over the body, so if a certain part of the body is targeted for imaging, thus some exogenous contrast agents are needed to reduce background interference. To overcome this problem, many exogenous contrast agents were proposed including gold nanoparticles (Pu et al., 2014; Zhang et al., 2020), iron nanoparticles, (Shi et al., 2021) and other inorganic nanomaterials (Kenry, Duan, & Liu, 2018); organic dyes, (Zha et al., 2020) semiconductor nanoparticles (Zhang et al., 2021), and other organic nanomaterials (Li, Deng, et al., 2020). In addition, there are other ways to reduce background interference and enhance photoacoustic imaging. Since light penetrates the tissue, currently many contrast agents were designed to absorb NIR to avoid the strong light scattering phenomenon and enhance photoacoustic imaging effects. Apart from this, some organic contrast agents doped into inorganic metal materials. In this way, the thermal expansion process can be enhanced through the heat conduction of the metal resulting in the enhancement of photoacoustic imaging effects.

In this chapter, we have summarized the existing tumor treatment methods, the principles of photoacoustic imaging, contrast agents, and applications of photoacoustic imaging were discussed. On the one hand, we have developed an efficient way to construct oxygen-embedded nonacene with regular zigzag edges (O-Nonacene) (Wang et al., 2019). It is worth noting that compared with many other O-annulated polycyclic aromatic hydrocarbons, O-embedded quinoidal nonacene exhibit excellent stability, strong intermolecular interaction, greatly expanded π-conjugation, narrow HOMO-LUMO gap, large red shift absorption and a quenching photoluminescence due to the partial interruption of electronic communication. Besides, we also show a kind of photoacoustic imaging to guide tumor PTT treatment by using molecule Nano(O-Nonacene)-PEG. On the other hand, due to the doping of metal particles, the property of heat conduction and dissipation can be enhanced, and as a result, photoacoustic imaging effects can ultimately be enhanced. Thus, we incorporated $Zn_{0.4}Fe_{2.6}O_4$ into Nano(O-Nonacene)-PEG to obtain the effect of enhancing the photoacoustic signal through the thermal conduction of metal.

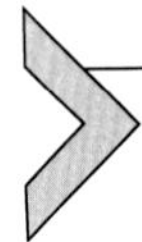

2. Design of Nano(O-Nonacene)-PEG and Nano(O-Nonacene@$Zn_{0.4}Fe_{2.6}O_4$)-PEG

Nano(O-Nonacene)-PEG is composed of O-Nonacene, $PEG\text{-}NH_2$ (5k), and Poly(styrene-*co*-maleic anhydride) (PSMA). Among them, O-Nonacene has a strong absorption peak in NIR-I. PSMA is a surfactant that can encapsulate O-Nonacene to enhance the water solubility of O-Nonacene. After PSMA modifies O-Nonacene to form nanoparticles, the surface leaks COOH, namely Nano(O-Nonacene)-COOH, which shows negative charges. Further modification of $PEG\text{-}NH_2$ (5k), namely Nano(O-Nonacene)-PEG by amino-carboxyl coupling, can not only enhance biocompatibility but also enhance tumor targeting ability. After that, $Zn_{0.4}Fe_{2.6}O_4$ is wrapped in Nano(O-Nonacene)-PEG, and the photoacoustic property can be further enhanced by enhancing heat conduction.

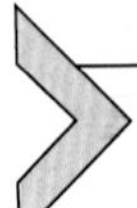

3. Synthesis of Nano(O-Nonacene)-PEG

3.1 Synthesis of O-Nonacene (Fig. 1)

1. Preheat an oil bath.
2. 18.2 mmol of hydroquinone, 18.2 mmol of 2,6-dichlorobenzaldehyde, and 50 mL of acetic acid were added into 100 mL of xylene.
3. The mixture was purged by nitrogen for 10 min at ambient temperature.
4. 0.2 mL of concentrated sulfuric acid was slowly added.
5. The reaction mixture was heated up to 140 °C and continued to heat for 30 h under the nitrogen atmosphere.
6. After cooling down, the crude product was precipitated by water and washed with methanol/water.
7. The obtained product was re-dissolved into dichloromethane, and 0.5 g of $SnCl_2$ was added under stirring.

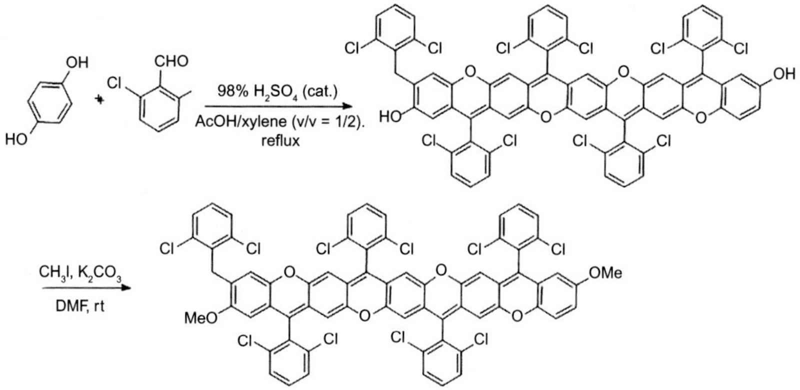

Fig. 1 Synthesis of O-Nonacene.

8. After 10 min of stirring, the product was collected by filtration and further purified by flash column chromatography on silica gel (petroleum ether/tetrahydrofuran = 3:2) to afford O-Nonacene-OH.
9. Next, 1.0 mmol of O-Nonacene-OH, K_2CO_3, and 2.4 mmol of methyl iodide were added into dry 40 mL of *N*, *N*-Dimethylformamide (DMF).
10. After stirring for several hours, the reaction solution was washed with brine, and dried by anhydrous Na_2SO_4.
11. After filtration, the organic solvents were evaporated to dryness.
12. To obtain the pure compound, the as-prepared O-Nonacene was purified by flash column chromatography.

3.2 Synthesis of Nano(O-Nonacene)-PEG (Fig. 2)

1. Prepare a 5 mL serum bottle in advance.
2. 100 μg of O-Nonacene and 5 mg of Poly(styrene-*co*-maleic anhydride) (PSMA) were dissolved in 1 mL of tetrahydrofuran (THF) and then placed in a serum bottle.
3. Put the serum bottle into the ultrasonic instrument and sonicate for 2 min.
4. Fill 9 mL water into a 50 mL serum bottle.
5. 1 mL of THF solution was quickly injected into 9 mL water.
6. Ultrasound for another 1 min.
7. Then add Na_2CO_3 aqueous solution (10 mg/mL, 0.5 mL) into the above solution.
8. Ultrasound for another 5 min.
9. THF was removed by rotary evaporation (45 °C, 100 rpm).

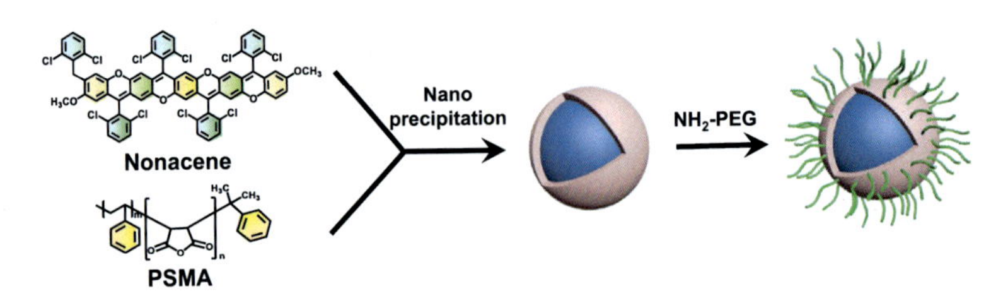

Fig. 2 Synthesis of Nano(O-Nonacene)-PEG.

10. Wash the obtained Nano(O-Nonacene)-COOH by ultrafiltration three times (4000 rpm, 3 min).
11. Dilute to 2 mL with water for future use.
12. To modification of PEG, add 500 μg of PEG-NH_2 (5 k) and 150 μg of 1-Ethyl-3-(3-dimethylaminopropyl) carbodiimide hydrochloride (EDC • HCl) to the above solution under stirring.
13. After stirring overnight, the obtained Nano(O-Nonacene)-PEG was purified by ultrafiltration and stored at 4 °C for future use.

3.3 Synthesis of Nano(PCPDTBT)-PEG

1. Prepare a 5 mL serum bottle ahead.
2. 100 μg of poly[2,6-(4,4-bis(2-ethylhexyl)-4H-cyclopenta[2,1-b;3,4-b′]-dithiophene)-alt-4,7(2,1,3-benzothiadiazole)] (PCPDTBT) and 5 mg of Poly(styrene-*co*-maleic anhydride) (PSMA) were dissolved in 1 mL of tetrahydrofuran (THF), and then placed in a serum bottle.
3. Other processes were consistent with the synthesis of Nano(O-Nonacene)-PEG (steps 2–12) above.

3.4 Synthesis of Nano(O-Nonacene@$Zn_{0.4}Fe_{2.6}O_4$)-PEG

1. Prepare to heat the oil bath to 300 °C in advance.
2. Prepare a 50 mL three-neck round-bottom flask.
3. $ZnCl_2$ (27 mg) and $Fe(acac)_3$ (65 mg) were placed in the 50 mL three-neck round-bottom flask.
4. Add 20 mL of the trioctylamine solution containing 5 mL oleic acid and 20 mL oleylamine to the above system.
5. The reaction mixture was stirred at 300 °C for 1 h.
6. Cool the above solution to room temperature.
7. Next, add excess ethanol to obtain 12 nm $Zn_{0.4}Fe_{2.6}O_4$ nanoparticles.
8. Followed by centrifugation.

9. 12 nm of $Zn_{0.4}Fe_{2.6}O_4$ nanoparticles were prepared.
10. The concentration of metals was quantified by ICP-MS.
11. 100 μg of O-Nonacene, 20 μg of $Zn_{0.4}Fe_{2.6}O_4$, and 5 mg of poly(styrene-co-maleic anhydride) (PSMA) were dissolved in 1 mL of tetrahydrofuran (THF).
12. Other processes are consistent with the synthesis of Nano(O-Nonacene)-PEG (steps 2–12) above.

3.5 Synthesis of Nano(O-Nonacene@$Zn_{0.4}Fe_{2.6}O_4$)-PEG (Fig. 3)

1. 20 μg of $Zn_{0.4}Fe_{2.6}O_4$ and 5 mg of poly(styrene-co-maleic anhydride) (PSMA) were dissolved in 1 mL of tetrahydrofuran (THF).
2. Other processes are consistent with the synthesis of Nano(O-Nonacene)-PEG (steps 2–12) above.

4. Characterization of Nano(O-Nonacene)-PEG

After synthesizing nanomaterials, their properties need to be characterized. First, determine the particle size and surface potential to prove the successful synthesis of the material. Second, explore the optical properties of particles for photoacoustic imaging. Finally, a toxicity experiment of the particles is carried out to prove that the material is non-toxic for post-stage live tumor imaging and treatment.

4.1 TEM of nanoparticles

This experiment proved the size and morphology of the synthesized particles. First, we characterized the size and morphology of Nano(O-Nonacene)-PEG. Secondly, $Zn_{0.4}Fe_{2.6}O_4$ nanoparticles were synthesized and characterized by TEM. Finally, $Zn_{0.4}Fe_{2.6}O_4$ and Nano(O-Nonacene)-PEG were co-doped to synthesize nanoparticles. The morphology proved the successful doping of $Zn_{0.4}Fe_{2.6}O_4$. (Fig. 4).

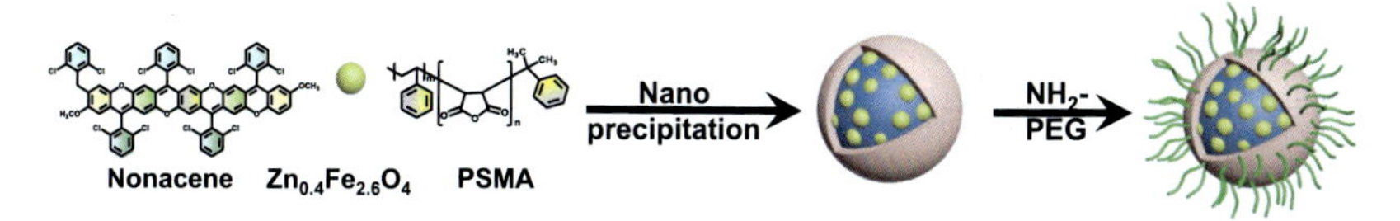

Fig. 3 Scheme of synthetic procedure for Nano(O-Nonacene@$Zn_{0.4}Fe_{2.6}O_4$)-PEG.

4.1.1 Equipment

Transmission electron microscope (TEM) (JEM-2100F (JEOL))
Micropipettes and tips (P10)
Infrared baking lamp
Copper mesh

4.1.2 Reagents

Nano(O-Nonacene)-PEG
$Zn_{0.4}Fe_{2.6}O_4$-PEG
Nano(O-Nonacene@$Zn_{0.4}Fe_{2.6}O_4$)-PEG
Solvents: Ultrapure water

4.1.3 Procedure

1. Ultrasound the synthesized Nano(O-Nonacene)-PEG for 10 min.
2. Take a large clean dish, place a piece of filter paper, use tweezers to grab the edge of the copper mesh, and place it on the filter paper.
3. Take 10 μL of ultrasonicated Nano(O-Nonacene)-PEG and slowly drop it on the copper mesh.
4. Place the large dish under the infrared baking lamp and bake at 40 °C for 30 min, until the sample is dry.
5. Submit the obtained copper mesh sample to a dedicated TEM tester to obtain accurate size and surface topography images.
6. The same procedure is used to obtain the size and surface morphology of $Zn_{0.4}Fe_{2.6}O_4$-PEG and Nano(O-Nonacene@$Zn_{0.4}Fe_{2.6}O_4$)-PEG as well.

4.1.4 Results

It can be seen from the TEM image that the synthesized Nano(O-Nonacene)-PEG has a spherical structure. Fig. 4C shows that $Zn_{0.4}Fe_{2.6}O_4$ has been successfully packaged into Nano(O-Nonacene)-PEG.

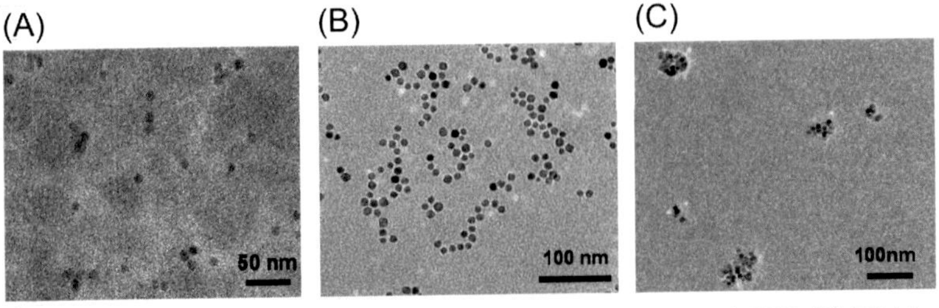

Fig. 4 TEM of nanoparticles. (A) TEM image of Nano(O-Nonacene)-PEG. (B) TEM image of $Zn_{0.4}Fe_{2.6}O_4$-PEG. (C) TEM image of Nano(O-Nonacene@$Zn_{0.4}Fe_{2.6}O_4$)-PEG.

4.2 DLS and zeta potential of nanoparticles

This experiment identifies the size and zeta potential of each compound. These data were crucial because they can characterize and indicate the formation of particles, $Zn_{0.4}Fe_{2.6}O_4$ doped into Nano(O-Nonacene), and the success of subsequent surface modification with PEG. As a result, successful modification of PEG helps to enhance the biocompatibility of particles. Below, we describe our characterization of Nano(O-Nonacene), Nano(O-Nonacene)-PEG, $Zn_{0.4}Fe_{2.6}O_4$-PEG, and Nano(O-Nonacene@$Zn_{0.4}Fe_{2.6}O_4$)-PEG. The properties of these four compounds are described in Table 1.

4.2.1 *Equipment*

Dynamic light scattering (DLS) (Malvern Zetasizer Nano ZS90 (Malvern))
Cuvette and corresponding caps
Potential cup
Micropipettes and tips (P1000)

4.2.2 *Reagents*

Nano(O-Nonacene)-COOH
Nano(O-Nonacene)-PEG
$Zn_{0.4}Fe_{2.6}O_4$-PEG
Nano(O-Nonacene@$Zn_{0.4}Fe_{2.6}O_4$)-PEG
Solvents: Ultrapure water

4.2.3 *Procedure*

1. Warm up the equipment of DLS at least 10 min in advance of the experiment.
2. Set up the experiment to collect the size of Nano(O-Nonacene)-PEG, scan three rounds.
3. Meanwhile, Prepare the solution, disperse 10 μL of Nano(O-Nonacene)-PEG (1 mg/mL) in 1 mL of water and sonicate for 5 min.

Table 1 Summary of DLS and Zeta.

Compound	DLS	Zeta
Nano(O-Nonacene)-COOH	–	−45.3 mV
Nano(O-Nonacene)-PEG	21 nm	−0.1 mV
$Zn_{0.4}Fe_{2.6}O_4$-PEG	43.8 nm	−12.9 mV
Nano(O-Nonacene@$Zn_{0.4}Fe_{2.6}O_4$)-PEG	122 nm	–

4. Add the above aqueous buffer to a cuvette (0.5–1 mL) and cap the cuvette to minimize solvent evaporation.
5. Collect size data.
6. Repeat steps 2–5 for $Zn_{0.4}Fe_{2.6}O_4$-PEG and Nano(O-Nonacene @$Zn_{0.4}Fe_{2.6}O_4$)-PEG.
7. Then, set up the experiment to collect zeta potential of Nano(O-Nonacene)-COOH, scan three rounds.
8. Meanwhile, prepare the solution, disperse 10 μL of Nano(O-Nonacene)-COOH (1 mg/mL) in 1 mL of water and sonicate for 5 min.
9. Add the above aqueous buffer to a potential cup (1 mL) and ensure that the liquid can completely immerse the electrodes on both sides.
10. Collect zeta potential data.
11. Repeat steps 7–10 for Nano(O-Nonacene)-PEG and $Zn_{0.4}Fe_{2.6}O_4$-PEG.

4.2.4 Results

Comparing Nano(O-Nonacene)-PEG and Nano(O-Nonacene)-COOH, after PEGylation, we can find that the zeta potential is changed from −45.3 to −0.1 mV, which proves that PEG has been successfully bonded on the particle surface. Apart from this, after packing $Zn_{0.4}Fe_{2.6}O_4$ into Nano(O-Nonacene)-PEG, the particle size increased from 21 to 122 nm, proving that $Zn_{0.4}Fe_{2.6}O_4$ has been successfully packed.

4.3 UV characterization of Nano(O-Nonacene)-PEG

This experiment identifies each compound's maximum absorption wavelength. This data is critical because it can initially determine whether the material can be used for further photoacoustic imaging, and it can also determine the concentration of the material. Below, we describe our characterization of Nano(O-Nonacene)-PEG, Nano(PCPDTBT)-PEG and Au nanorods as an example (Fig. 5).

4.3.1 Equipment

UV–visible range spectrophotometer
Quartz cuvettes with corresponding cuvette caps
Micropipettes and tips (P10, P200, P1000)

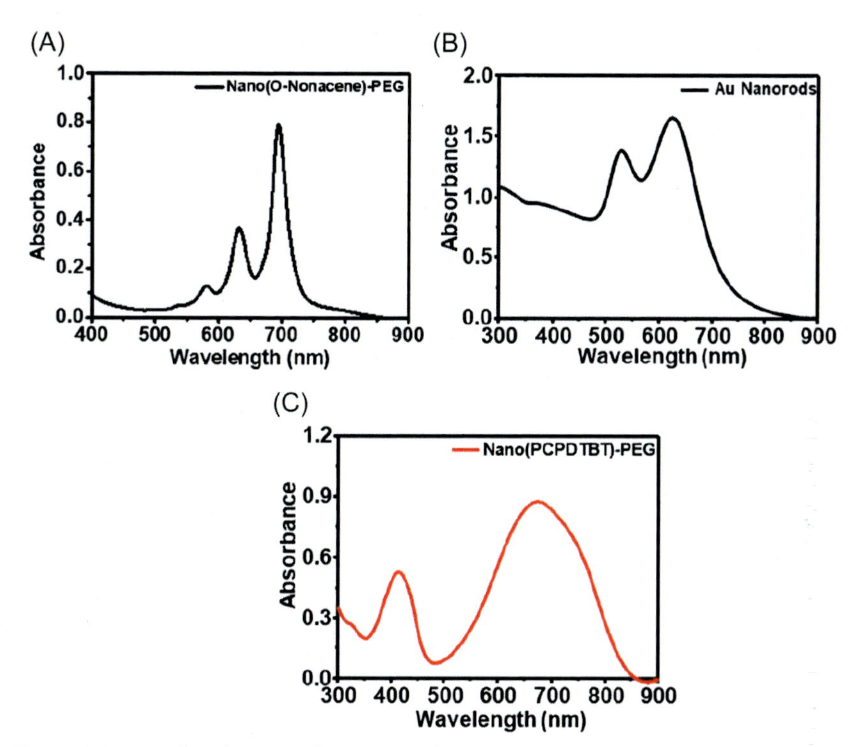

Fig. 5 UV–vis absorbance of nanoparticles. (A) UV–vis absorbance of Nano(O-Nonacene)-PEG. (B) UV–vis absorbance of Au Nanorods. (C) UV–vis absorbance of Nano(PCPDTBT)-PEG.

4.3.2 Reagents

Nano(O-Nonacene)-PEG
Nano(PCPDTBT)-PEG
Au nanorods
Solvents: Ultrapure water

4.3.3 Procedure

1. Warm up the spectrophotometer at least 15 min in advance of the measurement.
2. Set up the collection range from 300 to 900 nm, with the increasement of 0.5 nm.
3. Meanwhile, prepare a stock solution of Nano(O-Nonacene)-PEG.
4. Add 200 μL of ultrapure water to a quartz cuvette.

5. Scan the quartz cuvette in step 4 by using the baseline function.
6. Pour out the water and add in the stock solution of Nano(O-Nonacene)-PEG.
7. Put the cuvette back to ensure that the cuvette is placed in the same position and direction as scanning the baseline.
8. Collect an absorbance spectrum of Nano(O-Nonacene)-PEG.
9. Make sure the peak of absorbance lays between 0.1 and 1.0 a.u. If not, concentrate or dilute the Nano(O-Nonacene)-PEG solution.
10. Fully clean the quartz cuvette.
11. Repeat steps 6–10 for Nano(PCPDTBT)-PEG and Au nanorods.

4.4 Fluorescence and photodynamic effect of Nano(O-Nonacene)-PEG

Based on the Jablonski energy diagram, the energy that light delivered to chromophore forming excited state dissipates via the three following pathways: (1) fluorescence emission from the singlet excited state; (2) intersystem crossing to a triplet excited state to enable the generation of phosphorescence and/or reactive oxygen species (e.g., singlet oxygen); and (3) non-radiation in the form of thermal deactivation. Based on this, we verified the fluorescence and photodynamic effects of Nano(O-Nonacene)-PEG (Fig. 6).

4.4.1 Equipment

660 nm laser
Fluorescence spectrometer
Micro cuvette
Micropipettes and tips (P10, P200)

4.4.2 Reagents

Nano(O-Nonacene)-PEG
Nano(PCPDTBT)-PEG
Solvents: Ultrapure water

4.4.3 Procedure

1. For fluorescence, warm up the spectrophotometer at least 10 min in advance of the measurement.
2. Setting parameters.
   (a) EX: 680 nm
   (b) EM: 780–900 nm

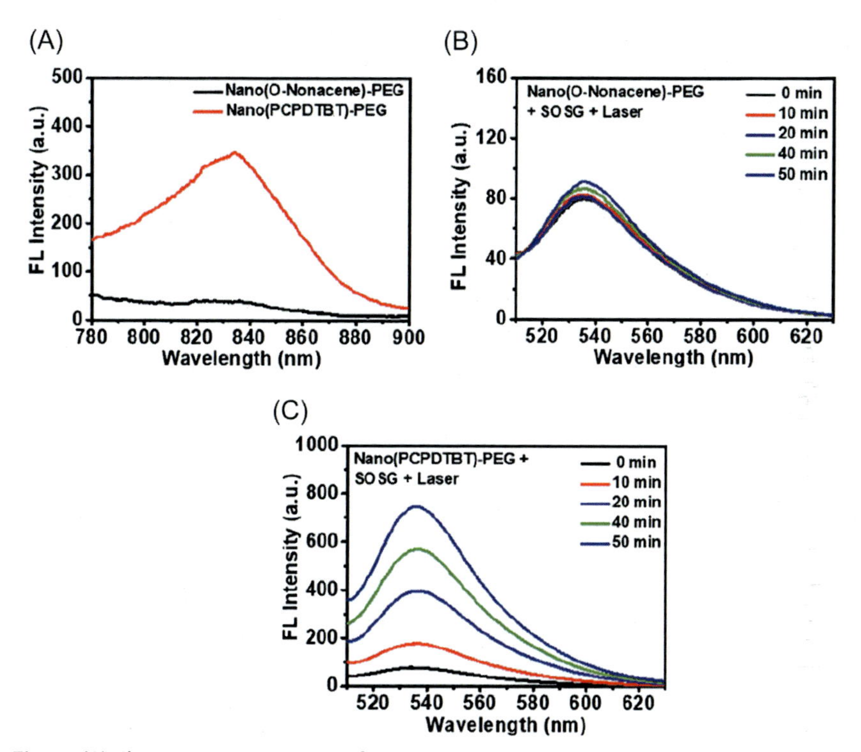

Fig. 6 (A) Fluorescence spectrum of Nano(O-Nonacene)-PEG and Nano(PCPDTBT)-PEG (excitation at 680 nm). (B and C) Generation of singlet oxygen of Nano(O-Nonacene)-PEG and Nano(PCPDTBT)-PEG measured by the fluorescence intensity of SOSG, upon designated time ($t = 0$–50 min) irradiation of a 660 nm laser.

3. Add 200 μL of Nano(O-Nonacene)-PEG to cuvette.
4. Collect fluorescence spectrum.
5. Repeat steps 2–4 for Nano(PCPDTBT)-PEG consistent with the absorption of Nano(O-Nonacene)-PEG at 680 nm.
6. For photodynamic, warm up the spectrophotometer at least 10 min in advance of the measurement.
7. Setting parameters.
   (a) EX: 494 nm
   (b) EM: 510–630 nm
8. Prepare Nano(PCPDTBT)-PEG and Nano(O-Nonacene)-PEG solutions with the same absorption at 660 nm.
9. Add 196 μL of Nano(O-Nonacene)-PEG and 4 μL commercialized SOSG (167 μM) to cuvette, mix thoroughly by pipetting up and down 10 times, and then collect the initial fluorescence spectrum.

10. Take out the cuvette for lighting.
11. Set 660 nm laser parameters and fix the sample so that the laser can irradiate the sample more accurately.
    (a) Power: 0.8 W
    (b) The distance between the fiber and the center of the centrifuge tube: 1 cm
12. Turn on the laser for 10 min.
13. Turn off the laser.
14. Put the cuvette into the fluorescence instrument again.
15. Collect fluorescence spectrum.
16. Repeat step 11–15 for another four times.
17. Wash the cuvette thoroughly.
18. Repeat step 9–17 for Nano(PCPDTBT)-PEG and ultrapure water.

4.4.4 Results

According to the law of conservation of energy, electrons transition from the ground state to the excited state. The excited state is an unstable state. There are several ways in the process of returning from the excited state to the ground state. The first is to release in the form of fluorescence through a radiation transition, and the second is to transfer energy to the surrounding oxygen to generate active oxygen through the way of intersystem crossing, and the third is to release it in a thermal way through non-radiative transition. Therefore, we verified the fluorescence and photodynamic effects of Nano (O-Nonacene)-PEG. From Fig. 6A, we can see that compared to Nano (PCPDTBT)-PEG, Nano(O-Nonacene)-PEG has almost no fluorescence. From Fig. 6B and C, we can see that compared to Nano(PCPDTBT)-PEG, Nano(O-Nonacene)-PEG has almost no active oxygen generation, so we predict that Nano(O-Nonacene)-PEG will have good photothermal and photoacoustic signals.

4.5 Photothermal studies of Nano(O-Nonacene)-PEG in solution

In order to verify the photothermal effect of Nano(O-Nonacene)-PEG, we compared Nano(O-Nonacene)-PEG with commercial Nano(PCPDTBT)-PEG and traditional inorganic material Au nanorods. Proved that this effect can be better used for subsequent imaging to guide tumor treatment (Fig. 7).

4.5.1 Equipment

660 nm laser
UV–visible range spectrophotometer

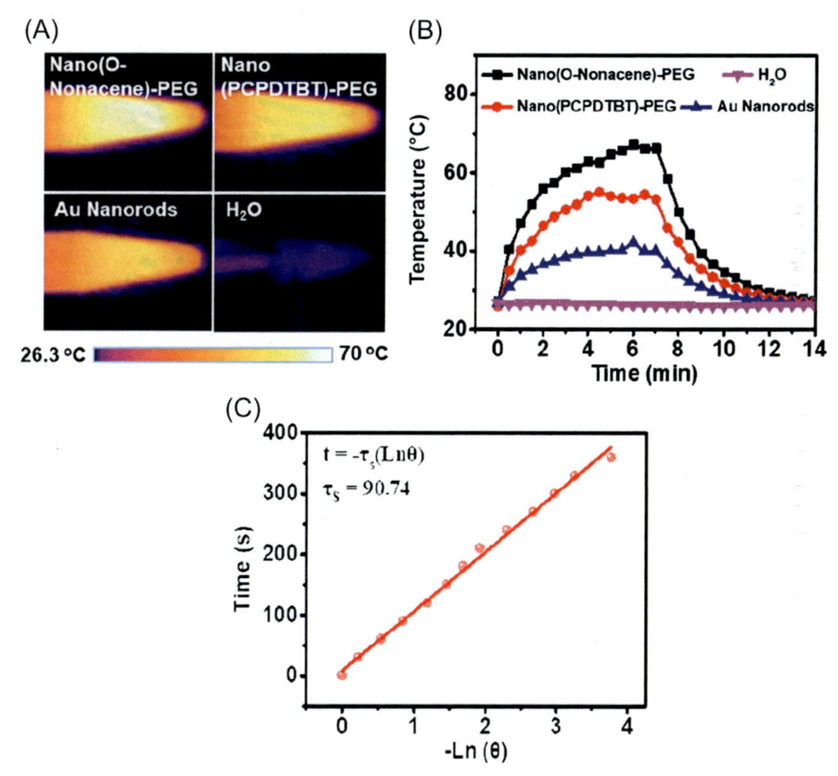

Fig. 7 Photothermal properties of Nano(O-Nonacene)-PEG under 660 nm laser irradiation ($1\,W/cm^2$). (A) Thermal images of OD_{660}-matched Nano(O-Nonacene)-PEG, Nano(PCPDTBT)-PEG and Au nanorods at 7 min of laser irradiation. (B) Temperature curves of OD_{660}-matched Nano(O-Nonacene)-PEG, Nano(PCPDTBT)-PEG, Au nanorods and H_2O under laser irradiation for 7 min and cooling down spontaneously. (C) The cooling time toward the negative natural logarithm of the temperature driving force, which is obtained from the cooling stage of Nano(O-Nonacene)-PEG.

Quartz cuvettes with corresponding cuvette caps
Thermal Imager
Micropipettes and tips (P100)
300 μL Eppendorf tubes (EP tube)

4.5.2 Reagents

Nano(O-Nonacene)-PEG
Nano(PCPDTBT)-PEG
Au nanorods
solvents: Ultrapure water

4.5.3 Procedure

1. Warm up the spectrophotometer at least 15 min in advance of the experiment.
2. Use the method of 4.2 to determine the absorption of the Nano(O-Nonacene)-PEG solution, and make sure the absorption of Au nanorods and Nano(PCPDTBT)-PEG at 660 nm is consistent with that of Nano(O-Nonacene)-PEG.
3. Take 100 μL of Nano(O-Nonacene)-PEG solution in a 300 μL EP tube and close the cap of the EP tube.
4. Set 660 nm laser parameters and fix the sample so that the laser can irradiate the sample more accurately.
   (a) Power: 1 W
   (b) The distance between the fiber and the center of the centrifuge tube: 1 cm
5. Turn on the thermal imager and record the initial temperature of the sample.
6. Turn on the laser and record the sample temperature in real time (record a temperature every 30 s) and acquire photothermal images simultaneously.
7. Turn off the laser until the sample no longer heats up significantly.
8. Record the sample cooling data (record a temperature every 30 s) until the sample drops to room temperature.
9. Repeat steps 3–8 for pure solvent, Nano(PCPDTBT)-PEG and Au nanorods. (The heating time and cooling time are consistent with Nano(O-Nonacene)-PEG).
10. Use the following formula to calculate the photothermal conversion efficiency of Nano(O-Nonacene)-PEG, Nano(PCPDTBT)-PEG and Au nanorods.

4.5.4 Photothermal conversion efficiency (η) (Chechetka et al., 2017) of Nano(O-Nonacene)-PEG

The η was calculated by equation:

$$\eta = \frac{hS(T_{Max} - T_{Sur}) - Q_{Dis}}{I(1 - 10^{-A_{660}})} \tag{1}$$

hS can be calculated by the follow equation:

$$hs = \frac{m_D C_D}{\tau_s} \tag{2}$$

$$\theta = \frac{T - T_{sur}}{T_{max} - T_{sur}} \tag{3}$$

$$t = -\tau_s \ln(\theta) \tag{4}$$

h: the heat transfer coefficient

S: the surface area of the container

T_{max}: the plateau temperature ($T_{max} = 66.3\,°C$)

T_{sur}: the surrounding temperature ($T_{sur} = 27.5\,°C$)

Q_{dis}: the energy of solvent and the same cell under laser irradiation. ($Q_{dis} = 0.00107$)

I: the lase power ($1\,W/cm^2$)

A_{660}: the absorbance at 660 nm ($A_{660} = 0.4$)

mD: the mass of solution (mD = 0.1 g)

CD: the heat capacity of water ($CD = 4.2\,J\,g^{-1}\,K^{-1}$)

According to the obtained data, the time constant (τ_s) was 90.74, the photothermal conversion efficiency of Nano(O-Nonacene)-PEG was calculated to be 30.5%. In the same way, the photothermal conversion efficiency of Nano(PCPDTBT)-PEG and Au nanorods were 20.6% and 10.8%, respectively.

4.5.5 Results

Due to the strong NIR absorption and the absence of fluorescence and photodynamic properties of Nano(O-Nonacene)-PEG, we further studied the photothermal effect of Nano(O-Nonacene)-PEG. From Fig. 7A and B, we can find that compared to Nano(PCPDTBT)-PEG and Au nanorods, under the same absorption at 660 nm, Nano(O-Nonacene)-PEG has better photothermal effects.

4.6 Photostability of Nano(O-Nonacene)-PEG in solution

Under laser irradiation, many organic materials will undergo photobleaching, which will lead to ineffective tumor imaging and tumor phototherapy. Therefore, the light stability of the material is very important. Here we compare the photostability of Nano(O-Nonacene)-PEG with commercial ICG (Fig. 8).

4.6.1 Equipment

660 nm laser

IR thermal camera

Micropipettes and tips (P100)

300 μL Eppendorf tubes (EP tube)

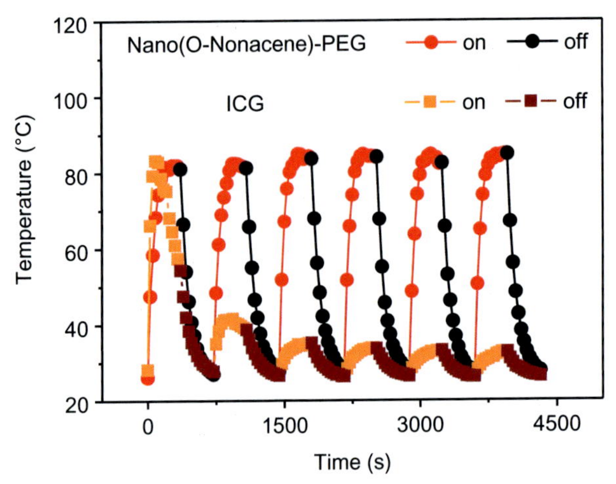

Fig. 8 The thermal curves of Nano(O-Nonacene)-PEG and ICG under six on-and-off cycles laser irradiation.

4.6.2 Reagents

Nano(O-Nonacene)-PEG
ICG
Solvents: Ultrapure water

4.6.3 Procedure

1. Take 100 μL of solution Nano(O-Nonacene)-PEG into a 300 μL EP tube.
2. Set 660 nm laser parameters and fix the sample so that the laser can irradiate the sample more accurately.
   - **(a)** Power: 1 W
   - **(b)** The distance between the fiber and the center of the centrifuge tube: 1 cm
3. Turn on the thermal imager and record the initial temperature of the sample.
4. Turn on the laser and record the sample temperature in real time (record a temperature every 30 s).
5. Turn off the laser until the sample no longer heats up significantly.
6. Record the sample cooling data (record a temperature every 30 s) until the sample drops to room temperature.
7. Repeat steps 4–6 six times, the heating and cooling time is the same as the first round.

8. Adjust the ICG concentration to make sure the maximum temperature is as same as that of Nano(O-Nonacene)-PEG solution under the same laser conditions.
9. Repeat steps 1–7 for ICG.

4.6.4 Results

As shown in the Fig. 8, under the first round of illumination, the ICG has been destroyed, and during the subsequent illumination process, the temperature can only rise to about 30 °C at most. However, the maximum temperature of Nano(O-Nonacene)-PEG remained almost unchanged during the six rounds of light heating and cooling. This result shows that Nano (O-Nonacene)-PEG has good light stability.

4.7 Photoacoustic studies in solution

Here, we performed photoacoustic imaging of Nano(O-Nonacene)-PEG and compared the photoacoustic signals of Nano(PCPDTBT)-PEG, Au nanorods and Nano(O-Nonacene)-PEG under the same absorption. In addition, it is proved that doping $Zn_{0.4}Fe_{2.6}O_4$ into Nano(O-Nonacene)-PEG can enhance the photoacoustic imaging effect by enhancing heat conduction (Fig. 9).

4.7.1 Equipment

UV–visible range spectrophotometer
Quartz cuvettes with corresponding cuvette caps
Micropipettes and tips (P100, P200, P1000)
Photoacoustic tomographer (InVision 256-TF imaging system (iThera Medical, Germany))
300 μL Eppendorf tubes
Inductively coupled plasma mass spectrometry (ICP-MS)

4.7.2 Reagents

Nano(O-Nonacene)-PEG
Nano(PCPDTBT)-PEG
$Zn_{0.4}Fe_{2.6}O_4$-PEG
Nano(O-Nonacene@$Zn_{0.4}Fe_{2.6}O_4$)-PEG
Au nanorods
Solvents: Ultrapure water

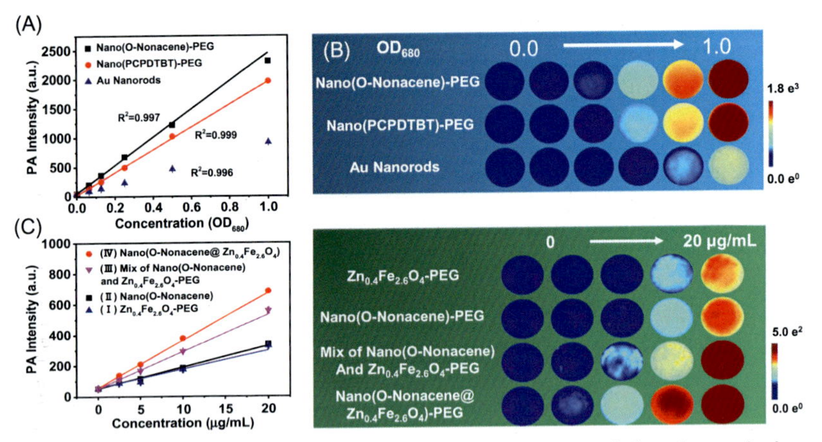

Fig. 9 Photoacoustic properties of Nano(O-Nonacene)-PEG and the elevated photoacoustic signals via incorporating magnetic $Zn_{0.4}Fe_{2.6}O_4$ nanoparticles at 680 nm laser excitation. (A) Photoacoustic signal intensities of OD_{680}-matched Nano(O-Nonacene)-PEG, Nano(PCPDTBT)-PEG and Au nanorods. (B) Corresponding photoacoustic images of (A). (C) Photoacoustic intensity (680 nm excitation) of various nanoparticles under various concentrations of Nano(O-Nonacene)-PEG. (D) Corresponding photoacoustic images of (C).

4.7.3 Procedure

1. Use an UV–visible range spectrophotometer to determine that Nano(PCPDTBT)-PEG, Au nanorods and Nano(O-Nonacene)-PEG have the same absorption at 680 nm. ($OD_{680}=1$, 0.5, 0.25, 0.125, 0.0625).
2. Take 300 μL of Nano(O-Nonacene)-PEG with $OD_{680}=1$, put them in a 300 μL Eppendorf tube, ensure that there are no bubbles in the Eppendorf tube.
3. Turn on the photoacoustic instrument, set the response parameters, and wait for the instrument to heat up to 25 °C.
4. Put the Eppendorf tube into the photoacoustic special bracket and put it in the photoacoustic device.
5. Adjust the parameters of the photoacoustic instrument to ensure that the image is clear and without ghost.
6. Adjust the position of the sample to find the position with the best photoacoustic signal.
7. Scan the photoacoustic spectrum of 680–980 nm with the intervals of 5 nm.
8. After the experiment, reconstruct the image, compare all obtained photoacoustic images after unified processing.

9. Repeat steps 2–8 for Nano(PCPDTBT)-PEG, Au nanorods.
10. For enhanced photoacoustic signal, determine the concentration of $Zn_{0.4}Fe_{2.6}O_4$ in $Zn_{0.4}Fe_{2.6}O_4$-PEG and Nano(O-Nonacene@$Zn_{0.4}Fe_{2.6}O_4$)-PEG by ICP-MS.
11. Repeat steps 2–8 for $Zn_{0.4}Fe_{2.6}O_4$-PEG and Nano (O-Nonacene@ $Zn_{0.4}Fe_{2.6}O_4$)-PEG at the same concentrations of $Zn_{0.4}Fe_{2.6}O_4$ or O-Nonacene.

4.7.4 Results

In view of the good photothermal performance of Nano (O-Nonacene)-PEG, we further tested the photoacoustic imaging properties of Nano(O-Nonacene)-PEG. As shown in Fig. 9A and B, under the same 680 nm absorption, Nano(O-Nonacene)-PEG has better photoacoustic performance than Nano(PCPDTBT)-PEG and Au nanorods. Because metal has good thermal conductivity, we doped $Zn_{0.4}Fe_{2.6}O_4$ into Nano (O-Nonacene)-PEG and tested its photoacoustic effect. As shown in Fig. 9C and D, under the same concentration of $Zn_{0.4}Fe_{2.6}O_4$-PEG and Nano(O-Nonacene)-PEG, Nano(O-Nonacene@$Zn_{0.4}Fe_{2.6}O_4$)-PEG shows better photoacoustic effect than Nano(O-Nonacene)-PEG, $Zn_{0.4}Fe_{2.6}O_4$-PEG and simple mixed Nano(O-Nonacene)-PEG and $Zn_{0.4}Fe_{2.6}O_4$-PEG.

4.8 Investigating the cytotoxicity of Nano(O-Nonacene)-PEG

Since Nano(O-Nonacene)-PEG is used for in vivo photoacoustic imaging and photothermal therapy, proving the dark toxicity of Nano(O-Nonacene)-PEG is vital important, which paves its way to be better used for later imaging and treatment (Fig. 10).

4.8.1 Equipment

96-well plate
Incubator suitable for mammalian cell culture, with 5% CO_2 atmosphere
microscope
Enzyme-labeled instrument
Shaker

4.8.2 Reagents

4 T1 cells
Dulbecco's Modified Eagle's medium (DMEM) (containing 1% of penicillin-streptomycin (PS), 10% of fetal bovine serum (FBS))
Nano(O-Nonacene)-PEG (0, 0.1, 0.2, 0.4, 0.6, 0.8, and 1 mg/mL in PBS)

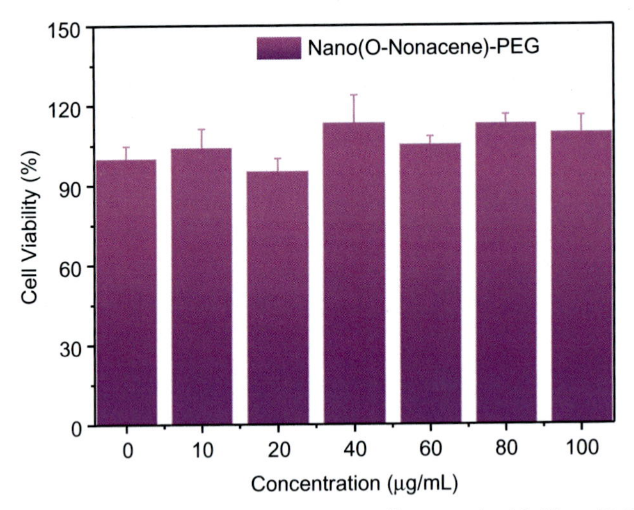

Fig. 10 Relative cellular viability of 4T1 cancer cells treated with Nano(O-Nonacene)-PEG for 24 h.

4.8.3 Procedure

1. Warm the modified DMEM media in a 37 °C water bath.
2. Add DPBS to the outermost circle of the well plate (200 μL/well).
3. Add 42 wells with 4T1 cells (200 μL/well, $1 * 10^4$ cells/well of media).
4. Put the plate in the incubator and incubate the plate until the wells approach 50–60% confluence.
5. Meanwhile, prepare the solutions of Nano(O-Nonacene)-PEG.
6. Dilute to 0, 10, 20, 40, 60, 80, 100 μg/mL with DMEM respectively before use.
7. Throw away the original DMEM in each well, and then add the above diluted solutions (Six wells in parallel for each concentration, 200 μL/well).
8. Put the plate in the incubator and incubate the plate, in the dark, for 24 h.
9. Throw away the original solution in each well, wash each well once with DPBS.
10. Then add 200 μL of DMEM containing MTT (0.5 mg/mL) to each well.
11. Put the plate in the incubator and incubate the plate for another 4 h.
12. Discard the DMEM, add 150 μL of DMSO, shake the shaker at low speed for 10 min to completely dissolve the crystals.
13. Open the lid of the well plate and put it in the Enzyme-labeled instrument.
14. Enzyme-labeled instrument measures the absorption of 96-well plate at 490 nm.

4.8.4 Results

The ideal photothermal reagent should have mild dark toxicity and can effectively kill tumor cells under light. Therefore, we verified the dark toxicity of Nano(O-Nonacene)-PEG in this experiment. By incubating Nano (O-Nonacene)-PEG with 4 T1 cells for 24 h, we found that Nano(O-Nonacene)-PEG has almost no dark toxicity, indicating that Nano (O-Nonacene)-PEG can be furtherly used for in vivo imaging and treatment.

5. Characterization of Nano(O-Nonacene)-PEG in vivo

The previous in vitro experiments proved that Nano(O-Nonacene)-PEG has good photoacoustic imaging and photothermal effects. And after $Zn_{0.4}Fe_{2.6}O_4$ doping, there is an obvious photoacoustic and photothermal enhancement. In the next part, we prove that after intratumoral injection, Nano(O-Nonacene)-PEG has a good photothermal imaging property. At the same time, through tail vein injection, Nano(O-Nonacene)-PEG can also be effectively enriched in the tumor.

Female BALB/c mice (5–7 weeks) were obtained from Hunan SJA Laboratory Animal Co., Ltd. All animal experiments were approved by the Institutional Animal Care and Use Committee of Hunan University.

5.1 Establishment of tumor model

5.1.1 Equipment

Sterile insulin syringes
Electric shaver, for hair removal
Bottle depilatory cream, for hair removal

5.1.2 Animals

Female BALB/c mouse, 5–7 weeks old

5.1.3 Reagents

4 T1 cells
PBS

5.1.4 Procedure

1. Use shaving machine and depilatory cream to remove all the hair of the mouse except the head.
2. 25 μL of PBS containing 1×10^6 of 4 T1 cells was injected into the right back of each BALB/c mice.
3. Use when the tumor grows to $100\,mm^3$.

5.2 In vivo photoacoustic study of Nano(O-Nonacene)-PEG

Good tumor enrichment can guide tumor treatment, so we verified the tumor targeting ability of Nano(O-Nonacene)-PEG after tail vein injection (Fig. 11).

5.2.1 Equipment

Photoacoustic tomographer (InVision 256-TF imaging system (iThera Medical, Germany))
Vaporizer equipped with isoflurane and oxygen (for anesthesia)
Cling film, for fixing mice
Ultrasound gel

5.2.2 Animals

Female BALB/c mouse, 5–7 weeks old

5.2.3 Reagents

Nano(O-Nonacene)-PEG
PBS

5.2.4 Procedure

1. Turn on the photoacoustic instrument, set the response parameters, and wait for the instrument to heat up to 37 °C.
2. Anesthetize the mouse in an isoflurane/oxygen chamber, then maintain an anesthetized state using a stream of isoflurane through a nose cone for the remainder of the experiment.

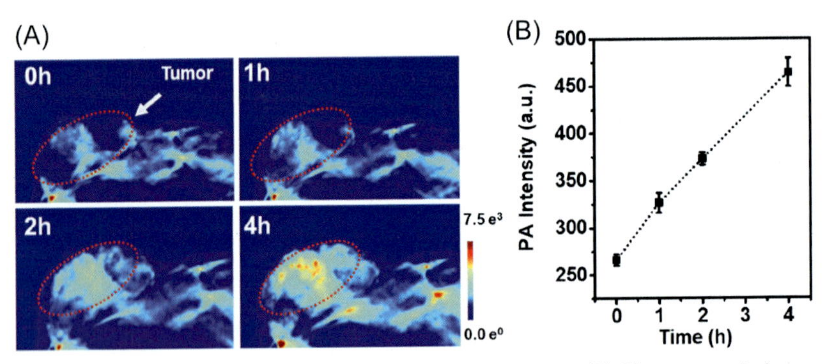

Fig. 11 Photoacoustic imaging of mice bearing 4T1 tumor. (A) Photoacoustic images (680 nm excitation) after i.v. injection of Nano(O-Nonacene)-PEG. (red cycle indicated tumor). (B) Corresponding photoacoustic intensity of tumor areas (A).

3. Lay the plastic wrap on the photoacoustic bracket, and spread the ultrasonic couplant gel evenly on the corresponding position on the plastic wrap.
4. Use a sterile insulin needle to inject 25 μL of Nano(O-Nonacene)-PEG (2 mg/mL) into the tail vein.
5. Anesthetize the mouse and place it on the coated ultrasound gel and fix the mouse's nose in an isoflurane airflow tube.
6. Apply ultrasound gel to the mouse to make the mouse tumor position 1 cm above and below and the corresponding position of the abdomen in the uniform ultrasound gel.
7. Put the special bracket in a photoacoustic instrument for imaging.
8. Adjust and find the location of the mouse tumor.
9. Adjust the parameters of the photoacoustic instrument to ensure that the image is clear and there is no ghost.
10. Scan the photoacoustic spectrum of 680–750 nm with the intervals of 5 nm.
11. Collect photoacoustic spectra at 1, 2, and 4 h after injection. (In order to reduce the experimental error, the mice should not be moved in the photoacoustic instrument).
12. After the experiment, reconstruct the image, unify the data of 0, 1, 2, and 4 h, and compare them to obtain the photoacoustic enrichment data.

5.2.5 Results

After verifying the photoacoustic solution of Nano(O-Nonacene)-PEG, we verified the enrichment effect of Nano(O-Nonacene)-PEG in vivo by tail vein injection. From the Fig. 11A and B, we can see that after tail vein injection, the photoacoustic signal at the tumor site gradually increases over time, which proves that the material can be enriched in the tumor via enhanced permeability and retention (EPR) effect, which provides the imaging guidance for our future treatment.

5.3 In vivo photothermal imaging of Nano(O-Nonacene)-PEG

This experiment is to verify that Nano(O-Nonacene)-PEG through intratumoral injection can well heat up the tumor site under the irradiation of laser, so as to obtain photothermal imaging of the tumor site and pave the way to tumor photothermal treatment (Fig. 12).

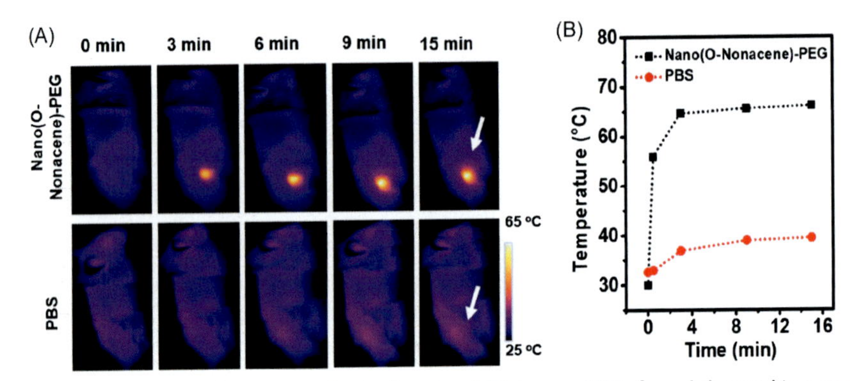

Fig. 12 Photothermal imaging of mice bearing 4 T1 tumor. (A) Infrared thermal images after i.t. injection of Nano(O-Nonacene)-PEG and PBS under laser irradiation (660 nm, 0.7 W/cm^2). (B) Corresponding temperature of tumor areas (A).

5.3.1 *Equipment*

660 nm laser

IR thermal camera

Vaporizer equipped with isoflurane and oxygen (for anesthesia) sterile insulin syringes

Special tape for mice, for fix mice

5.3.2 *Animals*

Female BALB/c mouse, 5–7 weeks old

5.3.3 *Reagents*

Nano(O-Nonacene)-PEG

PBS

5.3.4 *Procedure*

1. Anesthetize the mouse in an isoflurane/oxygen chamber, then maintain an anesthetized state using a stream of isoflurane through a nose cone for the remainder of the experiment.
2. Use a sterile insulin syringes to intratumorally inject 50 μL of Nano (O-Nonacene)-PEG (1.5 mg/mL).
3. Set 660 nm laser parameters and fix the mouse so that the laser can irradiate the tumor accurately.
4. Power: 0.7 W.

5. The distance between the fiber and the center of the tumor: 1 cm.
6. Turn on the IR thermal camera to record the real-time temperature of mouse tumors.
7. Half an hour after the injection of Nano(O-Nonacene)-PEG, turn on the laser and irradiate tumor for 15 min, and record the temperature every minute by IR thermal camera, collect 16 series of temperature data and images.
8. Change to another mouse, intratumorally inject 50 μL of PBS into the tumor, repeat steps 6–8.

5.3.5 Results

According to the picture, after intratumor injection of Nano(O-Nonacene)-PEG, the temperature of the tumor site can rise to 64.6 °C under laser irradiation, but under the same conditions, the PBS group can only rise to less than 40 °C, which also proves that our materials have good photothermal imaging effect and the ability of visualize tumors during photothermal imaging.

6. Concluding remarks

In this article, we synthesized a photoacoustic imaging molecule Nano(O-Nonacene)-PEG, and proved that the incorporation of $Zn_{0.4}Fe_{2.6}O_4$ can enhance the photoacoustic imaging effect of Nano (O-Nonacene)-PEG by enhancing heat conduction. At the same time, Nano(O-Nonacene)-PEG has a good photothermal effect, which can effectively guide tumor treatment during the photoacoustic imaging process. Therefore, we provide a new material for photoacoustic imaging. We believe this will be a very effective phototheranostic agent.

References

Boros, E., & Packard, A. B. (2019). Radioactive transition metals for imaging and therapy. *Chemical Reviews*, *119*(2), 870–901. https://doi.org/10.1021/acs.chemrev.8b00281.

Chechetka, S. A., Yu, Y., Zhen, X., Pramanik, M., Pu, K., & Miyako, E. (2017). *Nature Communications*, *8*, 15432–15450. https://doi.org/10.1038/ncomms15432.

Chitgupi, U., Nyayapathi, N., Kim, J., Wang, D., Sun, B., Li, C., et al. (2019). Surfactant-stripped micelles for NIR-II photoacoustic imaging through 12 cm of breast tissue and whole human breasts. *Advanced Materials*, *31*(40), 1902279. https://doi.org/10.1002/adma.201902279.

Debroye, E., & Parac-Vogt, T. N. (2014). Towards polymetallic lanthanide complexes as dual contrast agents for magnetic resonance and optical imaging. *Chemical Society Reviews*, *43*(23), 8178–8192. https://doi.org/10.1039/C4CS00201F.

Fan, Q., Cheng, K., Hu, X., Ma, X., Zhang, R., Yang, M., et al. (2014). Transferring biomarker into molecular probe: Melanin nanoparticle as a naturally active platform for multimodality imaging. *Journal of the American Chemical Society*, *136*(43), 15185–15194. https://doi.org/10.1021/ja505412p.

Fu, Q., Zhu, R., Song, J., Yang, H., & Chen, X. (2019). Photoacoustic imaging: Contrast agents and their biomedical applications. *Advanced Materials*, *31*(6), 1805875. https://doi.org/10.1002/adma.201805875.

Haedicke, K., Agemy, L., Omar, M., Berezhnoi, A., Roberts, S., Longo-Machado, C., et al. (2020). High-resolution optoacoustic imaging of tissue responses to vascular-targeted therapies. *Nature Biomedical Engineering*, *4*(3), 286–297. https://doi.org/10.1038/s41551-020-0527-8.

Huang, J., & Pu, K. (2020). Activatable molecular probes for second near-infrared fluorescence, chemiluminescence, and photoacoustic imaging. *Angewandte Chemie International Edition*, *59*(29), 11717–11731. https://doi.org/10.1002/anie.202001783.

Jiang, X., Du, B., Tang, S., Hsieh, J.-T., & Zheng, J. (2019). Photoacoustic imaging of nanoparticle transport in the kidneys at high temporal resolution. *Angewandte Chemie International Edition*, *58*(18), 5994–6000. https://doi.org/10.1002/anie.201901525.

Kenry, Duan, Y., & Liu, B. (2018). Recent advances of optical imaging in the second near-infrared window. *Advanced Materials*, *30*(47), 1802394. https://doi.org/10.1002/adma.201802394.

Knox, H. J., & Chan, J. (2018). Acoustogenic probes: A new frontier in photoacoustic imaging. *Accounts of Chemical Research*, *51*(11), 2897–2905. https://doi.org/10.1021/acs.accounts.8b00351.

Li, S., Deng, Q., Zhang, Y., Li, X., Wen, G., Cui, X., et al. (2020). Rational design of conjugated small molecules for superior photothermal theranostics in the NIR-II biowindow. *Advanced Materials*, *32*(33), 2001146. https://doi.org/10.1002/adma.202001146.

Li, Q., Li, S., He, S., Chen, W., Cheng, P., Zhang, Y., et al. (2020). An activatable polymeric reporter for near-infrared fluorescent and photoacoustic imaging of invasive cancer. *Angewandte Chemie International Edition*, *59*(18), 7018–7023. https://doi.org/10.1002/anie.202000035.

Li, X., Park, E.-Y., Kang, Y., Kwon, N., Yang, M., Lee, S., et al. (2020). Supramolecular phthalocyanine assemblies for improved photoacoustic imaging and photothermal therapy. *Angewandte Chemie International Edition*, *59*(22), 8630–8634. https://doi.org/10.1002/anie.201916147.

Liu, J.-N., Bu, W., & Shi, J. (2017). Chemical design and synthesis of functionalized probes for imaging and treating tumor hypoxia. *Chemical Reviews*, *117*(9), 6160–6224. https://doi.org/10.1021/acs.chemrev.6b00525.

Men, X., Wang, F., Chen, H., Liu, Y., Men, X., Yuan, Y., et al. (2020). Ultrasmall semiconducting polymer dots with rapid clearance for second near-infrared photoacoustic imaging and photothermal cancer therapy. *Advanced Functional Materials*, *30*(24), 1909673. https://doi.org/10.1002/adfm.201909673.

Ni, D., Bu, W., Ehlerding, E. B., Cai, W., & Shi, J. (2017). Engineering of inorganic nanoparticles as magnetic resonance imaging contrast agents. *Chemical Society Reviews*, *46*(23), 7438–7468. https://doi.org/10.1039/C7CS00316A.

Nie, L., & Chen, X. (2014). Structural and functional photoacoustic molecular tomography aided by emerging contrast agents. *Chemical Society Reviews*, *43*(20), 7132–7170. https://doi.org/10.1039/C4CS00086B.

Ong, S. Y., Zhang, C., Dong, X., & Yao, S. Q. (2021). Recent advances in polymeric nanoparticles for enhanced fluorescence and photoacoustic imaging. *Angewandte Chemie International Edition*. https://doi.org/10.1002/anie.202101964. n/a(n/a).

Owens, E. A., Henary, M., El Fakhri, G., & Choi, H. S. (2016). Tissue-specific near-infrared fluorescence imaging. *Accounts of Chemical Research*, *49*(9), 1731–1740. https://doi.org/10.1021/acs.accounts.6b00239.

Pellico, J., Gawne, P. J., & de Rosales, T. M. R. (2021). Radiolabelling of nanomaterials for medical imaging and therapy. *Chemical Society Reviews*, *50*(5), 3355–3423. https://doi.org/10.1039/D0CS00384K.

Pu, K., Shuhendler, A. J., Jokerst, J. V., Mei, J., Gambhir, S. S., Bao, Z., et al. (2014). Semiconducting polymer nanoparticles as photoacoustic molecular imaging probes in living mice. *Nature Nanotechnology*, *9*(3), 233–239. https://doi.org/10.1038/nnano.2013.302.

Shi, L., Wang, Y., Zhang, C., Zhao, Y., Lu, C., Yin, B., et al. (2021). An acidity-unlocked magnetic nanoplatform enables self-boosting ROS generation through upregulation of lactate for imaging-guided highly specific chemodynamic therapy. *Angewandte Chemie International Edition*. https://doi.org/10.1002/anie.202014415. n/a(n/a).

Wang, X., Pang, Y., Ku, G., Xie, X., Stoica, G., & Wang, L. V. (2003). Noninvasive laser-induced photoacoustic tomography for structural and functional in vivo imaging of the brain. *Nature Biotechnology*, *21*(7), 803–806. https://doi.org/10.1038/nbt839.

Wang, Y., Qiu, S., Xie, S., Zhou, L., Hong, Y., Chang, J., et al. (2019). Synthesis and characterization of oxygen-embedded quinoidal pentacene and nonacene. *Journal of the American Chemical Society*, *141*(5), 2169–2176. https://doi.org/10.1021/jacs.8b13884.

Wang, L. V., & Yao, J. (2016). A practical guide to photoacoustic tomography in the life sciences. *Nature Methods*, *13*(8), 627–638. https://doi.org/10.1038/nmeth.3925.

Wang, Z., Zhan, M., Li, W., Chu, C., Xing, D., Lu, S., et al. (2021). Photoacoustic cavitation-ignited reactive oxygen species to amplify peroxynitrite burst by photosensitization-free polymeric nanocapsules. *Angewandte Chemie International Edition*, *60*(9), 4720–4731. https://doi.org/10.1002/anie.202013301.

Weber, J., Beard, P. C., & Bohndiek, S. E. (2016). Contrast agents for molecular photoacoustic imaging. *Nature Methods*, *13*(8), 639–650. https://doi.org/10.1038/nmeth.3929.

Wilson, K., Homan, K., & Emelianov, S. (2012). Biomedical photoacoustics beyond thermal expansion using triggered nanodroplet vaporization for contrast-enhanced imaging. *Nature Communications*, *3*(1), 618. https://doi.org/10.1038/ncomms1627.

Yan, R., Hu, Y., Liu, F., Wei, S., Fang, D., Shuhendler, A. J., et al. (2019). Activatable NIR fluorescence/MRI bimodal probes for in vivo imaging by enzyme-mediated fluorogenic reaction and self-assembly. *Journal of the American Chemical Society*, *141*(26), 10331–10341. https://doi.org/10.1021/jacs.9b03649.

Zeng, L., Ma, G., Lin, J., & Huang, P. (2018). Photoacoustic probes for molecular detection: Recent advances and perspectives. *Small*, *14*(30), 1800782. https://doi.org/10.1002/smll.201800782.

Zha, M., Lin, X., Ni, J.-S., Li, Y., Zhang, Y., Zhang, X., et al. (2020). An ester-substituted semiconducting polymer with efficient nonradiative decay enhances NIR-II photoacoustic performance for monitoring of tumor growth. *Angewandte Chemie International Edition*, *59*(51), 23268–23276. https://doi.org/10.1002/anie.202010228.

Zhang, Y., He, S., Chen, W., Liu, Y., Zhang, X., Miao, Q., et al. (2021). Activatable polymeric nanoprobe for near-infrared fluorescence and photoacoustic imaging of T lymphocytes. *Angewandte Chemie International Edition*, *60*(11), 5921–5927. https://doi.org/10.1002/anie.202015116.

Zhang, Z., Xu, W., Kang, M., Wen, H., Guo, H., Zhang, P., et al. (2020). An all-round athlete on the track of phototheranostics: Subtly regulating the balance between radiative and nonradiative decays for multimodal imaging-guided synergistic therapy. *Advanced Materials*, *32*(36), 2003210. https://doi.org/10.1002/adma.202003210.

CHAPTER SEVENTEEN

A general strategy to optimize the performance of aza-BODIPY-based probes for enhanced photoacoustic properties

Anuj K. Yadav, Rodrigo Tapia Hernandez, and Jefferson Chan*
Department of Chemistry and Beckman Institute for Advanced Science and Technology, University of Illinois at Urbana – Champaign, Urbana, IL, United States
*Corresponding author: e-mail address: jeffchan@illinois.edu

Contents

Methods in Enzymology, Volume 657
ISSN 0076-6879
https://doi.org/10.1016/bs.mie.2021.06.022

Abstract

In this chapter, we describe a generalizable strategy to obtain a high PA output platform that is optimized for ratiometric imaging. Our approach entails conformationally restricting pendant aryl rings on the aza-BODIPY core to enhance orbital overlap which consequently increases the extinction coefficient. This strategy can potentially be applied to other dye platforms to enhance their signal intensity. We provide detailed protocols for the synthesis, *in vitro* characterization, and *in vivo* application.

1. Introduction

Photoacoustic (PA) imaging is a light in, sound out technology with deep tissue (up to 10 cm) imaging capability and excellent spatial resolution (Wang & Yao, 2016). This has encouraged development of small molecule-based contrast agents and acoustogenic probes. The first small molecule-based dye developed for PA imaging was aza-BODIPY (Li, Zhang, Smaga, Hoffman, & Chan, 2015). This is because aza-BODIPYs absorb maximally in the near infrared (NIR) range, exhibit excellent photostability, and can be readily functionalized (Knox & Chan, 2018). However, a robust and generalizable approach toward optimization of the photophysical properties of aza-BODIPYs by chemical tuning was lacking. The ability to chemically tune has the potential to profoundly impact the photophysical properties and generate optimized platforms.

For an ideal PA probe to be suitable for *in vivo* imaging three criteria must be satisfied. First, the absorption of the probe must be in NIR window for deep tissue penetration. Second, the signals produced by the probe must be strong enough to overcome the background signals produced by endogenous species (Weber, Beard, & Bohndiek, 2016). The strength of signal is dependent on high extinction coefficient and low quantum yield of the probe. Third, the signals between probe and its product must be well separated with minimal overlap, i.e., the probe should be well separated from its product's PA signal ($\Delta\lambda$) for resolving the ratiometric signals (Knox, Kim, Zhu, & Chan, 2018; Li et al., 2019; Lu et al., 2018; Reinhardt, Zhou, Jorgensen, Partipilo, & Chan, 2018; Roberts et al., 2018; Yin et al., 2019; Zhang, Smaga, Satyavolu, Chan, & Lu, 2017; Zhou et al., 2018). Ratiometric imaging (in comparison to intensity-based turn-on or turn-off) is ideally preferred for *in vivo* imaging because it accounts for any experimental variability in probe distribution amongst different tissue types.

In this regard, we have explored conformational restrictions to pendant aromatic substituents of the aza-BODIPY core structure. Such restrictions allow better orbital overlaps thus increasing the extinction coefficient, which in turn enhances the contrast agents PA signal. Moreover, it also red-shifts the absorbance maxima—a favorable property for *in vivo* PA imaging. Finally, due to the absence of any additional substituents the molecular weight of contrast agents does not change substantially. Thus, maintaining the physical properties intact as the original unrestricted structure.

In this chapter, we have described a generalizable strategy based on restricted bond rotations of pendant substituents to develop CRaB (conformationally restricted aza-BODIPY) platforms with improved PA intensity and $\Delta\lambda$. We then describe the rational of our design, synthetic strategies and comparisons of *in vitro* photophysical properties with unrestricted dye. Finally, we have synthesized and compared optimized PA probes for hypoxia, nitric oxide (NO) and copper (Cu) imaging with unrestricted probes *in vitro*. Lastly, we have compared the performance of the CRaB hypoxia probe with the unrestricted analog *in vivo* (Zhou, Knox, Liu, Zhao, & Chan, 2019).

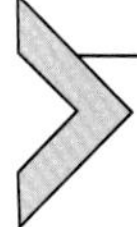

2. Strategy for optimization of aza-BODIPY platform by conformational restrictions

Generally, there are two approaches to diversify the aza-BODIPY platform. First, we can extend the conjugation to aza-BODIPY core structure by appending new aromatic rings at various positions (Ge & O'Shea, 2016; Sheng et al., 2017). These substitutions also increases the molecular weight of the dye, which may render them less biocompatible. Another approach is to enhance coplanarity between the aza-BODIPY core structure and pendant aryl substituents by conformational restrictions (Jiao et al., 2014; Leen et al., 2011; Lu, Mack, Yang, & Shen, 2014; Zhao & Carreira, 2005). This strategy does not significantly increase the molecular weight and enhances the electron delocalization to provide desired and favorable photophysical properties.

2.1 Design of conformationally restricted aza-BODIPY platforms

We hypothesized that the molar extinction coefficient will increase owing to greater orbital overlap by conformationally restricting the free rotation of the pendant aryl rings on the aza-BODIPY core with an ethylene bridge, (Fig. 1).

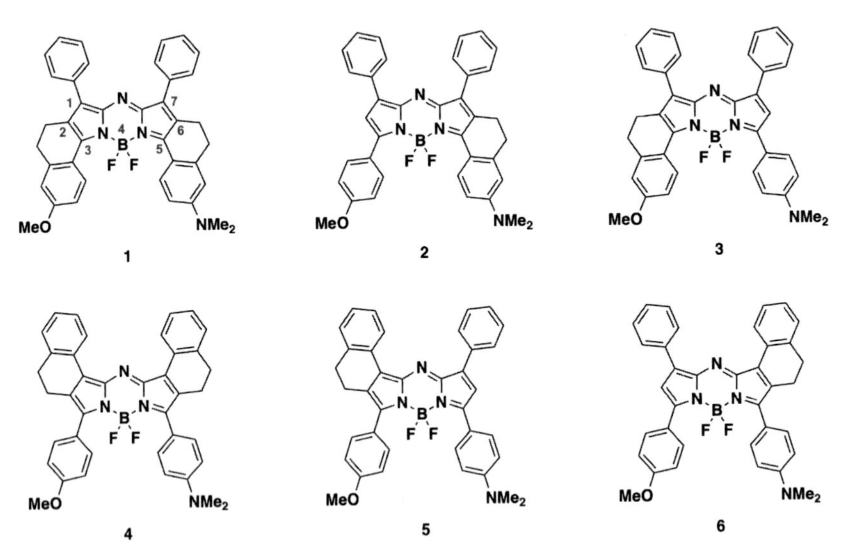

Fig. 1 Structures of conformationally restricted designed compounds 1–6. *Reprinted (adapted) with permission from Zhou, E. Y., Knox, H. J., Liu, C., Zhao, W. & Chan, J. (2019). A conformationally restricted aza-BODIPY platform for stimulus-responsive probes with enhanced photoacoustic properties.* Journal of the American Chemical Society, 141, *17601–17609. Copyright (2019) American Chemical Society.*

This, in turn, will result in a higher PA signal since the molar extinction coefficient is an indicator of how much light is being harvested by the molecule. For compound **1**, both bottom aryl rings at positions 3 and 5 are restricted. Compounds **2** and **3** each feature one restricted ring at positions 5 and 3, respectively. For compound **4**, we focused on restricting both of the top aryl rings at positions 1 and 7. Finally, for compounds **5** and **6**, restriction occurs at the aryl rings at position 1 and 7, respectively.

2.2 Photophysical characterizations and comparisons with unrestricted dye

To evaluate the impact of conformation restriction on the photophysical properties of compounds 1–6, we compared their wavelength of maximum absorbance (λ_{abs}) and emission (λ_{em}), molar extinction coefficient (ε), fluorescent quantum yield (Φ), and fold PA signal with respect to red-Hyp-1 (nonrestricted parent compound). All photophysical properties are summarized below in Table 1.

Equipment

1 UV–Vis spectrophotometer
1 Quartz cuvette with corresponding cuvette cap
3 Micropipettes and tips (P10, P200, P1000)

Table 1 Summary of photophysical properties of red-Hyp-1 and analogs measured in $CHCl_3$. Φ is measured with respect to ICG (DMSO) and fold PA is relative to red-Hyp-1.

Compounds	λ_{abs} (nm)	λ_{em} (nm)	ε ($M^{-1} \cdot cm^{-1}$)	ϕ	Fold PA
red-Hyp-1	760	798	5.40×10^4	0.043	1.00
1	796	813	1.72×10^5	0.122	2.42
2	759	791	1.57×10^5	0.195	2.26
3	789	822	1.35×10^5	0.054	2.36
4	757	803	7.95×10^4	0.047	1.00
5	759	785	9.05×10^4	0.195	0.99
6	782	818	8.92×10^4	0.061	1.24

Reprinted (adapted) with permission from Zhou, E. Y., Knox, H. J., Liu, C., Zhao, W. & Chan, J. (2019). A conformationally restricted aza-BODIPY platform for stimulus-responsive probes with enhanced photoacoustic properties. *Journal of the American Chemical Society, 141*, 17601–17609. Copyright (2019) American Chemical Society.

Reagents

3 aliquots of red-Hyp-1 (Zhou, Knox, Reinhardt, Partipilo, & Chan, 2020)
3 aliquots of compounds 1–6
Organic solvents: $CHCl_3$ and DMSO
Aqueous buffer: 50% ACN in PBS, adjusted to pH 7.4

Procedure

1. Turn spectrophotometer on for at least 15 min before the measurements to warm up the instrument
2. Set the instrument to scan from 400 to 900 nm with a step-size of 0.5 nm
3. Prepare stock solution of the dye being analyzed
4. Add a known volume of $CHCl_3$ or aqueous buffer to a quartz cuvette (0.5–1 mL) and use cap to minimize the solvent evaporation
5. Use the baseline function to account for background
6. Add a known volume of the dye at the desired concentration to the cuvette and mix well by pipetting up and down five times and recap the cuvette
7. Measure the absorbance spectrum
8. Repeat steps 5–7 to obtain at least five scans so that the peak absorbance value falls between 0.1 and 1.0 a.u
9. Continue steps 2–8 with each batch of dyes (A, B, and C) to obtain a total of 15 data points for each compound
10. Record the wavelength of maximum absorbance for each compound and plot absorbance vs concentration to determine the extinction coefficient using Beer-Lambert Law

2.3 Photophysical characterizations of protonated compounds and comparison with unrestricted dye: Red-Hyp-1

To evaluate each dye for potential acoustogenic probe development, we utilize acidic conditions (1% TFA) to protonate the *N,N*-diethyl moiety and use this as a proxy for an *N*-oxide-based hypoxia probe. This is based on the premise that protonation and trigger installation at this position will result in a similar blue-shift in the λ_{abs}. This allows us to approximate the $\Delta\lambda$ which is defined as the λ_{abs} before and after protonation. The experiment was performed as described above in Section 2.2. The results are summarized below in Table 2.

2.4 Photophysical characterizations in tissue phantom

For *in vitro* PA studies, the application of tissue phantoms are generally preferred due to the following reasons: (i) it mimics light scattering in tissues and (ii) it enables access to larger area of sample for irradiation by the excitation laser. The tissue phantom is prepared according to the published protocol (Zhou et al., 2020).

Equipment

1 Photoacoustic tomographer (Nexus 128, Endra Life Sciences)
1 Tissue-mimicking phantom (Zhou et al., 2020)
2 FEP tubes: FEP tubing (0.08 in. diameter, 5 cm length)
4 FEP caps: FEP tubing (0.12 in. diameter, 0.25 cm length)

Table 2 Measurement of absorbances (in $CHCl_3$) of protonated red-HyP analogs as proxy for *N*-oxide-based hypoxia probe.

Compounds	λabs (nm)	λ_{abs} (1% TFA) (nm)	Δλ (nm)
red-Hyp-1	760	669	91
1	796	719	77
2	759	699	60
3	789	686	103
4	757	675	82
5	759	697	62
6	782	690	92

Reprinted (adapted) with permission from Zhou, E. Y., Knox, H. J., Liu, C., Zhao, W. & Chan, J. (2019). A conformationally restricted aza-BODIPY platform for stimulus-responsive probes with enhanced photoacoustic properties. *Journal of the American Chemical Society, 141*, 17601–17609. Copyright (2019) American Chemical Society.

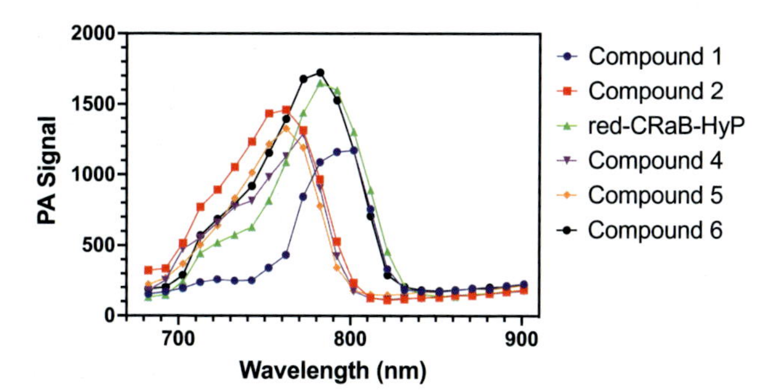

Fig. 2 PA spectra acquired in tissue phantom for compounds 1, 2, 4, 5, 6 and red-CRaB-Hyp (**3**). *Reprinted (adapted) with permission from Zhou, E. Y., Knox, H. J., Liu, C., Zhao, W. & Chan, J. (2019). A conformationally restricted aza-BODIPY platform for stimulus-responsive probes with enhanced photoacoustic properties.* Journal of the American Chemical Society, 141, *17601–17609. Copyright (2019) American Chemical Society.*

Reagents

Organic solvent: $CHCl_3$

3 aliquots each of red-Hyp-1, compounds 1–6 (10 μM in $CHCl_3$)

Procedure

1. Warm the photoacoustic tomographer for at least 30 min before any measurements are recorded
2. Prepare 10 μM solution from the aliquots in $CHCl_3$ and carefully transfer 200 μL into a FEP tube using P200 micropipette
3. Seal one edge of the tube by folding the edge on to itself and securing with a FEP cap. Thread the uncapped edge through the tissue mimicking phantom and seal the edge as previously
4. Repeat steps 2 and 3 with another FEP tube so that the phantom contains two samples of the dye
5. Acquire PA spectrum of the phantom by placing in the bowl with enough deionized (Milli-Q) water just to submerge tube. Place a weight over the phantom to prevent it from floating
6. Repeat steps 2–5 for each compound
7. Plot PA signal *vs* wavelength (nm) (Fig. 2)

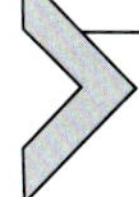

3. Selection of conformationally restricted aza-BODIPY platform

The optimized aza-BODIPY platform for PA imaging is selected after comparing the photophysical properties of compounds 1–6. Based on the ε

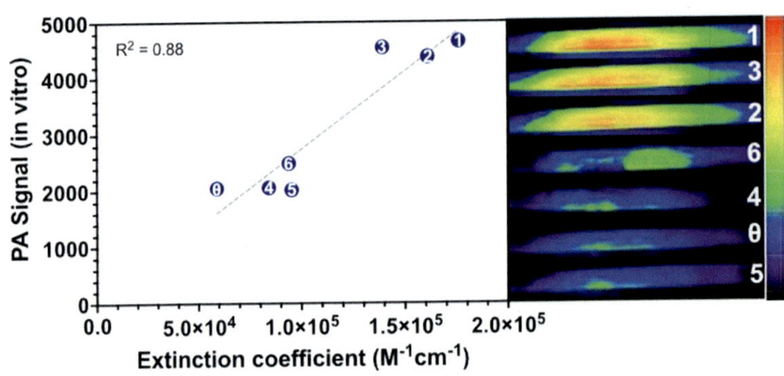

Fig. 3 Plot of extinction coefficient *vs in vitro* PA signal (10 μM, $CHCl_3$, in tissue-mimicking phantom) of red-HyP-1 (indicated by the θ) and compounds 1–6 showing linear relationship. Representative *in vitro* PA images in tissue mimicking phantom are displayed on the right (in decreasing order of intensity). *Reprinted (adapted) with permission from Zhou, E. Y., Knox, H. J., Liu, C., Zhao, W. & Chan, J. (2019). A conformationally restricted aza-BODIPY platform for stimulus-responsive probes with enhanced photoacoustic properties.* Journal of the American Chemical Society, 141, *17601–17609. Copyright (2019) American Chemical Society.*

values and PA signal strength, it is apparent that compounds 1–3 are more sensitive than 4–6 (Fig. 3). Among compounds 1–3, 3 has the largest Δλ which is ideal for enhancing the ratiometric capabilities of an acoustogenic probe. Thus, compound 3 was chosen for the development of PA probes for *in vivo* imaging.

4. Design of CRaB probes

After identifying compound 3 as the most sensitive aza-BODIPY platform, we designed three probes based on our previous work to demonstrate generalizability of our approach. First, installation of an *N*-oxide functional group allows us to develop CRaB-Hyp, a probe for hypoxia. Second, appending a Cu(II)-responsive picolinic ester trigger allows us to access CRaB-OMe-APC, a probe for Cu(II) detection. Finally, we incorporate an *N*-nitroso functional group to afford CRaB-photoNOD, which can deliver an equivalent of nitric oxide upon irradiation of NIR light.

5. Synthesis of CRaB probes

Procedures to synthesize the CRaB-based probes mentioned above require an understanding of standard organic chemistry techniques and access to standard laboratory equipment.

5.1 Synthesis of CRaB-HyP-1 and red-CRaB-HyP-1

red-CRaB-HyP and CRaB-HyP were synthesized over 10 steps as follows (Scheme 1):

Compound 7

1. Add styrene (22.9 mL, 200 mmol, 1 eq.) to a 250 mL round-bottom flask equipped with a magnetic stir bar and cap with a rubber septum

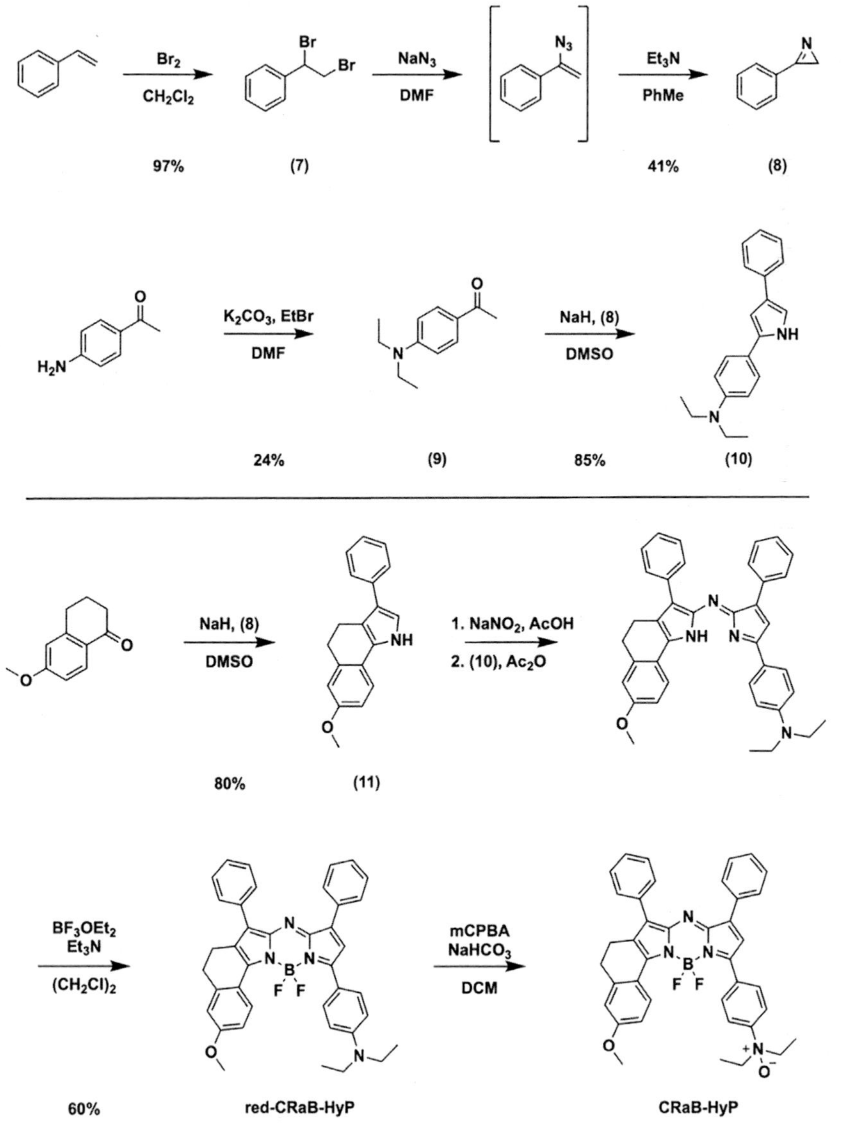

Scheme 1 Synthetic route to CRaB-HyP and red-CRaB-HyP.

2. Measure 400 mL of dichloromethane with a graduated cylinder. Briefly remove the rubber septum from the reaction vessel, pour the solvent into the flask with continuous magnetic stirring
3. Equip the flask with an addition funnel
4. Cool the mixture to 0 °C in an ice-bath
5. Transfer bromine (12.3 mL, 240 mmol, 1.2 eq.) into the addition funnel and add to the flask dropwise
6. Allow the red mixture to warm to room temperature and stir for 1 h or until reaction is complete by TLC
7. Cool the reaction to 0 °C in an ice-bath, dilute with water and slowly add saturated sodium thiosulfate until complete decolorization is observed
8. Transfer the mixture into a separatory funnel and isolate the organic layer
9. Dry the organic layer over sodium sulfate and concentrate by rotary evaporation into a white powder
10. Use directly without purification

Compound 8

1. Measure 15.0 mL of toluene and 2.0 mL of triethylamine with graduated cylinders and add to a 100 mL round-bottom flask equipped with a magnetic stir bar
2. Preheat the flask to 100 °C, this will be used for step 10
3. Pre-pack a silica column with 5% ethyl acetate/hexanes. This will be used immediately after step 10
4. Add (1,2-dibromoethyl)benzene (**7**, 2.64 g, 10 mmol, 1.0 eq.) to a 100 mL round-bottom flask equipped with a magnetic stir bar
5. Equip an addition funnel to the flask with a rubber septum. Flush the funnel and the reaction vessel with nitrogen
6. Transfer 40.0 mL of anhydrous DMF to the addition funnel. Add the solvent to the flask. While flowing nitrogen into the funnel quickly remove the funnel and replace it with a rubber septum to maintain the flask free of air
7. Measure 1.95 g of sodium azide (30 mmol, 3.0 eq.). Do not measure using a metal spatula. Add the sodium azide into the flask and stir the mixture vigorously for 5 h or until complete consumption of **7** is observed by TLC (Hexanes)
8. Dilute the reaction mixture with water, transfer into a separatory funnel, add diethyl ether and allow the layers to separate. Isolate the organic layer and wash with an equal volume of brine

9. Dry the organic layer over sodium sulfate and concentrate by rotary evaporation into an oil
10. Transfer the oil into the preheated round-bottom flask described in step 2, increase the temperature to 110 °C and let the reaction stir for 40 min
11. Cool the reaction rapidly in an ice-bath. Transfer the material directly into the pre-packed silica column from step 3 and purify the product with 5% ethyl acetate/Hexanes to produce a pale-yellow oil

Compound 9

1. Add 4-aminoacetophenone (1.00 g, 7.50 mmol, 1.0 eq.), potassium carbonate (2.28 g, 16.5 mmol, 2.2 eq.), DMF (7.5 mL) and bromoethane (1.2 mL, 16.5 mmol, 2.2 eq.) into a 50 mL round-bottom flask equipped with a magnetic stir bar and a rubber septum
2. Heat the reaction mixture to 60 °C and allow to stir for 24 h
3. Cool the reaction to room temperature and dilute the mixture with water
4. Transfer the mixture into a separatory funnel, extract the material with diethyl ether by isolating the organic layer
5. Dry the organic layer over sodium sulfate and concentrate by rotary evaporation
6. Purify the product through a silica column with a gradient 15–30% ethyl acetate/Hexanes with 0.1% triethylamine to give a pale oil that slowly solidifies

Compound 10

1. Carefully add NaH (60% wt., 246 mg, 6.15 mmol, 3.4 eq.) and 1.5 mL of anhydrous DMSO to a 25 mL round-bottom flask equipped with a magnetic stir bar and a rubber septum
2. Purge the reaction flask with nitrogen
3. Add 4-diethylaminoacetophenone (**9**, 350 mg, 1.83 mmol, 1.0 eq.) to the flask and stir for 20 min
4. Add 3-phenyl-2*H*-azirine (**8**, 679 mg, 5.8 mmol, 3.4 eq.) to the reaction vessel and stir at room temperature for 8 h
5. Dilute the mixture with water
6. Transfer the mixture into a separatory funnel, extract the material with diethyl ether by isolating the organic layer
7. Wash the organic layer with equal volume of brine
8. Dry the organic layer over sodium sulfate and concentrate by rotary evaporation into an orange solid
9. Purify the product through a neutral alumina column with a gradient of 15–35% ethyl acetate/Hexanes resulting in a beige solid

Compound 11

1. Carefully add NaH (60% wt., 278 mg, 6.96 mmol, 1.20 eq.) and 5.8 mL of anhydrous DMSO to a 25 mL round-bottom flask equipped with a magnetic stir bar and a rubber septum
2. Purge the reaction flask with nitrogen
3. Add 6-methoxy-1-tetralone (1.02 g, 5.80 mmol, 1.00 eq.) to the flask and stir at room temperature for 30 min
4. Cool the reaction vessel to 0 °C in an ice-bath and add 3-phenyl-2*H*-azirine (**8**, 0.679 g, 5.80 mmol, 1.00 eq.) dropwise
5. Allow the reaction to warm to room temperature and stir for 1 h or until completion by TLC
6. Dilute the mixture with water and saturated sodium bicarbonate
7. Transfer the mixture into a separatory funnel, add diethyl ether to separate the organics from the aqueous layer. Isolate the organic layer
8. Dry the organic layer over sodium sulfate and concentrate by rotary evaporation
9. Purify the product through a silica column with a gradient of 10–15% ethyl acetate/Hexanes resulting in a beige solid

Red-CRaB-HyP

1. Add 7-methoxy-3-phenyl-4,5-dihydro-1*H*-benzo[*g*]indole (**11**, 83 mg, 0.30 mmol, 1 eq.) in a 25 mL round-bottom flask equipped with a magnetic stir bar and a rubber septum
2. Purge the reaction flask with nitrogen
3. Add sodium nitrite (21 mg, 0.30 mmol, 1 eq.) and acetic acid (3 mL, 10 mL per mmol pyrrole) in the flask, allow the mixture to stir at room temperature for 20 min
4. Add *N*,*N*-diethyl-4-(4-phenyl-1*H*-pyrrol-2-yl)aniline (**10**, 87 mg, 0.30 mmol, 1 eq.) and acetic anhydride (1.2 mL, 4 mL per mmol of pyrrole) into the mixture and stir at room temperature for 20 min
5. Heat the reaction to 80 °C in an oil-bath and stir for 30 min
6. Cool the reaction to room temperature and dilute with cold saturated solution of sodium bicarbonate, filter the material by vacuum filtration to isolate the heterodimer as a dark solid
7. Suspend the solid in dichloromethane (20 mL per mmol of dimer) then add triethylamine (17 eq.) and $BF_3.OEt_2$ (19 eq.). Allow the mixture to stir at room temperature for 20 min
8. Heat the reaction to 80 °C and stir for 30 min
9. Allow the reaction to cool to room temperature, dilute the mixture with water and transfer the mixture to a separatory funnel

10. Extract the material with dichloromethane, isolate the organic layer
11. Dry the organic layer over sodium sulfate and concentrate by rotary evaporation
12. Purify the product through a silica column with 75% dichloromethane/Hexanes with 0.1% Et_3N

CRaB-HyP

1. Add red-CRaB-Hyp-1 (100 mg, 0.16 mmol, 1.0 eq.) in a 25 mL round-bottom flask equipped with a magnetic stir bar and a rubber septum
2. Dissolve the material with dichloromethane (3.2 mL) and add sodium bicarbonate (14.8 mg, 0.176 mmol, 1.1 eq.)
3. Cool the reaction vessel to 0 °C in an ice bath
4. Add mCPBA (77% wt., 39.6 mg, 0.176 mmol, 1.1 eq.) to the reaction mixture and allow the reaction to stir at 0 °C for 5 min
5. Allow the reaction to warm to room temperature and allow the reaction to stir for another 25 min
6. Dilute the reaction mixture with a saturated solution of sodium bicarbonate, transfer the mixture to a separatory funnel, and extract the material with ethyl acetate and isolate the organic layer
7. Dry the organic layer over sodium sulfate and concentrate by rotary evaporation
8. Purify the material through a silica column with 10% methanol/dichloromethane

5.2 Synthesis of CRaB analog of photoNOD and rNOD

r-CRaB-NOD and CRaB-photoNOD were synthesized over 11 steps as follows (Scheme 2):

Compound 12

1. Add 4′-aminoacetophenone (6.7 g, 50 mmol, 1.0 eq.) to a 250 mL round-bottom flask equipped with a magnetic stir bar and a rubber septum
2. Add di-tert-butyl decarbonate (13.1 g, 60 mmol, 1.2 eq.) to the flask
3. Flush the reaction vessel with nitrogen
4. Dissolve the material with 60 mL of 1,4-dioaxane, heat the reaction to 100 °C and stir at that temperature for 8.5 h
5. Cool to room temperature and concentrate by rotary evaporation into an oil that should slowly crystallize into a solid
6. Wash the solid with 25% ethyl acetate/Hexanes, vacuum filtrate the solid to produce one batch of pure product

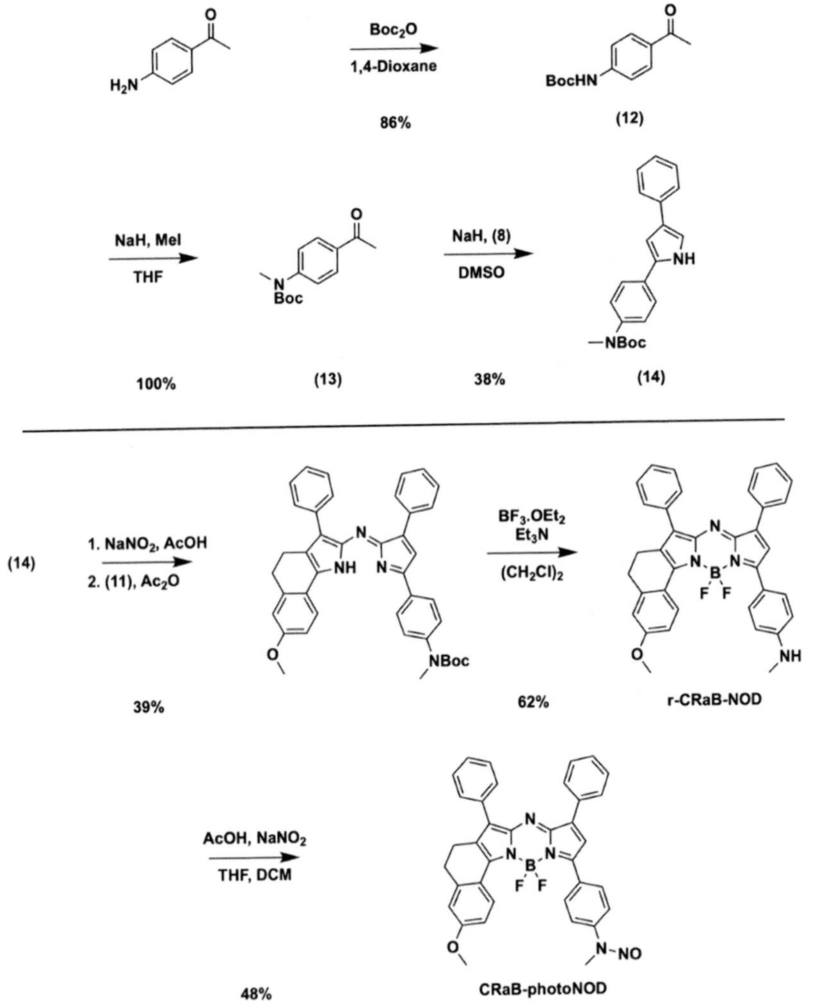

Scheme 2 Synthetic route to CRaB-photoNOD and r-CRaB-NOD.

7. Collect the filtrate and concentrate into a solid, wash the solid with 20% ethyl acetate/Hexanes and filtrate to afford a second batch of pure product

Compound 13

1. Add *tert*-butyl (4-acetylphenyl)carbamate (**12**, 3.0 g, 12.8 mmol, 1.0 eq.) to a 250 mL round-bottom flask equipped with a magnetic stir bar and a rubber septum

2. Flush the reaction vessel with nitrogen
3. Dissolve the material with 40 mL of anhydrous THF
4. Cool the reaction to 0 °C in an ice-bath, add sodium hydride (60% wt. dispersion in mineral oil) (0.61 g, 15.3 mmol, 1.2 eq.) portion-wise over 10 min. After addition is complete, stir for 20 min at 0 °C
5. Add methyl iodide (1.57 mL, 31.9 mmol, 2.5 eq.), warm to room temperature and stir for 40 min or until completion as observed by TLC (15% ethyl acetate/Hexanes)
6. Concentrate the reaction mixture by rotary evaporation
7. Dilute the mixture with ethyl acetate and transfer to a separatory funnel
8. Wash the mixture with brine (3 ×) and isolate the organic layer
9. Dry the organic layer over sodium sulfate and concentrate by rotary evaporation to obtain a yellow oil

Compound 14

1. Add sodium hydride (60% wt. dispersion in mineral oil) (48 mg, 1.2 mmol, 1.2 eq.) and dissolve with 1.0 mL of anhydrous DMSO at room temperature in a 25 mL round-bottom flask equipped with a magnetic stir bar and a rubber septum
2. Add *tert*-butyl (4-acetylphenyl)(methyl)carbamate (**13**, 250 mg, 1.0 mmol, 1 eq.) to the mixture and stir for 1 h at room temperature
3. Add 3-phenyl-2*H*-azirine (**8**, 140 mg, 1.20 mmol, 1.2 eq.) to the reaction mixture and stir for 1 h
4. Dilute the reaction mixture with water and transfer into a separatory funnel. Extract the material with diethyl ether by isolating the organic layer
5. Wash the organic layer with brine and isolate the organic layer
6. Dry the organic layer over sodium sulfate and concentrate by rotary evaporation
7. Purify the material through a silica column with a gradient of 15–30% ethyl acetate/Hexanes to obtain a pale green oil, dry the oil under vacuum to obtain a yellow foam

r-CRaB-NOD

1. Add 7-methoxy-3-phenyl-4,5-dihydro-1*H*-benzo[*g*]indole (**11**, 78.1 mg, 0.284 mmol, 1.0 eq.), sodium nitrite (19.8 mg, 0.287 mmol, 1.0 eq.), and acetic acid (2.9 mL) into a 25 mL round-bottom flask equipped with a magnetic stir bar and a rubber septum
2. Stir at room temperature for 20 min

3. Add *tert*-butyl methyl(4-(40phenyl-1*H*-pyrrol-2-yl)phenyl)carbamate (**14,** 99.5 mg, 0.286 mmol, 1.0 eq.) and acetic anhydride into the flask. Stir at room temperature for 20 min
4. Heat the reaction to 80 °C in an oil-bath and stir at this temperature for 30 min
5. Cool the reaction to room temperature and dilute the mixture with a cold saturated solution of sodium bicarbonate
6. Filter the mixture to obtain the heterodimer as a dark solid
7. Purify the material through an alumina column using 3% ethyl acetate/toluene
8. Add the purified heterodimer (31.5 mg, 0.050 mmol, 1.0 eq.) into a 25 mL round-bottom flask equipped with a magnetic stir bar and a rubber septum
9. Suspend the material in 1.0 mL dichloromethane, add triethylamine (0.12 mL) and BF_3OEt_2 (0.12 mL) and stir at room temperature for 20 min
10. Heat the reaction mixture to 80 °C and stir for 30 min
11. Allow the mixture to cool to room temperature, dilute with water and transfer into a separatory funnel. Extract the material with dichloromethane by isolating the organic layer
12. Wash the organic layer with saturated sodium bicarbonate and isolate the organic layer
13. Dry the organic layer over sodium sulfate and concentrate by rotary evaporation
14. Purify the material through a silica column with 3% ethyl acetate/toluene

CRaB-photoNOD

1. Add r-CRaB-NOD (16.4 mg, 0.028 mmol, 1.0 eq.) in a 25 mL round-bottom flask equipped with a magnetic stir bar and a rubber septum
2. Dissolve the material with acetic acid (1.0 mL), anhydrous THF (2.0 mL), and dichloromethane (1.0 mL).
3. Cool the reaction mixture to 0 °C in an ice bath
4. Add sodium nitrite (9.7 mg, 0.14 mmol, 5.0 eq.) and allow the mixture to stir at the same temperature for 1 h
5. Dilute the reaction mixture with a saturated solution of sodium bicarbonate and transfer to a separatory funnel. Extract the material with ethyl acetate by isolating the organic layer
6. Dry the organic layer over sodium sulfate and concentrate by rotary evaporation to obtain a green film

5.3 Synthesis of CRaB analog of OMe-APC and t-OMe-APC

t-CRaB-OMe-APC and CRaB-OMe-APC were synthesized over 13 steps as follows (Scheme 3):

Compound 15

1. Add 2,6-dichlorophenol (3.26 g, 20 mmol, 1.0 eq.) to a 100 mL round-bottom flask equipped with a magnetic stir bar and a rubber septum

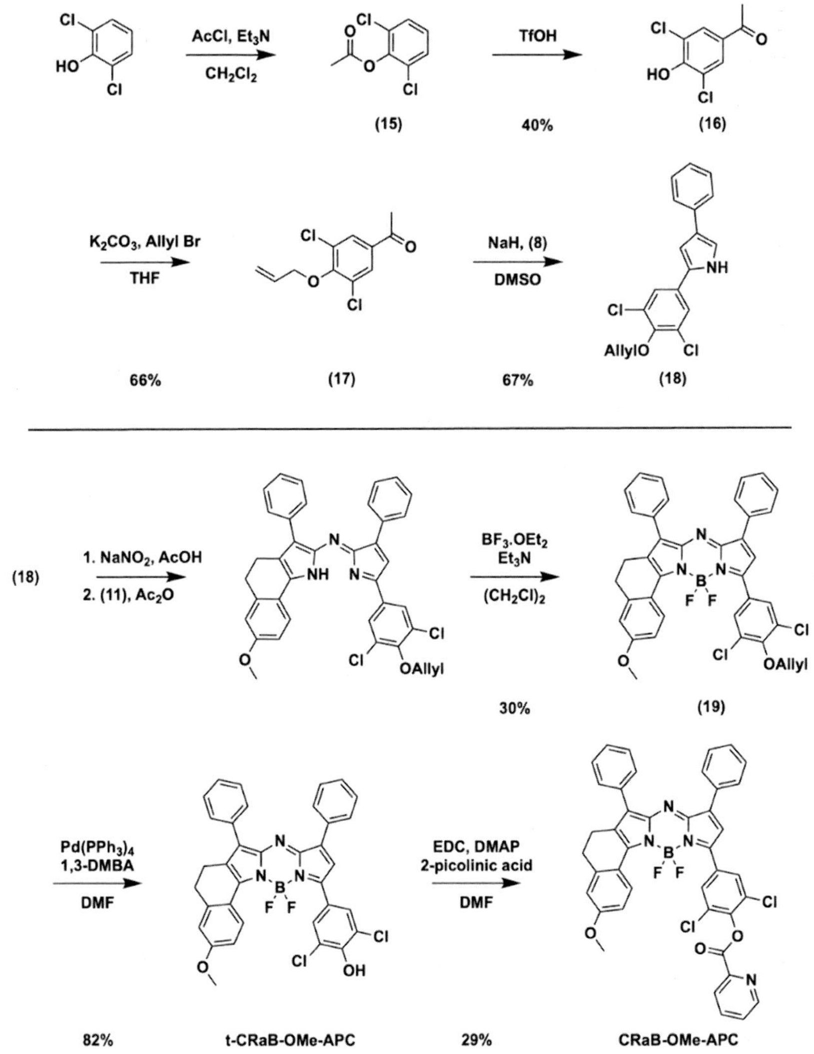

Scheme 3 Synthetic route to t-CRaB-OMe-APC and CRaB-OMe-APC.

2. Add 36 mL of dichloromethane and triethylamine (7.0 mL, 50 mmol, 2.5 eq.) and cool the reaction mixture to 0 °C in an ice-bath
3. Add acetyl chloride (1.7 mL, 24 mmol, 1.2 eq.) dropwise, warm the thick yellow slurry to room temperature and stir for 4 h or until completion as observed by TLC
4. Wash the reaction mixture with saturated sodium bicarbonate to quench the reaction, transfer to a separatory funnel and isolate the organic layer
5. Dry the organic layer over sodium sulfate and concentrate by rotary evaporation to obtain a yellow oil

Compound 16

1. Add 2,6-dichlorophenyl acetate (**15**, 450 mg, 1.55 mmol, 1.0 eq.) to a 50 mL round-bottom flask equipped with a magnetic stir bar and a rubber septum
2. Add triflic acid (7 mL) to the flask, heat the reaction mixture to 40 °C and stir for 18 h
3. Cool the reaction to 0 °C in an ice-bath and basify with a saturated sodium carbonate solution. Transfer the solution to a separatory funnel and wash with ethyl acetate to remove residual starting material by isolating the aqueous layer
4. Acidify the aqueous layer with 1 M aq. HCl. Transfer to a separatory funnel and extract the material with ethyl acetate by isolating the organic layer
5. Dry the organic layer over sodium sulfate and concentrate by rotary evaporation

Compound 17

1. Add 3,5-dichloro-5-hydroxyacetophenone (**16**, 500 mg, 2.44 mmol, 1.0 eq.) to a 25 mL round-bottom flask equipped with a magnetic stir bar and a rubber septum
2. Add potassium carbonate (674 mg, 4.88 mmol, 2.0 eq.), 2.4 mL of acetonitrile, and allyl bromide (337 μL, 3.90 mmol, 1.6 eq.)
3. Attach a reflux condenser to the flask and reflux the reaction for 4.5 h
4. Concentrate the crude material by rotary evaporation and purify the material through a silica column with 10% ethyl acetate/Hexanes to obtain a clear oil

Compound 18

1. Add sodium hydride (60% wt. dispersion in mineral oil) (48 mg, 1.20 mmol, 1.0 eq.) and anhydrous DMSO to a 25 mL round-bottom flask equipped with a magnetic stir bar and a rubber septum
2. Add 1-(4-(allyloxy)-3,5-dichlorophenyl)ethan-1-one (**17**, 245 mg, 1.00 mmol, 1.0 eq.) and allow the mixture to stir for 40 min at room temperature
3. Add 3-phenyl-2*H*-azirine (**8,** 140 mg, 1.20 mmol, 1.0 eq.) to the flask and allow to stir for 1 h
4. Dilute the reaction mixture with water, transfer to a separatory funnel and extract the material with diethyl ether by isolating the organic layer
5. Dry the organic layer over sodium sulfate, concentrate the material by rotary evaporation
6. Purify the material through a silica column with a gradient of 8% to 15% ethyl acetate/hexanes to obtain a beige solid

Compound 19

1. Add 7-methoxy-3-phenyl-4,5-dihydro-1*H*-benzo[*g*]indole (**11**, 156 mg, 0.567 mmol, 1.0 eq.), sodium nitrite (39.2 mg, 0.568 mmol, 1.0 eq.), and acetic acid (6.5 mL) to a 25 mL round-bottom flask equipped with a magnetic stir bar and a rubber septum. Allow the mixture to stir for 20 min at room temperature
2. Add 2-(4-(allyloxy)-3,5-dichlorophenyl)-4-phenyl-1*H*-pyrrole (**18**, 195 mg, 0.568 mmol, 1.0 eq.) and acetic anhydride (2.6 mL) to the flask. Stir for 20 min at room temperature
3. Heat the reaction mixture to 80 °C and stir for another 30 min
4. Cool to room temperature and dilute with a cold saturated solution of sodium bicarbonate
5. Filter the material to isolate the heterodimer as a dark solid
6. Transfer the solid to a 25 mL round-bottom flask equipped with a magnetic stir bar and a rubber septum
7. Suspend the solid in 6 mL of dichloromethane
8. Add triethylamine (680 μL) and BF_3OEt_2 (680 μL) to the flask and stir the mixture for 20 min at room temperature
9. Heat the reaction mixture to 80 °C and stir for another 30 min
10. Allow the reaction to cool to room temperature and dilute with water. Transfer to a separatory funnel and extract the material with ethyl acetate by isolating the organic layer

11. Wash the organic layer with saturated sodium bicarbonate
12. Isolate the organic layer and dry it over sodium sulfate and concentrate by rotary evaporation
13. Purify the material through an alumina column with a gradient from 50% dichloromethane/Hexanes to 75% dichloromethane/Hexanes

t-CRaB-OMe-APC

1. Add compound **19** (34 mg, 0.050 mmol, 1.0 eq.), 1,3-dimethylbabituric acid (10.2 mg, 0.065 mmol, 1.3 eq.), and $Pd(PPh_3)_4$ (11.1 mg, 0.0096 mmol, 0.2 eq.) to a 25 mL two-neck round-bottom flask equipped with a magnetic stir bar and rubber septa
2. Dry the solids under vacuum and flush with nitrogen
3. Add 1.0 mL of anhydrous DMF and stir for 1 h
4. Pour the reaction mixture into brine and transfer to a separatory funnel
5. Extract the material with ethyl acetate by isolating the organic layer
6. Dry the organic layer over sodium sulfate and concentrate it by rotary evaporation
7. Purify the product through a silica column with 0.1% AcOH in dichloromethane

CRaB-OMe-APC

1. Add t-CRaB-OMe-APC (100 mg, 0.157 mmol, 1.0 eq.) to a 25 mL round-bottom flask equipped with a magnetic stir bar and a rubber septum
2. Flush the reaction vessel with nitrogen
3. Add 2-picolinic acid (77.5 mg, 0.630 mmol, 4.0 eq.), 4-(dinmethylamino) pyridine (3.4 mg, 0.0278 mmol, 0.18 eq.), EDC-HCl (46.1 mg, 0.241 mmol, 1.5 eq.), and 7.8 mL of anhydrous DMF
4. Stir for 1 h at room temperature
5. Dilute the reaction mixture with brine and transfer to a separatory funnel
6. Extract the material with ethyl acetate by isolating the organic layer
7. Wash the organic layer with brine and 1% aq. HCl, isolate the organic layer
8. Dry the organic layer over sodium sulfate and concentrate by rotary evaporation
9. Purify the product through a silica column with a gradient of 30% to 60% to 100% ethyl acetate/Hexanes with 0.1% AcOH

6. *In vitro* characterization of CRaB analogs

After synthesizing the CRaB-based probes for hypoxia, Cu, and NO delivery, we evaluated their photophysical properties *in vitro* (Table 3). In all

Table 3 Measurement of PA wavelengths (λ_{red} and λ_{blue}) for each probe-product pair in $CHCl_3$ and in 1:1 acetonitrile/PBS mixture.

		$CHCl_3$		1:1 MeCN/PBS	
Probe	**Product**	λ_{blue}	λ_{red}	λ_{blue}	λ_{red}
HyP-1	red-HyP-1	680	770	680	770
photoNOD-1	rNOD-1	680	730	680	740
Ome-APC	t-Ome-APC	680	750	680	750
CRaB-HyP	red-CRaB-HyP	690	780	680	790
CRaB-photoNOD	r-CRaB-NOD	700	760	680	770
CRaB-Ome-APC	t-CRaB-Ome-APC	680	780	680	780

Reprinted (adapted) with permission from Zhou, E. Y., Knox, H. J., Liu, C., Zhao, W. & Chan, J. (2019). A conformationally restricted aza-BODIPY platform for stimulus-responsive probes with enhanced photoacoustic properties. *Journal of the American Chemical Society, 141*, 17601–17609). Copyright (2019) American Chemical Society.

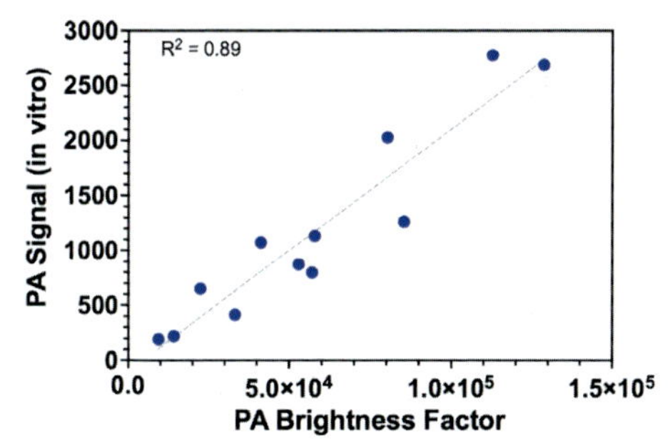

Fig. 4 Plot of *in vitro* PA signal vs calculated PABF for all 12 compounds. *Reprinted (adapted) with permission from Zhou, E. Y., Knox, H. J., Liu, C., Zhao, W. & Chan, J. (2019). A conformationally restricted aza-BODIPY platform for stimulus-responsive probes with enhanced photoacoustic properties.* Journal of the American Chemical Society, 141, *17601–17609. Copyright (2019) American Chemical Society.*

cases the restricted probes have higher sensitivities compared to unrestricted ones. Finally, when *in vitro* PA signals are plotted against PA brightness factor ($PABF = \varepsilon(1\text{-}\Phi)$) it shows a linear relationship (Fig. 4).

7. *In vivo* characterization of CRaB-Hyp

Once the *in vitro* characterization was completed, the CRaB-Hyp probe and red-CRaB-Hyp product was used for *in vivo* imaging. To perform *in vivo* experiments, training in ethics and practice of handling animals and institutional approvals are required. All the *in vivo* experiments are performed upon approval from the Institutional Animal Care and Use Committee of the University of Illinois at Urbana – Champaign, following the principles outlined by the American Physiological Society on research animal use. For this study male and female BALB/c mice between 6 and 8 weeks old are acquired from a breeding colony.

7.1 Identification of optimal wavelengths for monitoring CRaB-Hyp probes *in vivo*

Since the solubility of CRaB-Hyp and red-CRaB-Hyp may be different in tissues compared to *in vitro* solvent, we conducted the following experiment to identify the optimal wavelengths of the dye for *in vivo* ratiometric imaging.

Equipment

1 Photoacoustic tomographer (Nexus 128, Endra Life Sciences)
1 Vaporizer equipped with isoflurane and oxygen (for anesthesia)
4 sterile Eppendorf tubes
4 sterile insulin syringes
1 electric shaver, for hair removal
1 bottle depilatory cream, for hair removal

Animals

1 Female BALB/c mouse, 6–8 weeks old

Reagents

CRaB-Hyp and Hyp-1
Red-CRaB-Hyp and red-Hyp-1
Sterile saline containing 15% DMSO by volume

Procedure

1. Turn on the photoacoustic tomographer for at least 30 min before imaging
2. Use isoflurane/oxygen chamber to anesthetize the mouse and to keep it anesthetized, place it under a stream of isofluorene through a nose cone for the remainder of the experiment
3. Shave and remove residual hair from both flanks of the mouse using depilatory cream and washing with warm damp gauze

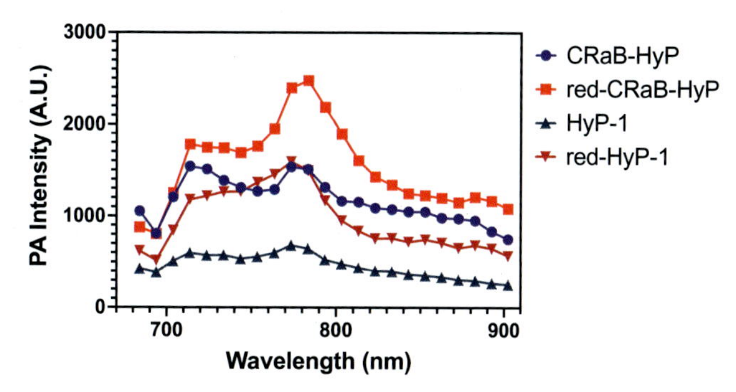

Fig. 5 *In vivo* PA spectrum of Hyp-1, red-Hyp-1, CRaB-Hyp and red-CRaB-Hyp are recorded after 1 h of probe injection into subcutaneous space. 710 nm and 770 nm are selected as λ_{blue} and λ_{red}, respectively. *Reprinted (adapted) with permission from Zhou, E. Y., Knox, H. J., Liu, C., Zhao, W. & Chan, J. (2019). A conformationally restricted aza-BODIPY platform for stimulus-responsive probes with enhanced photoacoustic properties.* Journal of the American Chemical Society, 141, *17,601–17,609. Copyright (2019) American Chemical Society.*

4. Formulate 50 μM CRaB-Hyp in the saline/DMSO solution under sterile conditions and inject 50 μL subcutaneously in one flank
5. Acquire the PA images of the flank after placing the mouse in the tomographer bowl (continuous mode, 6 s rotation time), scanning from 680 to 900 nm at 10 nm intervals
6. Repeat steps 2–5 for red-CRaB-Hyp by injecting the dye into the opposite flank
7. Process the data after reconstruction of three-dimensional images by selecting equal volume ROIs, and quantify the mean PA signal from each ROI, using Horos DICOM viewer
8. Plot PA signal *vs* excitation wavelength to determine the optimal maximum wavelength required for *in vivo* imaging (Fig. 5)

7.2 Application and comparison of CRaB-Hyp/red-CRaB-Hyp with unrestricted probes Hyp-1/red-Hyp-1 *in vivo*

The goal of this experiment is to show that ratiometric fold turn-on from CRaB analogs are higher than the unrestricted HyP probes. We generate 4T1 tumors in mice flanks and image after injecting probes at various time points.

Equipment

1 Photoacoustic tomographer (Nexus 128, Endra Life Sciences)

1 Vaporizer equipped with isoflurane and oxygen (for anesthesia)

24 Insulin syringes

Cells and animals

4T1 cells

6 male and female BALB/c mice, 6–8 weeks old

Reagents

Matrigel

Serum-free RPMI 1640

CRaB-Hyp and Hyp-1

Sterile saline

Trypsin

Procedure

1. Culture 4T1 cells under standard conditions and at least 3h before implantation, transfer Matrigel to an ice-bath for thawing
2. Count cells after trypsinization and prepare a 1:1 mixture of cell suspension (5×10^5) in Matrigel (50 μL) and implant in both flanks of mice
3. Weigh out mice and measure the volume of the tumors, which is 350–650 mm^3 after approximately 3-weeks
4. Formulate CRaB-Hyp and Hyp-1 (50 μM) in sterile saline containing 15% DMSO and injected intratumorally
5. Acquire PA images from 710 to 770 nm by placing the mice in a bowl
6. Generate three-dimensional reconstructed images for processing the data by selecting equal-volume regions of interest (ROIs) from each image. Quantify the mean PA signal in each ROI at both wavelengths and different timepoints, using the Horos DICOM viewer or equivalent software
7. Calculate the ratiometric fold turn-on of CRaB-Hyp to red-CRaB-Hyp by using the equation: Ratiometric fold turn-on $= (PA_{lred}/PA_{lblue})_{final}/(PA_{lred}/PA_{lblue})_{initial}$
8. Plot the data to obtain fold turn-on (Fig. 6)

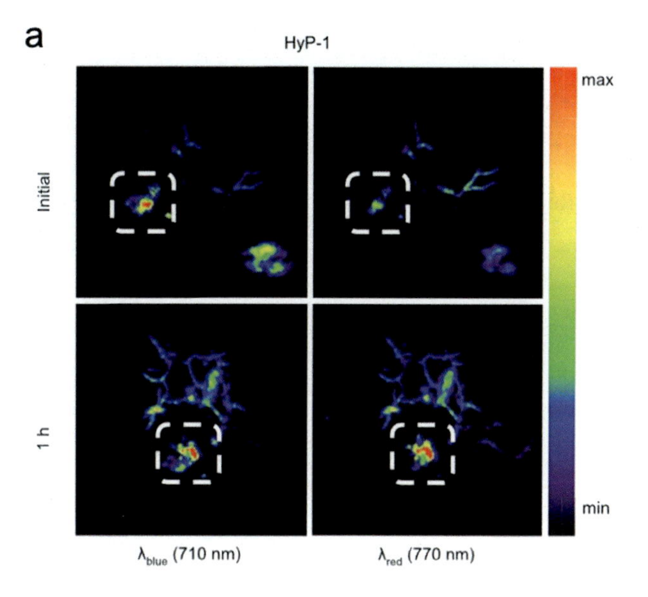

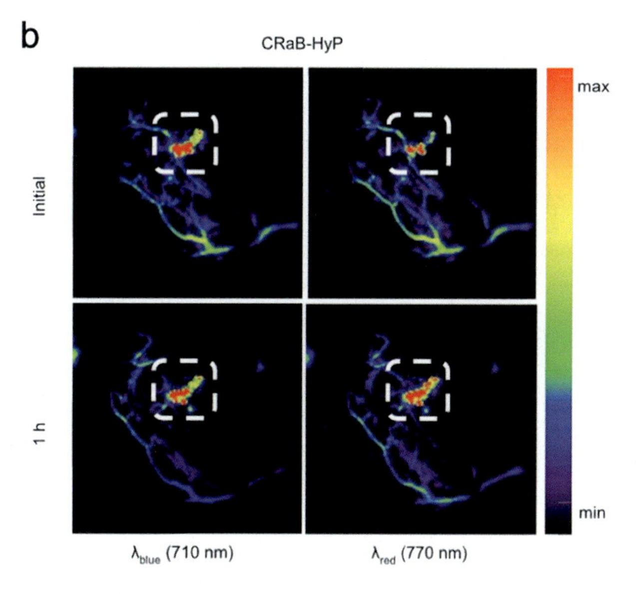

Fig. 6 Representative PA images of hypoxic tumors just after local injection and after 1 h with compounds (A) HyP-1 and (B) CRaB-HyP at λ_{blue} and λ_{red}. White boxes indicate pooled dye ROIs quantified for further analysis. (C) Ratiometric fold turn-on of HyP-1 *vs* CRaB-HyP in hypoxic tumors at various time points. Data are presented as mean ± SD ($n = 6$). Statistical analysis between HyP-1 and CRaB-HyP was performed using two-way ANOVA. Ratiometric fold turn-on was compared at each time point using Sidak's multiple comparison test ($\alpha = 0.05$); $^{*}P < 0.05$ and $^{**}P < 0.01$. *Reprinted (adapted) with permission from Zhou, E. Y., Knox, H. J., Liu, C., Zhao, W. & Chan, J. (2019). A conformationally restricted aza-BODIPY platform for stimulus-responsive probes with enhanced photoacoustic properties.* Journal of the American Chemical Society, 141, *17601–17609. Copyright (2019) American Chemical Society.*

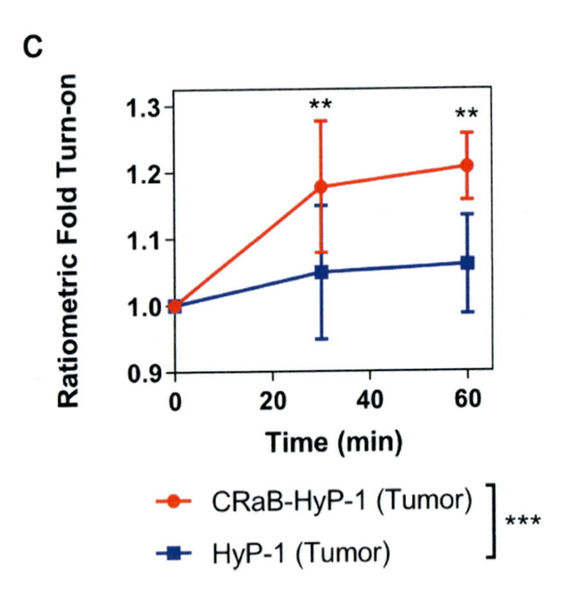

Fig. 6—Cont'd

8. Concluding remarks

In summary, we have discussed a generalizable strategy to optimize the aza-BODIPY platform for development of PA imaging probes of enhanced signal intensity and favorable physical properties. More importantly, we have provided a logical approach which may potentially be applicable to other dye platforms to obtain optimized and sensitive PA imaging agents.

References

Ge, Y., & O'Shea, D. F. (2016). Azadipyrromethenes: From traditional dye chemistry to leading edge applications. *Chemical Society Reviews*, *45*(14), 3846–3864.

Jiao, L., et al. (2014). Conformationally restricted aza-dipyrromethene boron difluorides (Aza-BODIPYs) with high fluorescent quantum yields. *Chemistry- An Asian Journal*, *9*(3), 805–810.

Knox, H. J., & Chan, J. (2018). Acoustogenic probes: A new frontier in photoacoustic imaging. *Accounts of Chemical Research*, *51*(11), 2897–2905.

Knox, H. J., Kim, T. W., Zhu, Z., & Chan, J. (2018). Photophysical tuning of *N* -oxide-based probes enables Ratiometric photoacoustic imaging of tumor hypoxia. *ACS Chemical Biology*, *13*(7), 1838–1843.

Leen, V., et al. (2011). 1,7-Disubstituted boron Dipyrromethene (BODIPY) dyes: Synthesis and spectroscopic properties. *The Journal of Organic Chemistry*, *76*(20), 8168–8176.

Li, H., Zhang, P., Smaga, L. P., Hoffman, R. A., & Chan, J. (2015). Photoacoustic probes for Ratiometric imaging of copper(II). *Journal of the American Chemical Society*, *137*(50), 15628–15631.

Li, X., et al. (2019). A small-molecule probe for ratiometric photoacoustic imaging of hydrogen sulfide in living mice. *Chemical Communications*, *55*(42), 5934–5937.

Lu, H., Mack, J., Yang, Y., & Shen, Z. (2014). Structural modification strategies for the rational design of red/NIR region BODIPYs. *Chemical Society Reviews*, *43*(13), 4778–4823.

Lu, X., et al. (2018). Enhancing hydrophilicity of photoacoustic probes for effective ratiometric imaging of hydrogen peroxide. *Journal of Materials Chemistry B*, *6*(27), 4531–4538.

Reinhardt, C. J., Zhou, E. Y., Jorgensen, M. D., Partipilo, G., & Chan, J. (2018). A Ratiometric Acoustogenic probe for *in Vivo* imaging of endogenous nitric oxide. *Journal of the American Chemical Society*, *140*(3), 1011–1018.

Roberts, S., et al. (2018). Calcium sensor for photoacoustic imaging. *Journal of the American Chemical Society*, *140*(8), 2718–2721.

Sheng, W., et al. (2017). [*a*]-phenanthrene-fused BF_2 azadipyrromethene (AzaBODIPY) dyes as bright near-infrared fluorophores. *The Journal of Organic Chemistry*, *82*(19), 10341–10349.

Wang, L. V., & Yao, J. (2016). A practical guide to photoacoustic tomography in the life sciences. *Nature Methods*, *13*, 627–638.

Weber, J., Beard, P. C., & Bohndiek, S. E. (2016). Contrast agents for molecular photoacoustic imaging. *Nature Methods*, *13*, 639–650.

Yin, L., et al. (2019). Quantitatively visualizing tumor-related protease activity *in Vivo* using a ratiometric photoacoustic probe. *Journal of the American Chemical Society*, *141*(7), 3265–3273.

Zhang, J., Smaga, L. P., Satyavolu, N. S. R., Chan, J., & Lu, Y. (2017). DNA aptamer-based Activatable probes for photoacoustic imaging in living mice. *Journal of the American Chemical Society*, *139*(48), 17225–17228.

Zhou, E. Y., et al. (2018). Near-infrared photoactivatable nitric oxide donors with integrated photoacoustic monitoring. *Journal of the American Chemical Society*, *140*(37), 11686–11697.

Zhou, E. Y., Knox, H. J., Liu, C., Zhao, W., & Chan, J. (2019). A conformationally restricted aza-BODIPY platform for stimulus-responsive probes with enhanced photoacoustic properties. *Journal of the American Chemical Society*, *141*(44), 17601–17609.

Zhou, E. Y., Knox, H. J., Reinhardt, C. J., Partipilo, G., & Chan, J. (2020). Near-infrared photoactivatable nitric oxide donors with photoacoustic readout. *Methods in Enzymology*, *641*, 113–147. https://doi.org/10.1016/bs.mie.2020.05.003.

Zhao, W., & Carreira, E. M. (2005). Conformationally restricted Aza-Bodipy: A highly fluorescent, stable, near-infrared-absorbing dye. *Angewandte Chemie International Edition*, *44*(11), 1677–1679.

CHAPTER EIGHTEEN

Functionalized contrast agents for multimodality photoacoustic microscopy, optical coherence tomography, and fluorescence microscopy molecular retinal imaging

Van Phuc Nguyen[a,b], Wei Qian[c], Xueding Wang[d,*], and Yannis M. Paulus[a,d,*]

[a]Department of Ophthalmology and Visual Sciences, University of Michigan, Ann Arbor, MI, United States
[b]NTT-Hitech Institutes, Nguyen Tat Thanh University, Ho Chi Minh City, Vietnam
[c]IMRA America Inc, Ann Arbor, MI, United States
[d]Department of Biomedical Engineering, University of Michigan, Ann Arbor, MI, United States
*Corresponding authors: e-mail address: xdwang@umich.edu; ypaulus@med.umich.edu

Contents

Methods in Enzymology, Volume 657
ISSN 0076-6879
https://doi.org/10.1016/bs.mie.2021.06.038

Abstract

Near-infrared (NIR) targeting contrast agents have been investigated as great photo-absorbers to improve photoacoustic microscopy (PAM), OCT, and fluorescence imaging contrast for visualization of various diseases. In ophthalmology, a limited number of NIR contrast agents have been approved for clinical use. Recently, gold nanoparticles with different size and shapes have been developed for molecular imaging. This chapter provides the principles of multimodality PAM, OCT, and fluorescence imaging as well as a brief overview of contrast agents for optical imaging. A detailed protocol for the fabrication of discrete colloidal gold nanoparticles (GNPs), synthesis of functionalized RGD-conjugated chain-like GNP (CGNP) clusters labeled with indocyanine green (ICG) fluorescence dye (ICG@CGNP clusters-RGD), and validation of the synthesized nanoparticles to evaluate newly developed blood vessels in the retina, named choroidal neovascularization (CNV), is described. Using RGD peptide, ICG@CGNPs clusters-RGD can bind integrin which is expressed on activated endothelial cells and newly developed CNV. The targeting efficiency of nanoparticles is monitored by multimodality PAM, OCT, and fluorescence imaging longitudinally.

1. Introduction

Gold nanoparticles (GNPs) have been explored as theranostic agents in the diagnosis and treatment of various diseases such as cancer, angiogenesis, and eye diseases. GNPs have unique optical and physicochemical properties that are dominated by the size, morphology, and concentration of the nanoparticles (Chen, Si, de la Zerda, Jokerst, & Myung, 2021; Chen, Zhao, Yoon, Gambhir, & Emelianov, 2019; Kim et al., 2017; Nguyen et al., 2020, 2019, 2021; Weber, Beard, & Bohndiek, 2016). GNPs allow covalent surface modifications *via* the formation of gold–sulfur bonds for optimizing biocompatibility such as polyethylene glycol (PEG) functionalization, stability using silica encapsulation, and active targeting using

Arg-Gly-Asp (RGD), F3, and Gastrin Releasing Peptide Receptor (GRPR) targeting peptides (Chen et al., 2019; Nguyen et al., 2021; Qin, Zong, & Kopelman, 2014; Weber et al., 2016). In addition, GNPs have a high surface-to-volume ratio, allowing a large number of targeting moieties to be attached on the surface of single particle, increasing the probability of target-binding.

The optical properties of GNPs are derived from their localized surface plasmon resonance (LSPR). The conduction electrons of a GNP oscillate relative to the core when exposed to light of an appropriate wavelength and rapidly convert a substantial part of the oscillation energy into heat, resulting in acoustic signal generation. The frequency of the LSPR determines the peak absorption and scattering wavelength of the GNPs. The peak absorption and scattering wavelength of GNPs can be tuned by changing the morphology, size, or surface-to-volume ratio of GNPs (Chen et al., 2019; De Silva Indrasekara, Johnson, Odion, & Vo-Dinh, 2018). The absorption peak of GNPs can be shifted from the visible window (i.e., 520 nm) to NIR (I and II) windows (i.e., 650–1415 nm) by modifying the core size and shape of the GNPs (Chen et al., 2019; De Silva Indrasekara et al., 2018; Nguyen et al., 2021; Si et al., 2019, 2018). For examples, LSPR of gold nanorods (GNRs) can be shifted to NIR window through varying their aspect ratio (Fig. 1A) (Li et al., 2013). Another example using gold nanostars (GNS) or hollow gold nanoparticles (nanocages or nanoshells) (De Silva Indrasekara et al., 2018; Jain & El-Sayed, 2007), the LSPR can be conveniently changed not only by changing the core size of GNPs but also by changing the number of branches (GNS) or shell thickness (thinner shells lead to a red shift) as shown in Fig. 1B–E (De Silva Indrasekara et al., 2018). Recently, our group has reported novel chain-like CNP clusters (CGNP clusters) that shifted the peak absorption of colloidal GNPs from 520 to 650 nm while keeping the colloidal GNPs at the smallest size of 20 nm (Fig. 1F) (Nguyen et al., 2021).

Several synthesis methods have been developed to fabricate and functionalize GNPs with diverse sizes and shapes. Both physical and chemical methods are widely used to fabricate GNPs. Chemical methods using different precursors are used to synthesize GNPs. However, the synthesized GNPs are associated with poor colloidal stability and easy aggregation. To improve stability of the GNPs, surfactant such as cetyltrimethylammonium bromide (CTAB) is usually used. Unfortunately, the use of this hazardous chemical reagent increases the toxicity and can induce cellular necrosis *in vitro* (Alkilany & Murphy, 2010; Jia et al., 2020; Murphy et al., 2008). To reduce

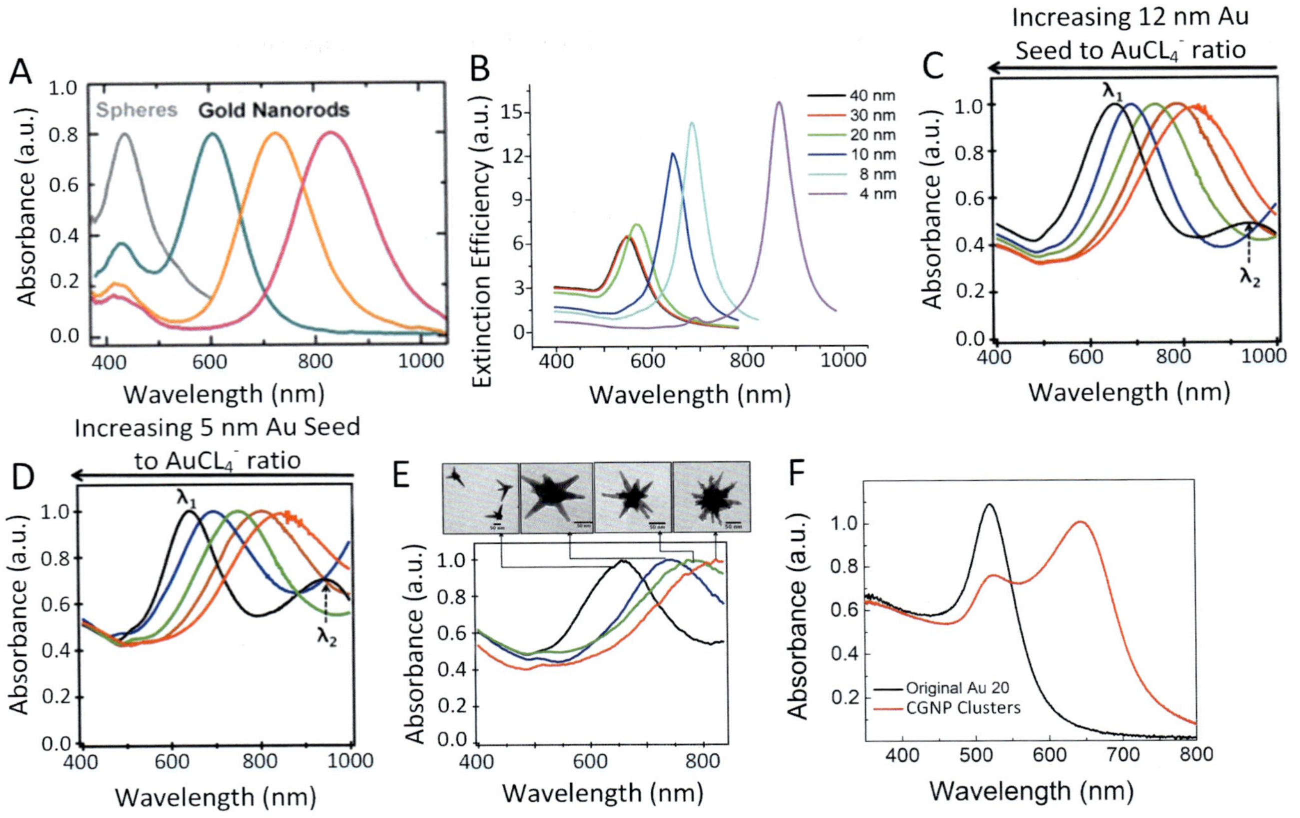

Fig. 1 See figure legend on opposite page.

cytotoxicity, increase stability of synthesized GNPs, as well as increase the conjugation capacity for targeting peptides or for the drug release vehicles, the surface of the GNPs is coated with PEG or silica (Park, Dumani, Arsiwala, Emelianov, & Kane, 2018; Sua & Jokerst, 2017). The GNPs modified with high molecular weight PEG (>5000 Da) are more stable and less toxic than that of the one conjugated with PEG having lower molecular weight under 5000 Da (Zhai et al., 2015; Zhang et al., 2009). In addition, the colloidal stability and cytotoxicity issues of GNPs could be solved by producing them using physical methods, including femtosecond pulsed laser ablation (Liu, Hu, Che, Chen, & Pan, 2007; Mafuné, Kohno, Takeda, & Kondow, 2001), lithography, and high-energy irradiation (Treguer et al., 1998; Zhang & Wang, 2008). The fabricated GNPs is ultrapure and colloidal stable without requiring any surfactants or stabilizers, allowing for improved application of GNPs in medicine.

The application of GNPs in ophthalmology has especially attracted attention recently. GNPs have been applied to improve visualization of retinal and choroidal vessels, retinal neovascularization (RNV), choroidal neovascularization (CNV), and ocular tumors (Kim et al., 2017; Nguyen et al., 2020; Nguyen, Li, Qian, et al., 2019; Nguyen et al., 2021). Recently, GNPs were used to label stem cells/photoreceptor precursors cells for tissue regeneration (Chemla et al., 2019; Kubelick, Snider, Ethier, & Emelianov, 2019). A study reported by de la Zerda et al. has described that GNRs could

Fig. 1 Optical properties of GNPs. (A) UV–Vis absorption spectra of GNRs of different aspect ratios. (B) LSPR of gold nanoshells shifted to the NIR range by modifying the shell thickness relative to the core size of 80 nm diameter. (C and D) LSPR absorption spectra of GNS turned to NIR range by adjusting the core size and the amount of gold seed precursors. (E) Absorption spectra of GNS shifted to NIR range by adjusting number or branches and branching length. (F) Comparison absorption spectra between colloidal GNPs and CGNP clusters. The absorption peak shifted from visible (520 nm) to NIR (650 nm) through clustering GNP monomers without changing the size of GNPs. *Panel (A) adapted with permission from Li, J., Guo, H. & Li, Z.-Y. Microscopic and macroscopic manipulation of gold nanorod and its hybrid nanostructures. Photonics Research 1, 28–41 (2013); Panel (B) adapted with permission from Jain, P. K. & El-Sayed, M. A. Universal scaling of plasmon coupling in metal nanostructures: Extension from particle pairs to nanoshells. Nano Letters 7, 2854–2858 (2007); Panel (C, D, and E) adapted with permission from De Silva Indrasekara, A. S., Johnson, S. F., Odion, R. A. & Vo-Dinh, T. Manipulation of the geometry and modulation of the optical response of surfactant-free gold nanostars: A systematic bottom-up synthesis. ACS Omega 3, 2202–2210 (2018).*

increase optical coherence tomography (OCT) image contrast (de la Zerda et al., 2015). Our group has reported that 20 nm colloidal GNPs could improve both photoacoustic microscopy (PAM) and OCT image contrast (Nguyen, Li, Qian, et al., 2019). To improve visualization of RNV and CNV as well as distinguish them from the surrounding microvasculature, GNS, GNR, and CGNP clusters have been investigated. These GNPs were conjugated with RGD peptides for targeted delivery to the locations of CNV or RNV after intravenous injection for up to 5 days post injection. Among these GNPs, CGNP clusters were fabricated by physical methods and were not toxic to various cells (HeLa, RPE, and ARPE-19 cells) *in vitro* and *in vivo* at tested concentrations.

In this chapter, we highlight the criteria for selecting RGD as a targeting peptide used in our study. We then provide a brief principle of PAM, OCT, and FM and the requirements for ocular imaging. Finally, we will provide a detailed protocol of synthesizing CGNP clusters and applying them for visualization of RNV and CNV *in vivo* using PAM, OCT, and FM.

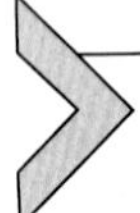

2. Requirements of ophthalmology exogenous contrast agent design

In order to improve imaging contrast, exogenous contrast agent must provide appropriate photophysical and biological properties so as to generate strong signal for imaging and be able to specifically target the area of disease. The ideal contrast agents rely on several characteristics (Nguyen & Paulus, 2018): (1) a high molar extinction coefficient to maximize the amount of light absorbed; (2) a sharp peak to ensure unambiguous identification by spectral unmixing even at low molar concentrations; (3) LSPR absorption peak in the NIR or second NIR window (620–1410 nm) to maximize penetration depth by avoiding the strong absorption of intrinsic chromophores, allowing for detection of the agent deep in biological tissues; (4) high photostability to ensure that target features are not varied by light illumination; (5) highly efficient conversion of heat energy to produce acoustic waves; (6) having a specific targeting moiety such as peptides (i.e., RGD, F3, GRPR), adherens, antibodies, aptamers, or proteins; (7) high biocompatibility to minimize internal cytotoxicity to neural tissues. Fig. 2 shows the schematic diagram of multimodality image contrast agents in the eyes.

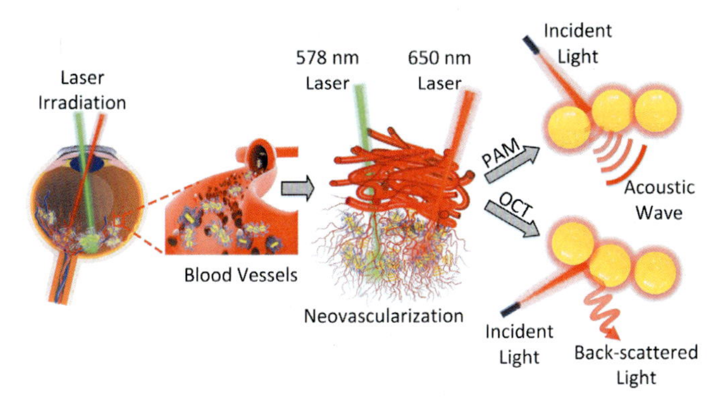

Fig. 2 Schematic illustration of GNPs as a multimodality PAM and OCT image contrast agents for molecular imaging of retinal tissues. GNPs with and without targeting molecules can be administrated into the eye. GNPs can produce strong back-scattered light or acoustic signal when illuminated with laser beam having appropriate wavelength. This signal can be detected by an OCT photodiode to form OCT image or ultrasound detection to reconstruct PA image. Note that the wavelength of 578 nm is used to observe hemoglobin in the vessels and 650 nm is used to detect extravasation of GNPs at the targeted vessels, allowing for discrimination of neovascularization.

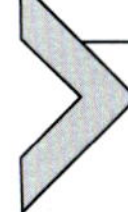

3. Multimodality PAM, OCT, and fluorescence imaging

3.1 Principle of PAM, OCT, and fluorescence imaging

PAM is an emerging and noninvasive imaging tool based on the optical energy conversion from light to sound (Beard, 2011; Nguyen et al., 2021; Tian, Zhang, Mordovanakis, Wang, & Paulus, 2017; Zhang et al., 2020). PAM imaging is considered as a hybrid imaging technique by combining the high contrast, great resolution, and spectroscopic-based specificity of optical imaging with the high penetration depth of ultrasound imaging. In PAM, a beam output from a short pulse laser is used to irradiate the specimen and light energy penetrates into these specimens depending on the optical wavelength. Some light energy is absorbed by chromophores and partially converted into heat. Due to a rapid localized temperature increase, pressure waves are generated and propagated as acoustic waves, which are termed photoacoustic waves.

Another optical imaging technology based on the back-scattering of the incident light is OCT. OCT was first described in 1991 and is a noninvasive imaging technology that produces high resolution images of the internal microstructure of living tissue (Alamouti & Funk, 2003; Budenz et al., 2007; Huang et al., 1991; Nassif et al., 2004). OCT is widely used to diagnose and monitor numerous ophthalmologic diseases (Adhi & Duker, 2013; De Carlo, Romano, Waheed, & Duker, 2015; Hee et al., 1996; Ishibazawa et al., 2015; Jia et al., 2014; Regatieri, Branchini, Carmody, Fujimoto, & Duker, 2012; Yi et al., 2015). To capture OCT images with high resolution and high contrast, a low-coherence light source is used to excite the sample. This light is split into two pathways, a reference arm and sample arm, and evaluated with an interferometer (Huang et al., 1991). A part of the excitation light will be absorbed by the tissue, and some light will be scattered back to the light source. The backscattered light and reference beam will combine to generate an interference pattern, which is recorded by the photodetector. The reflection index based light echoes vs the depth profiles can be determined from the recorded interference pattern. To achieve better penetration depth in tissue, near-infrared light with the central wavelength of 850, 900, and 1310 nm are often used as a light source for OCT.

Fluorescence imaging relies on the emitted photons of a fluorescent dye when it is excited by an appropriate excitation wavelength. The depth of this imaging modality is limited by the excitation and emission wavelengths.

To better visualize different structural and functional information of biological tissues, PAM, OCT, and FM can be integrated to form a multimodality imaging tool as shown in Fig. 3 (Nguyen et al., 2018, 2019, 2020; Nguyen, Li, Zhang, Wang, & Paulus, 2018; Nguyen, Li, Zhang, Wang, & Paulus, 2019; Nguyen & Paulus, 2018; Nguyen et al., 2021; Tian, Zhang, Nguyen, Wang, & Paulus, 2018; Zhang et al., 2020; Zhao et al., 2018). By using the multimodal imaging system, the image of biological tissues can be obtained from each modality and coregistered on the same orthogonal imaging plane. The OCT system can provide supplemental information for PAM and FM such as location, structure, and thickness of the tissue using cross-sectional B-scans. OCT can also be used to guide surgical procedures such as subretinal delivery of vascular endothelial growth factor (VEGF) into the subretinal space in the retina or as an alignment tool to guide PAM. Thus, the position and structure of the tissues can be more easily observed.

3.2 Requirement for retinal molecular imaging

Retinal tissue is fragile and extremely sensitive to the light that excites the eye. Intense laser illumination may induce thermal damage, thermoacoustic damage, and photochemical damage to the retinal tissue (Kuo et al., 2010;

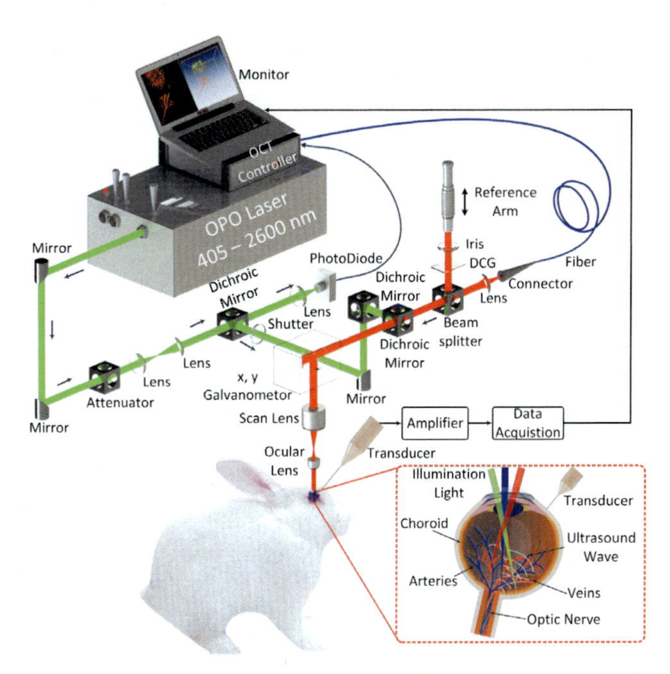

Fig. 3 Schematic diagram of the custom-built multimodality PAM and OCT imaging system.

Organisciak & Vaughan, 2010). To avoid damaging sensitive neural tissue, the light fluence delivered into the eye must be below the American National Standards Institute (ANSI) safety limit standard (ANSI Z136.1, 2007; Tian et al., 2017). Fast acquisition time is mandatory to achieve high-resolution imaging while minimizing motion artifacts since the eye can frequently move and scan and has a very short fixation time (~500 ms) (Robinson, 1964). Lastly, a noninvasive, noncontact or minimally invasive imaging modality is highly desirable to decrease systemic side effects such as nausea, vomiting, allergic reactions, and patient discomfort by the administration of exogenous contrast agents.

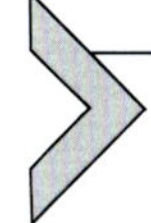

4. Physical production of ultrapure spherical colloidal GNPs

We first produced raw capping agent-free spherical colloidal GNPs used for the fabrication of CGNP clusters *via* a physical method of femtosecond pulsed laser ablation (PLA) of a gold target as previously described in the literatures (Liu et al., 2007; Liu, Hu, Murakami, & Che, 2012).

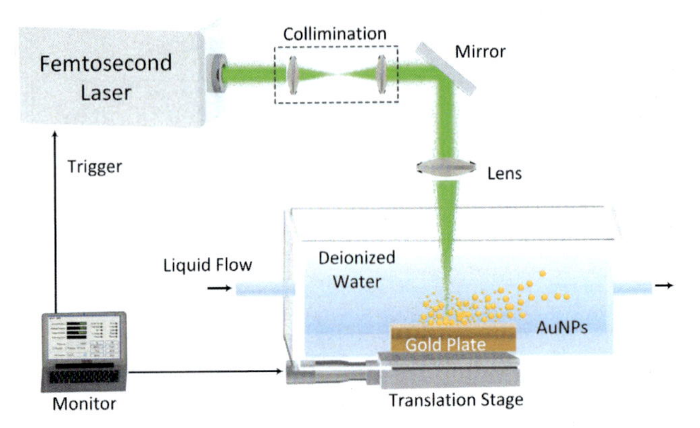

Fig. 4 Ultrapure colloidal gold nanoparticles fabrication. Colloidal gold nanoparticles were produced by femtosecond pulsed laser ablation of a gold target in flowing deionized water. *This schematic diagram was redesigned with permission from Qian, W., Murakami, M., Ichikawa, Y. & Che, Y. Highly efficient and controllable PEGylation of gold nanoparticles prepared by femtosecond laser ablation in water. The Journal of Physical Chemistry C 115, 23293–23298 (2011).*

This method uses tightly focused micro-joule (μJ) femtosecond laser pulses to produce nanoparticles and the size/size distribution of generated nanoparticles can be precisely controlled by optimizing laser parameters, such as wavelength, pulse fluence, duration, and repetition rate as shown in Fig. 4. The GNPs produced using the PLA method are naturally negatively charged and no capping agents and stabilizing ligands are required for maintaining their colloidal stability. This unique feature of having capping-agent free surface for the GNPs produced this way compared with chemically synthesized GNPs allows versatile surface modification to obtain controllable surface chemistry (Qian et al., 2011).

4.1 Equipment

1. Ytterbium-doped femtosecond fiber laser (FCPA μJewel D-1000, IMRA America, Ann Arbor, MI)
2. Bulk gold target
3. XYZ precise linear translation stage

4.2 Procedure

Briefly, the ytterbium-doped femtosecond fiber laser (FCPA μJewel D-1000, IMRA America, Ann Arbor, MI) operating at 1.045 μm delivered pulsed laser at a repetition rate of 100 kHz with 10 μJ pulse fluence and 700 fs pulse duration. The emitted laser beam was first focused by an objective lens and then reflected by a scanning mirror to the surface of the bulk gold target, which was submerged in flowing deionized water (18 MΩcm). The size of the laser spot on the gold target was estimated to be 50 μm and its position was precisely controlled by the scanning mirror. A translation stage was employed to produce relative movements between the laser beam and the gold sample in the ablation process. During the pulsed laser ablation, GNPs were partially oxidized by oxygen present in solution. These Au-O compounds were hydroxylated, followed by a proton transfer to give a surface of Au-O^- as described by Sylvestre et al. (2004). Colloidal GNPs with an average diameter of 20 nm were produced and used in our experiments. The generated GNPs have a narrow size distribution and have an absorption peak at 520 nm due to the LSPR.

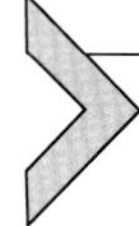

5. Synthesis of indocyanine green (ICG)-labeled and Arginine(*R*)-Glycine(G)-Aspartic(D) (RGD) peptide-conjugated CGNP clusters (ICG@CGNP clusters-RGD)

ICG@CGNP clusters-RGD can be synthesized in four steps from spherical GNP monomers with diameter of 20 nm. The following protocol aims to provide step-by-step procedures for the synthesis of high-quality samples of ICG@CGNP clusters-RGD with high reproducibility as contrast agents for PAM and OCT imaging applications. These procedures require understanding of standard chemistry techniques and the availability of basic laboratory equipment. All procedures are performed at room temperature.

5.1 Self-assembly of spherical GNP monomers into CGNP clusters

5.1.1 Equipment

1. Conical centrifuge tubes (Falcon 15 and 50 mL)
2. Glass vials (2 mL capacity)
3. Centrifuge machine
4. UV–Vis spectrophotometer
5. UV–Vis cuvettes (10 × 10 mm light path)
6. Analytical balance
7. Graduated cylinder
8. Vortex
9. Pipettes and pipette tips

5.1.2 Reagents

1. 5 mL of 20 nm diameter spherical GNP monomers with optical density (OD) 10 at 520 nm
2. Pentapeptide with amino acid sequence of Cys (C)-Ala (A)-Leu (L)-Asn (N)-Asn (N) (≥3 mg)
3. Cysteamine ((≥2 mg)
4. Deionized water (DI H_2O, 18.0 MΩ-cm) ((≥100 mL)

5.1.3 Procedure

1. Prepare 1 mL of pentapeptide CALNN stock solution with concentration of 5 mM
   a. Weigh 2.67 mg of powdered CALNN peptide into a glass vial
   b. Add 1 mL of DI H_2O into the glass vial using a pipette and mix the reagents well by pipetting up and down for several times
   c. Label the glass vial as "5 mM CALNN"
2. Prepare 1 mL of cysteamine stock solution with concentration of 20 mM
   a. Weigh 1.54 mg of powdered CALNN peptide into a glass vial
   b. Add 1 mL of DI H_2O into the glass vial using a pipette and mix the reagents well by pipetting up and down for several times
   c. Label the glass vial as "10 mM cysteamine"
3. Transfer entire 5 mL of 20 nm diameter spherical GNP monomers with OD 10 at 520 nm from its original container into a Falcon 50 mL centrifuge tube.
4. Measure 45 mL of DI H_2O using a graduated cylinder. Pour the DI H_2O directly into the centrifuge tube and mix by gently vortexing for several seconds.
5. Add 20 μL of 5 mM CALNN peptide solution to the centrifuge tube so as to achieve a defined molar ratio of 2000:1 between CALNN peptides and spherical GNP monomers and immediately mix by gently vortexing for several seconds.
6. Cover the centrifuge tube and keep the mixture inside undisturbed for 2 h at room temperature to enable sufficient conjugation of CALNN peptides to the spherical GNP monomers *via* gold-sulfur bonds.
7. Add 4.5 μL of 20 mM cysteamine solution to the centrifuge tube so as to achieve a defined molar ratio of 1800:1 between cysteamine molecules and spherical GNP monomers and immediately mix by gently vortexing for several seconds.

8. Cover the centrifuge tube and keep the mixture inside undisturbed at room temperature until the observation of significant color change from red-pink to blue, which serves as a clear evidence of a successful self-assembly of spherical GNP monomers into CGNP clusters. NOTE: This color change typically occurs between 24h to serval days after addition of cysteamine molecules.
9. Spin down CGNP clusters at 1000g for 1h to a pellet using a centrifuge machine and remove supernatant as much as possible using a pipette.
10. Add 4mL of DI H_2O into the centrifuge tube to redisperse the pellet and transfer the pellet to a Falcon 15mL centrifuge tube.
11. Measure the OD of the preadjustment sample of CGNP clusters:
    a. Transfer 100μL of CGNP cluster solution into a UV–Vis cuvette using a pipette
    b. Add 900μL of DI H_2O into the cuvette using a pipette and mix well by pipetting up and down for several times to make a final sample with 10× dilution
    c. Insert the cuvette into a UV–Vis spectrophotometer and record an absorption spectrum from 350 to 800nm. The obtained colloidal solution of CGNP clusters should present a characteristic band around 650nm
12. Adjust the final volume of CGNP cluster solution with adding DI H_2O to obtain OD 10 at 650nm

5.2 PEGylation of CGNP clusters

5.2.1 Equipment

1. Conical centrifuge tubes (Falcon 15mL)
2. Glass vials (2mL capacity)
3. Analytical balance
4. Vortex
5. Pipettes and pipette tips

5.2.2 Reagents

1. 5mL of CGNP cluster solution with OD 10 at 650nm
2. Thiol/sulfhydryl (-SH) functionalized methoxy polyethylene glycol with molecular weight of 2000g/mol (mPEG 2000-SH, ≥3mg)
3. Deionized water (DI H_2O, 18.0 MΩ-cm) ((≥10mL)

5.2.3 Procedure

1. Prepare 1 mL of mPEG 2000-SH stock solution with concentration of 1 mM
   a. Weigh 2.0 mg of powdered mPEG 2000-SH into a glass vial
   b. Add 1 mL of DI H_2O into the glass vial using a pipette and mix the reagents well by pipetting up and down for several times
   c. Label the glass vial as "1 mM mPEG 2000-SH"
2. Transfer entire 5 mL of CGNP cluster solution with OD 10 at 650 nm from its original container into a Falcon 15 mL centrifuge tube.
3. Add 20 μL of 1 mM mPEG 2000-SH solution to the centrifuge tube and immediately mix by gently vortexing for several seconds.
4. Cover the centrifuge tube and keep the mixture inside undisturbed for 2 h at room temperature to enable sufficient conjugation of mPEG 2000-SH molecules to the CGNP clusters *via* gold-sulfur bonds. NOTE: After 2 h reaction, 5 mL of PEGylated CGNP clusters with OD 10 around 650 nm is formed.

5.3 Conjugation of RGD peptide onto PEGylated CGNP clusters

5.3.1 Equipment

1. Conical centrifuge tubes (Falcon 15 and 50 mL)
2. Glass vials (2 mL capacity)
3. Analytical balance
4. Vortex
5. Pipettes and pipette tips

5.3.2 Reagents

1. 5 mL of PEGylated CGNP cluster solution with OD 10 at 650 nm
2. RGD peptide with amino acid sequence of RGDRGDRGDRGDPGC ($\geq$2 mg)
3. Deionized water (DI H_2O, 18.0 MΩ-cm) (($\geq$10 mL)

5.3.3 Procedure

1. Prepare 1 mL of RGD stock solution with concentration of 1 mM
   a. Weigh 1.59 mg of powdered RGD into a glass vial
   b. Add 1 mL of DI H_2O into the glass vial using a pipette and mix the reagents well by pipetting up and down for several times
   c. Label the glass vial as "1 mM RGD"

2. Transfer entire 5 mL of PEGylated CGNP cluster solution with OD 10 at 650 nm from its original container into a Falcon 15 mL centrifuge tube.
3. Add 50 μL of 1 mM RGD solution to the centrifuge tube and immediately mix by gently vortexing for several seconds.
4. Cover the centrifuge tube and keep the mixture inside undisturbed for 2 h at room temperature to enable sufficient conjugation RGD peptide to the CGNP clusters *via* gold-sulfur bonds. NOTE: After 2 h of reaction, 5 mL of PEGylated CGNP clusters conjugated with RGD peptide (CGNP clusters-RGD) with OD 10 around 650 nm is formed.

5.4 Synthesis of ICG-labeled and RGD peptide-conjugated CGNP clusters (ICG@CGNP clusters-RGD)

5.4.1 Equipment

1. Conical centrifuge tubes (Falcon 15 and 50 mL)
2. Glass vials (2 mL capacity)
3. Plastic container (125 mL capacity)
4. Centrifuge machine
5. UV–Vis spectrophotometer
6. UV–Vis cuvettes (10 × 10 mm light path)
7. Analytical balance
8. Graduated cylinder
9. Vortex
10. Pipettes and pipette tips

5.4.2 Reagents

1. 5 mL of CGNP clusters-RGD with OD 10 around 650 nm
2. ICG and thiol (-SH) heterofunctionalized polyethylene glycol with molecular weight of 2000 g/mol (ICG-PEG 2000-SH) (≥3 mg)
3. 100 mM borate buffer (pH 8.2)
4. Bovine serum albumin (BSA) (≥500 mg)
5. Deionized water (DI H_2O, 18.0 MΩ-cm) ((≥150 mL)

5.4.3 Procedure

1. Prepare 1 mL of ICG-PEG 2000-SH stock solution with concentration of 1 mM
   a. Weigh 2.0 mg of powdered ICG-PEG 2000-SH into a glass vial
   b. Add 1 mL of DI H_2O into the glass vial using a pipette and mix the reagents well by pipetting up and down for several times
   c. Label the glass vial as "1 mM ICG-PEG 2000-SH"

2. Prepare 100 mL of storage buffer stock solution (4 mM borate buffer containing 5 mg/ml BSA, pH 8.2)
   a. Add 4 mL of 100 mM borate buffer (pH 8.2) into a plastic container using a pipette
   b. Measure 96 mL of DI H_2O using a graduated cylinder. Pour the DI H_2O directly into the plastic container and mix by gently inverting the container for several times
   c. Weigh 500 mg of powdered BSA into the plastic container
   d. Label the plastic container as "storage buffer"
3. Transfer entire 5 mL of CGNP clusters-RGD with OD 10 around 650 nm from its original container into a Falcon 50 mL centrifuge tube.
4. Add 70 μL of 1 mM ICG-PEG 2000-SH to the centrifuge tube and immediately mix by gently vortexing for several seconds.
5. Cover the centrifuge tube and keep the mixture inside undisturbed for 2 h at room temperature to enable sufficient conjugation of ICG-PEG 2000-SH to the CGNP clusters-RGD *via* gold-sulfur bonds.
6. Measure 45 mL of storage buffer using a graduated cylinder. Pour the storage buffer directly into the centrifuge tube and mix by gently vortexing for several seconds.
7. Spin down ICG@CGNP clusters-RGD at 1000 g for 1 h to a pellet using a centrifuge machine and remove supernatant as much as possible using a pipette.
8. Add 4 mL of storage buffer into the centrifuge tube to redisperse the pellet and transfer the pellet to a Falcon 15 mL centrifuge tube.
9. Measure the OD of the preadjustment sample of ICG@CGNP clusters-RGD:
   a. Transfer 100 μL of ICG@CGNP clusters-RGD solution into a UV–Vis cuvette using a pipette
   b. Add 900 μL of storage buffer into the cuvette using a pipette and mix well by pipetting up and down for several times to make a final sample with 10 × dilution
   c. Insert the cuvette into a UV–Vis spectrophotometer and record an absorption spectrum from 350 to 800 nm
10. Adjust the final volume of ICG@CGNP clusters-RGD solution with adding storage buffer to obtain OD 10 around 650 nm.
11. Store the conjugate of ICG@CGNP clusters-RGD at 4 °C until use. DO NOT FREEZE.

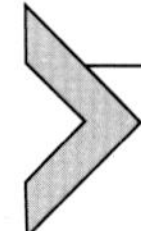

6. Characterization of CGNP clusters

6.1 Optical properties

Prior to the application of the synthesized ICG@CGNP clusters-RGD *in vivo*, photophysical properties of the GNPs were assessed including the optical properties, stability, and biocompatibility. The presence of RGD, ICG, and PEG were characterized using Fourier transform infrared spectroscopy (FTIR) analysis. Transmission electron microscopy (TEM) is used to visualize the morphology of the colloidal GNPs and CGNP clusters. UV–vis spectrophotometry is used to determine the absorption spectrum of the NPs as well as the stability of GNPs over time. Below, we described a detailed characterization protocol of GNPs and CGNP clusters-RGD. The optical properties these GNPs were summarized in Table 1.

6.1.1 Equipment

1. UV–vis spectrophotometer
2. Quartz cuvettes with corresponding cuvette caps
3. Nano-ZS90 Zetasizer
4. Fourier transform infrared spectroscopy (FTIR) spectrometer
5. Micropipettes and tips (P10, P200, P1000)

6.1.2 Reagents

1. GNPs suspension solution
2. CGNP clusters suspension solution
3. Deionized water (DI H2O, 18.0 MΩ-cm)

6.1.3 Procedure

1. Warm up the spectrometer for 15 min before measurement
2. Select the absorbance spectra from 350 to 800 nm with an interval of 1 nm

Table 1 Optical properties of GNPs.

Nanoparticles	Size (nm)	Peak wavelength of LSPR (nm)	Zeta value (mV)	Ratio surface area:volume
GNPs	20	520	-28 ± 1	0.3
CGNPs-RGD	64	650	-41 ± 2	0.9

3. Prepare 3 quartz cuvettes: 1 filled with DI water as blank and the other 2 filled with 1 mL of GNPs and CGNP clusters, respectively, at a final concentration of 0.1 mg/mL
4. Put the blank cuvette into the spectrometer and run baseline function
5. Replace the blank cuvette by a new cuvette containing the GNP sample
6. Run the spectrometer and collect the absorption spectrum
7. Repeat step 5–6 for CGNP cluster sample
8. Plot the absorption spectrum using Origin software and determine the absorption peak wavelengths of the GNP and CGNP clusters samples (Fig. 5).

6.2 Stability and photostability of CGNP clusters

The stability and photostability of CGNP clusters is very important for molecular imaging. If the synthesis of GNPs was not stable at room or *in vivo* temperature or under laser illumination, it would cause the *in vivo* imaging signal to change, leading to difficulty evaluating the targeting objects. Thus, it is essential to evaluate the stability and photostability of CGNP clusters. The absorption spectra of CGNP clusters were obtained at different time points with and without laser illumination.

6.2.1 Equipment

1. UV–vis spectrometer
2. Quartz cuvettes with corresponding cuvette caps
3. Micropipettes and tips (P10, P200, P1000)
4. 96-wells tissue culture plate

6.2.2 Reagents

1. CGNP cluster suspension solution
2. Deionized water (DI H_2O, 18.0 MΩ-cm)

6.2.3 Procedure

1. Prepare stock solution of CGNP clusters at a final concentration of 0.04 mg/mL in DI water
2. Add 50 μL of CGNP cluster solution into 96-wells tissue culture plate. A total of 5 wells were added with CGNP clusters
3. Illuminate the sample with laser fluence of 0 (without laser), 0.005, 0.01, 0.02, and 0.04 mJ/cm^2
4. Measure the absorption spectrum of CGNP clusters after laser illumination follow the procedure in Section 6.1 and plot the data (Fig. 6)

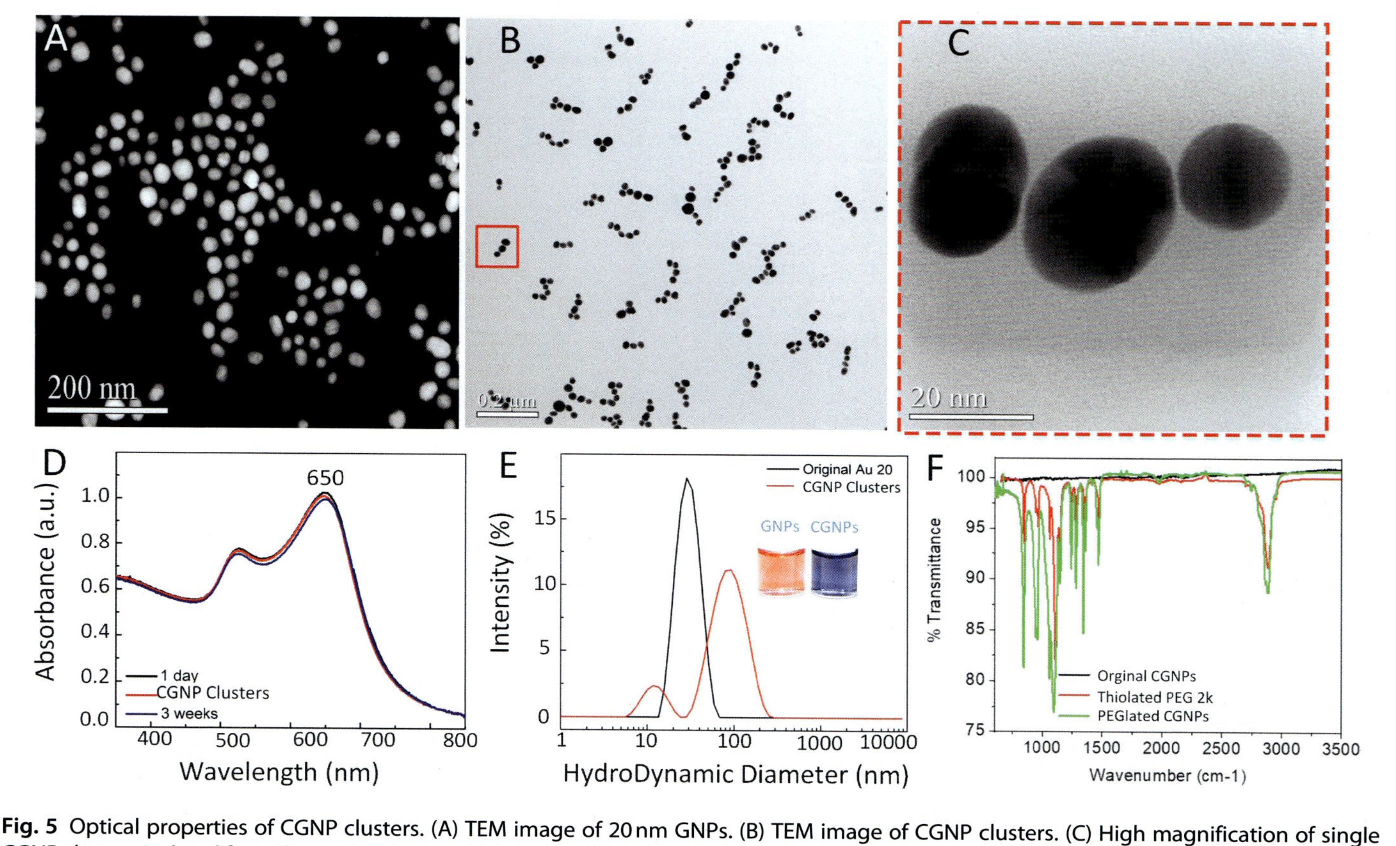

Fig. 5 Optical properties of CGNP clusters. (A) TEM image of 20 nm GNPs. (B) TEM image of CGNP clusters. (C) High magnification of single CGNP clusters isolated from the red box in panel (B). (D) Stability of CGNP clusters over time. (E) Nanoparticle size distribution characterized by dynamic light scattering (DLS). (F) FTIR spectra. *Adapted with permission from Nguyen, V. P., Qian, W., Li, Y., Liu, B., Aaberg, M., Henry, J., Zhang, W., Wang, X. & Paulus, Y. M. Chain-like gold nanoparticle clusters for multimodal photoacoustic microscopy and optical coherence tomography enhanced molecular imaging. Nature Communications 12, 1–14 (2021).*

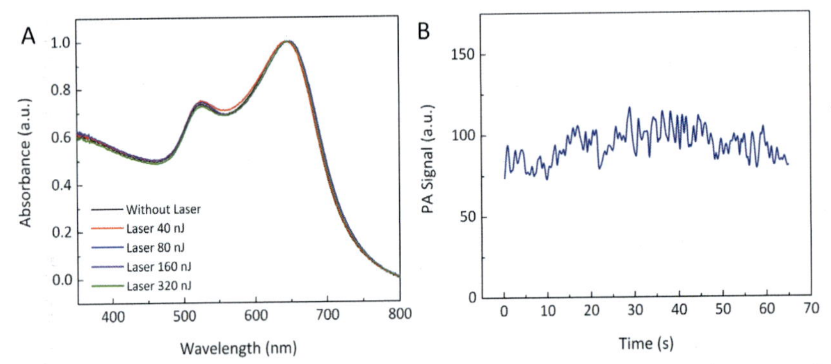

Fig. 6 Photostability analysis. (A) UV–vis absorption spectra of CGNP clusters under laser illumination at different pulse energy. The absorption peak wavelength of the treated samples did not change, confirming that CGNP clusters had great photostability. (B) PA signal amplitudes as a function of laser irradiation times. During the laser irradiation, PA signal varied less than 2%. *Adapted with permission from Nguyen, V. P., Qian, W., Li, Y., Liu, B., Aaberg, M., Henry, J., Zhang, W., Wang, X. & Paulus, Y. M. Chain-like gold nanoparticle clusters for multimodal photoacoustic microscopy and optical coherence tomography enhanced molecular imaging. Nature Communications 12, 1–14 (2021).*

6.3 Cytotoxicity analysis

CGNP clusters were designed as multimodal contrast agent for PAM, OCT, and FM imaging to visualize newly developed neovascularization in the eye of living rabbits. Therefore, it is important to evaluate the safety of the NPs to avoid any side effects that may damage healthy retinal tissues. Cytotoxicity of GNPs were examined on different cell lines (HeLa, Bovine retinal endothelial cells (BRECs), bovine brain endothelial cells (b.End3), and Raw 264.7) using multiple methods including MTT assay, confocal microscope analysis, flow cytometry, and photoacoustic microscopy.

6.3.1 Equipment

1. 96-well tissue culture plate
2. 25 mm^2 tissue culture plate with coated glass bottom
3. Tissue culture facilities: Tissue culture hood, microscope, and an incubator with 5% CO_2 and temperature maintained at 38 °C
4. Confocal laser scanning microscope

6.3.2 Reagent

1. HeLa, Bovine retinal endothelial cells (BRECs), bovine brain endothelial cells (b.End3) BRECs, RAW 264.7, ARPE-19 cells

2. Dulbecco's modified eagle's medium (DMEM) medium supplemented with 10% FBS and antibiotics was utilized as the culture medium for the b.End3, Raw 264.7, and HeLa cells
3. Fibronectin
4. MCDB-131 supplemented with 10% FBS, 1.18 g sodium bicarbonate, 20 ng/mL EGF, 200 mg EndoGRO, 90 mg heparin, 1 mL tylosin, and 10 mL antibiotics/antimycotics, were prepared as the culture medium for the BREC cells
5. 0.25% trypsin-EDTA solution
6. Dimethyl sulfoxide (DMSO)
7. Hoechst 33342, propidium iodide (PI), and Annexin-V FITC
8. Cell scrapper (for harvesting RAW 264.7)
9. Microplate reader

6.3.2.1 MTT procedure

1. Preheat the completed media in 37 °C water bath
2. Coat the surface of cell culture plate with fibronectin for 4 h (only use for BREC cells)
3. Prepare 2 96-wells culture plate and seed 48 wells with 100 mL of cells at density of 10^4 cells/mL on each plate
4. Incubate for 24 h at 37 °C and 5% CO_2
5. Prepare a stock solution of CGNP clusters at a final concentration of 12.5, 25, 50, 100, 200, 400, and 500 μg/mL in media and MTT reagent (1 mg/mL) in medium
6. After 24 h, gently discard media and replace fresh media contained CGNPs
7. Incubate for 24 h and 48 h
8. Replace media with 50 μL of MTT reagent in media
9. Cover the plate with foil and keep in the dark for 4 h
10. Add 100 μL of DMSO into each well
11. Keep the plate in room temperature for 20 min
12. Measure the optical density (OD) at 570 nm
13. Quantify the relative cell viability using the following formula:

$$P = \frac{\text{OD experimental group}}{\text{OD control group}} \times 100$$

14. Plot the data (Fig. 7A–D)

6.3.2.2 Flow cytometry analysis procedure

1. Seed 18 cells in 24-wells plates with 1 mL of cell solution at density of 1×10^5 cells/mL

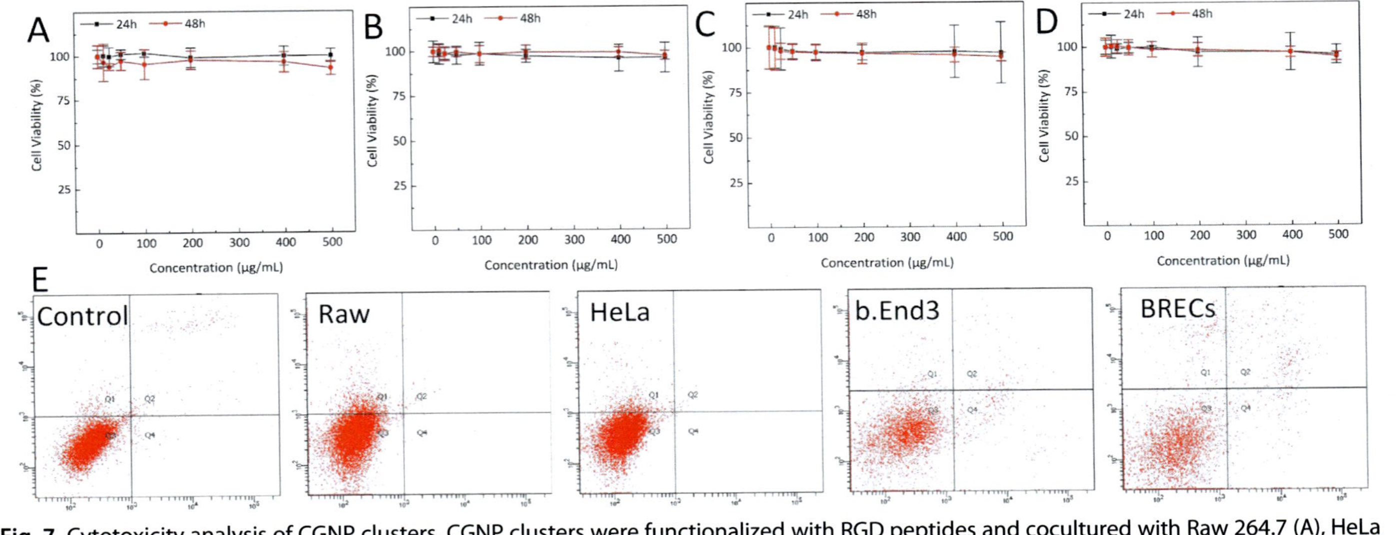

Fig. 7 Cytotoxicity analysis of CGNP clusters. CGNP clusters were functionalized with RGD peptides and cocultured with Raw 264.7 (A), HeLa (B), b.End3 cells (C), and BRECs (D). (E) Flow cytometry analysis. CGNP clusters-RGD did not cause any severe intracellular necrosis. This implies that CGNP clusters are safe for the cells at the treated concentrations. *Adapted with permission from Nguyen, V. P., Qian, W., Li, Y., Liu, B., Aaberg, M., Henry, J., Zhang, W., Wang, X. & Paulus, Y. M. Chain-like gold nanoparticle clusters for multimodal photoacoustic microscopy and optical coherence tomography enhanced molecular imaging. Nature Communications 12, 1–14 (2021).*

2. Incubate for 24 h at at 37 °C in a humidified atmosphere of 5% CO_2 or until the cells reach 80% confluence
3. The cells were cultured with a final concentration of 200 μg/mL of CGNPs in media and incubated for 24 h and 48 h
4. Harvest the treated cells using trypsin or cell scraper (for RAW 264.7).
5. Centrifuge the harvested cells at 1500 rpm for 3 min and discard the media
6. Re-suspend the cell in 500 μL of 1 × binding buffer
7. Add 5 μL (10 μg/mL) of Annexin-V FITC and PI solution into the cell suspension
8. Incubate for 15 min in a dark environment
9. Dilute the cell suspension with 1 mL cold PBS
10. Transfer the solution to 1.5 mL glass tube for flow cytometer analysis
11. Prepare negative control samples by placing a 50% harvested cells in 56 °C water bath
12. Separate into three samples: 1 unstaining, 1 sample stain with 5 μL (10 μg/mL) FITC, and 1 sample stain with 5 μL (10 μg/mL) PI. Observe the analysis data on 4 quadrants: Lower left portion (Q1) denotes viable cells (Annexin-V^-/PI^-); lower right (Q2) is early apoptotic cells (Annexin-V^+/PI^-), upper right (Q3) represents late apoptotic cells (Annexin-V^+/PI^+), and upper left (Q4) indicates necrotic cells (Annexin-V^-/PI^+) (Fig. 7E).

6.3.2.3 Cellular uptake procedure

1. Seed cells in 35 mm^2 microplates (glass bottom) with 2 mL of cells at density of 2×10^5 cells/mL
2. Incubate for 24 h at at 37 °C in a humidified atmosphere of 5% CO_2
3. Replace media with fresh media contain ICG@CGNP clusters-RGD at a final concentration of 200 μg/mL
4. Incubate for an additional 24 h
5. Wash the cells with cold PBS three times
6. Add 500 μL of 1 × binding buffer to the cells and incubate for 15 min
7. Add 5 μL of Annexin-V FITC and 5 μL of PI and incubate for an additional 15 min in the dark
8. Wash the cells with cold PBS three times
9. Fix the cells with 2.5% formaldehyde and incubate for 20 min at 37 °C
10. Wash the cells with PBS two times
11. Stain the fixed cells with 300 μL of 10 μg/mL Hoechst 33342 solution and maintain for 20 min

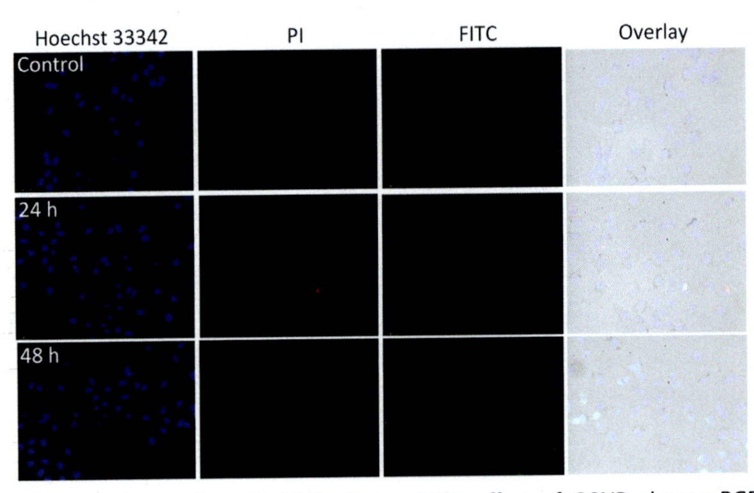

Fig. 8 Cytotoxicity studies of CGNP clusters-RGD: effect of CGNP clusters-RGD on bovine retinal endothelial cells (BRECs) at incubation times of 24 and 48 h, respectively. Confocal fluorescence microscopy images of the cells stained with Hoechst (33342), propidium iodide (PI), and Annexin-V FITC after treatment with CGNP clusters-RGD at concentration of 50 μg/mL. The right images are the overlaid fluorescence and bright-field images. Blue fluorescent color shows the morphology of the cells' nuclei stained with Hoechst 33342. The dead cells were stained with PI and displayed red color. Apoptotic cells were stained with FITC and displayed green color on the image. The fluorescence images were obtained at emission wavelengths of 461 nm for Hoechst 33342, 530 for FITC, and 617 nm for PI, under laser excitation wavelengths at 350 nm for Hoechst 33342, 470 for FITC, and 535 nm for PI, respectively. *Adapted with permission from Nguyen, V. P., Qian, W., Li, Y., Liu, B., Aaberg, M., Henry, J., Zhang, W., Wang, X. & Paulus, Y. M. Chain-like gold nanoparticle clusters for multimodal photoacoustic microscopy and optical coherence tomography enhanced molecular imaging. Nature Communications 12, 1–14 (2021).*

12. Wash the cells three times with PBS
13. Capture the confocal laser scanning image using three channels: red (PI), green (FITC) and blue (Hoechst 33342) (Fig. 8).

6.3.2.4 Single cell detection using PAM procedure

1. Warm the laser system for 15–30 min
2. Fixed cell samples prepared in Section 6.3.2.3 were used to image with the PAM imaging system
3. Place the sample on the stable platform with micro linear XY stages
4. Place the ultrasound transducer and adjust its position to avoid removing cell samples

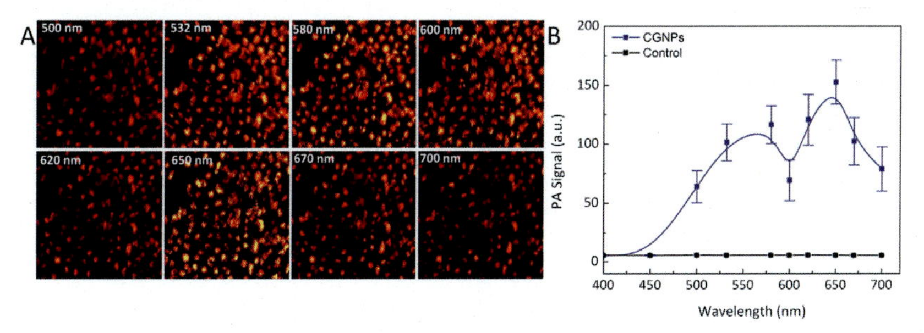

Fig. 9 *In vitro* PAM of labeled cells with CGNP clusters.-RGD (A) Maximum projection intensity (MIP) PAM image of HeLa cells treated with CGNP clusters. (B) Graph of PA signal as a function of excitation wavelength ranging from 400 to 700 nm. Error bar represents the standard deviation measured from 20 independent regions of interest on PAM images. *Adapted with permission from Nguyen, V. P., Qian, W., Li, Y., Liu, B., Aaberg, M., Henry, J., Zhang, W., Wang, X. & Paulus, Y. M. Chain-like gold nanoparticle clusters for multimodal photoacoustic microscopy and optical coherence tomography enhanced molecular imaging. Nature Communications 12, 1–14 (2021).*

5. Observe the photoacoustic signal (PA) on the oscilloscope
6. Adjust the position of the transducer to maximize the PA signal
7. Adjust the laser energy to ensure the PA is not saturated
8. Acquire the PA image
9. Change the optical wavelength from 500 to 710 nm and repeat step 7–8
10. Measure the PA signal amplitude at each wavelength and plot the PAM images and PAM spectrum as a function of wavelengths (Fig. 9)

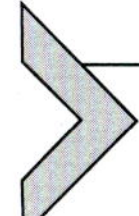

7. *In vitro* PAM and OCT evaluation of CGNP clusters-RGD and photostability analysis

To determine the optimal concentration of CGNP clusters for *in vivo* imaging as well as the optimal optical wavelength for PAM, we employed *in vitro* PAM and OCT imaging of the phantoms. In addition, the photostability of CGNP clusters under multiple short pulsed laser illumination was also characterized.

7.1 Equipment

1. Multimodal PAM and OCT imaging system (Fig. 3)
2. Silicone tube with an inner diameter of 0.30 mm and outer diameter of 0.64 mm

3. Capillaries glass tubes (inner diameter = 0.30 mm and outer diameter = 0.54 mm)
4. Optical adhesive
5. Coverslip
6. Degassed water tank
7. Stabilization platform
8. 1 mL insulin syringe with a 30-gauge needle

7.2 Reagent

1. CGNP clusters-RGD stock solution (5 mg/mL)
2. Human blood
3. Degassed water

7.3 Procedure

1. Turn ON the laser system
2. Dilute CGNP-clusters RGD stock solution to different concentrations of 0.005, 0.01, 0.02, 0.04, and 0.08 mg/mL
3. Mix blood and CGNP clusters-RGD solution to achieve final concentration of 0.02, 0.04, and 2.5 mg/mL
4. Fill the tubes with CGNP clusters-RGD or mixed blood and CGNP clusters-RGD
5. Seal both the distal ends of each tube with optical adhesive
6. Mount the phantom samples on cover glass
7. Place the samples on the degassed water tank (for OCT, the sample was placed in air)
8. Connect the ultrasound transducer and find the maximum photoacoustic signal
9. Acquire the PAM image at different optical wavelengths
10. Measure the PA signal amplitudes as a function of concentration and wavelength using region of interest (ROI) analysis
11. For photostability, the sample was illuminated with 65,000 short pulsed lasers at an energy of 80 nJ. The PA signal amplitude was recorded
12. Plot the PA signal amplitude and OCT signal intensity for each condition (Fig. 10)

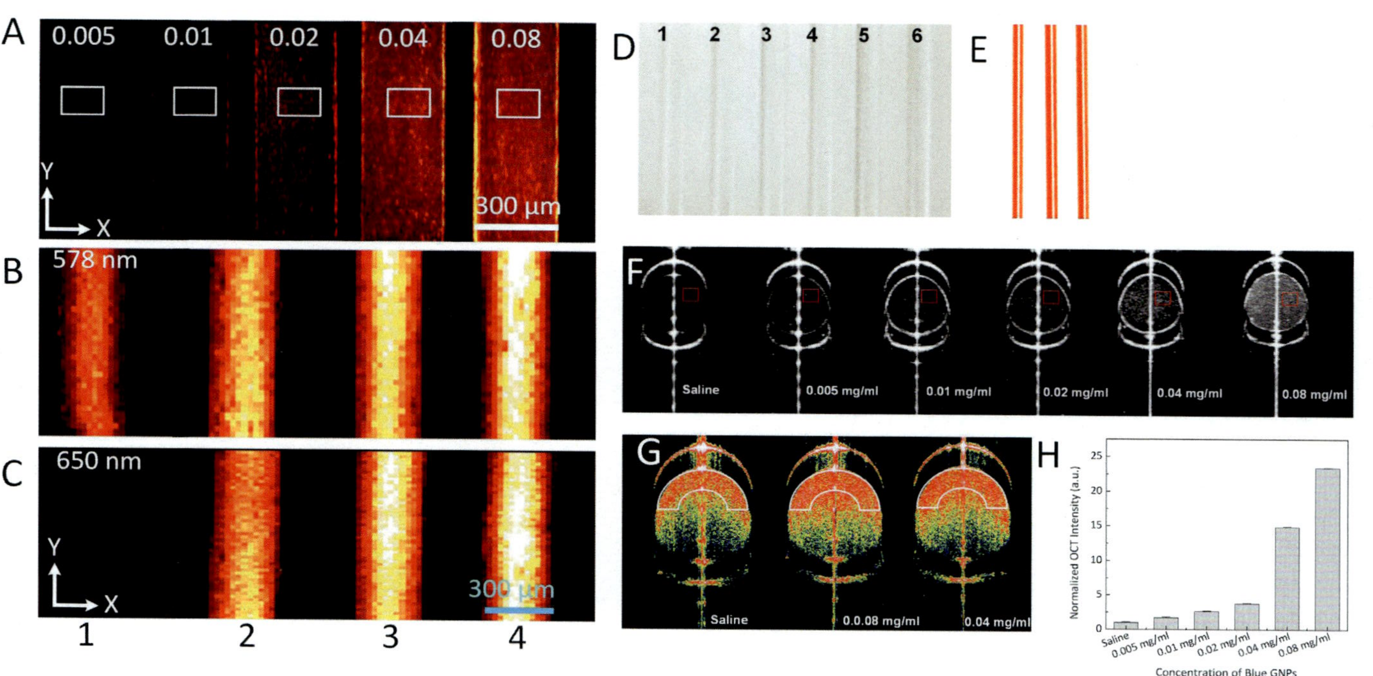

Fig. 10 *In vitro* quantitative analysis of photoacoustic and OCT response of CGNP clusters: (A) PA image of phantom made from silicone tubes filled with CGNP clusters at different concentrations. PA images of blood filled phantom containing CGNP clusters with different final concentrations of 0 mg/mL (sample 1), 0.02 (sample 2), 0.04 (sample 3), and 0.08 mg/mL (sample 4) acquired at 578 nm (B) and 650 nm (C). (D) Photograph of phantom filled with CGNP clusters. (E) Photograph of phantoms filled with fresh rabbit blood and mixture of rabbit blood and CGNP clusters at various concentrations. (F-E) corresponding B-scan OCT image. (H) Plot of the OCT signal intensity as a function of CGNPs concentrations. *Adapted with permission from Nguyen, V. P., Qian, W., Li, Y., Liu, B., Aaberg, M., Henry, J., Zhang, W., Wang, X. & Paulus, Y. M. Chain-like gold nanoparticle clusters for multimodal photoacoustic microscopy and optical coherence tomography enhanced molecular imaging. Nature Communications 12, 1–14 (2021).*

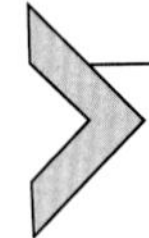

8. *In vivo* multimodal PAM and OCT retinal molecular imaging

The feasibility of CGNP clusters-RGD to improve PAM and OCT image contrast has been tested *in vitro*. In addition, evaluation of the targeting efficiency of CGNP clusters-RGD in living animals is important. In this section, biodistribution of CGNP clusters at the newly developed blood vessels is validated in two different clinically-relevant rabbit models. First, we validate the application of CGNP clusters-RGD to target choroidal neovascularization (CNV) induced by retinal vein occlusion. Then, we demonstrate that CGNP clusters-RGD can be used as an effective targeting contrast agent to monitor CNV with subretinal injection of VEGF-165 with the accumulation of CGNP clusters at CNV that changes over time.

For *in vivo* experiments in large animals like rabbits, extensive training must be done before working on the animal including training in the ethics, laser safety training, rabbit handling, subretinal injection method, intravenous injection, intramuscular injection, and optical imaging modalities such as fluorescein angiography and indocyanine green angiography. All rabbit studies should follow the guidelines of the ARVO (The Association for Research in Vision and Ophthalmology) Statement on the care and use of laboratory animals in Ophthalmic and Vision Research. The experimental protocol should be approved by an appropriate Institutional Animal Care and Use Committee (IACUC). New Zealand White rabbits that were 2–3 months old and weighed 1.8–2.8 kg were obtained by generous donation from the University of Michigan Center for Advanced Models and Translational Sciences and Therapeutics (CAMTraST).

8.1 Application of CGNP clusters-RGD for visualization of CNV in rabbits with retinal vein occlusion model

The hypothesis of this study is to demonstrate that CGNP clusters-RGD can bind at the location of CNV and generate strong PA signal and OCT that can help to distinguish CNV from the surrounding retinal blood vessels which has strong intrinsic PA signal in the visible wavelength (i.e., 578 nm). Because hemoglobin in blood vessels has very low PA signal in the NIR window, the blood vessels have low signal on the PA image obtained in the NIR wavelength. In contrast, CGNP clusters-RGD has strong absorption in the NIR window which generates significant image contrast. In addition, RGD peptide can bind to integrin receptors which are expressed at CNV, allowing for precise evaluation of the margin of

CNV. We apply CGNP clusters in the eye because no existing imaging technique can visualize CNV at an early stage to date. We create the CNV model using laser photocoagulation. Once the CNV appears and is stable, the animal receives CGNP clusters and is followed with different imaging modalities. After treatment with CGNP clusters-RGD, the morphology and margin of CNV is distinguished with multiple wavelength PAM imaging. In addition, the CNV position is confirmed by OCT imaging.

8.1.1 Equipment

1. Custom-built multimodality PAM and OCT imaging system
2. 50-degree color fundus photography (Topcon 50EX, Topcon Corporation, Tokyo, Japan)
3. Laser photocoagulation system (Vitra 532 nm, Quantel Medical, Cournon d'Auvergne, France)
4. Contact lens (Volk H-R Wide Field, laser spot 2 × magnification, Volk Optical Inc., Mentor, OH, USA)
5. Pulse oximeter (Smiths Medical, MN, USA)
6. Water-circulating blanket (TP-700, Stryker Corporation, Kalamazoo, MI)
7. Custom-made stabilization platforms

8.1.2 Animal

Three New Zealand white rabbits that were 2–3 months old.

8.1.3 Reagent

1. Ketamine (40 mg/kg IM, 100 mg/mL)
2. Xylazine (5 mg/kg IM, 100 mg/mL)
3. Vaporized isoflurane anesthetic (Surgivet, MN, USA)
4. Tropicamide 1% ophthalmic
5. Phenylephrine hydrochloride 2.5% ophthalmic
6. Tetracaine 0.5% ophthalmic
7. Artificial tear (Systane, Alcon Inc., TX, USA)
8. Gonak Hypromellose Ophthalmic Demulcent Solution 2.5% (Akorn, Lake Forest, IL, USA)
9. Rose Bengal (5 mg/mL)

8.1.4 Procedure

1. Anesthetize the rabbit using ketamine and xylazine
2. Dilate the rabbit's pupil with a drop of tropicamide 1% and phenylephrine 2.5% ophthalmic
3. Cover untreated eye to avoid dehydration
4. Monitor the animal vitals (mucous membrane color, heart rate, respiratory rate, and rectal temperature were monitored)

5. Turn on the laser system and set up the treatment parameters (power = 150 mW, beam size = 75 μm in aerial diameter, irradiation time = 0.5 s)
6. When the pupil is fully dilated, apply a drop of tetracaine
7. Add Gonak gel on the contact lens
8. Place the contact lens on the cornea of the rabbit eye
9. Determine the target blood vessels under the slit lamp
10. Inject Rose Bengal into the rabbit *via* the marginal ear vein *via* intravenous injection
11. Illuminate 20 shots of laser at a distance of a half to one-disc diameter from the optic disc margin
12. Increase the laser power up to 300 mW and illuminate for further 20 shots at the same position to prevent blood vessel reperfusion
13. Observe the treated area under color fundus photography
14. Change to fluorescein angiography (FA) imaging
15. Perform I.V. injection of 0.2 mL of fluorescein sodium and acquire FA images immediately during the transit phase after intravenous injection
16. Acquire late phase FA at least every minute for a period of at least 15 min
17. Apply terramycin ophthalmic ointment to the treated eye and cover the eye with tape
18. Administrate a dose of meloxicam under the rabbit skin to reduce discomfort
19. Visualize the development of CNV at day 28 post laser photocoagulation using FA imaging and color fundus photography
20. Warm the OPO laser system
21. Transfer the rabbit to the PAM and OCT imaging system
22. Place the rabbit body and heat on two different platforms to minimize motion artifacts
23. Maintain the rabbit's body temperature using heat blanket
24. Maintain anesthesia using vaporized isoflurane
25. Adjust the rabbit head to find the location of CNV using the CCD camera integrated on the OCT system
26. Obtain baseline PAM at 578 and 650 nm and OCT images (2D and 3D)
27. Perform intravenous injection of 0.4 mL, 5 mg/mL of CGNP clusters-RGD suspension solution into the rabbit with a 1 mL syringe, 27-gauge needle in the marginal ear vein
28. Acquire PAM at different wavelengths ranging from 500 to 700 nm, OCT, color fundus photography, and FA images at different time points after injection
29. Plot the PAM and OCT signals over time (Figs. 11 and 12)

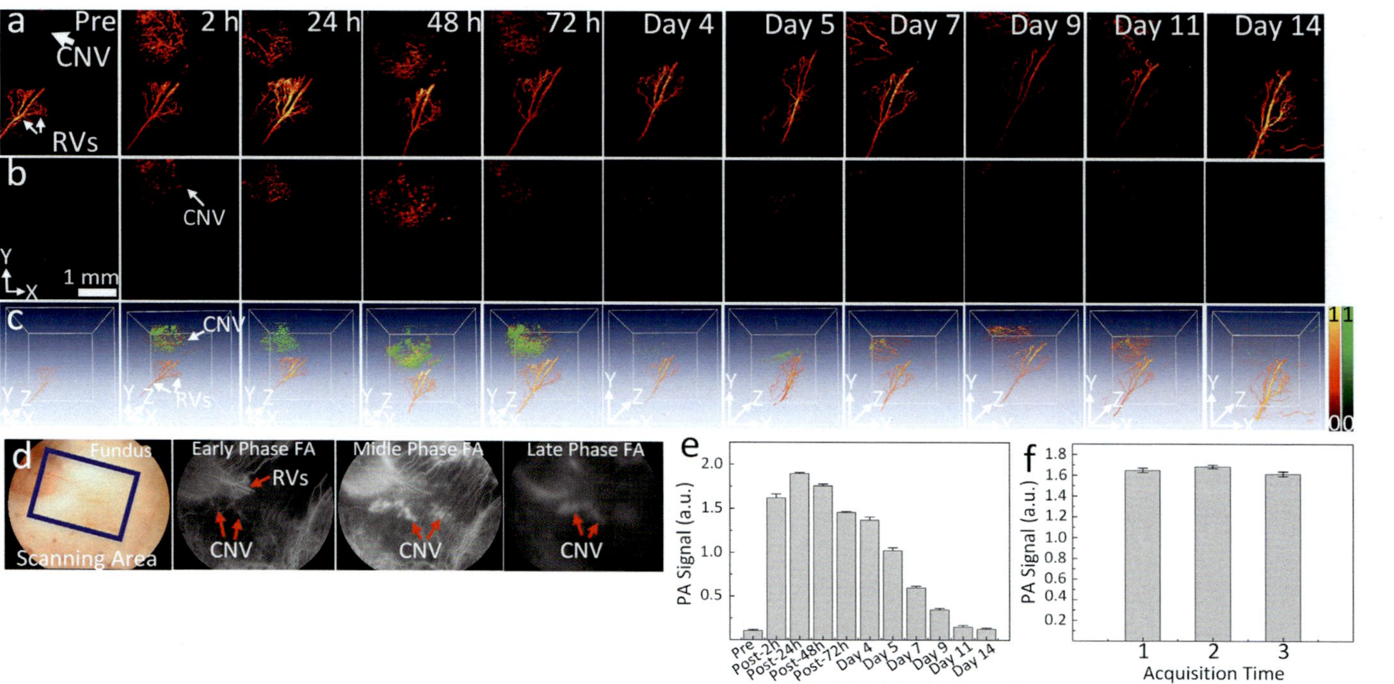

Fig. 11 *In vivo* PAM visualization of CGNP clusters accumulated at CNV. (A and B) PAM images of CNV before and after the injection of 0.5 mL CGNP clusters-RGD at concentration of 2.5 mg/mL acquired along the selected area outlined in fundus image (D) under nanosecond pulsed laser illumination at wavelength of 578 and 650 nm, respectively. (C) Overlay 3D images showed the distribution of CGNP clusters-RGD accumulated at CNV location in rabbit retina (pseudo-green color). (E) Rabbit injected with CGNP clusters-RGD exhibited significantly higher PA signal than preinjection. Note that the peak PA signal occurred at 24 h post injection. Then, the PA signals gradually decreased over time. (F) *In vivo* photostability of CGNP clusters-RGD. The error bars in e and f represent standard error of the average PA signal measured from three different animals ($N = 3$), $P < 0.05$. *Adapted with permission from Nguyen, V. P., Qian, W., Li, Y., Liu, B., Aaberg, M., Henry, J., Zhang, W., Wang, X. & Paulus, Y. M. Chain-like gold nanoparticle clusters for multimodal photoacoustic microscopy and optical coherence tomography enhanced molecular imaging. Nature Communications 12, 1–14 (2021).*

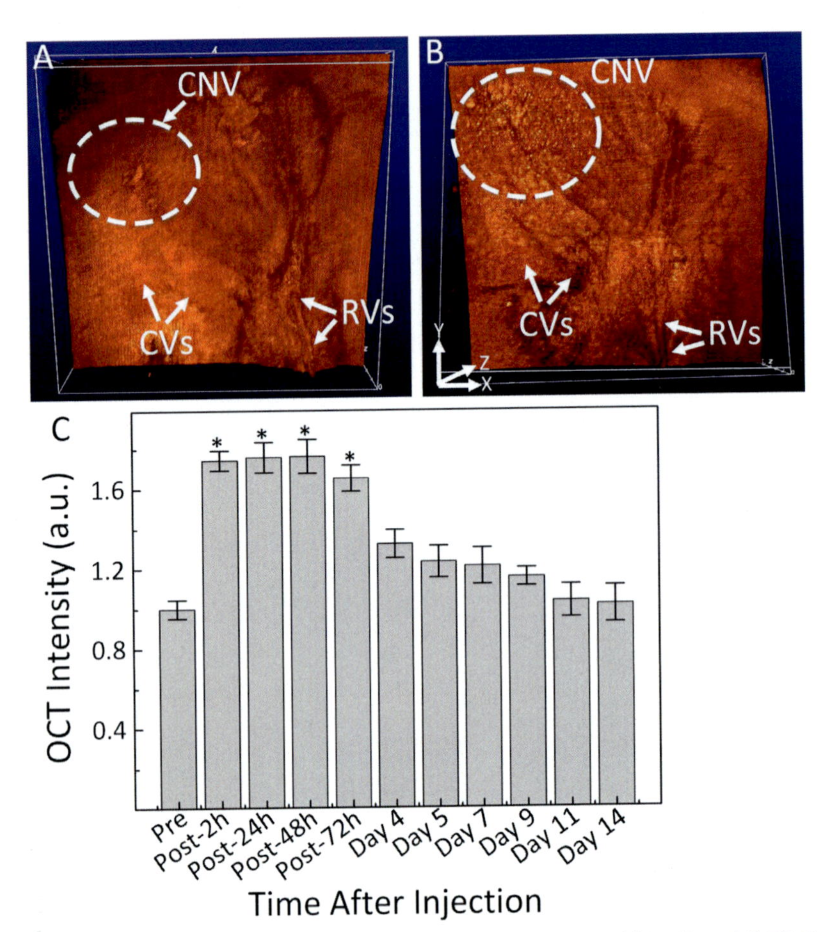

Fig. 12 *In vivo* contrast enhancement of OCT images of CNV in rabbits. (A and B) 3D OCT images before and after the injection of CGNP clusters-RGD. (C) Quantitative measurement OCT signal over times. *Adapted with permission from Nguyen, V. P., Qian, W., Li, Y., Liu, B., Aaberg, M., Henry, J., Zhang, W., Wang, X. & Paulus, Y. M. Chain-like gold nanoparticle clusters for multimodal photoacoustic microscopy and optical coherence tomography enhanced molecular imaging. Nature Communications 12, 1–14 (2021).*

8.2 Application of CGNP clusters for visualization of choroidal neovascularization in rabbit with subretinal injection of VEGF model

The goal of this study is to validate the potential application of CGNP clusters-RGD for targeting different disease models as well as to ensure this method can be repeated.

8.2.1 Equipment

1. Custom-built multimodal PAM and OCT imaging system
2. 50-degree color fundus photography (Topcon 50EX, Topcon Corporation, Tokyo, Japan)

3. Operating microscope
4. Hamilton syringe with 30G needle
5. Ophthalmic surgical toolkit including eyelid speculum, forceps, and scissors

8.2.2 Animal

Three New Zealand white rabbits that were 2–3 months old.

8.2.3 Reagent

1. Matrigel
2. VEGF (100 μg/mL)
3. Silicone contact lens
4. Gonak Hypromellose Ophthalmic Demulcent Solution 2.5% (Akorn, Lake Forest, IL, USA)
5. 26G sharp disposable presterilized needle

8.2.4 Procedure

1. Thaw Matrigel and VEGF solution
2. Mix 20 μL of Matrigel with 7.5 μL of VEGF
3. Anesthetize the rabbit and dilate the pupil at least 30 min before subretinal injection
4. Place the rabbit head under a dissecting microscope
5. Position the head onto its side so that the eye that will be injected is facing the ceiling
6. Remove the superior rectus muscle using scissors
7. Make a scleral tunnel 3.5 mm posterior to the limbus using the 26G sharp disposable pre-sterilized needle
8. Fill a drop of Gonak gel into a contact lens
9. Place the lens on the cornea
10. Insert the tip of the syringe containing 27.5 μL mixed Matrigel and VEGF with the 30G needle through the hole
11. Gently push the tip through the eye until observing the needle tip approach the retinal tissue under the operating microscope
12. Inject the mixed solution slowly into the subretinal space
13. Retract the syringe slowly
14. Monitor the injection area with color fundus photography, FA, PAM, and OCT
15. Five-day postinjection, CNV is noted to develop and monitored by color fundus photography, FA, PAM, and OCT
16. Repeat step 25–28 in Section 8.1 for a period of 14 days
17. Plot the PAM and OCT signals over time (Fig. 13)

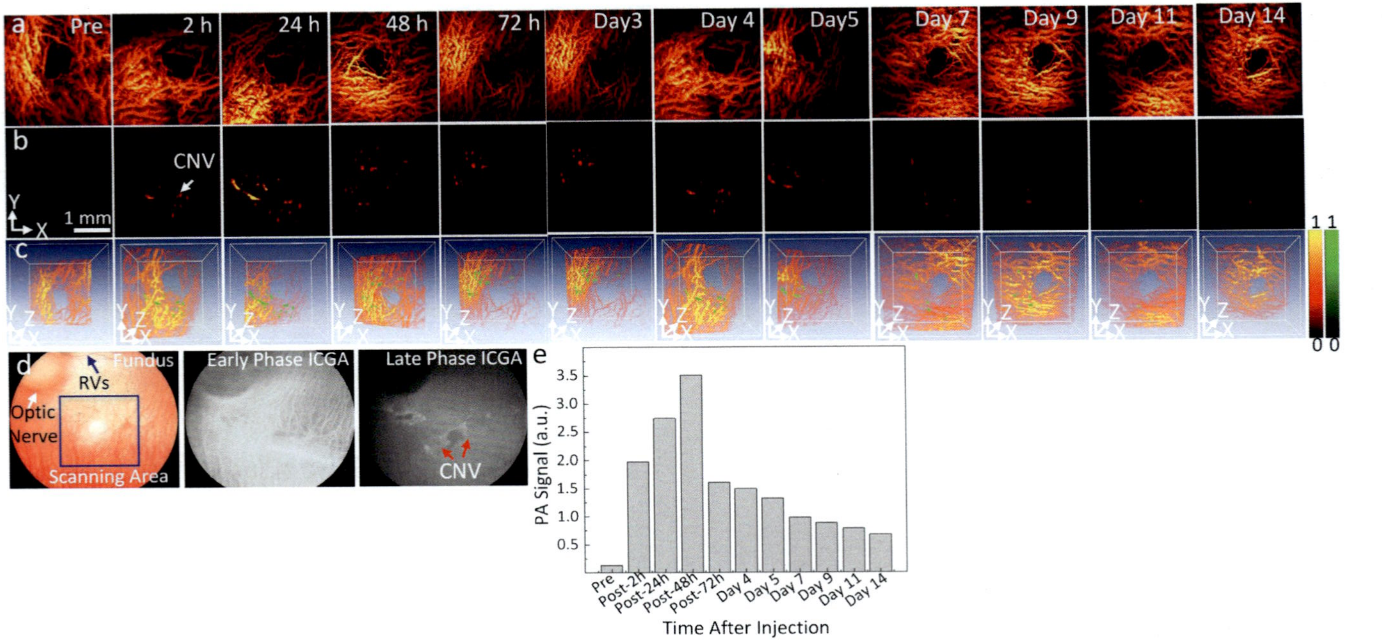

Fig. 13 Study of contrast-enhanced PA detection of CNV model in living rabbits induced by subretinal injection. (A and B) PAM images of CNV obtained along the selected region shown in color fundus image d at 578 and 650 nm, respectively, before and after the injection of CGNP clusters-RGD (0.5 mL at concentration of 2.5 mg/mL). The CNV was obviously distinguished at 650 nm. (C) Fused 3D visualization of the data set shown in panels (A and B). The retinal vasculature is shown in red and the distribution of CGNP clusters-RGD is shown in green. (D) ICGA confirming location of CNV. (E) PA signals measured for each time points. The error bars represent the standard deviation of the mean PA signals ($N=3$). *Adapted with permission from Nguyen, V. P., Qian, W., Li, Y., Liu, B., Aaberg, M., Henry, J., Zhang, W., Wang, X. & Paulus, Y. M. Chain-like gold nanoparticle clusters for multimodal photoacoustic microscopy and optical coherence tomography enhanced molecular imaging. Nature Communications 12, 1–14 (2021).*

9. Conclusions

In this chapter, we describe the synthesis of novel ultrapure CGNP clusters and validate their potential application as dual PAM and OCT imaging contrast agents that can be used for the study of retinal pathologies in living large animal eyes. The synthesized CGNP clusters have several benefits: (1) CGNP clusters shift the optical absorption spectra from visible region (520 nm) to near infrared region (650 nm) while keeping the GNPs at a small size; (2) Chain-like structures can be disassembled into NP monomers, resulting in improved clearance; (3) Conjugation of RGD onto CGNP clusters allow them to bind to molecular targets (integrin receptors) which are overexpressed in neovascularization. This biocompatible exogenous contrast agent provides a unique nanoprobes for visualization of the microvasculature. This manuscript describes in detail the synthesis and validation of the CGNP clusters in different disease models in living animals as well as the details of the custom-built multimodal imaging setup. We believe that the application of these novel nanoprobes for multimodality PAM, OCT, and FM imaging provides a promising technology for the evaluation of numerous disease pathologies.

References

Adhi, M., & Duker, J. S. (2013). Optical coherence tomography–current and future applications. *Current Opinion in Ophthalmology*, *24*, 213.

Alamouti, B., & Funk, J. (2003). Retinal thickness decreases with age: An OCT study. *British Journal of Ophthalmology*, *87*, 899–901.

Alkilany, A. M., & Murphy, C. J. (2010). Toxicity and cellular uptake of gold nanoparticles: What we have learned so far? *Journal of Nanoparticle Research*, *12*, 2313–2333.

ANSI Z136.1. (2007). *https://www.lia.org/store/product/ansi-z1361-2014-safe-use-lasers-electronic-version*.

Beard, P. (2011). Biomedical photoacoustic imaging. *Interface Focus*, *1*, 602–631.

Budenz, D. L., Anderson, D. R., Varma, R., Schuman, J., Cantor, L., Savell, J., et al. (2007). Determinants of normal retinal nerve fiber layer thickness measured by stratus OCT. *Ophthalmology*, *114*, 1046–1052.

Chemla, Y., Betzer, O., Markus, A., Farah, N., Motiei, M., Popovtzer, R., et al. (2019). Gold nanoparticles for multimodal high-resolution imaging of transplanted cells for retinal replacement therapy. *Nanomedicine*, *14*, 1857–1871.

Chen, F., Si, P., de la Zerda, A., Jokerst, J. V., & Myung, D. (2021). Gold nanoparticles to enhance ophthalmic imaging. *Biomaterials Science*, *9*, 367–390.

Chen, Y.-S., Zhao, Y., Yoon, S. J., Gambhir, S. S., & Emelianov, S. (2019). Miniature gold nanorods for photoacoustic molecular imaging in the second near-infrared optical window. *Nature Nanotechnology*, *14*, 465–472.

De Carlo, T. E., Romano, A., Waheed, N. K., & Duker, J. S. (2015). A review of optical coherence tomography angiography (OCTA). *International Journal of Retina and Vitreous*, *1*, 5.

de la Zerda, A., Prabhulkar, S., Perez, V. L., Ruggeri, M., Paranjape, A. S., Habte, F., et al. (2015). Optical coherence contrast imaging using gold nanorods in living mice eyes. *Clinical & Experimental Ophthalmology*, *43*, 358–366.

De Silva Indrasekara, A. S., Johnson, S. F., Odion, R. A., & Vo-Dinh, T. (2018). Manipulation of the geometry and modulation of the optical response of surfactant-free gold nanostars: A systematic bottom-up synthesis. *ACS Omega*, *3*, 2202–2210.

Hee, M. R., Baumal, C. R., Puliafito, C. A., Duker, J. S., Reichel, E., Wilkins, J. R., et al. (1996). Optical coherence tomography of age-related macular degeneration and choroidal neovascularization. *Ophthalmology*, *103*, 1260–1270.

Huang, D., Swanson, E. A., Lin, C. P., Schuman, J. S., Stinson, W. G., Chang, W., et al. (1991). Optical coherence tomography. *Science*, *254*, 1178–1181.

Ishibazawa, A., Nagaoka, T., Takahashi, A., Omae, T., Tani, T., Sogawa, K., et al. (2015). Optical coherence tomography angiography in diabetic retinopathy: A prospective pilot study. *American Journal of Ophthalmology*, *160*, 35–44.e31. https://doi.org/10.1016/j.ajo.2015.04.021.

Jain, P. K., & El-Sayed, M. A. (2007). Universal scaling of plasmon coupling in metal nanostructures: Extension from particle pairs to nanoshells. *Nano Letters*, *7*, 2854–2858.

Jia, Y., Bailey, S. T., Wilson, D. J., Tan, O., Klein, M. L., Flaxel, C. J., et al. (2014). Quantitative optical coherence tomography angiography of choroidal neovascularization in age-related macular degeneration. *Ophthalmology*, *121*, 1435–1444.

Jia, Y. P., Shi, K., Liao, J. F., Peng, J. R., Hao, Y., Qu, Y., et al. (2020). Effects of cetyltrimethylammonium bromide on the toxicity of gold nanorods both in vitro and in vivo: Molecular origin of cytotoxicity and inflammation. *Small Methods*, *4*, 1900799.

Kim, H., Van Phuc Nguyen, P. M., Jung, M. J., Kim, S. W., Oh, J., & Kang, H. W. (2017). Doxorubicin-fucoidan-gold nanoparticles composite for dual-chemo-photothermal treatment on eye tumors. *Oncotarget*, *8*, 113719.

Kubelick, K. P., Snider, E. J., Ethier, C. R., & Emelianov, S. (2019). Development of a stem cell tracking platform for ophthalmic applications using ultrasound and photoacoustic imaging. *Theranostics*, *9*, 3812.

Kuo, T.-R., Hovhannisyan, V. A., Chao, Y.-C., Chao, S.-L., Chiang, S.-J., Lin, S.-J., et al. (2010). Multiple release kinetics of targeted drug from gold nanorod embedded polyelectrolyte conjugates induced by near-infrared laser irradiation. *Journal of the American Chemical Society*, *132*, 14163–14171.

Li, J., Guo, H., & Li, Z.-Y. (2013). Microscopic and macroscopic manipulation of gold nanorod and its hybrid nanostructures. *Photonics Research*, *1*, 28–41.

Liu, B., Hu, Z., Che, Y., Chen, Y., & Pan, X. (2007). Nanoparticle generation in ultrafast pulsed laser ablation of nickel. *Applied Physics Letters*, *90*, 044103.

Liu, B., Hu, Z., Murakami, M., & Che, Y. (2012). *Google patents.*

Mafuné, F., Kohno, J.-y., Takeda, Y., & Kondow, T. (2001). Dissociation and aggregation of gold nanoparticles under laser irradiation. *The Journal of Physical Chemistry B*, *105*, 9050–9056.

Murphy, C. J., Gole, A. M., Stone, J. W., Sisco, P. N., Alkilany, A. M., Goldsmith, E. C., et al. (2008). Gold nanoparticles in biology: Beyond toxicity to cellular imaging. *Accounts of Chemical Research*, *41*, 1721–1730.

Nassif, N., Cense, B., Park, B., Pierce, M., Yun, S., Bouma, B., et al. (2004). In vivo high-resolution video-rate spectral-domain optical coherence tomography of the human retina and optic nerve. *Optics Express*, *12*, 367–376.

Nguyen, V. P., Li, Y., Aaberg, M., Zhang, W., Wang, X., & Paulus, Y. M. (2018). In vivo 3D imaging of retinal neovascularization using multimodal photoacoustic microscopy and optical coherence tomography imaging. *Journal of Imaging*, *4*, 150.

Nguyen, V. P., Li, Y., Folz, J., Henry, J., Aaberg, M., Zhang, W., et al. (2019). *Frontiers in optics. FM5F. Vol. 5.* Optical Society of America.

Nguyen, V.-P., Li, Y., Henry, J., Zhang, W., Aaberg, M., Jones, S., et al. (2020). Plasmonic gold nanostar-enhanced multimodal photoacoustic microscopy and optical coherence tomography molecular imaging to evaluate choroidal neovascularization. *ACS Sensors*, *5*, 3070–3081.

Nguyen, V. P., Li, Y., Qian, W., Liu, B., Tian, C., Zhang, W., et al. (2019). Contrast agent enhanced multimodal photoacoustic microscopy and optical coherence tomography for imaging of rabbit choroidal and retinal vessels in vivo. *Scientific Reports*, *9*, 1–17.

Nguyen, V. P., Li, Y., Zhang, W., Wang, X., & Paulus, Y. M. (2018). Multi-wavelength, en-face photoacoustic microscopy and optical coherence tomography imaging for early and selective detection of laser induced retinal vein occlusion. *Biomedical Optics Express*, *9*, 5915–5938.

Nguyen, V. P., Li, Y., Zhang, W., Wang, X., & Paulus, Y. M. (2019). High-resolution multimodal photoacoustic microscopy and optical coherence tomography image-guided laser induced branch retinal vein occlusion in living rabbits. *Scientific Reports*, *9*, 1–14.

Nguyen, V. P., & Paulus, Y. M. (2018). Photoacoustic ophthalmoscopy: Principle, application, and future directions. *Journal of Imaging*, *4*, 149.

Nguyen, V. P., Qian, W., Li, Y., Liu, B., Aaberg, M., Henry, J., et al. (2021). Chain-like gold nanoparticle clusters for multimodal photoacoustic microscopy and optical coherence tomography enhanced molecular imaging. *Nature Communications*, *12*, 1–14.

Organisciak, D. T., & Vaughan, D. K. (2010). Retinal light damage: Mechanisms and protection. *Progress in Retinal and Eye Research*, *29*, 113–134. https://doi.org/10.1016/j.preteyeres.2009.11.004.

Park, J. H., Dumani, D. S., Arsiwala, A., Emelianov, S., & Kane, R. S. (2018). Tunable aggregation of gold-silica janus nanoparticles to enable contrast-enhanced multi-wavelength photoacoustic imaging in vivo. *Nanoscale*, *10*, 15365–15370.

Qian, W., Murakami, M., Ichikawa, Y., & Che, Y. (2011). Highly efficient and controllable PEGylation of gold nanoparticles prepared by femtosecond laser ablation in water. *The Journal of Physical Chemistry C*, *115*, 23293–23298.

Qin, M., Zong, H., & Kopelman, R. (2014). Click conjugation of peptide to hydrogel nanoparticles for tumor-targeted drug delivery. *Biomacromolecules*, *15*, 3728–3734.

Regatieri, C. V., Branchini, L., Carmody, J., Fujimoto, J. G., & Duker, J. S. (2012). Choroidal thickness in patients with diabetic retinopathy analyzed by spectral-domain optical coherence tomography. *Retina (Philadelphia, Pa.)*, *32*, 563–568. https://doi.org/10.1097/IAE.0b013e31822f5678.

Robinson, D. (1964). The mechanics of human saccadic eye movement. *The Journal of Physiology*, *174*, 245–264.

Si, P., Shevidi, S., Yuan, E., Yuan, K., Lautman, Z., Jeffrey, S. S., et al. (2019). Gold nanobipyramids as second near infrared optical coherence tomography contrast agents for in vivo multiplexing studies. *Nano Letters*, *20*, 101–108.

Si, P., Yuan, E., Liba, O., Winetraub, Y., Yousefi, S., SoRelle, E. D., et al. (2018). Gold nanoprisms as optical coherence tomography contrast agents in the second near-infrared window for enhanced angiography in live animals. *ACS Nano*, *12*, 11986–11994.

Sua, S., & Jokerst, J. V. (2017). Silica/gold hybrid nanoparticles for imaging and therapy. In *Hybrid Nanomaterials: Design, Synthesis, and Biomedical Applications* (p. 355). Taylor Francis Group.

Sylvestre, J.-P., Poulin, S., Kabashin, A. V., Sacher, E., Meunier, M., & Luong, J. H. (2004). Surface chemistry of gold nanoparticles produced by laser ablation in aqueous media. *The Journal of Physical Chemistry B*, *108*, 16864–16869.

Tian, C., Zhang, W., Mordovanakis, A., Wang, X., & Paulus, Y. M. (2017). Noninvasive chorioretinal imaging in living rabbits using integrated photoacoustic microscopy and optical coherence tomography. *Optics Express*, *25*, 15947–15955.

Tian, C., Zhang, W., Nguyen, V. P., Wang, X., & Paulus, Y. M. (2018). Novel photoacoustic microscopy and optical coherence tomography dual-modality chorioretinal imaging in living rabbit eyes. *JoVE (Journal of Visualized Experiments)*, *132*, e57135.

Treguer, M., de Cointet, C., Remita, H., Khatouri, J., Mostafavi, M., Amblard, J., et al. (1998). Dose rate effects on radiolytic synthesis of gold – silver bimetallic clusters in solution. *The Journal of Physical Chemistry B*, *102*, 4310–4321.

Weber, J., Beard, P. C., & Bohndiek, S. E. (2016). Contrast agents for molecular photoacoustic imaging. *Nature Methods*, *13*, 639–650.

Yi, J., Liu, W., Chen, S., Backman, V., Sheibani, N., Sorenson, C. M., et al. (2015). Visible light optical coherence tomography measures retinal oxygen metabolic response to systemic oxygenation. *Light: Science & Applications*, *4*, e334.

Zhai, J., Hinton, T. M., Waddington, L. J., Fong, C., Tran, N., Mulet, X., et al. (2015). Lipid–PEG conjugates sterically stabilize and reduce the toxicity of phytantriol-based lyotropic liquid crystalline nanoparticles. *Langmuir*, *31*, 10871–10880.

Zhang, W., Li, Y., Yu, Y., Derouin, K., Qin, Y., Nguyen, V. P., et al. (2020). Simultaneous photoacoustic microscopy, spectral-domain optical coherence tomography, and fluorescein microscopy multi-modality retinal imaging. *Photoacoustics*, *20*, 100194.

Zhang, G., & Wang, D. (2008). Fabrication of heterogeneous binary arrays of nanoparticles via colloidal lithography. *Journal of the American Chemical Society*, *130*, 5616–5617.

Zhang, G., Yang, Z., Lu, W., Zhang, R., Huang, Q., Tian, M., et al. (2009). Influence of anchoring ligands and particle size on the colloidal stability and in vivo biodistribution of polyethylene glycol-coated gold nanoparticles in tumor-xenografted mice. *Biomaterials*, *30*, 1928–1936.

Zhao, H., Wang, G., Lin, R., Gong, X., Song, L., Li, T., et al. (2018). Three-dimensional hessian matrix-based quantitative vascular imaging of rat iris with optical-resolution photoacoustic microscopy in vivo. *Journal of Biomedical Optics*, *23*(046006).

Printed in the United States
by Baker & Taylor Publisher Services